全栈开发技术丛书

SSM+Spring Boot+Vue. js 3

全栈开发从入门到实战 微课视频版

陈　恒　李正光　主　编

楼偶俊　刁建华　副主编

清華大學出版社

北京

内容简介

本书从 Spring、Spring MVC 和 MyBatis 的基础知识讲起，从而让读者无难度地学习 Spring Boot。为更好地帮助读者巩固学习，本书分阶段安排三个完整的综合案例：基于 SSM + JSP 的名片管理系统、基于 Spring Boot + MyBatis + Thymeleaf 的电子商务平台，以及基于 Spring Boot + Vue 3 + MyBatis 的人事管理系统。

全书共 16 章，内容涵盖 Spring，Spring MVC，MyBatis，名片管理系统的设计与实现（SSM + JSP），Spring Boot 的入门、核心、Web 开发、数据访问，电子商务平台的设计与实现（Spring Boot + MyBatis + Thymeleaf）、Spring Boot 的安全控制、异步消息、热部署与单元测试、应用的监控，Vuc 3 基础、进阶，人事管理系统的设计与实现（Spring Boot + Vue 3 + MyBatis）。书中实例侧重实用性、通俗易懂，使读者能够快速掌握 SSM、Spring Boot 以及 Vue 3 的基础知识、编程技巧以及完整的开发体系，为大型项目开发打下坚实的基础。

本书可作为大学计算机及相关专业的教材或教学参考书，也可作为 Java 技术的培训教材。

图书在版编目（CIP）数据

SSM + Spring Boot + Vue.js 3 全栈开发从入门到实战：微课视频版 / 陈恒，李正光主编. —北京：清华大学出版社，2022.1（2025.1 重印）

（全栈开发技术丛书）

ISBN 978-7-302-59850-3

Ⅰ. ①S… Ⅱ. ①陈… ②李… Ⅲ. ①网页制作工具-程序设计 Ⅳ. ①TP393.092.2

中国版本图书馆 CIP 数据核字（2021）第 280326 号

策划编辑：魏江江
责任编辑：王冰飞
封面设计：刘 键
责任校对：时翠兰
责任印制：杨 艳

出版发行：清华大学出版社
网 址：https://www.tup.com.cn, https://www.wqxuetang.com
地 址：北京清华大学学研大厦 A 座 邮 编：100084
社 总 机：010-83470000 邮 购：010-62786544
投稿与读者服务：010-62776969, c-service@tup.tsinghua.edu.cn
质量反馈：010-62772015, zhiliang@tup.tsinghua.edu.cn
课件下载：https://www.tup.com.cn, 010-83470236
印 装 者：三河市龙大印装有限公司
经 销：全国新华书店
开 本：185mm×260mm **印 张**：32.25 **字 数**：823 千字
版 次：2022 年 3 月第 1 版 **印 次**：2025 年 1 月第 9 次印刷
印 数：15501～17500
定 价：99.80 元

产品编号：091883-01

前言

党的二十大报告指出：教育、科技、人才是全面建设社会主义现代化国家的基础性、战略性支撑。必须坚持科技是第一生产力、人才是第一资源、创新是第一动力，深入实施科教兴国战略、人才强国战略、创新驱动发展战略，开辟发展新领域新赛道，不断塑造发展新动能新优势。高等教育与经济社会发展紧密相连，对促进就业创业、助力经济社会发展、增进人民福祉具有重要意义。

时至今日，在脚本语言和敏捷开发大行其道之时，基于 Spring 框架的 Java EE 开发显得烦琐许多，开发者经常遇到两个非常头疼的问题：①大量的配置文件；②与第三方框架整合。Spring Boot 的出现颠覆了 Java EE 开发，可以说具有划时代意义。Spring Boot 的目标是帮助开发者编写更少的代码实现所需功能，遵循“约定优于配置”原则，从而使开发者只需很少的配置，或者使用默认配置就可以快速搭建项目。虽然 Spring Boot 给开发者带来了开发效率，但 Spring Boot 并不是什么新技术，完全是一个基于 Spring 的应用。例如 Spring Boot 的最大优点——自动配置是通过 Spring 的@Conditional 注解实现的，所以读者在学习 Spring Boot 前，最好快速学习 Spring、Spring MVC 的基础知识。另外，本书第三阶段主要学习基于 Vue 3 的前端开发，建议读者拥有 HTML + CSS + JavaScript 的基础知识。

本书系统介绍 SSM、Spring Boot 和 Vue 3 的重要内容，分三个阶段：第一阶段为 SSM 框架整合开发（第 1～4 章），内容包括 Spring、Spring MVC、MyBatis，以及基于 SSM + JSP 的案例开发；第二阶段为 Spring Boot 框架开发（第 5～13 章），内容包括 Spring Boot 的入门、核心、Web 开发、数据访问、安全控制、异步消息、热部署与单元测试、应用监控，以及基于 Spring Boot + MyBatis + Thymeleaf 的案例开发；第三阶段为 Vue 3 前端框架开发（第 14～16 章），内容包括 Vue 3 基础、进阶，以及基于 Spring Boot + Vue 3 + MyBatis 的案例开发。本书的重点不是简单地介绍基础知识，而是精心设计了大量实例和案例。读者通过本书可以快速地掌握 SSM、Spring Boot 以及 Vue 3 的实践应用，提高 Java EE 应用的开发能力。

全书内容分三个阶段，共 16 章，具体如下。

第一阶段：SSM 框架整合开发

第 1 章：Spring，包括 Spring 开发环境的构建、Spring IoC、Spring AOP、Spring Bean 以及 Spring 的数据库编程等内容。

第 2 章：Spring MVC，包括 Spring MVC 的工作原理、Spring MVC 的工作环境、基于注解的控制器、表单标签库与数据绑定、JSON 数据交互以及 Spring MVC 的基本配置等内容。

第 3 章：MyBatis，包括 MyBatis 的工作原理、SSM 框架整合开发、核心配置文件、SQL 映射文件、级联查询、动态 SQL 以及 MyBatis 的缓存机制等内容。

第 4 章：名片管理系统的设计与实现（SSM + JSP），本章内容是对第 1～3 章学习的巩固。

第二阶段：Spring Boot 框架开发

第 5 章：Spring Boot 入门，包括 Spring Boot 特性、Maven 简介、使用 Spring Tool Suite

（STS）快速构建 Spring Boot 应用以及使用 IntelliJ IDEA 快速构建 Spring Boot 应用等内容。

第 6 章：Spring Boot 核心，包括核心注解、基本配置、自动配置原理以及条件注解等内容。

第 7 章：Spring Boot 的 Web 开发，包括 Spring Boot 的 Web 开发支持、Thymeleaf 模板引擎、JSON 数据交互、文件上传与下载、异常统一处理以及对 JSP 的支持等内容。

第 8 章：Spring Boot 的数据访问，包括 Spring Data JPA、Spring Boot 整合 MyBatis、REST、MongoDB、Redis、数据缓存 Cache 等内容。

第 9 章：电子商务平台的设计与实现（Spring Boot + MyBatis + Thymeleaf），本章内容是对第 5～8 章学习的巩固。

第 10 章：Spring Security 的安全控制，包括 Spring Security 快速入门以及基于 Spring Data JPA 的 Spring Boot Security 操作实例等内容。

第 11 章：Spring Boot 的异步消息，讲解企业级系统间异步消息通信，包括消息模型、JMS 与 AMQP 企业级消息代理、Spring Boot 对异步消息的支持以及异步消息通信实例等内容。

第 12 章：Spring Boot 的热部署与单元测试，包括模板引擎的热部署、使用 spring-boot-devtools 进行热部署以及 Spring Boot 的单元测试等内容。

第 13 章：Spring Boot 应用的监控，包括端点的分类与测试、自定义端点以及自定义 HealthIndicator 等内容。

第三阶段：Vue 3 前端框架开发

第 14 章：Vue 3 基础，包括 Vue 3 的安装、Vue 3 的生命周期、插值与表达式、计算属性、指令、在 Vue 3 中动态使用样式、组件以及自定义指令等内容。

第 15 章：Vue 3 进阶，包括 render 函数、组合 API、webpack、Vue CLI、路由 vue-router 以及状态管理与 Vuex 等内容。

第 16 章：人事管理系统的设计与实现（Spring Boot + Vue 3 + MyBatis），本章内容是对本书整体学习的巩固。

为便于教学，本书提供丰富的配套资源，包括教学大纲、教学课件、电子教案、程序源码、习题答案、在线作业和微课视频。

资源下载提示

课件等资源：扫描封底的“课件下载”二维码，在公众号“书圈”下载。
素材（源码）等资源：扫描目录上方的二维码下载。
在线作业：扫描封底的作业系统二维码，登录网站在线做题及查看答案。
视频等资源：扫描封底的文泉云盘防盗码，再扫描书中相应章节中的二维码，可以在线学习。

由于编者水平有限，书中难免会有不足之处，敬请广大读者批评指正。

编　者
2022 年 1 月

源码下载

目录

第 1 章 Spring

第2章 Spring MVC

第3章 MyBatis

第4章 名片管理系统的设计与实现（SSM+JSP）

第 5 章 Spring Boot 入门

第 6 章 Spring Boot 核心

第 7 章 Spring Boot 的 Web 开发

第 8 章 Spring Boot 的数据访问

第 9 章 电子商务平台的设计与实现(Spring Boot + MyBatis + Thymeleaf)

第 10 章 Spring Boot 的安全控制

第 11 章 Spring Boot 的异步消息

第 12 章 Spring Boot 的热部署与单元测试

第 13 章 Spring Boot 应用的监控

第 14 章 Vue 3 基础

第 15 章 Vue 3 进阶

第 16 章 人事管理系统的设计与实现（Spring Boot + Vue 3 + MyBatis）

第1章 Spring

本章重点讲解 Spring 的基础知识。通过本章的学习，了解 Spring 的体系结构，理解 Spring IoC 与 AOP 的基本原理，了解 Spring Bean 的生命周期、实例化以及作用域，掌握 Spring 的事务管理。

主要内容

- ❖ Spring 开发环境的构建
- ❖ Spring IoC
- ❖ Spring AOP
- ❖ Spring Bean
- ❖ Spring 的数据库编程

Spring 是当前主流的 Java 开发框架，为企业级应用开发提供了丰富的功能。掌握 Spring 框架的使用，已是 Java 开发者必备的技能之一。本章将学习如何使用 Eclipse 开发 Spring 程序，不过在此之前需要构建 Spring 的开发环境。

1.1　Spring 概述

视频讲解

1.1.1　Spring 的由来

Spring 是一个轻量级 Java 开发框架，最早由 Rod Johnson 创建，目的是解决企业级应用开发的业务逻辑层和其他各层的耦合问题。它是一个分层的 JavaSE/EEfull-stack（一站式）轻量级开源框架，为开发 Java 应用程序提供全面的基础架构支持。Spring 负责基础架构，因此 Java 开发者可以专注于应用程序的开发。

1.1.2　Spring 的体系结构

Spring 的功能模块被有组织地分散到约 20 个模块中，这些模块分布在核心容器（Core Container）层、数据访问/集成（Data Access/Integration）层、Web 层、面向切面的编程（Aspect Oriented Programming，AOP）模块、植入（Instrumentation）模块、消息传输（Messaging）和测试（Test）模块中，如图 1.1 所示。

❶ Core Container

Spring 的 Core Container 是其他模块建立的基础，由 Beans（spring-beans）、Core（spring-core）、Context（spring-context）和 Expression（spring-expression，Spring 表达式语言）等模块组成。

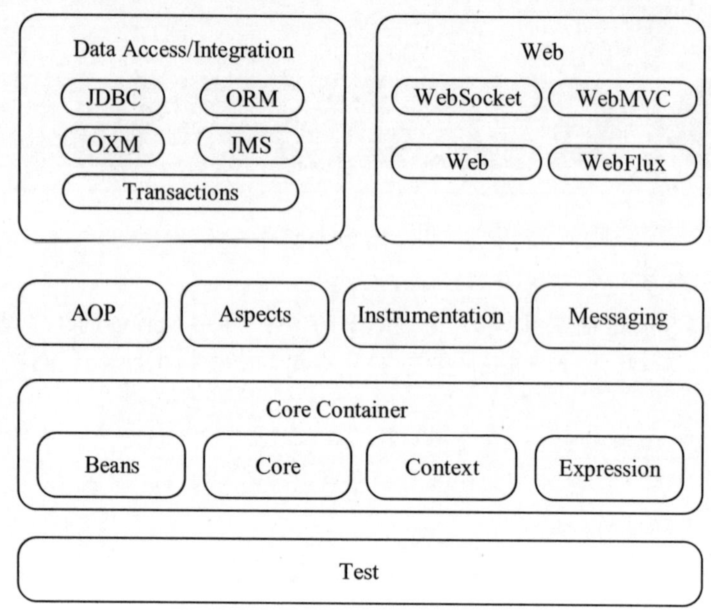

图 1.1　Spring 的体系结构

spring-beans 模块：提供了 BeanFactory，是工厂模式的一个经典实现，Spring 将管理对象称为 Bean。

spring-core 模块：提供了框架的基本组成部分，包括控制反转（Inversion of Control，IoC）和依赖注入（Dependency Injection，DI）功能。

spring-context 模块：建立在 spring-beans 和 spring-core 模块基础上，提供一个框架式的对象访问方式，是访问定义和配置的任何对象媒介。

spring-expression 模块：提供了强大的表达式语言支持运行时查询和操作对象图。这是对 JSP 2.1 规范中规定的统一表达式语言（Unified EL）的扩展。该语言支持设置和获取属性值、属性分配、方法调用、访问数组、集合和索引器的内容、逻辑和算术运算、变量命名以及从 Spring 的 IoC 容器中以名称检索对象；它还支持列表投影、选择以及常见的列表聚合。

❷ AOP 和 Instrumentation

Spring 框架中与 AOP 和 Instrumentation 相关的模块有 AOP（spring-aop）模块、Aspects（spring-aspects）模块以及 Instrumentation（spring-instrument）模块。

spring-aop 模块：提供了一个符合 AOP 要求的面向切面的编程实现，允许定义方法拦截器和切入点，将代码按照功能进行分离，以便干净地解耦。

spring-aspects 模块：提供了与 AspectJ 的集成功能，AspectJ 是一个功能强大且成熟的 AOP 框架。

spring-instrument 模块：提供了类植入（Instrumentation）支持和类加载器的实现，可以在特定的应用服务器中使用。Instrumentation 提供了一种虚拟机级别支持的 AOP 实现方式，使得开发者无须对 JDK 做任何升级和改动，就可以实现某些 AOP 的功能。

❸ Messaging

Spring 4.0 以后新增了 Messaging（spring-messaging）模块，该模块提供了对消息传递体系结构和协议的支持。

❹ Data Access/Integration

数据访问/集成层由 JDBC（spring-jdbc）、ORM（spring-orm）、OXM（spring-oxm）、JMS（spring-jms）和 Transactions（spring-tx）模块组成。

spring-jdbc 模块：提供了一个 JDBC 的抽象层，消除了烦琐的 JDBC 编码和数据库厂商特有的错误代码解析。

spring-orm 模块：为流行的对象关系映射（Object-Relational Mapping）API 提供集成层，包括 JPA 和 Hibernate。使用 spring-orm 模块，可以将这些 O/R 映射框架与 Spring 提供的所有其他功能结合使用，例如声明式事务管理功能。

spring-oxm 模块：提供了一个支持对象/XML 映射的抽象层实现，如 JAXB、Castor、JiBX 和 XStream。

spring-jms 模块（Java Messaging Service）：指 Java 消息传递服务，包含用于生产和使用消息的功能。自 Spring 4.1 后，提供了与 spring-messaging 模块的集成。

spring-tx 模块（事务模块）：支持用于实现特殊接口和所有 POJO（普通 Java 对象）类的编程和声明式事务管理。

❺ Web

Web 层由 Web（spring-web）、WebMVC（spring-webmvc）、WebSocket（spring-websocket）和 WebFlux（spring-webflux）模块组成。

spring-web 模块：提供了基本的 Web 开发集成功能，例如多文件上传功能、使用 Servlet 监听器初始化一个 IoC 容器以及 Web 应用上下文。

spring-webmvc 模块：也称为 Web-Servlet 模块，包含用于 Web 应用程序的 Spring MVC 和 REST Web Services 实现。 Spring MVC 框架提供了领域模型代码和 Web 表单之间的清晰分离，并与 Spring Framework 的所有其他功能集成。本书后续章节将会详细讲解 Spring MVC 框架。

spring-websocket 模块：Spring 4.0 后新增的模块，它提供了 WebSocket 和 SockJS 的实现，主要是与 Web 前端的全双工通信的协议。

Web Flux 模块：spring-webflux 是一个新的非阻塞函数式 Reactive Web 框架，可以用来建立异步的、非阻塞、事件驱动的服务，并且扩展性非常好。（该模块是 Spring 5 的新增模块。）

❻ Test

Test（spring-test）模块：支持使用 JUnit 或 TestNG 对 Spring 组件进行单元测试和集成测试。

1.2 Spring 开发环境的构建

视频讲解

使用 Spring 框架开发应用前，应先搭建其开发环境。本书前 4 章使用的开发环境是基于 Eclipse 平台的 Java Web 应用的开发环境。

1.2.1 使用 Eclipse 开发 Java Web 应用

为了提高开发效率，通常需要安装 IDE（集成开发环境）工具。Eclipse 是一个可用于开发 Web 应用的 IDE 工具。

登录https://www.eclipse.org/，打开如图 1.2 所示的 Eclipse 官方主页。

图 1.2　Eclipse 官方主页

单击图 1.2 中的 Download 按钮，打开如图 1.3 所示的 Eclipse IDE 2020-12 下载页面。

图 1.3　Eclipse IDE 2020-12 下载页面

单击图 1.3 中的 Download Packages 超链接，打开如图 1.4 所示的 Eclipse IDE 2020-12 版本选择页面。

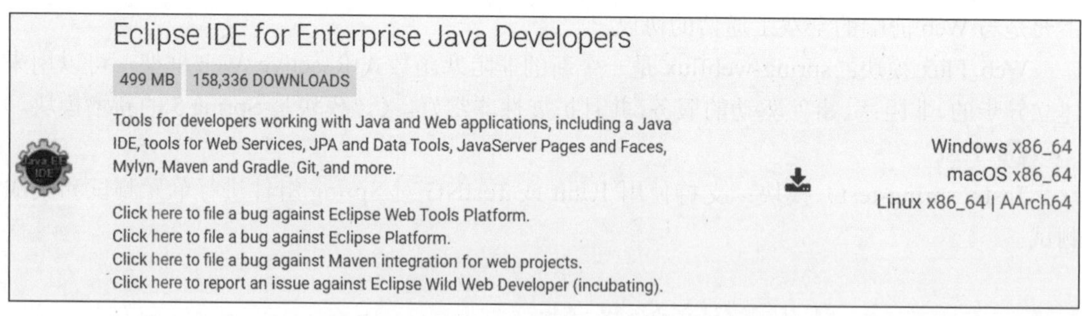

图 1.4　Eclipse IDE 2020-12 版本选择页面

根据操作系统的位数，下载相应的 Eclipse。本书采用的 Eclipse 是 Windows 64-bit（eclipse-jee-2020-12-R-win32-x86_64.zip）。

eclipse-jee-2020-12-R-win32-x86_64 自带 Java SE 15，因此，使用此版本的 Eclipse 开发 Java Web 应用，仅需对 Web 服务器和 Eclipse 进行一些必要的设置。

❶ Web 服务器

目前，比较常用的 Web 服务器包括 Tomcat、JRun、Resin、WebSphere、WebLogic 等，本书采用的是 Tomcat 9.0。

登录 Apache 软件基金会的官方网站 http://jakarta.Apache.org/tomcat，下载 Tomcat 9.0 的免安装版（本书采用的 Tomcat 是 apache-tomcat-9.0.30-windows-x64.zip）。登录网站后，首先在 Download 中选择 Tomcat 9，然后在 Binary Distributions 的 Core 中选择相应版本即可。

将下载的 apache-tomcat-9.0.30-windows-x64.zip 解压到某个目录下，例如解压到 E:\Java soft，解压缩后将出现如图 1.5 所示的目录结构。

名称	修改日期	类型	大小
bin	2019/12/7 16:43	文件夹	
conf	2020/1/30 11:20	文件夹	
lib	2019/12/7 16:43	文件夹	
logs	2020/1/30 11:20	文件夹	
temp	2019/12/7 16:43	文件夹	
webapps	2019/12/7 16:43	文件夹	
work	2020/1/30 11:20	文件夹	
BUILDING.txt	2019/12/7 16:43	文本文档	20 KB
CONTRIBUTING.md	2019/12/7 16:43	MD 文件	6 KB
LICENSE	2019/12/7 16:43	文件	57 KB
NOTICE	2019/12/7 16:43	文件	3 KB
README.md	2019/12/7 16:43	MD 文件	4 KB
RELEASE-NOTES	2019/12/7 16:43	文件	7 KB
RUNNING.txt	2019/12/7 16:43	文本文档	17 KB

图 1.5　Tomcat 目录结构

❷ 安装 Eclipse

Eclipse 下载完成后，解压到自己设置的路径下，即可完成安装。安装 Eclipse 后，双击 Eclipse 安装目录下的 eclipse.exe 文件，启动 Eclipse。

❸ 集成 Tomcat

启动 Eclipse，选择 Window/Preferences 菜单项，在弹出的对话框中选择 Server/Runtime Environments 命令。在弹出的对话框中，单击 Add 按钮，弹出如图 1.6 所示的 New Server Runtime Environment 对话框，在此可以配置各种版本的 Web 服务器。

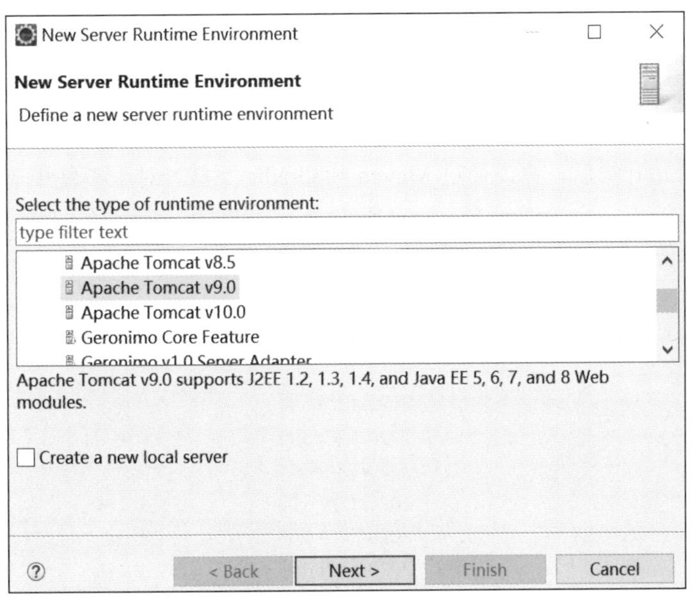

图 1.6　Tomcat 配置对话框

在图 1.6 中选择 Apache Tomcat v9.0 服务器版本，单击 Next 按钮，进入如图 1.7 所示对话框。

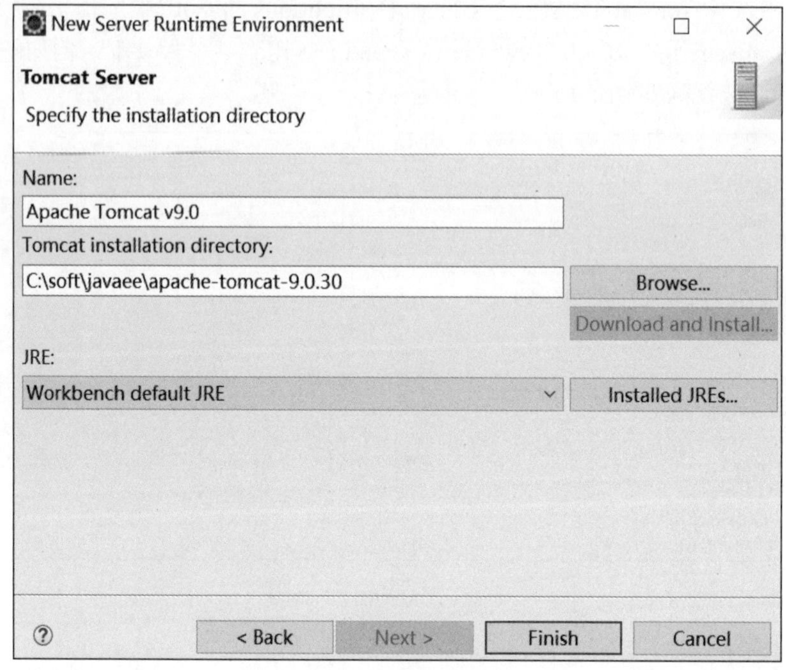

图 1.7　选择 Tomcat 目录

在图 1.7 中单击 Browse 按钮，选择 Tomcat 的安装目录，单击 Finish 按钮即可完成 Tomcat 配置。

至此，可以使用 Eclipse 创建 Dynamic Web Project（Java Web 应用），并在 Tomcat 下运行。

1.2.2　Spring 的下载及目录结构

使用 Spring 框架开发应用程序时，除了引用 Spring 自身的 JAR 包外，还需要引用 commons.logging 的 JAR 包。

❶ Spring 的 JAR 包

Spring 官方网站升级后，建议通过 Maven 和 Gradle 下载。对于不使用 Maven 和 Gradle 下载的开发者，本书给出一个 Spring Framework JAR 官方直接下载路径：https://repo.spring.io/ui/native/release/org/springframework/spring。本书采用的是 spring-5.3.2-dist.zip。将下载到的 ZIP 文件解压缩，解压缩后的目录结构如图 1.8 所示。

名称	修改日期	类型	大小
docs	2020/12/9 6:11	文件夹	
libs	2020/12/9 6:11	文件夹	
schema	2020/12/9 6:11	文件夹	
license.txt	2020/12/9 6:03	文本文档	16 KB
notice.txt	2020/12/9 6:03	文本文档	1 KB
readme.txt	2020/12/9 6:03	文本文档	1 KB

图 1.8　spring-framework-5.3.2 的目录结构

图 1.8 中，docs 目录包含 Spring 的 API 文档和开发规范。

图 1.8 中，libs 目录包含开发 Spring 应用所需要的 JAR 包和源代码。该目录下有三类 JAR 文件，其中，以-5.3.2.jar 结尾的文件是 Spring 框架 class 的 JAR 包，即开发 Spring 应用所需要的 JAR 包；以-5.3.2-javadoc.jar 结尾的文件是 Spring 框架 API 文档的压缩包；以-5.3.2-sources.jar 结尾的文件是 Spring 框架源文件的压缩包。在 libs 目录中，有 4 个基础包：spring-core-5.3.2.jar、spring-beans-5.3.2.jar、spring-context-5.3.2.jar 和 spring-expression-5.3.2.jar，分别对应 Spring 核心容器的 4 个模块：spring-core 模块、spring-beans 模块、spring-context 模块和 spring-expression 模块。

图 1.8 中，schema 目录包含开发 Spring 应用所需要的 schema 文件，这些 schema 文件定义了 Spring 相关配置文件的约束。

❷ commons.logging 的 JAR 包

Spring 框架依赖于 Apache Commons Logging 组件，该组件的 JAR 包可以通过官方网站 http://commons.apache.org/proper/commons-logging/download_logging.cgi 下载，本书下载的是 commons-logging-1.2-bin.zip，解压缩后，即可找到 commons-logging-1.2.jar。

对于 Spring 框架的初学者，开发 Spring 应用时，只需要将 Spring 的 4 个基础包和 commons-logging-1.2.jar 复制到 Web 应用的 WEB-INF/lib 目录下。如果不明白需要哪些 JAR 包，可以将 Spring 的 libs 目录中的 spring-XXX-5.3.2.jar 全部复制到 WEB-INF/lib 目录下。

1.2.3　第一个 Spring 入门程序

本节通过一个简单的入门程序向读者演示 Spring 框架的使用过程。

【例 1-1】Spring 框架的使用过程。

具体实现步骤如下。

❶ 使用 Eclipse 创建 Web 应用并导入相关 JAR 包

使用 Eclipse 创建一个名为 ch1_1 的 Dynamic Web Project，并将 Spring 的 4 个基础包和第三方依赖包 commons-logging-1.2.jar 复制到 ch1_1 的 WEB-INF/lib 目录中，如图 1.9 所示。

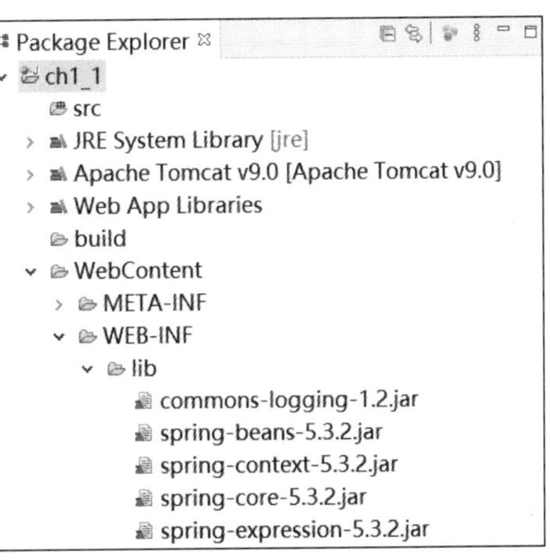

图 1.9　Web 应用 ch1_1 导入的 JAR 包

> **注意：** 在讲解 Spring MVC 框架前，本书的实例并没有真正运行 Web 应用。创建 Web 应用的目的是方便添加相关 JAR 包。

❷ 创建接口 TestDao

Spring 解决的是业务逻辑层和其他各层的耦合问题，因此它将面向接口的编程思想贯穿整个应用系统。

在 src 目录下，创建一个 dao 包，并在 dao 包中创建接口 TestDao，接口中定义一个 sayHello() 方法，代码如下：

```
package dao;
public interface TestDao {
    public void sayHello();
}
```

❸ 创建接口 TestDao 的实现类 TestDaoImpl

在 dao 包下创建接口 TestDao 的实现类 TestDaoImpl，代码如下：

```
package dao;
public class TestDaoImpl implements TestDao{
    @Override
    public void sayHello() {
        System.out.println("Hello, Study hard!");
    }
}
```

❹ 创建配置文件 applicationContext.xml

在 src 目录下，创建名为 config 的包，并在该包中创建 Spring 的配置文件 applicationContext.xml，在配置文件中使用实现类 TestDaoImpl 创建一个 id 为 test 的 Bean，代码如下：

```
<?xml version="1.0" encoding="UTF-8"?>
<beans xmlns="http://www.springframework.org/schema/beans"
    xmlns:xsi="http://www.w3.org/2001/XMLSchema-instance"
    xsi:schemaLocation="http://www.springframework.org/schema/beans
        http://www.springframework.org/schema/beans/spring-beans.xsd">
    <!-- 将指定类 TestDaoImpl 配置给 Spring，让 Spring 创建其实例 -->
    <bean id="test" class="dao.TestDaoImpl" />
</beans>
```

注意： 配置文件的名称可以自定义，但习惯上命名为 applicationContext.xml，有时也命名为 beans.xml。配置文件信息不需要读者手写，可以从 Spring 的帮助文档中复制（首先使用浏览器打开\spring-framework-5.3.2\docs\reference\html\index.html，在页面中单击超链接 Core，在 1.2.1.Configuration Metadata 小节下即可找到配置文件的约束信息）。

❺ 创建测试类

在 src 目录下，创建一个 test 包，并在 test 包中创建 Test 类，代码如下：

```
package test;
import org.springframework.context.ApplicationContext;
import org.springframework.context.support.ClassPathXmlApplicationContext;
import dao.TestDao;
public class Test {
    public static void main(String[] args) {
```

```
        //初始化 Spring 容器 ApplicationContext，加载配置文件
        //@SuppressWarnings 抑制警告的关键字，有泛型未指定类型
        @SuppressWarnings("resource")
        ApplicationContext appCon =
        new ClassPathXmlApplicationContext("config/applicationContext.xml");
        //通过容器获取 test 实例
        TestDao tt = (TestDao)appCon.getBean("test");//test 为配置文件中的 id
        tt.sayHello();
    }
}
```

执行上述 main()方法后，将在控制台输出“Hello, Study hard!”。上述 main()方法中并没有使用 new 运算符创建 TestDaoImpl 类的对象，而是通过 Spring 容器获取实现类 TestDaoImpl 的对象，这就是 Spring IoC 的工作机制。

1.3　Spring IoC

视频讲解

1.3.1　Spring IoC 的基本概念

IoC（控制反转）是一个比较抽象的概念，是 Spring 框架的核心，用来消减计算机程序的耦合问题。依赖注入（DI）是 IoC 的另外一种说法，只是从不同的角度，描述相同的概念。下面通过实际生活中的一个例子解释 IoC 和 DI。

人们需要一件东西时，第一反应就是找东西，例如想吃面包。在没有面包店和有面包店两种情况下，你会怎么做？在没有面包店时，最直观的做法可能是你按照自己的口味制作面包，也就是一个面包需要主动制作。然而，时至今日，各种面包店的盛行，你不想制作面包时，可以把自己的口味告诉店家，一会儿就可以吃到可口的面包了。注意，你并没有制作面包，而是由店家制作，但是完全符合你的口味。

上面只是列举了一个非常简单的例子，但包含了控制反转的思想，即把制作面包的主动权交给店家。下面通过面向对象编程思想，继续探讨这两个概念。

当某个 Java 对象（调用者，比如你）需要调用另一个 Java 对象（被调用者，即被依赖对象，比如面包）时，在传统编程模式下，调用者通常会采用“new 被调用者”的代码方式创建对象（比如你自己制作面包）。这种方式会增加调用者与被调用者之间的耦合性，不利于后期代码的升级与维护。

当 Spring 框架出现后，对象的实例不再由调用者创建，而是由 Spring 容器（比如面包店）创建。Spring 容器会负责控制程序之间的关系（比如面包店负责控制你与面包的关系），而不是由调用者的程序代码直接控制。这样，控制权由调用者转移到 Spring 容器，控制权发生了反转，这就是 Spring 的控制反转。

从 Spring 容器角度来看，Spring 容器负责将被依赖对象赋值给调用者的成员变量，相当于为调用者注入它所依赖的实例，这就是 Spring 的依赖注入，主要目的是解耦，体现一种“组合”的理念。

综上所述，控制反转是一种通过描述（在 Spring 中可以是 XML 或注解）并通过第三方去产生或获取特定对象的方式。在 Spring 中实现控制反转的是 IoC 容器，其实现方法是依赖注入。

▶ 1.3.2 Spring 的常用注解

在 Spring 框架中，尽管使用 XML 配置文件可以很简单地装配 Bean（如例 1-1），但如果应用中有大量的 Bean 需要装配时，会导致 XML 配置文件过于庞大，不方便以后的升级维护。因此，更多时候推荐开发者使用注解（annotation）的方式去装配 Bean。

Spring 框架基于 AOP 编程（1.4 节）实现注解解析，因此，在使用注解编程时，需要导入 spring-aop-5.3.2.jar 包。

在 Spring 框架中定义了一系列的注解，常用注解如下。

❶ 声明 Bean 的注解

1）@Component

该注解是一个泛化的概念，仅仅表示一个组件对象（Bean），可以作用在任何层次上，没有明确的角色。

2）@Repository

该注解用于将数据访问层（DAO）的类标识为 Bean，即注解数据访问层 Bean，其功能与@Component()相同。

3）@Service

该注解用于标注一个业务逻辑组件类（Service 层），其功能与@Component()相同。

4）@Controller

该注解用于标注一个控制器组件类（Spring MVC 的 Controller），其功能与@Component()相同。

❷ 注入 Bean 的注解

1）@Autowired

该注解可以对类的成员变量、方法及构造方法进行标注，完成自动装配的工作。通过@Autowired 的使用消除 setter 和 getter 方法。默认按照 Bean 的类型进行装配。

2）@Resource

该注解与@Autowired 功能一样。区别在于，该注解默认按照名称装配注入，只有当找不到与名称匹配的 Bean 时才会按照类型装配注入；而@Autowired 默认按照 Bean 的类型进行装配，如果想按照名称装配注入，则需要结合@Qualifier 注解一起使用。

@Resource 注解有两个属性：name 和 type。name 属性指定 Bean 的实例名称，即按照名称来装配注入；type 属性指定 Bean 类型，即按照 Bean 的类型进行装配。

3）@Qualifier

该注解与@Autowired 注解配合使用。当@Autowired 注解需要按照名称装配注入时，则需要结合该注解一起使用，Bean 的实例名称由@Qualifier 注解的参数指定。

▶ 1.3.3 基于注解的依赖注入

Spring IoC 容器（ApplicationContext）负责创建和注入 Bean。Spring 提供使用 XML 配置、注解、Java 配置以及 groovy 配置实现 Bean 的创建和注入。本书前 4 章尽量使用注解（@Component、@Repository、@Service 以及@Controller 等业务 Bean 的配置）和 XML 配置方式。

下面通过一个简单实例向读者演示基于注解的依赖注入的使用过程。

【例 1-2】基于注解的依赖注入的使用过程。该实例的具体要求是：在 Controller 层中，依赖注入 Service 层；在 Service 层中，依赖注入 DAO 层。

具体实现步骤如下。

❶ 使用 Eclipse 创建 Web 应用并导入 JAR 包

使用 Eclipse 创建一个名为 ch1_2 的 Dynamic Web Project，并将 Spring 的 4 个基础包、spring-aop-5.3.2.jar 和第三方依赖包 commons-logging-1.2.jar 复制到 ch1_2 的 WEB-INF/lib 目录中。

❷ 创建 DAO 层

在 src 目录中，创建名为 annotation.dao 的包，并在该包中创建 TestDao 接口和 TestDaoImpl 实现类，并将实现类 TestDaoImpl 使用@Repository 注解标注为数据访问层。

TestDao 的代码如下：

```
package annotation.dao;
public interface TestDao {
    public void save();
}
```

TestDaoImpl 的代码如下：

```
package annotation.dao;
import org.springframework.stereotype.Repository;
@Repository("testDaoImpl")
/**相当于@Repository，但如果在 service 层使用@Resource(name="testDaoImpl")，
   testDaoImpl 不能省略。testDaoImpl 为 IoC 容器中的对象名**/
public class TestDaoImpl implements TestDao{
    @Override
    public void save() {
        System.out.println("testDao save");
    }
}
```

❸ 创建 Service 层

在 src 目录中，创建名为 annotation.service 的包，并在该包中创建 TestService 接口和 TestServiceImpl 实现类，并将实现类 TestServiceImpl 使用@Service 注解标注为业务逻辑层。

TestService 的代码如下：

```
package annotation.service;
public interface TestService {
    public void save();
}
```

TestServiceImpl 的核心代码如下：

```
@Service("testServiceImpl")//相当于@Service
public class TestServiceImpl implements TestService{
    @Resource(name="testDaoImpl")
    /**相当于@Autowired，@Autowired 默认按照 Bean 类型注入**/
    private TestDao testDao;
    @Override
    public void save() {
        testDao.save();
```

```
        System.out.println("testService save");
    }
}
```

❹ 创建 Controller 层

在 src 目录中，创建名为 annotation.controller 的包，并在该包中创建 TestController 类，并将 TestController 类使用@Controller 注解标注为控制器层。

TestController 的核心代码如下：

```
@Controller
public class TestController {
    @Autowired
    private TestService testService;
    public void save() {
        testService.save();
        System.out.println("testController save");
    }
}
```

❺ 创建配置文件

使用注解时，在 Spring 的配置文件中需要使用“<context:component-scan base-package="Bean 所在的包路径"/>”语句扫描使用注解的包，Spring IoC 容器根据 XML 配置文件的扫描信息，提供包以及子包中使用注解的类的实例供应用程序使用。

在 src 目录中，创建名为 config 的包，并在该包中创建名为 applicationContext.xml 的配置文件。

applicationContext.xml 的代码如下：

```
<?xml version="1.0" encoding="UTF-8"?>
<beans xmlns="http://www.springframework.org/schema/beans"
    xmlns:xsi="http://www.w3.org/2001/XMLSchema-instance"
    xmlns:context="http://www.springframework.org/schema/context"
    xsi:schemaLocation="http://www.springframework.org/schema/beans
       http://www.springframework.org/schema/beans/spring-beans.xsd
       http://www.springframework.org/schema/context
       http://www.springframework.org/schema/context/spring-context.xsd">
    <!-- 扫描annotation包及其子包中的注解 -->
    <context:component-scan base-package="annotation"/>
</beans>
```

❻ 创建测试类

在 src 目录中，创建名为 annotation.test 的包，并在该包中创建测试类 TestAnnotation，核心代码如下：

```
public class TestAnnotation {
    public static void main(String[] args) {
        @SuppressWarnings("resource")
        ApplicationContext appCon =
        new ClassPathXmlApplicationContext("config/applicationContext.xml");
        TestController tt = (TestController)appCon.getBean("testController");
        tt.save();
    }
}
```

1.4 Spring AOP

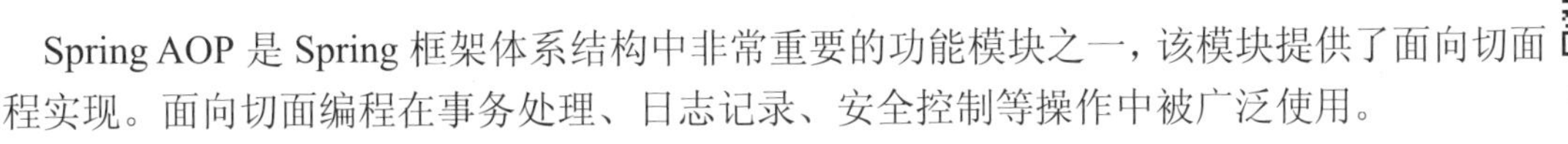

Spring AOP 是 Spring 框架体系结构中非常重要的功能模块之一，该模块提供了面向切面编程实现。面向切面编程在事务处理、日志记录、安全控制等操作中被广泛使用。

▶ 1.4.1 Spring AOP 的基本概念

❶ AOP 的概念

AOP，即面向切面编程。它与 OOP（Object-Oriented Programming，面向对象编程）相辅相成，提供了与 OOP 不同的抽象软件结构的视角。在 OOP 中，以类作为程序的基本单元，而 AOP 中的基本单元是 Aspect（切面）。Struts 2 的拦截器设计就是基于 AOP 的思想，是个比较经典的应用。

在业务处理代码中，通常都有日志记录、性能统计、安全控制、事务处理、异常处理等操作。尽管使用 OOP 可以通过封装或继承的方式达到代码的重用，但仍然存在同样的代码分散到各个方法中。因此，采用 OOP 处理日志记录等操作，不仅增加了开发者的工作量，而且提高了升级维护的困难。为了解决此类问题，AOP 思想应运而生。AOP 采取横向抽取机制，即将分散在各个方法中的重复代码提取出来，然后在程序编译或运行阶段，再将这些抽取出来的代码应用到需要执行的地方。这种横向抽取机制采用传统的 OOP 是无法办到的，因为 OOP 实现的是父子关系的纵向重用。但是 AOP 不是 OOP 的替代品，而是 OOP 的补充，它们是相辅相成的。

在 AOP 中，横向抽取机制的类与切面的关系如图 1.10 所示。

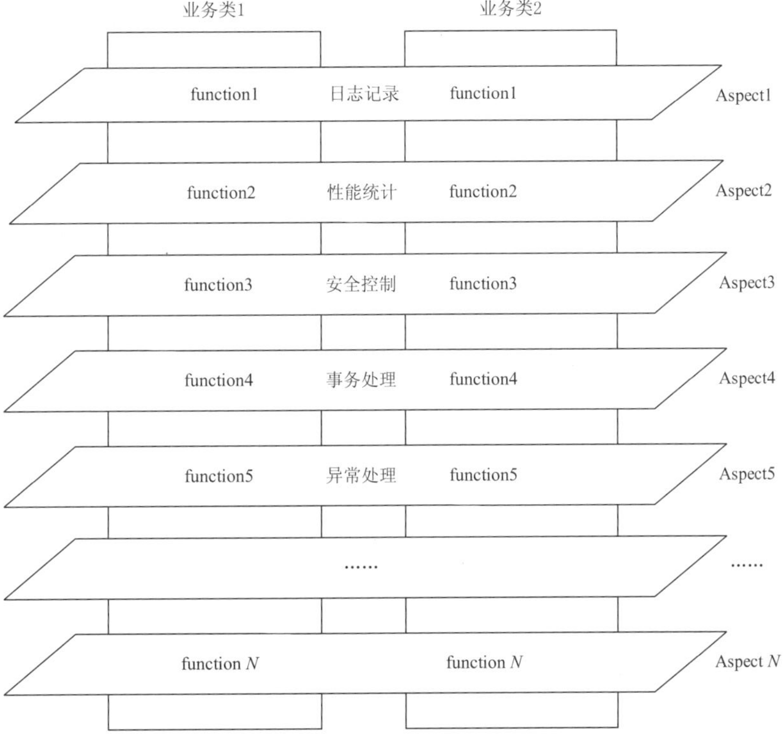

图 1.10 AOP 中类与切面的关系

从图 1.10 可以看出，通过切面 Aspect 分别在业务类 1 和业务类 2 中加入了日志记录、性能统计、安全控制、事务处理、异常处理等操作。

❷ AOP 的术语

在 Spring AOP 框架中，涉及以下常用术语。

1）切面

切面（Aspect）是指封装横切到系统功能（如事务处理）的类。

2）连接点

连接点（Joinpoint）是指程序运行中的一些时间点，如方法的调用或异常的抛出。

3）切入点

切入点（Pointcut）是指那些需要处理的连接点。在 Spring AOP 中，所有的方法执行都是连接点，而切入点是一个描述信息，它修饰的是连接点，通过切入点确定哪些连接点需要被处理。切面、连接点和切入点的关系如图 1.11 所示。

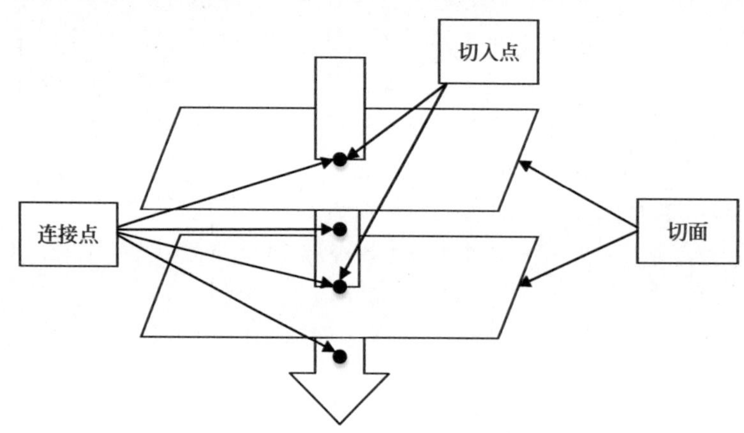

图 1.11　切面、连接点和切入点的关系

4）通知（增强处理）

通知是由切面添加到特定的连接点（满足切入点规则）的一段代码，即在定义好的切入点处所要执行的程序代码。可以将其理解为切面开启后，切面的方法。因此，通知是切面的具体实现。

5）引入

引入（Introduction）允许在现有的实现类中添加自定义的方法和属性。

6）目标对象

目标对象（Target Object）是指所有被通知的对象。如果 AOP 框架使用运行时代理的方式（动态的 AOP）来实现切面，那么通知对象总是一个代理对象。

7）代理

代理（Proxy）是通知应用到目标对象之后，被动态创建的对象。

8）组入

组入（Weaving）是将切面代码插入目标对象上，从而生成代理对象的过程。根据不同的实现技术，AOP 组入有三种方式：

（1）编译器组入，需要有特殊的 Java 编译器。

（2）类装载器组入，需要有特殊的类装载器。

（3）动态代理组入，在运行期为目标类添加通知生成子类。Spring AOP 框架默认采用动态代理组入，而 AspectJ（基于 Java 语言的 AOP 框架）采用编译器组入和类装载器组入。

1.4.2 基于注解开发 AspectJ

基于注解开发 AspectJ 要比基于 XML 配置开发 AspectJ 便捷许多，所以在实际开发中推荐使用注解方式。在讲解 AspectJ 之前，先了解一下 Spring 的通知类型。根据 Spring 中通知在目标类方法的连接点位置，通知可以分为以下 6 种类型。

❶ 环绕通知

环绕通知是在目标方法执行前和执行后实施增强，可以应用于日志记录、事务处理等。

❷ 前置通知

前置通知是在目标方法执行前实施增强，可应用于权限管理等。

❸ 后置返回通知

后置返回通知是在目标方法成功执行后实施增强，可应用于关闭流、删除临时文件等。

❹ 后置（最终）通知

后置通知是在目标方法执行后实施增强，与后置返回通知不同的是，不管是否发生异常都要执行该通知，可应用于释放资源。

❺ 异常通知

异常通知是在方法抛出异常后实施增强，可以应用于异常处理、日志记录等。

❻ 引入通知

引入通知是在目标类中添加一些新的方法和属性，可以应用于修改目标类（增强类）。

AspectJ 注解如表 1.1 所示。

表 1.1 AspectJ 注解

注解名称	描　述
@Aspect	用于定义一个切面，注解在切面类上
@Pointcut	用于定义切入点表达式。在使用时，需要定义一个切入点方法。该方法是一个返回值 void，且方法体为空的普通方法
@Before	用于定义前置通知。在使用时，通常为其指定 value 属性值，该值可以是已有的切入点，也可以直接定义切入点表达式
@AfterReturning	用于定义后置返回通知。在使用时，通常为其指定 value 属性值，该值可以是已有的切入点，也可以直接定义切入点表达式
@Around	用于定义环绕通知。在使用时，通常为其指定 value 属性值，该值可以是已有的切入点，也可以直接定义切入点表达式
@AfterThrowing	用于定义异常通知。在使用时，通常为其指定 value 属性值，该值可以是已有的切入点，也可以直接定义切入点表达式。另外，还有一个 throwing 属性用于访问目标方法抛出的异常，该属性值与异常通知方法中同名的形参一致
@After	用于定义后置（最终）通知。在使用时，通常为其指定 value 属性值，该值可以是已有的切入点，也可以直接定义切入点表达式

下面通过一个实例讲解基于注解开发 AspectJ 的过程。

【例 1-3】基于注解开发 AspectJ 的过程。该实例的具体要求是：首先，在 DAO 层的实现类中，定义 save、modify 和 delete 三个待增强的方法；然后，使用@Aspect 注解定义一个切面，在该切面中定义各类型通知，增强 DAO 层中的 save、modify 和 delete 方法。

具体实现步骤如下。

❶ 使用 Eclipse 创建 Web 应用并导入 JAR 包

使用 Eclipse 创建一个名为 ch1_3 的 Dynamic Web Project，除了将 Spring 的 4 个基础包、spring-aop-5.3.2.jar 和第三方依赖包 commons-logging-1.2.jar 复制到 ch1_3 的 WEB-INF/lib 目录中外，还需要将 Spring 为 AspectJ 框架提供的实现 spring-aspects-5.3.2.jar 以及 AspectJ 框架所提供的规范包 aspectjweaver-xxx.jar 复制到 WEB-INF/lib 目录中。

AspectJ 框架所提供的规范包 aspectjweaver-xxx.jar 可通过地址 http://mvnrepository.com/artifact/org.aspectj/aspectjweaver 下载，本书使用的是 aspectjweaver-1.9.6.jar。

❷ 创建接口及实现类

在 src 目录中，创建一个名为 aspectj.dao 的包，并在该包中创建接口 TestDao 和接口实现类 TestDaoImpl。该实现类作为目标类，在切面类中对其方法进行增强处理。使用注解@Repository 将目标类 aspectj.dao.TestDaoImpl 注解为目标对象。

TestDao 的代码如下：

```
package aspectj.dao;
public interface TestDao {
    public void save();
    public void modify();
    public void delete();
}
```

TestDaoImpl 的代码如下：

```
package aspectj.dao;
import org.springframework.stereotype.Repository;
@Repository("testDao")
public class TestDaoImpl implements TestDao{
    @Override
    public void save() {
        System.out.println("保存");
    }
    @Override
    public void modify() {
        System.out.println("修改");
    }
    @Override
    public void delete() {
        System.out.println("删除");
    }
}
```

❸ 创建切面类

在 src 目录中，创建一个名为 aspectj.annotation 的包，并在该包中创建切面类 MyAspect。在该类中，首先使用@Aspect 注解定义一个切面类，由于该类在 Spring 中是作为组件使用的，所以还需要使用@Component 注解；然后使用@Pointcut 注解定义切入点表达式，并通过定义方法表示切入点名称；最后在每个通知方法上添加相应的注解，并将切入点名称作为参数传递给需要执行增强的通知方法。

MyAspect 的核心代码如下：

```
/**
 * 切面类，在此类中编写各种类型通知
 */
@Aspect//@Aspect 声明一个切面
@Component//@Component 让此切面成为 Spring 容器管理的 Bean
public class MyAspect {
    /**
     * 定义切入点，通知增强哪些方法。
"execution(* aspectj.dao.*.*(..))" 是定义切入点表达式，
该切入点表达式的意思是匹配 aspectj.dao 包中任意类的任意方法的执行。
其中 execution()是表达式的主体，第一个*表示返回类型，使用*代表所有类型；
aspectj.dao 表示需要匹配的包名，后面第二个*表示类名，使用*代表匹配包中所有的类；
第三个*表示方法名，使用*表示所有方法；后面(..)表示方法的参数，其中“..”表示任意参数。
另外，注意第一个*与包名之间有一个空格。
     */
    @Pointcut("execution(* aspectj.dao.*.*(..))")
    private void myPointCut() {
    }
    /**
     * 前置通知，使用 Joinpoint 接口作为参数获得目标对象信息
     */
    @Before("myPointCut()")//myPointCut()是切入点的定义方法
    public void before(JoinPoint jp) {
        System.out.print("前置通知:模拟权限控制");
        System.out.println(",目标类对象:" + jp.getTarget()
        + ",被增强处理的方法:" + jp.getSignature().getName());
    }
    /**
     * 后置返回通知
     */
    @AfterReturning("myPointCut()")
    public void afterReturning(JoinPoint jp) {
        System.out.print("后置返回通知:" + "模拟删除临时文件");
        System.out.println(",被增强处理的方法:" + jp.getSignature().getName());
    }
    /**
     * 环绕通知
     * ProceedingJoinPoint 是 JoinPoint 子接口，代表可以执行的目标方法
     * 返回值类型必须是 Object
     * 必须有一个参数是 ProceedingJoinPoint 类型
     * 必须是 throws Throwable
     */
    @Around("myPointCut()")
    public Object around(ProceedingJoinPoint pjp) throws Throwable{
        //开始
        System.out.println("环绕开始：执行目标方法前，模拟开启事务");
        //执行当前目标方法
        Object obj = pjp.proceed();
```

```
        //结束
        System.out.println("环绕结束：执行目标方法后，模拟关闭事务");
        return obj;
    }
    /**
     * 异常通知
     */
    @AfterThrowing(value="myPointCut()",throwing="e")
    public void except(Throwable e) {
        System.out.println("异常通知:" + "程序执行异常" + e.getMessage());
    }
    /**
     * 后置(最终)通知
     */
    @After("myPointCut()")
    public void after() {
        System.out.println("最终通知：模拟释放资源");
    }
}
```

❹ 创建配置文件

在 src 目录中，创建一个名为 aspectj.config 的包，并在该包中创建配置文件 applicationContext.xml，在配置文件中指定需要扫描的包，使注解生效。同时，需要启动基于注解的 AspectJ 支持。

applicationContext.xml 的代码如下：

```
<?xml version="1.0" encoding="UTF-8"?>
<beans xmlns="http://www.springframework.org/schema/beans"
    xmlns:xsi="http://www.w3.org/2001/XMLSchema-instance"
    xmlns:aop="http://www.springframework.org/schema/aop"
    xmlns:context="http://www.springframework.org/schema/context"
    xsi:schemaLocation="http://www.springframework.org/schema/beans
        http://www.springframework.org/schema/beans/spring-beans.xsd
        http://www.springframework.org/schema/aop
        http://www.springframework.org/schema/aop/spring-aop.xsd
        http://www.springframework.org/schema/context
        http://www.springframework.org/schema/context/spring-context.xsd">
    <!-- 指定需要扫描的包，使注解生效 -->
    <context:component-scan base-package="aspectj"/>
    <!-- 启动基于注解的 AspectJ 支持 -->
    <aop:aspectj-autoproxy/>
</beans>
```

❺ 创建测试类

在 src 目录中，创建一个名为 aspectj.test 的包，并在该包中创建测试类 AspectjAOPTest。AspectjAOPTest 的核心代码如下：

```
public class AspectjAOPTest {
    public static void main(String[] args) {
        @SuppressWarnings("resource")
        ApplicationContext appCon =
```

```
        new ClassPathXmlApplicationContext("aspectj/config/
          applicationContext.xml");
        //从容器中，获取增强后的目标对象
        TestDao testDaoAdvice = (TestDao)appCon.getBean("testDao");
        //执行方法
        testDaoAdvice.save();
        System.out.println("==============");
        testDaoAdvice.modify();
        System.out.println("==============");
        testDaoAdvice.delete();
    }
}
```

❻ 运行测试类

运行测试类 AspectjAOPTest 的 main 方法，运行结果如图 1.12 所示。

```
Markers  Properties  Servers  Data Source Explorer  Snippets  Console
<terminated> AspectjAOPTest [Java Application] C:\soft\javaee\eclipse\plugins\org.eclipse.justj.op
环绕开始：执行目标方法前，模拟开启事务
前置通知：模拟权限控制，目标类对象：aspectj.dao.TestDaoImpl@2e2ff723，被增强处理的方法：save
保存
后置返回通知：模拟删除临时文件，被增强处理的方法：save
最终通知：模拟释放资源
环绕结束：执行目标方法后，模拟关闭事务
==============
环绕开始：执行目标方法前，模拟开启事务
前置通知：模拟权限控制，目标类对象：aspectj.dao.TestDaoImpl@2e2ff723，被增强处理的方法：modify
修改
后置返回通知：模拟删除临时文件，被增强处理的方法：modify
最终通知：模拟释放资源
环绕结束：执行目标方法后，模拟关闭事务
==============
环绕开始：执行目标方法前，模拟开启事务
前置通知：模拟权限控制，目标类对象：aspectj.dao.TestDaoImpl@2e2ff723，被增强处理的方法：delete
删除
后置返回通知：模拟删除临时文件，被增强处理的方法：delete
最终通知：模拟释放资源
环绕结束：执行目标方法后，模拟关闭事务
```

图 1.12 ch1_3 应用的运行结果

1.5 Spring Bean

视频讲解

在 Spring 应用中，Spring IoC 容器可以创建、装配和配置应用组件对象，这里的组件对象称为 Bean。

▶ 1.5.1 Bean 的实例化

在面向对象编程中，想使用某个对象时，需要事先实例化该对象。同样，在 Spring 框架中，想使用 Spring 容器中的 Bean，也需要实例化 Bean。Spring 框架实例化 Bean 有三种方式：构造方法实例化、静态工厂实例化和实例工厂实例化（其中，最常用的实例方法是构造方法实例化）。

下面通过一个实例 ch1_4 演示 Bean 的实例化过程。

【例 1-4】Bean 的实例化过程。该实例的具体要求是：分别使用构造方法、静态工厂和实例工厂实例化 Bean。

具体实现步骤如下。

❶ 使用 Eclipse 创建 Web 应用并导入相关 JAR 包

使用 Eclipse 创建一个名为 ch1_4 的 Dynamic Web Project，并将 Spring 的 4 个基础包和第三方依赖包 commons-logging-1.2.jar 复制到 ch1_4 的 WEB-INF/lib 目录中。

❷ 创建实例化 Bean 的类

在 src 目录中，创建一个名为 instance 的包，并在该包中创建 BeanClass、BeanInstanceFactory 以及 BeanStaticFactory 等实例化 Bean 的类。

BeanClass 的代码如下：

```
package instance;
public class BeanClass {
    public String message;
    public BeanClass() {
        message = "构造方法实例化 Bean";
    }
    public BeanClass(String s) {
        message = s;
    }
}
```

BeanInstanceFactory 的代码如下：

```
package instance;
public class BeanInstanceFactory {
    public BeanClass createBeanClassInstance() {
        return new BeanClass("调用实例工厂方法实例化 Bean");
    }
}
```

BeanStaticFactory 的代码如下：

```
package instance;
public class BeanStaticFactory {
    private static BeanClass beanInstance = new BeanClass("调用静态工厂方法
       实例化 Bean");
    public static BeanClass createInstance() {
       return beanInstance;
    }
}
```

❸ 创建配置文件

在 src 目录中，创建名为 config 的包，并在该包中创建名为 applicationContext.xml 的配置文件，在配置文件中分别使用 BeanClass、BeanInstanceFactory 以及 BeanStaticFactory 实例化 Bean。

applicationContext.xml 的代码如下：

```
<?xml version="1.0" encoding="UTF-8"?>
<beans xmlns="http://www.springframework.org/schema/beans"
```

```
    xmlns:xsi="http://www.w3.org/2001/XMLSchema-instance"
    xsi:schemaLocation="http://www.springframework.org/schema/beans
        http://www.springframework.org/schema/beans/spring-beans.xsd">
    <!-- 构造方法实例化 Bean -->
    <bean id="constructorInstance" class="instance.BeanClass"/>
    <!-- 静态工厂方法实例化 Bean，createInstance 为静态工厂类 BeanStaticFactory
       中的静态方法-->
    <bean id="staticFactoryInstance"
        class="instance.BeanStaticFactory" factory-method="createInstance"/>
    <!-- 配置工厂 -->
    <bean id="myFactory" class="instance.BeanInstanceFactory"/>
    <!-- 使用 factory-bean 属性指定配置工厂，使用 factory-method 属性指定使用工厂中
       哪个方法实例化 Bean-->
    <bean id="instanceFactoryInstance"
        factory-bean="myFactory" factory-method="createBeanClassInstance"/>
</beans>
```

❹ 创建测试类

在 src 目录中，创建 test 包，并在该包中创建测试类 TestInstance，核心代码如下：

```
public class TestInstance {
    public static void main(String[] args) {
        @SuppressWarnings("resource")
        ApplicationContext appCon =
          new ClassPathXmlApplicationContext("config/applicationContext.xml");
        //测试构造方法实例化 Bean
        BeanClass b1=(BeanClass)appCon.getBean("constructorInstance");
        System.out.println(b1+b1.message);
        //测试静态工厂方法实例化 Bean
        BeanClass b2=(BeanClass)appCon.getBean("staticFactoryInstance");
        System.out.println(b2+b2.message);
        //测试实例工厂方法实例化 Bean
        BeanClass b3 =(BeanClass)appCon.getBean("instanceFactoryInstance");
        System.out.println(b3+b3.message);
    }
}
```

❺ 运行测试类

运行测试类 TestInstance 的 main 方法，运行结果如图 1.13 所示。

```
Markers  Properties  Servers  Data Source Explorer
<terminated> TestInstance [Java Application] C:\soft\javaee
instance.BeanClass@13b6aecc构造方法实例化Bean
instance.BeanClass@158a8276调用静态工厂方法实例化Bean
instance.BeanClass@3c3d9b6b调用实例工厂方法实例化Bean
```

图 1.13 ch1_4 应用的运行结果

1.5.2 Bean 的作用域

在 Spring 中，不仅可以完成 Bean 的实例化，还可以为 Bean 指定作用域。在 Spring 中为 Bean 的实例定义了如表 1.2 所示的作用域，通过属性 scope 设定。

表 1.2　Bean 的作用域

作用域名称	描　述
singleton	默认的作用域，使用 singleton 定义的 Bean 在 Spring 容器中只有一个 Bean 实例
prototype	Spring 容器每次获取 prototype 定义的 Bean，容器都将创建一个新的 Bean 实例
request	在一次 HTTP 请求中，容器将返回一个 Bean 实例，不同的 HTTP 请求返回不同的 Bean 实例。仅在 Web Spring 应用程序上下文中使用
session	在一个 HTTP Session 中，容器将返回同一个 Bean 实例。仅在 Web Spring 应用程序上下文中使用
application	为每个 ServletContext 对象创建一个实例，即同一个应用共享一个 Bean 实例。仅在 Web Spring 应用程序上下文中使用
websocket	为每个 WebSocket 对象创建一个 Bean 实例。仅在 Web Spring 应用程序上下文中使用

在表 1.2 所示的 6 种作用域中，singleton 和 prototype 是最常用的两种，后面 4 种作用域仅使用在 Web Spring 应用程序上下文中。下面通过一个实例演示 Bean 的作用域。

【例 1-5】Bean 的作用域。该实例的具体要求是：在应用 ch1_4 中，分别定义作用域为 singleton 和 prototype 的两个 Bean。

具体实现步骤如下。

❶ 添加配置文件内容

在应用 ch1_4 的配置文件 applicationContext.xml 中，定义两个 Bean，一个 Bean 的作用域为 singleton，另一个 Bean 的作用域为 prototype。具体添加的配置内容如下：

```
<bean id="scope1" class="instance.BeanClass" scope="singleton"/>
<bean id="scope2" class="instance.BeanClass" scope="prototype"/>
```

❷ 创建测试类

在应用 ch1_4 的 test 包中，创建测试类 TestScope，在该测试类中分别获得 id 为 scope1 和 scope2 的 Bean 实例，核心代码如下：

```
public class TestScope {
    public static void main(String[] args) {
        @SuppressWarnings("resource")
        ApplicationContext appCon =
          new ClassPathXmlApplicationContext("config/applicationContext.xml");
        BeanClass b1=(BeanClass)appCon.getBean("scope1");
        System.out.println(b1);
        BeanClass b2=(BeanClass)appCon.getBean("scope1");
        System.out.println(b2);
        System.out.println("==========");
        BeanClass b3=(BeanClass)appCon.getBean("scope2");
        System.out.println(b3);
        BeanClass b4=(BeanClass)appCon.getBean("scope2");
        System.out.println(b4);
    }
}
```

❸ 运行测试类

运行测试类 TestScope 的 main 方法，运行结果如图 1.14 所示。

```
Markers  Properties  Servers
<terminated> TestScope [Java Applic
instance.BeanClass@13b6aecc
instance.BeanClass@13b6aecc
==========
instance.BeanClass@158a8276
instance.BeanClass@3c3d9b6b
```

图 1.14　TestScope 的运行结果

从图 1.14 的运行结果可知，两次获取 id 为 scope1 的 Bean 实例时，IoC 容器返回两个相同的 Bean 实例；而两次获取 id 为 scope2 的 Bean 实例时，IoC 容器返回两个不同的 Bean 实例。

1.5.3　Bean 的初始化和销毁

在实际工程应用中，经常需要在 Bean 使用之前或之后做些必要的操作，Spring 为 Bean 生命周期的操作提供了支持。在配置文件中定义 Bean 时，可以使用 init-method 和 destroy-method 属性对 Bean 进行初始化和销毁。下面通过一个实例演示 Bean 的初始化和销毁。

【例 1-6】Bean 的初始化和销毁。该实例的具体要求是：在应用 ch1_4 中，首先定义一个 MyService 类，在该类中定义初始化方法和销毁方法；然后在配置文件中，使用 init-method 和 destroy-method 属性对 MyService 的 Bean 对象进行初始化和销毁。

具体实现步骤如下。

❶ 创建 Bean 的类

在应用 ch1_4 的 src 目录中，创建一个名为 service 的包，并在该包中创建 MyService 类，具体代码如下：

```
package service;
public class MyService {
    public MyService(){
        System.out.println("执行构造方法");
    }
    public void initService() {
        System.out.println("initMethod");
    }
    public void destroyService() {
        System.out.println("destroyMethod");
    }
}
```

❷ 添加配置文件内容

在应用 ch1_4 的配置文件 applicationContext.xml 中，配置一个 id 为 beanLife 的 Bean，并使用 init-method 属性指定初始化方法，使用 destroy-method 属性指定销毁方法。具体添加的配置内容如下：

```
<!-- 配置 bean，使用 init-method 属性指定初始化方法，使用 destroy-method 属性指定销毁
    方法-->
<bean id="beanLife" class="service.MyService"
        init-method="initService" destroy-method="destroyService"/>
```

❸ 创建测试类

在应用 ch1_4 的 test 包中，创建测试类 TestInitAndDestroy，核心代码如下：

```
public class TestInitAndDestroy {
    public static void main(String[] args) {
        //为了方便演示销毁方法的执行，这里使用 ClassPathXmlApplicationContext
        ClassPathXmlApplicationContext appCon =
          new ClassPathXmlApplicationContext("config/applicationContext.xml");
        System.out.println("获得对象前");
        MyService blife=(MyService)appCon.getBean("beanLife");
        System.out.println("获得对象后" + blife);
        appCon.close();//关闭容器，销毁 Bean 对象
    }
}
```

❹ 运行测试类

运行测试类 TestInitAndDestroy 的 main 方法，运行结果如图 1.15 所示。

```
Markers  Properties  Servers  Snippets  Console  Progress  Coverage
<terminated> TestInitAndDestroy [Java Application] C:\soft\javaee\eclipse\plugins\org.eclip
执行构造方法
initMethod
获得对象前
获得对象后service.MyService@75db5df9
destroyMethod
```

图 1.15　Bean 的初始化和销毁

从图 1.15 可以看出，加载配置文件时，创建 Bean 对象，执行了 Bean 的构造方法和初始化方法 initService()；获得对象后，关闭容器时，执行了 Bean 的销毁方法 destroyService()。

视频讲解

1.6　Spring 的数据库编程

数据库编程是互联网编程的基础，Spring 框架为开发者提供了 JDBC 模板模式，即 jdbcTemplate，它可以简化许多代码，但在实际应用中 jdbcTemplate 并不常用。实际工程中用的是 Hibernate 框架和 MyBatis 框架进行数据库编程。本节仅简要介绍 Spring jdbcTemplate 的使用方法，而 Hibernate 框架和 MyBatis 框架的相关内容不属于本节的内容。

1.6.1　Spring JDBC 的 XML 配置

本节（1.6 节）主要使用 Spring JDBC 模块的 core 包和 dataSource 包。core 包是 JDBC 的核心功能包，包括常用的 JdbcTemplate 类；dataSource 包是访问数据源的工具类包。使用 Spring JDBC 操作数据库，需要对其进行配置。XML 配置文件示例代码如下：

```
<!-- 配置数据源 -->
<bean id="dataSource" class=
        "org.springframework.jdbc.datasource.DriverManagerDataSource">
        <!-- MySQL 数据库驱动 -->
        <property name="driverClassName" value="com.mysql.jdbc.Driver"/>
        <!-- 连接数据库的 URL -->
        <property name="url" value="jdbc:mysql://localhost:3306/
         springtest?characterEncoding=utf8"/>
```

```
        <!-- 连接数据库的用户名 -->
        <property name="username" value="root"/>
        <!-- 连接数据库的密码 -->
        <property name="password" value="root"/>
</bean>
<!-- 配置 JDBC 模板 -->
<bean id="jdbcTemplate" class="org.springframework.jdbc.core.JdbcTemplate">
        <property name="dataSource" ref="dataSource"/>
</bean>
```

上述示例代码中，配置 JDBC 模板时，需要将 dataSource 注入 jdbcTemplate，而在数据访问层（如 Dao 类）使用 jdbcTemplate 时，也需要将 jdbcTemplate 注入对应的 Bean 中。代码示例如下：

```
@Repository
public class TestDaoImpl implements TestDao{
    @Autowired
    //使用配置文件中的 JDBC 模板
    private JdbcTemplate jdbcTemplate;
    …
}
```

▶ 1.6.2 Spring JdbcTemplate 的常用方法

获取 JDBC 模板后，如何使用它是本节要讲述的内容。首先，需要了解 JdbcTemplate 类的常用方法——update()和 query()。

1）public int update(String sql,Object args[])

该方法可以对数据表进行增加、修改、删除等操作。使用 args[]设置 SQL 语句中的参数，并返回更新的行数。示例代码如下：

```
String insertSql = "insert into user values(null,?,?)";
Object param1[] = {"chenheng1", "男"};
jdbcTemplate.update(insertSql, param1);
```

2）public List<T> query (String sql, RowMapper<T> rowMapper, Object args[])

该方法可以对数据表进行查询操作。rowMapper 将结果集映射到用户自定义的类中（前提是自定义类中的属性与数据表的字段对应）。示例代码如下：

```
String selectSql ="select * from user";
RowMapper<MyUser> rowMapper = new BeanPropertyRowMapper<MyUser>
  (MyUser.class);
List<MyUser> list = jdbcTemplate.query(selectSql, rowMapper, null);
```

下面通过一个实例演示 Spring JDBC 的使用过程。

【例 1-7】Spring JDBC 的使用过程。该实例的具体要求是：首先在 MySQL 数据库中创建数据表 user；然后使用 Spring JDBC 对数据表 user 进行增删改查。

具体实现步骤如下。

❶ 使用 Eclipse 创建 Web 应用并导入相关 JAR 包

使用 Eclipse 创建一个名为 ch1_5 的 Dynamic Web Project，除了将 Spring 的 4 个基础包

和第三方依赖包 commons-logging-1.2.jar 复制到 ch1_5 的 WEB-INF/lib 目录中，还需要将 MySQL 数据库的连接驱动 JAR 包（mysql-connector-java-5.1.45-bin.jar）、Spring JDBC（spring-jdbc-5.3.2.jar）、Spring 事务处理（spring-tx-5.3.2.jar）以及 Spring AOP 的 JAR 包（spring-aop-5.3.2.jar）复制到 WEB-INF/lib 目录中。

❷ 创建配置文件

在 src 目录中，创建名为 config 的包，并在该包中创建名为 applicationContext.xml 的配置文件，在配置文件中配置数据源和 JDBC 模板，具体内容如下：

```
<?xml version="1.0" encoding="UTF-8"?>
<beans xmlns="http://www.springframework.org/schema/beans"
    xmlns:xsi="http://www.w3.org/2001/XMLSchema-instance"
    xmlns:context="http://www.springframework.org/schema/context"
    xsi:schemaLocation="http://www.springframework.org/schema/beans
       http://www.springframework.org/schema/beans/spring-beans.xsd
       http://www.springframework.org/schema/context
       http://www.springframework.org/schema/context/spring-context.xsd">
    <!-- 指定需要扫描的包(包括子包)，使注解生效 -->
    <context:component-scan base-package="dao"/>
    <context:component-scan base-package="service"/>
    <!-- 配置数据源 -->
    <bean id="dataSource" class=
             "org.springframework.jdbc.datasource.DriverManagerDataSource">
        <!-- MySQL 数据库驱动 -->
        <property name="driverClassName" value="com.mysql.jdbc.Driver"/>
        <!-- 连接数据库的 URL -->
        <property name="url" value="jdbc:mysql://localhost:3306/
             springtest?characterEncoding=utf8"/>
        <!-- 连接数据库的用户名 -->
        <property name="username" value="root"/>
        <!-- 连接数据库的密码 -->
        <property name="password" value="root"/>
    </bean>
    <!-- 配置 JDBC 模板 -->
    <bean id="jdbcTemplate" class="org.springframework.jdbc.core.JdbcTemplate">
        <property name="dataSource" ref="dataSource"/>
    </bean>
</beans>
```

❸ 创建数据表与实体类

使用 Navicat for MySQL 创建数据库 springtest，并在该数据库中创建数据表 user，数据表 user 的结构如图 1.16 所示。

名	类型	长度	小数点	允许空值 (	
uid	int	11	0	☐	🔑1
uname	varchar	20	0	☑	
usex	varchar	10	0	☑	

图 1.16 数据表 user 的结构

在应用 ch1_5 的 src 目录中，创建一个名为 entity 的包，并在该包中创建实体类 MyUser，具体代码如下：

```
package entity;
public class MyUser {
    private Integer uid;
    private String uname;
    private String usex;
    //省略 set 和 get 方法
    public String toString() {
        return "myUser [uid=" + uid +", uname=" + uname + ", usex=" + usex + "]";
    }
}
```

❹ 创建数据访问层

在应用 ch1_5 的 src 目录中，创建一个名为 dao 的包，在该包中创建数据访问接口 TestDao 和接口实现类 TestDaoImpl。在实现类 TestDaoImpl 中使用@Repository 注解标注此类为数据访问层，并使用@Autowired 注解依赖注入 JdbcTemplate。

TestDao 的核心代码如下：

```
public interface TestDao {
    public int update(String sql, Object[] param);
    public List<MyUser> query(String sql, Object[] param);
}
```

TestDaoImpl 的核心代码如下：

```
@Repository
public class TestDaoImpl implements TestDao{
     @Autowired
    //使用配置类中的 JDBC 模板
    private JdbcTemplate jdbcTemplate;
    /**
     * 更新方法，包括添加、修改、删除
     * param 为 sql 中的参数，如通配符？
     */
    @Override
    public int update(String sql, Object[] param) {
        return jdbcTemplate.update(sql, param);
    }
    /**
     * 查询方法
     * param 为 sql 中的参数，如通配符？
     */
    @Override
    public List<MyUser> query(String sql, Object[] param) {
        RowMapper<MyUser> rowMapper =
            new BeanPropertyRowMapper<MyUser>(MyUser.class);
        return jdbcTemplate.query(sql, rowMapper);
    }
}
```

❺ 创建业务逻辑层

在应用 ch1_5 的 src 目录中，创建一个名为 service 的包，在该包中创建数据访问接口

TestService 和接口实现类 TestServiceImpl。在实现类 TestServiceImpl 中使用@Service 注解标注此类为业务逻辑层，并使用@Autowired 注解依赖注入 TestDao。

TestService 的代码如下：

```
package service;
public interface TestService {
    public void testJDBC();
}
```

TestServiceImpl 的核心代码如下：

```
@Service
public class TestServiceImpl implements TestService{
    @Autowired
    public TestDao testDao;
    @Override
    public void testJDBC() {
        String insertSql = "insert into user values(null,?,?)";
        //数组 param 的值与 insertSql 语句中的?一一对应
        Object param1[] = {"chenheng1", "男"};
        Object param2[] = {"chenheng2", "女"};
        Object param3[] = {"chenheng3", "男"};
        Object param4[] = {"chenheng4", "女"};
        //添加用户
        testDao.update(insertSql, param1);
        testDao.update(insertSql, param2);
        testDao.update(insertSql, param3);
        testDao.update(insertSql, param4);
        //查询用户
        String selectSql ="select * from user";
        List<MyUser> list = testDao.query(selectSql, null);
        for(MyUser mu : list) {
            System.out.println(mu);
        }
    }
}
```

❻ 创建测试类

在应用 ch1_5 的 src 目录中，创建一个名为 test 的包，并在该包中创建测试类 TestJDBC，核心代码如下：

```
public class TestJDBC {
    public static void main(String[] args) {
        @SuppressWarnings("resource")
        ApplicationContext appCon =
          new ClassPathXmlApplicationContext("config/applicationContext.xml");
        TestService ts = (TestService)appCon.getBean("testServiceImpl");
        ts.testJDBC();
    }
}
```

❼ 运行测试类

运行测试类 TestJDBC 的 main 方法，运行结果如图 1.17 所示。

```
Markers  Properties  Servers  Data Source
<terminated> TestJDBC [Java Application] C:\soft\ja
myUser [uid=31, uname=chenheng1, usex=男]
myUser [uid=32, uname=chenheng2, usex=女]
myUser [uid=33, uname=chenheng3, usex=男]
myUser [uid=34, uname=chenheng4, usex=女]
```

图 1.17　ch1_5 应用的运行结果

1.6.3　基于@Transactional 注解的声明式事务管理

Spring 的声明式事务管理是通过 AOP 技术实现的事务管理，其本质是对方法前后进行拦截，然后在目标方法开始之前创建或者加入一个事务，在执行完目标方法之后根据执行情况提交或者回滚事务。

声明式事务管理最大的优点是不需要通过编程的方式管理事务，因而不需要在业务逻辑代码中掺杂事务处理的代码，只需相关的事务规则声明，便可以将事务规则应用到业务逻辑中。通常情况下，在开发中使用声明式事务处理，不仅因为其简单，更主要是因为这样使得纯业务代码不被污染，极大地方便了后期的代码维护。

和编程式事务管理相比，声明式事务管理唯一不足的地方是，最细粒度只能作用到方法级别，无法做到像编程式事务管理那样可以作用到代码块级别。但即便有这样的需求，也可以通过变通的方法进行解决，例如，可以将需要进行事务处理的代码块独立为方法。Spring 的声明式事务管理可以通过两种方式来实现，一种是基于 XML 的方式，另一种是基于@Transactional 注解的方式。

@Transactional 注解可以作用于接口、接口方法、类以及类方法上。当作用于类上时，该类的所有 public 方法都将具有该类型的事务属性，同时，也可以在方法级别使用该注解来覆盖类级别的定义。虽然@Transactional 注解可以作用于接口、接口方法、类以及类方法上，但是 Spring 小组建议不要在接口或者接口方法上使用该注解，因为只有在使用基于接口的代理时它才会生效。可以使用@Transactional 注解的属性定制事务行为，具体属性如表 1.3 所示。

表 1.3　@Transactional 注解的属性

属　性	属性值含义	默认值
propagation	Propagation 定义了事务的生命周期，主要有以下选项。 ① Propagation.REQUIRED：需要事务支持的方法 A 被调用时，如果没有事务则新建一个事务。当在方法 A 中调用另一个方法 B 时，方法 B 将使用相同的事务。如果方法 B 发生异常需要数据回滚时，整个事务数据回滚。 ② Propagation.REQUIRES_NEW：对于方法 A 和 B，在方法调用时，无论是否有事务都开启一个新的事务；方法 B 有异常不会导致方法 A 的数据回滚。 ③ Propagation.NESTED：和 Propagation.REQUIRES_NEW 类似，仅支持 JDBC，不支持 JPA 或 Hibernate。 ④ Propagation.SUPPORTS：方法调用时有事务就使用事务，没有事务就不创建事务。 ⑤ Propagation.NOT_SUPPORTED：强制方法在事务中执行，若有事务，在方法调用到结束阶段事务都将会被挂起。 ⑥ Propagation.NEVER：强制方法不在事务中执行，若有事务则抛出异常。 ⑦ Propagation.MANDATORY：强制方法在事务中执行，若无事务则抛出异常	Propagation. REQUIRED

续表

属　性	属性值含义	默认值
isolation	Isolation（隔离）决定了事务的完整性，可以设置多事务对相同数据下的处理机制，主要包含以下隔离级别（前提是当前数据库是否支持）。 ① Isolation.READ_UNCOMMITTED：对于在事务 A 中修改了一条记录但没有提交事务，在事务 B 中可以读取修改后的记录。可导致脏读、不可重复读以及幻读。 ② Isolation.READ_COMMITTED：只有当在事务 A 中修改了一条记录且提交事务后，事务 B 才可以读取提交后的记录，防止脏读，但可能导致不可重复读和幻读。 ③ Isolation.REPEATABLE_READ：不仅能实现 Isolation .READ_COMMITTED 的功能，而且还能阻止当事务 A 读取了一条记录，事务 B 将不允许修改该条记录；阻止脏读和不可重复读，但可出现幻读。 ④ Isolation.SERIALIZABLE：此级别下事务是顺序执行的，可以避免上述级别的缺陷，但开销较大。 ⑤ Isolation.DEFAULT：使用当前数据库的默认隔离级别。如 Oracle 和 SQL Server 是 READ_COMMITTED；MySQL 是 REPEATABLE_READ	Isolation.DEFAULT
timeout	timeout 指定事务过期时间，默认为当前数据库的事务过期时间	
readOnly	指定当前事务是否是只读事务	false
rollbackFor	指定哪个或哪些异常可以引起事务回滚（Class 对象数组，必须继承自 Throwable）	Throwable 的子类
rollbackForClassName	指定哪个或哪些异常可以引起事务回滚（类名数组，必须继承自 Throwable）	Throwable 的子类
noRollbackFor	指定哪个或哪些异常不可以引起事务回滚（Class 对象数组，必须继承自 Throwable）	Throwable 的子类
noRollbackForClassName	指定哪个或哪些异常不可以引起事务回滚（类名数组，必须继承自 Throwable）	Throwable 的子类

本节通过实例演示基于@Transactional 注解的声明式事务管理。

【例 1-8】基于@Transactional 注解的声明式事务管理。例 1-8 是通过修改例 1-7 中的代码实现的。

具体步骤如下。

❶ 添加配置文件内容

在配置文件中，使用<tx:annotation-driven>元素为事务管理器注册注解驱动器，同时为数据源添加事务管理器。添加后的配置文件内容如下：

```
<?xml version="1.0" encoding="UTF-8"?>
<beans xmlns="http://www.springframework.org/schema/beans"
    xmlns:xsi="http://www.w3.org/2001/XMLSchema-instance"
    xmlns:context="http://www.springframework.org/schema/context"
    xmlns:tx="http://www.springframework.org/schema/tx"
    xsi:schemaLocation="http://www.springframework.org/schema/beans
      http://www.springframework.org/schema/beans/spring-beans.xsd
      http://www.springframework.org/schema/context
      http://www.springframework.org/schema/context/spring-context.xsd
      http://www.springframework.org/schema/tx
      http://www.springframework.org/schema/tx/spring-tx.xsd">
```

```
    <!-- 指定需要扫描的包(包括子包),使注解生效 -->
    <context:component-scan base-package="dao"/>
    <context:component-scan base-package="service"/>
    <!-- 配置数据源 -->
    <bean id="dataSource" class=
            "org.springframework.jdbc.datasource.DriverManagerDataSource">
        <!-- MySQL 数据库驱动 -->
        <property name="driverClassName" value="com.mysql.jdbc.Driver"/>
        <!-- 连接数据库的 URL -->
        <property name="url" value="jdbc:mysql://localhost:3306/
          springtest?characterEncoding=utf8"/>
        <!-- 连接数据库的用户名 -->
        <property name="username" value="root"/>
        <!-- 连接数据库的密码 -->
        <property name="password" value="root"/>
    </bean>
    <!-- 配置 JDBC 模板 -->
    <bean id="jdbcTemplate" class="org.springframework.jdbc.core.JdbcTemplate">
        <property name="dataSource" ref="dataSource"/>
    </bean>
    <!-- 为数据源添加事务管理器 -->
    <bean id="txManager" class=
        "org.springframework.jdbc.datasource.DataSourceTransactionManager">
        <property name="dataSource" ref="dataSource" />
    </bean>
    <!-- 为事务管理器注册注解驱动 -->
    <tx:annotation-driven transaction-manager="txManager" />
</beans>
```

❷ 修改业务逻辑层

在实际开发中，通常通过 Service 层进行事务管理，因此需要为 Service 层添加@Transactional 注解。

添加@Transactional 注解后的 TestServiceImpl 类的核心代码如下：

```
@Service
@Transactional
public class TestServiceImpl implements TestService{
    @Autowired
    public TestDao testDao;
    @Override
    public void testJDBC() {
        String insertSql = "insert into user values(null,?,?)";
        //数组 param 的值与 insertSql 语句中的?一一对应
        Object param1[] = {"chenheng1", "男"};
        Object param2[] = {"chenheng2", "女"};
        Object param3[] = {"chenheng3", "男"};
        Object param4[] = {"chenheng4", "女"};
        String insertSql1 = "insert into user values(?,?,?)";
        Object param5[] = {1,"chenheng5", "女"};
        Object param6[] = {1,"chenheng6", "女"};
        //添加用户
        testDao.update(insertSql, param1);
```

```
        testDao.update(insertSql, param2);
        testDao.update(insertSql, param3);
        testDao.update(insertSql, param4);
        //添加两个 ID 相同的用户，出现唯一性约束异常,使事务回滚
        testDao.update(insertSql1, param5);
        testDao.update(insertSql1, param6);
    }
}
```

❸ 测试事务处理

首先清空数据表 user 中的数据，然后运行测试类 TestJDBC，发现数据表 user 中并没有数据，这是因为最后执行添加数据时主键重复，事务回滚，即回到程序运行的初始状态。

▶ 1.6.4 如何在事务处理中捕获异常

声明式事务处理的流程是：

（1）Spring 根据配置完成事务定义，设置事务属性。

（2）执行开发者的代码逻辑。

（3）如果开发者的代码产生异常（如主键重复）并且满足事务回滚的配置条件，则事务回滚；否则，事务提交。

（4）事务资源释放。

现在的问题是，如果开发者在代码逻辑中加入了 try…catch…语句，Spring 还能不能在声明式事务处理中正常得到事务回滚的异常信息？答案是不能。例如，将 1.6.3 节中 TestServiceImpl 实现类的 testJDBC 方法的代码修改如下：

```
@Override
public voidtestJDBC () {
    String insertSql="insert into user values(null,?,?)";
    //数组 param 的值与 insertSql 语句中的?一一对应
    Object param1[]={"chenheng1", "男"};
    Object param2[]={"chenheng2", "女"};
    Object param3[]={"chenheng3", "男"};
    Object param4[]={"chenheng4", "女"};
    String insertSql1="insert into user values(?,?,?)";
    Object param5[]={1,"chenheng5", "女"};
    Object param6[]={1,"chenheng6", "女"};
    try {
        //添加用户
        testDao.update(insertSql, param1);
        testDao.update(insertSql, param2);
        testDao.update(insertSql, param3);
        testDao.update(insertSql, param4);
        //添加两个 ID 相同的用户，出现唯一性约束异常,使事务回滚
        testDao.update(insertSql1, param5);
        testDao.update(insertSql1, param6);
        } catch (Exception e) {
            System.out.println("主键重复,事务回滚。");
        }
    }
```

这时，再运行测试类，发现主键重复但事务并没有回滚。这是因为默认情况下，Spring 只在发生未被捕获的 RuntimeException 时才事务回滚。现在，如何在事务处理中捕获异常呢？具体修改如下。

（1）将 TestServiceImpl 类中的@Transactional 注解修改为：

```
@Transactional(rollbackFor= {Exception.class})
//rollbackFor 指定回滚生效的异常类，多个异常类之间用逗号分隔；
//noRollbackFor 指定回滚失效的异常类
```

（2）在 catch 语句中添加 “throw new RuntimeException();” 语句。

注意： 在实际工程应用中，经常仅需要在 catch 语句中添加 “TransactionAspectSupport.currentTransactionStatus().setRollbackOnly();” 语句即可。也就是说，不需要修改@Transactional 注解和在 catch 语句中添加 “throw new RuntimeException();” 语句。

1.7 本章小结

本章讲解了 Spring IoC、AOP、Bean 以及事务管理等基础知识，目的是让读者在学习 Spring MVC 之前，对 Spring 有个基本了解。

习 题 1

1. Spring 的核心容器由哪些模块组成？
2. 如何找到 Spring 框架的官方 API？
3. 什么是 Spring IoC？什么是依赖注入？
4. 在配置文件中如何开启 Spring 对 AspectJ 的支持？又如何开启 Spring 对声明式事务的支持？
5. 什么是 Spring AOP？它与 OOP 是什么关系？

第 2 章　Spring MVC

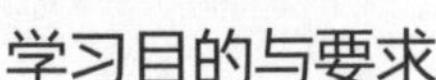

学习目的与要求

本章重点讲解 Spring MVC 的工作原理、控制器以及数据绑定。通过本章的学习，了解 Spring MVC 的工作原理，掌握 Spring MVC 应用的开发步骤。

主要内容

- ❖ Spring MVC 的工作原理
- ❖ Spring MVC 的工作环境
- ❖ 基于注解的控制器
- ❖ 表单标签库与数据绑定
- ❖ JSON 数据交互

MVC 思想将一个应用分成三个基本部分：Model（模型）、View（视图）和 Controller（控制器），让这三个部分以最低的耦合进行协同工作，从而提高应用的可扩展性及可维护性。Spring MVC 是一款优秀的基于 MVC 思想的应用框架，它是 Spring 提供的一个实现了 Web MVC 设计模式的轻量级 Web 框架。

2.1　Spring MVC 的工作原理

视频讲解

Spring MVC 框架是高度可配置的，包含多种视图技术，例如 JSP 技术、Velocity、Tiles、iText 和 POI。Spring MVC 框架并不关心使用的视图技术，也不会强迫开发者只使用 JSP 技术，但本章使用的视图是 JSP。

Spring MVC 框架主要由 DispatcherServlet、处理器映射、控制器、视图解析器、视图组成，其工作原理如图 2.1 所示。

从图 2.1 可总结出 Spring MVC 的工作流程如下。

（1）客户端请求提交到 DispatcherServlet。

（2）由 DispatcherServlet 控制器寻找一个或多个 HandlerMapping，找到处理请求的 Controller。

（3）DispatcherServlet 将请求提交到 Controller。

（4）Controller 调用业务逻辑处理后，返回 ModelAndView。

（5）DispatcherServlet 寻找一个或多个 ViewResolver 视图解析器，找到 ModelAndView 指定的视图。

（6）视图负责将结果显示到客户端。

图 2.1 中包含 4 个 Spring MVC 接口：DispatcherServlet、HandlerMapping、Controller 和 ViewResolver。

Spring MVC 所有的请求都经过 DispatcherServlet 来统一分发。DispatcherServlet 将请

求分发给 Controller 之前，需要借助于 Spring MVC 提供的 HandlerMapping 定位到具体的 Controller。

HandlerMapping 接口负责完成客户请求到 Controller 映射。

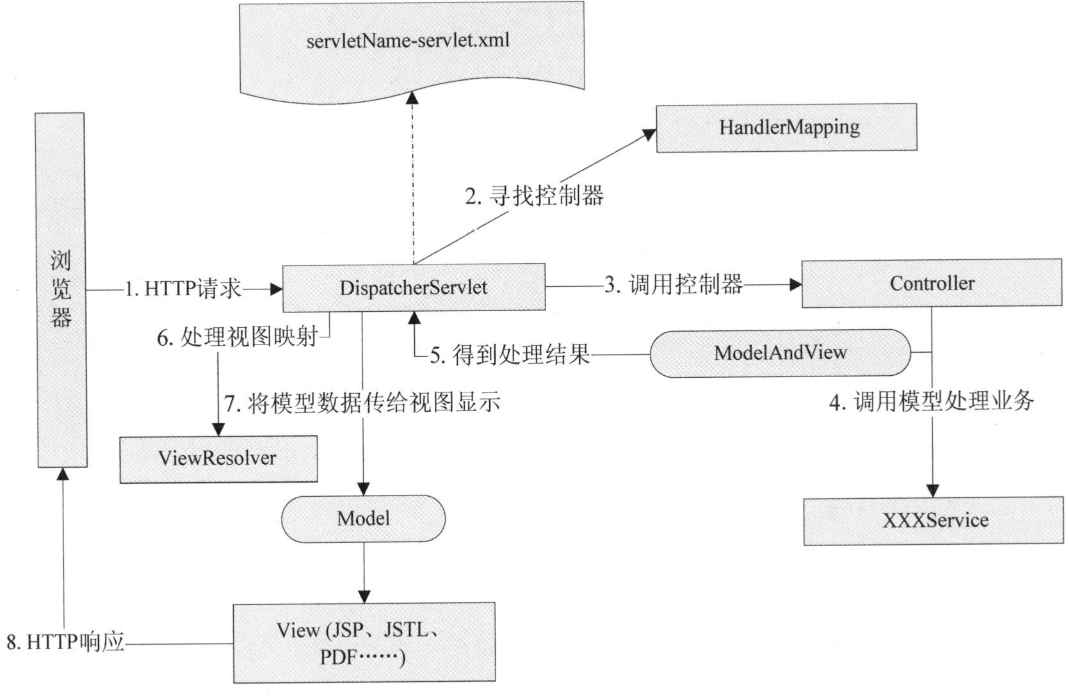

图 2.1　Spring MVC 工作原理图

Controller 接口将处理用户请求，这和 Java Servlet 扮演的角色是一致的。一旦 Controller 处理完用户请求，则返回 ModelAndView 对象给 DispatcherServlet 前端控制器，ModelAndView 中包含了模型（Model）和视图（View）。从宏观角度考虑，DispatcherServlet 是整个 Web 应用的控制器；从微观角度考虑，Controller 是单个 Http 请求处理过程中的控制器，而 ModelAndView 是 Http 请求过程中返回的模型（Model）和视图（View）。

ViewResolver 接口（视图解析器）在 Web 应用中负责查找 View 对象，从而将相应结果渲染给客户。

2.2　Spring MVC 的工作环境

视频讲解

2.2.1　Spring MVC 所需要的 JAR 包

在第 1 章 Java Web 开发环境的基础上，导入 Spring MVC 的相关 JAR 包，即可开发 Spring MVC 应用。

对于 Spring MVC 框架的初学者，开发 Spring MVC 应用时，只需将 Spring 的 4 个基础包、commons-logging-1.2.jar、注解时需要的 JAR 包 spring-aop-5.1.4.RELEASE.jar 和 Spring MVC 相关的 JAR 包（spring-web-5.3.2.jar 和 spring-webmvc-5.3.2.jar）复制到 Web 应用的 WEB-INF/lib 目录下即可。

2.2.2 使用 Eclipse 开发 Spring MVC 的 Web 应用

本节通过一个实例来演示 Spring MVC 入门程序的实现过程。

【例 2-1】Spring MVC 入门程序的实现过程。

具体实现步骤如下。

❶ 创建 Web 应用 ch2_1 并导入 JAR 包

创建 Web 应用 ch2_1，导入如 2.2.1 节所述的 JAR 包。

❷ 在 web.xml 文件中部署 Spring MVC 核心控制器 DispatcherServlet

在开发 Spring MVC 应用时，需要在 WEB-INF 目录下创建 web.xml 文件，并在该文件中部署 DispatcherServlet，示例代码如下：

```
<?xml version="1.0" encoding="UTF-8"?>
<web-app
xmlns:xsi="http://www.w3.org/2001/XMLSchema-instance"
xmlns="http://xmlns.jcp.org/xml/ns/javaee"
xsi:schemaLocation="http://xmlns.jcp.org/xml/ns/javaee
http://xmlns.jcp.org/xml/ns/javaee/web-app_4_0.xsd"
id="WebApp_ID"
version="4.0">
<!--配置 springmvcDispatcherServlet-->
<servlet>
    <servlet-name>springmvc</servlet-name>
    <servlet-class>org.springframework.web.servlet.DispatcherServlet
        </servlet-class>
    <load-on-startup>1</load-on-startup>
</servlet>
<servlet-mapping>
    <servlet-name>springmvc</servlet-name>
    <url-pattern>/</url-pattern>
</servlet-mapping>
</web-app>
```

上述 DispatcherServlet 的 servlet 对象 springmvc 初始化时，默认情况下，将在应用程序的 WEB-INF 目录下查找 Spring MVC 配置文件，该配置文件的命名规则是“servletName-servlet.xml”，例如 springmvc-servlet.xml。

另外，也可以将 Spring MVC 的配置文件存放在应用程序目录中的任何地方，但需要使用 servlet 的 init-param 元素加载配置文件，示例代码如下：

```
<!--配置 Spring MVC DispatcherServlet -->
<servlet>
    <servlet-name>springmvc</servlet-name>
    <servlet-class>org.springframework.web.servlet.DispatcherServlet
       </servlet-class>
    <init-param>
        <param-name>contextConfigLocation</param-name>
        <!-- classpath 是指到 src 目录查找配置文件 -->
        <param-value>classpath:config/springmvc.xml</param-value>
```

```
        <!-- 如果没有 classpath,是到 WebContent 目录找，示例如下 -->
        <!-- <param-value>/WEN-INF/spring-config/springmvc.xml</param-value> -->
    </init-param>
    <load-on-startup>1</load-on-startup>
</servlet>
<servlet-mapping>
    <servlet-name>springmvc</servlet-name>
    <url-pattern>/</url-pattern>
</servlet-mapping>
```

❸ 创建 Web 应用首页

在 ch2_1 应用的 WebContent 目录下，有个应用首页 index.jsp。index.jsp 的核心代码如下：

```
<body>
    没注册的用户，请<a href="index/register">注册</a>! <br>
    已注册的用户，去<a href="index/login">登录</a>!
</body>
```

❹ 创建 Controller 类

在 ch2_1 应用的 src 目录下，创建包 controller，并在该包中创建基于注解的名为 IndexController 的控制器类，该类中有两个处理请求方法，分别处理首页 index.jsp 中“注册（index/register）”和“登录（index/login）”超链接请求。IndexController 的代码如下：

```
package controller;
import org.springframework.stereotype.Controller;
import org.springframework.web.bind.annotation.RequestMapping;
/**"@Controller"表示 IndexController 的实例是一个控制器
 * @Controller 相当于@Controller("indexController")
 * 或@Controller(value = "indexController")
 */
@Controller
@RequestMapping("/index")
public class IndexController {
    @RequestMapping("/login")
    public String login() {
        /**login 代表逻辑视图名称，需要根据 Spring MVC 配置
         * 文件中 internalResourceViewResolver 的前缀和后缀找到对应的物理视图
         */
        return "login";
    }
    @RequestMapping("/register")
    public String register() {
        return "register";
    }
}
```

❺ 创建 Spring MVC 的配置文件

在 Spring MVC 中，使用扫描机制找到应用中所有基于注解的控制器类。所以，为了让控制器类被 Spring MVC 框架扫描到，需要在配置文件中声明 spring-context，并使用<context:component-scan/>元素指定控制器类的基本包（请确保所有控制器类都在基本包及其

子包下）。另外，需要在配置文件中定义 Spring MVC 的视图解析器（ViewResolver），示例代码如下：

```
<bean class="org.springframework.web.servlet.view.InternalResourceViewResolver"
        id="internalResourceViewResolver">
    <!-- 前缀 -->
    <property name="prefix" value="/WEB-INF/jsp/" />
    <!-- 后缀 -->
    <property name="suffix" value=".jsp" />
</bean>
```

上述视图解析器配置了前缀和后缀两个属性。因此，控制器类中视图路径仅需提供 register 和 login，视图解析器将会自动添加前缀和后缀。

在 ch2_1 应用的 src 目录下，创建名为 config 的包，并在该包中创建名为 springmvc.xml 的配置文件(此时需要在 web.xml 配置文件中指定 springmvc.xml 文件的具体位置，见步骤 2)，其代码如下：

```
<?xml version="1.0" encoding="UTF-8"?>
<beans xmlns="http://www.springframework.org/schema/beans"
    xmlns:xsi="http://www.w3.org/2001/XMLSchema-instance"
    xmlns:context="http://www.springframework.org/schema/context"
    xsi:schemaLocation="
     http://www.springframework.org/schema/beans
     http://www.springframework.org/schema/beans/spring-beans.xsd
        http://www.springframework.org/schema/context
        http://www.springframework.org/schema/context/spring-context.xsd">
    <!-- 使用扫描机制，扫描控制器类 -->
    <context:component-scan base-package="controller"/>
     <!-- 配置视图解析器 -->
     <bean class=
       "org.springframework.web.servlet.view.InternalResourceViewResolver"
           id="internalResourceViewResolver">
       <!-- 前缀 -->
       <property name="prefix" value="/WEB-INF/jsp/" />
       <!-- 后缀 -->
       <property name="suffix" value=".jsp" />
  </bean>
</beans>
```

❻ 应用的其他页面

IndexController 控制器的 register 方法处理成功后，跳转到/WEB-INF/jsp/register.jsp 视图；IndexController 控制器的 login 方法处理成功后，跳转到/WEB-INF/jsp/login.jsp 视图。因此，应用的/WEB-INF/jsp 目录下应有 register.jsp 和 login.jsp 页面，此两个 JSP 页面代码略。

❼ 发布并运行 Spring MVC 应用

在 Eclipse 中第 1 次运行 Spring MVC 应用时，需要将应用发布到 Tomcat。例如，运行 ch2_1 应用时，可以右击应用名称 ch2_1，选择 Run As/Run on Server，即完成发布并运行。

2.3 基于注解的控制器

视频讲解

在使用 Spring MVC 进行 Web 应用开发时，Controller 是 Web 应用的核心。Controller 实现类包含对用户请求的处理逻辑，是用户请求和业务逻辑之间的“桥梁”，是 Spring MVC 框架的核心部分，负责具体的业务逻辑处理。

▶ 2.3.1 Controller 注解类型

在 Spring MVC 中，使用 org.springframework.stereotype.Controller 注解类型声明某类的实例是一个控制器，例如 2.2.2 节中的 IndexController 控制器类。别忘了在 Spring MVC 的配置文件中使用<context:component-scan/>元素（见例 2-1）指定控制器类的基本包，进而扫描所有注解的控制器类。

▶ 2.3.2 RequestMapping 注解类型

在基于注解的控制器类中，可以为每个请求编写对应的处理方法。如何将请求与处理方法一一对应呢？需要使用 org.springframework.web.bind.annotation.RequestMapping 注解类型。

❶ 方法级别注解

方法级别注解示例代码如下：

```
@Controller
public class IndexController {
    @RequestMapping(value = "/index/login")
    public String login() {
        /**login 代表逻辑视图名称，需要根据 Spring MVC 配置中
         *internalResourceViewResolver 的前缀和后缀找到对应的物理视图
         */
        return "login";
    }
    @RequestMapping(value = "/index/register")
    public String register() {
        return "register";
    }
}
```

上述示例中有两个 RequestMapping 注解语句，它们都作用在处理方法上。注解的 value 属性将请求 URI 映射到方法，value 属性是 RequestMapping 注解的默认属性。如果就一个 value 属性，则可省略该属性。可以使用如下 URL 访问 login 方法（请求处理方法）：

```
http://localhost:xxx/yyyy/index/login
```

❷ 类级别注解

类级别注解示例代码如下：

```
@Controller
@RequestMapping("/index")
public class IndexController {
    @RequestMapping("/login")
    public String login() {
```

```
        return "login";
    }
    @RequestMapping("/register")
    public String register() {
        return "register";
    }
}
```

在类级别注解的情况下，控制器类中的所有方法都将映射为类级别的请求。可以使用如下 URL 访问 login 方法：

```
http://localhost:xxx/yyy/index/login
```

为了方便程序维护，建议开发者采用类级别注解，将相关处理放在同一个控制器类中，例如，对商品的增、删、改、查处理方法都可以放在一个名为 GoodsOperate 的控制类中。

2.3.3 编写请求处理方法

在控制器类中每个请求处理方法可以有多个不同类型的参数，以及一个多种类型的返回结果。

❶ 请求处理方法中常出现的参数类型

如果需要在请求处理方法中使用 Servlet API 类型，那么可以将这些类型作为请求处理方法的参数类型。Servlet API 参数类型示例代码如下：

```
@Controller
@RequestMapping("/index")
public class IndexController {
    @RequestMapping("/login")
    public String login(HttpSession session, HttpServletRequest request) {
        session.setAttribute("skey", "session 范围的值");
        request.setAttribute("rkey", "request 范围的值");
        return "login";
    }
}
```

除了 Servlet API 参数类型外，还有输入输出流、表单实体类、注解类型、与 Spring 框架相关的类型等。但特别重要的类型是 org.springframework.ui.Model 类型，该类型是一个包含 Map 的 Spring 框架类型。每次调用请求处理方法时，Spring MVC 都将创建 org.springframework.ui.Model 对象。Model 参数类型示例代码如下：

```
@Controller
@RequestMapping("/index")
public class IndexController {
    @RequestMapping("/register")
    public String register(Model model) {
    /*在视图中可以使用 EL 表达式${success}取出 model 中的值，有关 EL 的相关知识，请参
     考相关内容，不属于本书的范畴*/
        model.addAttribute("success", "注册成功");
        return "register";
    }
}
```

❷ 请求处理方法常见的返回类型

请求处理方法最常见的返回类型是代表逻辑视图名称的 String 类型。除了 String 类型外，还有 Model、View 以及其他任意的 Java 类型。

▶ 2.3.4　Controller 接收请求参数的常见方式

Controller 接收请求参数的方式有很多种，有的适合 get 请求方式，有的适合 post 请求方式，有的二者都适合。下面介绍几个常用的方式，读者可根据实际情况选择合适的接收方式。

❶ 通过实体 bean 接收请求参数

通过一个实体 bean 接收请求参数适用于 get 和 post 请求方式。需要注意的是，bean 的属性名称必须与请求参数名称相同。下面通过一个实例讲解通过实体 bean 接收请求参数。

【例 2-2】 通过实体 bean 接收请求参数。

具体步骤如下。

1）创建 Web 应用并导入 JAR 包

创建 Web 应用 ch2_2，导入如 2.2.1 节所述的 JAR 包。

2）创建视图文件

在应用 ch2_2 的/WEB-INF/目录下创建 jsp 文件夹，在 jsp 文件夹中创建 register.jsp、login.jsp 和 main.jsp 文件（main.jsp 的代码略）。

register.jsp 的核心代码如下：

```
<form action="user/register" method="post" name="registForm">
    <table>
        <tr>
            <td>姓名：</td>
            <td><input type="text" name="uname" value="${user.uname }"/></td>
        </tr>
        <tr>
            <td>密码：</td>
            <td><input type="password"  name="upass"/></td>
        </tr>
        <tr>
            <td>确认密码：</td>
            <td><input type="password"  name="reupass"/></td>
        </tr>
        <tr>
            <td colspan="2" align="center">
                <input type="submit" value="注册" />
            </td>
        </tr>
    </table>
</form>
```

login.jsp 的核心代码如下：

```
<form action="user/login" method="post">
    <table>
        <tr><td align="center" colspan="2">登录</td></tr>
        <tr>
```

```
                <td>姓名：</td>
                <td><input type="text" name="uname"></td>
            </tr>
            <tr>
                <td>密码：</td>
                <td><input type="password" name="upass"></td>
            </tr>
            <tr>
                <td colspan="2">
                    <input type="submit" value="提交" >
                    <input type="reset" value="重置" >
                </td>
            </tr>
        </table>
        ${messageError }
</form>
```

在应用 ch2_2 的/WebContent/目录下创建 index.jsp，核心代码如下：

```
<body>
    没注册的用户，请<a href="user/register">注册</a>! <br>
    已注册的用户，去<a href="user/login">登录</a>!
</body>
```

3）创建 POJO 实体类

在应用 ch2_2 的 src 目录下，创建包 pojo，并在该包中创建实体类 UserForm，代码如下：

```
package pojo;
public class UserForm {
    private String uname;       //与 form 表单的请求参数名称相同
    private String upass;
    private String reupass;
    //省略 getter 和 setter 方法
}
```

4）创建控制器类

在应用 ch2_2 的 src 目录下，创建包 controller，并在该包中创建控制器类 UserController 和 IndexController。IndexController 的代码与例 2-1 的相同，为节省篇幅，不再赘述。

UserController 的核心代码如下：

```
@Controller
@RequestMapping("/user")
public class UserController {
    /**
     * 处理登录，使用 UserForm 对象(实体 bean)user 接收登录页面提交的请求参数
     */
    @RequestMapping("/login")
    public String login(UserForm user, HttpSession session, Model model) {
            if("zhangsan".equals(user.getUname())
                    && "123456".equals(user.getUpass())) {
                session.setAttribute("u", user);
                return "main";//登录成功，跳转到 main.jsp
```

```
            }else{
                model.addAttribute("messageError", "用户名或密码错误");
                return "login";
            }
    }
    /**
     *处理注册，使用 UserForm 对象(实体 bean)user 接收注册页面提交的请求参数
     */
    @RequestMapping("/register")
    public String register(UserForm user, Model model) {
        if("zhangsan".equals(user.getUname())
                && "123456".equals(user.getUpass())) {
            return "login";//注册成功，跳转到 login.jsp
        }else{
            //在 register.jsp 页面上可以使用 EL 表达式取出 model 的 uname 值
            model.addAttribute("uname", user.getUname());
            return "register";//返回 register.jsp
        }
    }
}
```

5）创建 Web 与 Spring MVC 的配置文件

将例2-1的Web应用ch2_1的Web配置文件web.xml和Spring MVC配置文件springmvc.xml复制到应用 ch2_2 的相同位置。

6）发布并运行应用

右击应用 ch2_2，选择 Run As/Run on Server 发布并运行应用。

❷ 通过处理方法的形参接收请求参数

通过处理方法的形参接收请求参数，也就是直接把表单参数写在控制器类相应方法的形参中，即形参名称与请求参数名称完全相同。该接收参数方式适用于 get 和 post 请求方式。可以将例 2-2 的控制器类 UserController 中 register 方法的代码修改如下：

```
@RequestMapping("/register")
/**
 * 通过形参接收请求参数，形参名称与请求参数名称完全相同
 */
public String register(String uname, String upass, Model model) {
    if("zhangsan".equals(uname)
            && "123456".equals(upass)) {
        return "login";//注册成功，跳转到 login.jsp
    }else{
        //在 register.jsp 页面上可以使用 EL 表达式取出 model 的 uname 值
        model.addAttribute("uname", uname);
        return "register";//返回 register.jsp
    }
}
```

❸ 通过@RequestParam 接收请求参数

通过@RequestParam 接收请求参数适用于 get 和 post 请求方式。可以将例 2-2 的控制器类 UserController 中 register 方法的代码修改如下：

```
@RequestMapping("/register")
/**
 * 通过@RequestParam 接收请求参数
 */
public String register(@RequestParam String uname, @RequestParam String upass,
 Model model) {
    if("zhangsan".equals(uname)
            && "123456".equals(upass)) {
        return "login";//注册成功，跳转到 login.jsp
    }else{
        //在 register.jsp 页面上可以使用 EL 表达式取出 model 的 uname 值
        model.addAttribute("uname", uname);
        return "register";//返回 register.jsp
    }
}
```

通过@RequestParam 接收请求参数与通过处理方法的形参接收请求参数的区别是：当请求参数与接收参数名不一致时，通过处理方法的形参接收请求参数不会报 400 错误，而通过@RequestParam 接收请求参数会报 400 错误。

❹ 通过@ModelAttribute 接收请求参数

@ModelAttribute 注解放在处理方法的形参上时，用于将多个请求参数封装到一个实体对象中，从而简化数据绑定流程，而且自动暴露为模型数据用于视图页面展示时使用。而通过实体 bean 接收请求参数只是将多个请求参数封装到一个实体对象，并不能暴露为模型数据（需要使用 model.addAttribute 语句才能暴露为模型数据，数据绑定与模型数据展示可参考 2.4 节的内容）。

通过@ModelAttribute 注解接收请求参数适用于 get 和 post 请求方式。可以将例 2-2 的控制器类 UserController 中 register 方法的代码修改如下：

```
@RequestMapping("/register")
public String register(@ModelAttribute("user") UserForm user) {
    if("zhangsan".equals(user.getUname())
            && "123456".equals(user.getUpass())){
        return "login";//注册成功,跳转到 login.jsp
    }else{
        //使用@ModelAttribute("user")与 model.addAttribute("user", user)功能相同
        //在 register.jsp 页面上可以使用 EL 表达式${user.uname}取出 ModelAttribute
        //的 uname 值
        return "register";//返回 register.jsp
    }
}
```

2.3.5 重定向与转发

重定向是将用户从当前处理请求定向到另一个视图（如 JSP）或处理请求，以前的请求（request）中存放的信息全部失效，并进入一个新的 request 作用域；转发是将用户对当前处理的请求转发给另一个视图或处理请求，以前的 request 中存放的信息不会失效。

转发是服务器行为；重定向是客户端行为。具体工作流程如下。

转发过程：客户浏览器发送 http 请求，Web 服务器接受此请求，调用内部的一个方法在

容器内部完成请求处理和转发动作，将目标资源发送给客户；在这里，转发的路径必须是同一个 Web 容器下的 URL，其不能转向到其他的 Web 路径上去，中间传递的是自己的容器内的 request。在客户浏览器的地址栏中显示的仍然是其第一次访问的路径，也就是说客户是感觉不到服务器做了转发的。转发行为是浏览器只做了一次访问请求。

重定向过程：客户浏览器发送 http 请求，Web 服务器接受后发送 302 状态码响应及对应新的 location 给客户浏览器，客户浏览器发现是 302 响应，就自动再发送一个新的 http 请求，请求 URL 是新的 location 地址，服务器根据此请求寻找资源并发送给客户。在这里 location 可以重定向到任意 URL，既然是浏览器重新发出了请求，就没有什么 request 传递的概念了。在客户浏览器的地址栏中显示的是其重定向的路径，客户可以观察到地址的变化。重定向行为是浏览器做了至少两次的访问请求。

在 Spring MVC 框架中，控制器类中处理方法的 return 语句默认就是转发实现，只不过实现的是转发到视图。示例代码如下：

```
@RequestMapping("/register")
public String register() {
    return "register";//转发到 register.jsp
}
```

在 Spring MVC 框架中，重定向与转发的示例代码如下：

```
@Controller
@RequestMapping("/index")
public class IndexController {
    @RequestMapping("/login")
    public String login() {
        //转发到一个请求方法(同一个控制器类里，可省略/index/)
        return "forward:/index/isLogin";
    }
    @RequestMapping("/isLogin")
    public String isLogin() {
        //重定向到一个请求方法
        return "redirect:/index/isRegister";
    }
    @RequestMapping("/isRegister")
    public String isRegister() {
        //转发到一个视图
        return "register";
    }
}
```

在 Spring MVC 框架中，无论重定向还是转发，都需要符合视图解析器的配置，如果直接重定向到一个不需要 DispatcherServlet 的资源，如：

```
return "redirect:/html/my.html";
```

在 Spring MVC 配置文件中，需要使用 mvc:resources 配置，将不需要 DispatcherServlet 分发的请求释放，配置文件具体示例如下：

```
<?xml version="1.0" encoding="UTF-8"?>
<beans xmlns="http://www.springframework.org/schema/beans"
```

```
    xmlns:xsi="http://www.w3.org/2001/XMLSchema-instance"
    xmlns:context="http://www.springframework.org/schema/context"
    xmlns:mvc="http://www.springframework.org/schema/mvc"
    xsi:schemaLocation="
        http://www.springframework.org/schema/beans
        http://www.springframework.org/schema/beans/spring-beans.xsd
        http://www.springframework.org/schema/context
        http://www.springframework.org/schema/context/spring-context.xsd
        http://www.springframework.org/schema/mvc
        http://www.springframework.org/schema/mvc/spring-mvc.xsd
        ">
    <!-- 使用扫描机制，扫描控制器类 -->
    <context:component-scan base-package="controller"/>
    <mvc:annotation-driven />
    <!-- 使用resources过滤掉不需要dispatcher servlet的资源（即静态资源,如css、
      js、html、images）。
    使用resources时，必须使用annotation-driven，否则resources元素会阻止任意控制
      器被调用-->
    <!-- 允许WebContent/html目录下所有文件可见 -->
    <mvc:resources location="/html/" mapping="/html/**"></mvc:resources>
</beans>
```

▶ 2.3.6　应用@Autowired 进行依赖注入

在前面学习的控制器中，并没有体现 MVC 的 M 层，这是因为控制器既充当 C 层，又充当 M 层。这样设计程序的系统结构很不合理，应该将 M 层从控制器中分离出来。Spring MVC 框架本身就是一个非常优秀的 MVC 框架，它具有一个依赖注入的优点。可以通过 org.springframework.beans.factory.annotation.Autowired 注解类型将依赖注入一个属性（成员变量）或方法，例如：

```
@Autowired
public UserService userService;
```

在 Spring MVC 中，为了能被作为依赖注入，服务层的类必须使用 org.springframework.stereotype.Service 注解类型注明为@Service（一个服务）。另外，还需要在配置文件中使用<context:component-scan base-package="基本包"/>元素来扫描依赖基本包。下面将例 2-2 的 ch2_2 应用的“登录”和“注册”的业务逻辑处理分离出来，使用 Service 层实现。

首先，创建 service 包，在包中创建 UserService 接口和 UserServiceImpl 实现类。

UserService 接口的具体代码如下：

```
package service;
import pojo.UserForm;
public interface UserService {
    boolean login(UserForm user);
    boolean register(UserForm user);
}
```

UserServiceImpl 实现类的具体代码如下：

```
package service;
import org.springframework.stereotype.Service;
import pojo.UserForm;
//注解为一个服务
@Service
public class UserServiceImpl implements UserService{
    @Override
    public boolean login(UserForm user) {
        if("zhangsan".equals(user.getUname())
                && "123456".equals(user.getUpass()))
            return true;
        return false;
    }
    @Override
    public boolean register(UserForm user) {
        if("zhangsan".equals(user.getUname())
                && "123456".equals(user.getUpass()))
            return true;
        return false;
    }
}
```

其次，在配置文件中追加如下内容：

```
<context:component-scan base-package="service"/>
```

最后，修改控制器类 UserController，核心代码如下：

```
@Controller
@RequestMapping("/user")
public class UserController {
    //将服务层依赖注入属性 userService
    @Autowired
     public UserService userService;
    /**
     * 处理登录
     */
    @RequestMapping("/login")
    public String login(UserForm user, HttpSession session, Model model) {
        if(userService.login(user)){
            session.setAttribute("u", user);
            return "main";//登录成功,跳转到 main.jsp
        }else{
            model.addAttribute("messageError", "用户名或密码错误");
            return "login";
        }
    }
    /**
     *处理注册
     */
    @RequestMapping("/register")
    public String register(@ModelAttribute("user") UserForm user) {
        if(userService.register(user)){
```

```
            return "login";//注册成功,跳转到 login.jsp
        }else{
            return "register";//返回 register.jsp
        }
    }
}
```

2.3.7 @ModelAttribute

通过 org.springframework.web.bind.annotation.ModelAttribute 注解类型，可以实现如下两个功能。

❶ 绑定请求参数到实体对象（表单的命令对象）

该用法与 2.3.4 节的“通过@ModelAttribute 接收请求参数”一样：

```
@RequestMapping("/register")
public String register(@ModelAttribute("user") UserForm user) {
    if("zhangsan".equals(user.getUname())
            && "123456".equals(user.getUpass())){
        return "login";
    }else{
            return "register";
    }
}
```

上述代码中“@ModelAttribute("user") UserForm user”语句的功能有两个：一个是将请求参数的输入封装到 user 对象中；另一个是创建 UserForm 实例，以 user 为键值存储在 Model 对象中，与“model.addAttribute("user", user)”语句功能一样。如果没有指定键值，即“@ModelAttribute UserForm user”，那么创建 UserForm 实例时，以 userForm 为键值存储在 Model 对象中，与“model.addAttribute("userForm", user)”语句功能一样。

❷ 注解一个非请求处理方法

在控制器类中，被@ModelAttribute 注解的一个非请求处理方法将在每次调用该控制器类的请求处理方法前被调用。这种特性可以用来控制登录权限，当然控制登录权限的方法很多，例如拦截器、过滤器等。

使用该特性控制登录权限的示例代码如下：

```
public class BaseController {
    @ModelAttribute
    public void isLogin(HttpSession session) throws Exception {
        if(session.getAttribute("user") == null){
            throw new Exception("没有权限");
        }
    }
}
@Controller
@RequestMapping("/admin")
public class ModelAttributeController extends BaseController{
    @RequestMapping("/add")
    public String add(){
        return "addSuccess";
```

```
        }
        @RequestMapping("/update")
        public String update(){
            return "updateSuccess";
        }
        @RequestMapping("/delete")
        public String delete(){
            return "deleteSuccess";
        }
    }
```

上述 ModelAttributeController 类中的 add、update、delete 请求处理方法执行时，首先执行父类 BaseController 中的 isLogin 方法判断登录权限。

2.4　表单标签库与数据绑定

视频讲解

数据绑定是将用户参数输入值绑定到领域模型的一种特性。在 Spring MVC 的 Controller 和 View 参数数据传递中，所有 HTTP 请求参数的类型均为字符串。如果模型需要绑定的类型为 double 或 int，则需要手动进行类型转换；而有了数据绑定后，就不再需要手动将 HTTP 请求中的 String 类型转换为模型需要的类型。数据绑定的另一个好处是，当输入验证失败时，会重新生成一个 HTML 表单，无须重新填写输入字段。在 Spring MVC 中，为了方便、高效地使用数据绑定，还需要学习表单标签库。

2.4.1　表单标签库

表单标签库中包含了可以用在 JSP 页面中渲染 HTML 元素的标签。JSP 页面使用 Spring 表单标签库时，必须在 JSP 页面开头处声明 taglib 指令，指令代码如下：

```
<%@ taglib prefix="form" uri="http://www.springframework.org/tags/form" %>
```

表单标签库中有 form、input、password、hidden、textarea、checkbox、checkboxes、radiobutton、radiobuttons、select、option、options、errors 等标签。

form：渲染表单元素。

input：渲染<input type="text"/>元素。

password：渲染<input type="password"/>元素。

hidden：渲染<input type="hidden"/>元素。

textarea：渲染 textarea 元素。

checkbox：渲染一个<input type="checkbox"/>元素。

checkboxes：渲染多个<input type="checkbox"/>元素。

radiobutton：渲染一个<input type="radio"/>元素。

radiobuttons：渲染多个<input type="radio"/>元素。

select：渲染一个选择元素。

option：渲染一个选项元素。

options：渲染多个选项元素。

errors：在 span 元素中渲染字段错误。

❶ 表单标签

表单标签的语法格式如下：

```
<form:form modelAttribute="xxx" method="post" action="xxx">
    …
</form:form>
```

除了具有 HTML 表单元素属性外，表单标签还具有 acceptCharset、commandName、cssClass、cssStyle、htmlEscape 和 modelAttribute 等属性。各属性含义如下所示。

acceptCharset：定义服务器接受的字符编码列表。

commandName：暴露表单对象的模型属性名称，默认为 command。

cssClass：定义应用到 form 元素的 CSS 类。

cssStyle：定义应用到 form 元素的 CSS 样式。

htmlEscape：true 或 false，表示是否进行 HTML 转义。

modelAttribute：暴露 form backing object 的模型属性名称，默认为 command。

其中，commandName 和 modelAttribute 属性功能基本一致，属性值绑定一个 JavaBean 对象。假设控制器类 UserController 的方法 inputUser 是返回 userAdd.jsp 的请求处理方法，inputUser 方法的代码如下：

```
@RequestMapping(value = "/input")
public String inputUser(Model model) {
    …
    model.addAttribute("user", new User());
    return "userAdd";
}
```

userAdd.jsp 的表单标签代码如下：

```
<form:form modelAttribute="user"  method="post" action="user/save">
    …
</form:form>
```

注意：在 inputUser 方法中，如果没有 Model 属性 user，userAdd.jsp 页面就会抛出异常，因为表单标签无法找到在其 modelAttribute 属性中指定的 form backing object。

❷ input 标签

input 标签的语法格式如下：

```
<form:input path="xxx"/>
```

该标签除 cssClass、cssStyle、htmlEscape 属性外，还有一个最重要的属性——path。path 属性将文本框输入值绑定到 form backing object 的一个属性。示例代码如下：

```
<form:form modelAttribute="user"  method="post" action="user/save">
    <form:input path="userName"/>
</form:form>
```

上述代码将输入值绑定到 user 对象的 userName 属性。

❸ password 标签

password 标签的语法格式如下：

```
<form:password path="xxx"/>
```

该标签与 input 标签用法完全一致，为节省篇幅，不再赘述。

❹ hidden 标签

hidden 标签的语法格式如下：

```
<form:hidden path="xxx"/>
```

该标签与 input 标签用法基本一致，只不过它不可显示，不支持 cssClass 和 cssStyle 属性。

❺ textarea 标签

textarea 是一个支持多行输入的 input 元素，语法格式如下：

```
<form:textarea path="xxx"/>
```

该标签与 input 标签用法完全一致，为节省篇幅，不再赘述。

❻ checkbox 标签

checkbox 标签的语法格式如下：

```
<form:checkbox path="xxx" value="xxx"/>
```

多个 path 相同的 checkbox 标签，它们是一个选项组，允许多选。选项值绑定到一个数组属性。示例代码如下：

```
<form:checkbox path="friends" value="张三"/>张三
<form:checkbox path="friends" value="李四"/>李四
<form:checkbox path="friends" value="王五"/>王五
<form:checkbox path="friends" value="赵六"/>赵六
```

上述示例代码中复选框的值绑定到一个字符串数组属性 friends（String[] friends）。该标签的其他用法与 input 标签基本一致，为节省篇幅，不再赘述。

❼ checkboxes 标签

checkboxes 标签渲染多个复选框，是一个选项组，等价于多个 path 相同的 checkbox 标签。它有以下 3 个非常重要的属性。

items：用于生成 input 元素的 Collection、Map 或 Array。

itemLabel：items 属性中指定的集合对象的属性，为每个 input 元素提供 label。

itemValue：items 属性中指定的集合对象的属性，为每个 input 元素提供 value。

checkboxes 标签语法格式如下：

```
<form:checkboxes items="xxx"  path="xxx"/>
```

示例代码如下：

```
<form:checkboxes items="${hobbys}"  path="hobby" />
```

上述示例代码是将 model 属性 hobbys 的内容（集合元素）渲染为复选框。在 itemLabel 和 itemValue 缺省情况下，如果集合是数组，复选框的 label 和 value 相同；如果是 Map 集合，复选框的 label 是 Map 的值（value），复选框的 value 是 Map 的关键字（key）。

❽ radiobutton 标签

radiobutton 标签的语法格式如下：

```
<form:radiobutton path="xxx" value="xxx"/>
```

多个 path 相同的 radiobutton 标签，它们是一个选项组，只允许单选。

❾ radiobuttons 标签

radiobuttons 标签渲染多个 radio，是一个选项组，等价于多个 path 相同的 radiobutton 标签。radiobuttons 标签的语法格式如下：

```
<form:radiobuttons path="xxx" items="xxx"/>
```

该标签的 itemLabel 和 itemValue 属性与 checkboxes 标签的 itemLabel 和 itemValue 属性完全一样，但只允许单选。

❿ select 标签

select 标签的选项可以来自其属性 items 指定的集合，或者来自一个嵌套的 option 标签或 options 标签。语法格式如下：

```
<form:select path="xxx" items="xxx" />
```

或

```
<form:select path="xxx" items="xxx" >
    <option value="xxx">xxx</option>
</ form:select>
```

或

```
<form:select path="xxx">
    <form:options items="xxx"/>
</form:select>
```

该标签的 itemLabel 和 itemValue 属性与 checkboxes 标签的 itemLabel 和 itemValue 属性完全一样。

⓫ options 标签

options 标签生成一个 select 标签的选项列表。因此，需要与 select 标签一同使用，具体用法参见 select 标签。

⓬ errors 标签

errors 标签渲染一个或者多个 span 元素，每个 span 元素包含一个错误消息。它可以显示一个特定的错误消息，也可以显示所有错误消息。语法如下：

```
<form:errors path="*"/>
```

或

```
<form:errors path="xxx"/>
```

其中，“*”表示显示所有错误消息；“xxx”表示显示由“xxx”指定的特定错误消息。

2.4.2 数据绑定

为了让读者进一步学习数据绑定和表单标签，本节给出了一个应用实例 ch2_3。ch2_3 应用中实现了 User 类属性和 JSP 页面中表单参数的绑定，同时在 JSP 页面中分别展示了 input、textarea、checkbox、checkboxs、select 等标签。

【例 2-3】数据绑定和表单标签。

具体步骤如下。

❶ 创建应用并导入相关的 JAR 包

在 ch2_3 应用中需要使用 JSTL 标签，因此，不仅需要将 Spring MVC 相关 jar 包（如 2.2.1 节所示）复制到应用的 WEN-INF/lib 目录下，还需要从 Tomcat 的 webapps\examples\WEB-INF\lib 目录下将 JSTL 相关 jar 包（taglibs-standard-impl-1.2.5.jar 和 taglibs-standard-spec-1.2.5.jar）复制到应用的 WEN-INF/lib 目录下。

❷ 创建 Web 和 Spring MVC 配置文件

若在 WEB-INF 目录中创建 web.xml，为了避免中文乱码问题，则应该在 web.xml 中配置编码过滤器，同时 JSP 页面编码设置为 UTF-8，form 表单的提交方式为 post。

web.xml 的配置内容如下：

```
<?xml version="1.0" encoding="UTF-8"?>
<web-app xmlns:xsi="http://www.w3.org/2001/XMLSchema-instance"
    xmlns="http://xmlns.jcp.org/xml/ns/javaee"
    xsi:schemaLocation="http://xmlns.jcp.org/xml/ns/javaee
http://xmlns.jcp.org/xml/ns/javaee/web-app_4_0.xsd"
    id="WebApp_ID" version="4.0">
    <!--配置 Spring MVC DispatcherServlet -->
    <servlet>
        <servlet-name>springmvc</servlet-name>
        <servlet-class>org.springframework.web.servlet.DispatcherServlet
           </servlet-class>
        <init-param>
        <param-name>contextConfigLocation</param-name>
        <!-- classpath 是指到 src 目录查找配置文件 -->
        <param-value>classpath:config/springmvc.xml</param-value>
        </init-param>
    <load-on-startup>1</load-on-startup>
    </servlet>
    <servlet-mapping>
        <servlet-name>springmvc</servlet-name>
        <url-pattern>/</url-pattern>
    </servlet-mapping>
    <!-- 避免中文乱码 -->
    <filter>
        <filter-name>characterEncodingFilter</filter-name>
        <filter-class>org.springframework.web.filter.CharacterEncodingFilter
          </filter-class>
        <init-param>
            <param-name>encoding</param-name>
            <param-value>UTF-8</param-value>
        </init-param>
        <init-param>
            <param-name>forceEncoding</param-name>
            <param-value>true</param-value>
        </init-param>
    </filter>
    <filter-mapping>
```

```
        <filter-name>characterEncodingFilter</filter-name>
        <url-pattern>/*</url-pattern>
    </filter-mapping>
</web-app>
```

在 src 目录中，创建名为 config 的包，并在该包中创建 Spring MVC 的配置文件 springmvc.xml，具体代码如下：

```
<?xml version="1.0" encoding="UTF-8"?>
<beans xmlns="http://www.springframework.org/schema/beans"
    xmlns:xsi="http://www.w3.org/2001/XMLSchema-instance"
    xmlns:context="http://www.springframework.org/schema/context"
    xsi:schemaLocation="
    http://www.springframework.org/schema/beans
    http://www.springframework.org/schema/beans/spring-beans.xsd
        http://www.springframework.org/schema/context
        http://www.springframework.org/schema/context/spring-context.xsd">
    <!-- 使用扫描机制，扫描控制器类 -->
    <context:component-scan base-package="controller"/>
    <context:component-scan base-package="service"/>
    <!-- 配置视图解析器 -->
    <bean class="
      org.springframework.web.servlet.view.InternalResourceViewResolver"
          id="internalResourceViewResolver">
        <!-- 前缀 -->
        <property name="prefix" value="/WEB-INF/jsp/" />
        <!-- 后缀 -->
        <property name="suffix" value=".jsp" />
    </bean>
</beans>
```

❸ 创建 View 层

View 层包含两个 JSP 页面，一个是信息输入页面 userAdd.jsp，一个是信息显示页面 userList.jsp。在 ch2_3 应用的 WEB-INF/jsp/目录下，创建这两个 JSP 页面。

在 userAdd.jsp 页面中将 Map 类型的 hobbys 绑定到 checkboxes 上，将 String[]类型的 carrers 和 houseRegisters 绑定到 select 上，实现通过 option 标签对 select 添加选项，同时表单的 method 方法需指定为 post 来避免中文乱码问题。

在 userList.jsp 页面中使用 JSTL 标签遍历集合中的用户信息，JSTL 知识参见相关内容，不属于本书范畴。

userAdd.jsp 的核心代码如下：

```
<form:form modelAttribute="user"  method="post" action="user/save">
    <fieldset>
        <legend>添加一个用户</legend>
        <p>
            <label>用户名:</label>
            <form:input path="userName"/>
        </p>
        <p>
            <label>爱好:</label>
            <form:checkboxes items="${hobbys}"  path="hobby" />
```

```
        </p>
        <p>
            <label>朋友:</label>
            <form:checkbox path="friends" value="张三"/>张三
            <form:checkbox path="friends" value="李四"/>李四
            <form:checkbox path="friends" value="王五"/>王五
            <form:checkbox path="friends" value="赵六"/>赵六
        </p>
        <p>
            <label>职业:</label>
            <form:select path="carrer">
                <option/>请选择职业
                <form:options items="${carrers }"/>
            </form:select>
        </p>
        <p>
            <label>户籍:</label>
            <form:select path="houseRegister">
                <option/>请选择户籍
                <form:options items="${houseRegisters }"/>
            </form:select>
        </p>
        <p>
            <label>个人描述:</label>
            <form:textarea path="remark" rows="5"/>
        </p>
        <p id="buttons">
            <input id="reset" type="reset">
            <input id="submit" type="submit" value="添加">
        </p>
    </fieldset>
</form:form>
```

userList.jsp 的核心代码如下：

```
<h1>用户列表</h1>
<a href="<c:url value="user/input"/>">继续添加</a>
<table>
    <tr>
        <th>用户名</th>
        <th>兴趣爱好</th>
        <th>朋友</th>
        <th>职业</th>
        <th>户籍</th>
        <th>个人描述</th>
    </tr>
    <!-- JSTL 标签，请参考相关内容 -->
    <c:forEach items="${users}" var="user">
        <tr>
            <td>${user.userName }</td>
            <td>
              <c:forEach items="${user.hobby }" var="hobby">
                  ${hobby } 
```

```
            </c:forEach>
          </td>
          <td>
            <c:forEach items="${user.friends }" var="friend">
                ${friend } 
            </c:forEach>
          </td>
          <td>${user.carrer }</td>
          <td>${user.houseRegister }</td>
          <td>${user.remark }</td>
      </tr>
    </c:forEach>
</table>
```

❹ 创建领域模型

在应用中实现 User 类属性和 JSP 页面表单参数的绑定，User 类包含和表单参数名对应的属性，以及属性的 set 和 get 方法。在 ch2_3 应用的 src 目录下，创建包 pojo，并在该包中创建 User 类。

User 类的代码如下：

```
package pojo;
public class User {
    private String userName;
    private String[] hobby;//兴趣爱好
    private String[] friends;//朋友
    private String carrer;
    private String houseRegister;
    private String remark;
    //省略 setter 和 getter 方法
}
```

❺ 创建 Service 层

在 Service 层使用静态集合变量 users 模拟数据库存储用户信息，包括添加用户和查询用户两个功能方法。在 ch2_3 应用的 src 目录下，创建包 service，并在该包中创建 UserService 接口和 UserServiceImpl 实现类。

UserService 接口的核心代码如下：

```
public interface UserService {
    boolean addUser(User u);
    ArrayList<User> getUsers();
}
```

UserServiceImpl 实现类的核心代码如下：

```
@Service
public class UserServiceImpl implements UserService{
    //使用静态集合变量 users 模拟数据库
    private static ArrayList<User> users = new ArrayList<User>();
    @Override
    public boolean addUser(User u) {
        if(!"IT 民工".equals(u.getCarrer())){//不允许添加 IT 民工
```

```
            users.add(u);
            return true;
        }
        return false;
    }
    @Override
    public ArrayList<User> getUsers() {
        return users;
    }
}
```

❻ 创建 Controller 层

在 Controller 类的 UserController 中定义请求处理方法，其中包括处理 user/input 请求的 inputUser 方法，以及处理 user/save 请求的 addUser 方法。在 UserController 类中，通过@Autowired 注解注入 UserService 对象，实现对 user 对象的添加和查询等操作；通过 model 的 addAttribute 方法将 User 类对象、HashMap 类型的 hobbys 对象、String[]类型的 carrers 对象以及 String[]类型的 houseRegisters 对象传递给 View（userAdd.jsp）。在 ch2_3 应用的 src 目录下，创建包 controller，并在该包中创建 UserController 控制器类。

UserController 类的核心代码如下：

```
@Controller
@RequestMapping("/user")
public class UserController {
    @Autowired
    private UserService userService;
    @RequestMapping("/input")
    public String inputUser(Model model) {
        HashMap<String, String> hobbys = new HashMap<String, String>();
        hobbys.put("篮球", "篮球");
        hobbys.put("乒乓球", "乒乓球");
        hobbys.put("电玩", "电玩");
        hobbys.put("游泳", "游泳");
        model.addAttribute("user", new User());
        model.addAttribute("hobbys", hobbys);
        model.addAttribute("carrers", new String[] { "教师", "学生", "coding 搬运
          工", "IT 民工", "其他" });
        model.addAttribute("houseRegisters", new String[] { "北京", "上海",
          "广州", "深圳", "其他" });
        return "userAdd";
    }
    @RequestMapping("/save")
    public String addUser(@ModelAttribute User user, Model model) {
        if (userService.addUser(user)) {
            return "redirect:/user/list";
        } else {
            HashMap<String, String> hobbys = new HashMap<String, String>();
            hobbys.put("篮球", "篮球");
            hobbys.put("乒乓球", "乒乓球");
            hobbys.put("电玩", "电玩");
            hobbys.put("游泳", "游泳");
    /**这里不需要 model.addAttribute("user", newUser())，因为@ModelAttribute
```

```
        指定 form backing object*/
        model.addAttribute("hobbys", hobbys);
        model.addAttribute("carrers", new String[] { "教师", "学生", "coding 搬运
          工", "IT 民工", "其他" });
        model.addAttribute("houseRegisters", new String[] { "北京", "上海",
          "广州", "深圳", "其他" });
            return "userAdd";
        }
    }
    @RequestMapping("/list")
    public String listUsers(Model model) {
        List<User> users = userService.getUsers();
        model.addAttribute("users", users);
        return "userList";
    }
}
```

❼ 测试应用

通过网址 http://localhost:8080/ch2_3/user/input 测试应用，添加用户信息页面效果如图 2.2 所示。

如果在图 2.2 中，职业选择“IT 民工”时，添加失败。失败后返回到添加页面，输入过的信息不再输入，自动回填（必须结合 form 标签）。自动回填是数据绑定的一个优点。失败页面如图 2.3 所示。

图 2.2　添加用户信息页面

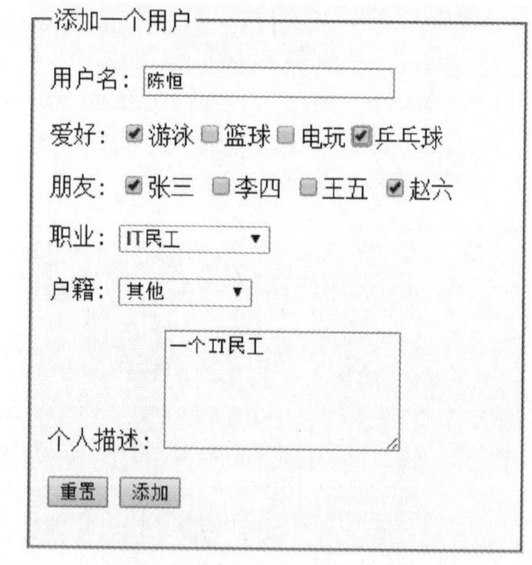

图 2.3　添加用户信息失败页面

在图 2.2 中输入正确信息，添加成功后，重定向到信息显示页面，效果如图 2.4 所示。

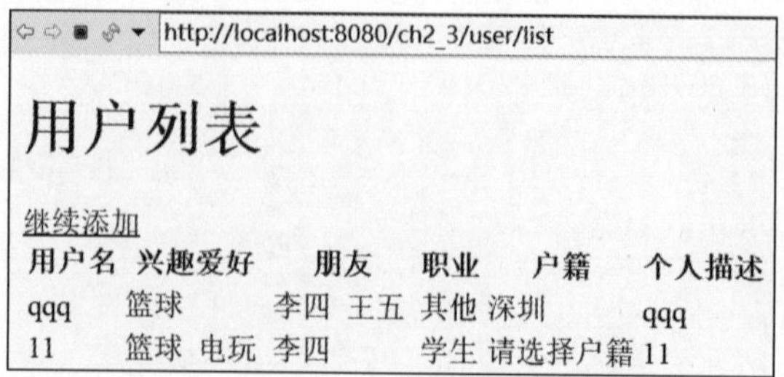

用户名	兴趣爱好	朋友	职业	户籍	个人描述
qqq	篮球	李四 王五	其他	深圳	qqq
11	篮球 电玩	李四	学生	请选择户籍	11

图 2.4　信息显示页面

2.5　JSON 数据交互

视频讲解

Spring MVC 在数据绑定的过程中，需要对传递数据的格式和类型进行转换，它既可以转换 String 等类型的数据，也可以转换 JSON 等其他类型的数据。本节将针对 Spring MVC 中 JSON 类型的数据交互进行讲解。

▶ 2.5.1　JSON 数据结构

JSON（JavaScript Object Notation，JS 对象标记）是一种轻量级的数据交换格式。与 XML 一样，JSON 也是基于纯文本的数据格式。它有两种数据结构。

❶ 对象结构

对象结构以“{”开始，以“}”结束，中间部分由 0 个或多个以英文“,”分隔的 key/value 对构成，key 和 value 之间以英文“:”分隔。对象结构的语法如下：

```
{
    key1:value1,
    key2:value2,
    …
}
```

其中，key 必须为 String 类型，value 可以是 String、Number、Object、Array 等数据类型。例如，一个 person 对象包含姓名、密码、年龄等信息，使用 JSON 的表示形式如下：

```
{
    "pname":"陈恒",
    "password":"123456",
    "page":40
}
```

❷ 数组结构

数组结构以“[”开始，以“]”结束，中间部分由 0 个或多个以英文“,”分隔的值的列表组成。数组结构的语法结构如下：

```
[
    value1,
    value2,
    …
]
```

上述两种（对象、数组）数据结构也可以分别组合构成更为复杂的数据结构。例如，一个 student 对象包含 sno、sname、hobby 和 college 对象，其 JSON 的表示形式如下：

```
{
    "sno":"201802228888",
    "sname":"陈恒",
    "hobby":["篮球","足球"],
    "college":{
        "cname":"清华大学",
        "city":"北京"
    }
}
```

▶ 2.5.2 JSON 数据转换

为实现浏览器与控制器类之间的 JSON 数据交互，Spring MVC 提供了 MappingJackson2HttpMessageConverter 实现类默认处理 JSON 格式请求响应。该实现类利用 Jackson 开源包读写 JSON 数据，将 Java 对象转换为 JSON 对象和 XML 文档，同时也可以将 JSON 对象和 XML 文档转换为 Java 对象。

Jackson 开源包及其描述如下。

- jackson-annotations.jar：JSON 转换的注解包。
- jackson-core.jar：JSON 转换的核心包。
- jackson-databind.jar：JSON 转换的数据绑定包。

以上 3 个 Jackson 的开源包在编写本书时，最新版本是 2.12.0，读者可通过网址 http://mvnrepository.com/artifact/com.fasterxml.jackson.core 下载得到。

在使用注解开发时，需要用到两个重要的 JSON 格式转换注解，分别是@RequestBody 和@ResponseBody。

@RequestBody：用于将请求体中的数据绑定到方法的形参中，该注解应用在方法的形参上。

@ResponseBody：用于直接返回 JSON 对象，该注解应用在方法上。

下面通过一个实例来演示 JSON 数据交互过程。在该实例中，针对返回实体对象、ArrayList 集合、Map<String, Object>集合以及 List<Map<String, Object>>集合分别处理。

【例 2-4】JSON 数据交互过程。

具体步骤如下。

❶ 创建 Web 应用并导入相关的 JAR 包

创建 Web 应用 ch2_4，除了导入如 2.2.1 节所示的 JAR 包外，还需将 JSON 相关的 3 个 JAR 包（jackson-annotations-2.12.0.jar、jackson-databind-2.12.0.jar 和 jackson-core-2.12.0.jar）复制到 WEB-INF/lib 目录中。

❷ 创建 Web 和 Spring MVC 配置文件

应用 ch2_4 的 Web 配置文件 web.xml 与例 2-3 一样，为节省篇幅，不再赘述。

在应用 ch2_4 的 src 目录下，创建名为 config 的包，并在该包中创建 Spring MVC 配置文件 springmvc.xml。springmvc.xml 的代码如下：

```
<?xml version="1.0" encoding="UTF-8"?>
<beans xmlns="http://www.springframework.org/schema/beans"
    xmlns:xsi="http://www.w3.org/2001/XMLSchema-instance"
    xmlns:context="http://www.springframework.org/schema/context"
    xmlns:mvc="http://www.springframework.org/schema/mvc"
    xsi:schemaLocation="
     http://www.springframework.org/schema/beans
     http://www.springframework.org/schema/beans/spring-beans.xsd
        http://www.springframework.org/schema/context
        http://www.springframework.org/schema/context/spring-context.xsd
        http://www.springframework.org/schema/mvc
        http://www.springframework.org/schema/mvc/spring-mvc.xsd">
    <context:component-scan base-package="controller"/>
    <mvc:annotation-driven />
```

```
    <mvc:resources location="/js/" mapping="/js/**"></mvc:resources>
</beans>
```

❸ 创建 JSP 页面，并引入 jQuery

首先从 jQuery 官方网站 http://jquery.com/download/下载 jQuery 插件 jquery-3.5.1.min.js，将其复制到 Web 项目开发目录的 WebContent/js 目录下，然后在 JSP 页面中，通过<script type="text/javascript" src="js/jquery-3.5.1.min.js"></script>代码将 jquery-3.5.1.min.js 引入当前页面中。

在应用 ch2_4 的 WebContent 目录下创建 JSP 文件 index.jsp，在该页面中使用 Ajax 向控制器异步提交数据，核心代码如下：

```
<script type="text/javascript" src="js/jquery-3.5.1.min.js"></script>
<script type="text/javascript">
    function testJson() {
        //获取输入的值 pname 为 id
        var pname = $("#pname").val();
        var password = $("#password").val();
        var page = $("#page").val();
        $.ajax({
            //请求路径
            url : "${pageContext.request.contextPath }/testJson",
            //请求类型
            type : "post",
            //data 表示发送的数据
            data : JSON.stringify({pname:pname,password:password,page:page}),
            //定义发送请求的数据格式为 JSON 字符串
            contentType : "application/json;charset=utf-8",
            //定义回调响应的数据格式为 JSON 字符串，该属性可以省略
            dataType: "json",
            //成功响应的结果
            success : function(data){
                if(data != null){
                    //返回一个 Person 对象
        //alert("输入的用户名:" + data.pname + ", 密码: " + data.password + ",
           年龄: " +  data.page);
                    //ArrayList<Person>对象
                    /**for(var i = 0; i < data.length; i++){
                        alert(data[i].pname);
                    }**/
                    //返回一个 Map<String, Object>对象
                    //alert(data.pname);//pname 为 key
                    //返回一个 List<Map<String, Object>>对象
                    for(var i = 0; i < data.length; i++){
                        alert(data[i].pname);
                    }
                }
            }
        });
    }
</script>
</head>
```

```
<body>
    <form action="">
        用户名：<input type="text" name="pname" id="pname"/><br>
        密码：<input type="password" name="password" id="password"/><br>
        年龄：<input type="text" name="page" id="page"/><br>
        <input type="button" value="测试" onclick="testJson()"/>
    </form>
</body>
</html>
```

❹ 创建实体类

在应用 ch2_4 的 src 目录下，创建名为 pojo 的包，在该包中创建 Person 实体类，代码如下：

```
package pojo;
public class Person {
    private String pname;
    private String password;
    private Integer page;
    //省略 set 和 get 方法
}
```

❺ 创建控制器类

在应用 ch2_4 的 src 目录下，创建名为 controller 的包，并在该包中创建 TestController 控制器类，在处理方法中使用@ResponseBody 和@RequestBody 注解进行 JSON 数据交互，核心代码如下：

```
@Controller
public class TestController {
    /**
     * 接收页面请求的 JSON 数据，并返回 JSON 格式结果
     */
    @RequestMapping("/testJson")
    @ResponseBody
    public List<Map<String, Object>> testJson(@RequestBody Person user) {
        //打印接收的 JSON 格式数据
        System.out.println("pname=" + user.getPname() +
                ", password=" + user.getPassword() + ",page=" + user.getPage());
        //返回 Person 对象
        //return user;
        /**ArrayList<Person> allp = new ArrayList<Person>();
        Person p1 = new Person();
        p1.setPname("陈恒 1");
        p1.setPassword("123456");
        p1.setPage(80);
        allp.add(p1);
        Person p2 = new Person();
        p2.setPname("陈恒 2");
        p2.setPassword("78910");
        p2.setPage(90);
        allp.add(p2);
        //返回 ArrayList<Person>对象
        return allp;
        **/
```

```
        Map<String, Object> map = new HashMap<String, Object>();
        map.put("pname", "陈恒2");
        map.put("password", "123456");
        map.put("page", 25);
        //返回一个Map<String, Object>对象
        //return map;
        //返回一个List<Map<String, Object>>对象
        List<Map<String,Object>> allp=new ArrayList<Map<String,Object>>();
        allp.add(map);
        Map<String, Object> map1 = new HashMap<String, Object>();
        map1.put("pname", "陈恒3");
        map1.put("password", "54321");
        map1.put("page", 55);
        allp.add(map1);
        return allp;
    }
}
```

❻ 测试应用

右击应用 ch2_4，选择 Run As/Run on Server 发布并运行应用。

2.6 拦截器

视频讲解

Spring MVC 的拦截器（Interceptor）与 Java Servlet 的过滤器（Filter）类似，它主要用于拦截用户的请求并做相应的处理，通常应用在权限验证、记录请求信息的日志、判断用户是否登录等功能上。

2.6.1 拦截器的定义

在 Spring MVC 框架中，定义一个拦截器可以通过两种方式：一种是通过实现 HandlerInterceptor 接口或继承 HandlerInterceptor 接口的实现类来定义；另一种是通过实现 WebRequestInterceptor 接口或继承 WebRequestInterceptor 接口的实现类来定义。本节以实现 HandlerInterceptor 接口的定义方式为例，讲解自定义拦截器的使用方法。示例代码如下：

```
public class TestInterceptor implements HandlerInterceptor{
    @Override
    public boolean preHandle(HttpServletRequest request, HttpServletResponse
        response, Object handler)
            throws Exception {
        System.out.println("preHandle 方法在控制器的处理请求方法前执行");
        /**返回 true 表示继续向下执行，返回 false 表示中断后续操作*/
        return true;
    }
    @Override
    public void postHandle(HttpServletRequest request, HttpServletResponse
        response, Object handler,
            ModelAndView modelAndView) throws Exception {
                System.out.println("postHandle 方法在控制器的处理请求方法调用之
                    后，解析视图之前执行");
```

```
    }
    @Override
    public void afterCompletion(HttpServletRequest request, HttpServletResponse
       response, Object handler, Exception ex)
            throws Exception {
        System.out.println("afterCompletion 方法在控制器的处理请求方法执行完成后
          执行，即视图渲染结束之后执行");
    }
}
```

在上述拦截器的定义中，实现了 HandlerInterceptor 接口，并实现了接口中的 3 个方法。有关 3 个方法的描述如下。

preHandle()方法：该方法在控制器的处理请求方法前执行，其返回值表示是否中断后续操作。返回 true 表示继续向下执行；返回 false 表示中断后续操作。

postHandle()方法：该方法在控制器的处理请求方法调用之后，解析视图之前执行。可以通过此方法对请求域中的模型和视图做进一步的修改。

afterCompletion()方法：该方法在控制器的处理请求方法执行完成后执行，即视图渲染结束后执行。可以通过此方法实现一些资源清理、记录日志信息等工作。

2.6.2 拦截器的配置

让自定义的拦截器生效，需要在 Spring MVC 的配置文件中进行配置，示例代码如下：

```
<!-- 配置拦截器 -->
<mvc:interceptors>
    <!-- 配置一个全局拦截器，拦截所有请求 -->
    <bean class="interceptor.TestInterceptor"/>
    <mvc:interceptor>
        <!-- 配置拦截器作用的路径 -->
        <mvc:mapping path="/**"/>
        <!-- 配置不需要拦截作用的路径 -->
        <mvc:exclude-mapping path=""/>
        <!-- 定义在<mvc:interceptor>元素中，表示匹配指定路径的请求才进行拦截 -->
        <bean class="interceptor.Interceptor1"/>
    </mvc:interceptor>
    <mvc:interceptor>
        <!-- 配置拦截器作用的路径 -->
        <mvc:mapping path="/gotoTest"/>
        <bean class="interceptor.Interceptor2"/>
    </mvc:interceptor>
</mvc:interceptors>
```

在上述示例代码中，<mvc:interceptors>元素用于配置一组拦截器，其子元素<bean>定义的是全局拦截器，即拦截所有的请求。<mvc:interceptor>元素中定义的是指定路径的拦截器，其子元素<mvc:mapping>用于配置拦截器作用的路径，该路径在其属性 path 中定义。如上述示例代码中，path 的属性值“/**”表示拦截所有路径，“/gotoTest”表示拦截所有以“/gotoTest”结尾的路径。如果在请求路径中包含不需要拦截的内容，可以通过<mvc:exclude-mapping>子元素进行配置。

需要注意的是，<mvc:interceptor>元素的子元素必须按照<mvc:mapping .../>、<mvc:exclude-mapping .../>、<bean .../>的顺序配置。

▶ 2.6.3　拦截器的执行流程

在配置文件中，如果只定义了一个拦截器，程序首先执行拦截器类中的 preHandle()方法。如果该方法返回 true，程序将继续执行控制器中处理请求的方法，否则中断执行。如果该方法返回 true，并且控制器中处理请求的方法执行后返回视图前，将执行 postHandle()方法；返回视图后，才执行 afterCompletion()方法。

在 Web 应用中，通常有多个拦截器同时工作，这时它们的 preHandle()方法将按照配置文件中拦截器的配置顺序执行，而它们的 postHandle()方法和 afterCompletion()方法则按照配置顺序的反序执行。下面通过一个应用 ch2_5 来演示拦截器的执行流程。

【例 2-5】多个拦截器的执行过程。

具体步骤如下。

❶ 创建 Web 应用并导入相关的 JAR 包

创建 Web 应用 ch2_5，将如 2.2.1 节所示的 JAR 包复制到 WEB-INF/lib 目录中。

❷ 创建 Web 配置文件 web.xml

在 WEB-INF 目录下，创建 web.xml 文件，该文件中的配置信息与例 2-1 相同，为节省篇幅，不再赘述。

❸ 创建控制器类

在 src 目录下，创建名为 controller 的包，并在该包中创建控制器类 InterceptorController，核心代码如下：

```
@Controller
public class InterceptorController {
    @RequestMapping("/gotoTest")
    public String gotoTest() {
        System.out.println("正在测试拦截器，执行控制器的处理请求方法中");
        return "test";
    }
}
```

❹ 创建拦截器类

在 src 目录下，创建一个名为 interceptor 的包，并在该包中创建拦截器类 TestInterceptor、Interceptor1 和 Interceptor2。TestInterceptor 的代码与 2.6.1 节的示例代码相同，为节省篇幅，不再赘述。

Interceptor1 类的核心代码如下：

```
public class Interceptor1 implements HandlerInterceptor{
    @Override
    public boolean preHandle(HttpServletRequest request, HttpServletResponse
       response, Object handler)
            throws Exception {
        System.out.println("Interceptor1 preHandle 方法执行");
        /**返回 true 表示继续向下执行，返回 false 表示中断后续的操作*/
        return true;
    }
    @Override
    public void postHandle(HttpServletRequest request, HttpServletResponse
       response, Object handler,
```

```
            ModelAndView modelAndView) throws Exception {
        System.out.println("Interceptor1 postHandle 方法执行");
    }
    @Override
    public void afterCompletion(HttpServletRequest request, HttpServletResponse
       response, Object handler, Exception ex)
            throws Exception {
        System.out.println("Interceptor1 afterCompletion 方法执行");
    }
}
```

Interceptor2 类的代码如下：

```
public class Interceptor2 implements HandlerInterceptor{
    @Override
    public boolean preHandle(HttpServletRequest request, HttpServletResponse
       response, Object handler)
            throws Exception {
        System.out.println("Interceptor2 preHandle 方法执行");
        /**返回 true 表示继续向下执行，返回 false 表示中断后续的操作*/
        return true;
    }
    @Override
    public void postHandle(HttpServletRequest request, HttpServletResponse
      response, Object handler,
            ModelAndView modelAndView) throws Exception {
        System.out.println("Interceptor2 postHandle 方法执行");
    }
    @Override
    public void afterCompletion(HttpServletRequest request, HttpServletResponse
       response, Object handler, Exception ex)
            throws Exception {
        System.out.println("Interceptor2 afterCompletion 方法执行");
    }
}
```

❺ 创建配置文件 springmvc.xml

在 src 目录下，创建名为 config 的包，并在该包中创建配置文件 springmvc.xml。配置文件包括一个全局拦截器和两个局部拦截器，具体代码如下：

```
<?xml version="1.0" encoding="UTF-8"?>
<beans xmlns="http://www.springframework.org/schema/beans"
    xmlns:xsi="http://www.w3.org/2001/XMLSchema-instance"
    xmlns:context="http://www.springframework.org/schema/context"
    xmlns:mvc="http://www.springframework.org/schema/mvc"
    xsi:schemaLocation="
    http://www.springframework.org/schema/beans
    http://www.springframework.org/schema/beans/spring-beans.xsd
       http://www.springframework.org/schema/context
       http://www.springframework.org/schema/context/spring-context.xsd
       http://www.springframework.org/schema/mvc
       http://www.springframework.org/schema/mvc/spring-mvc.xsd">
    <!-- 使用扫描机制，扫描控制器类 -->
    <context:component-scan base-package="controller"/>
```

```
    <!-- 配置视图解析器 -->
    <bean class=
      "org.springframework.web.servlet.view.InternalResourceViewResolver"
           id="internalResourceViewResolver">
      <!-- 前缀 -->
      <property name="prefix" value="/WEB-INF/jsp/" />
      <!-- 后缀 -->
      <property name="suffix" value=".jsp" />
    </bean>
    <mvc:interceptors>
        <!-- 配置一个全局拦截器，拦截所有请求 -->
         <bean class="interceptor.TestInterceptor"/>
        <mvc:interceptor>
            <!-- 配置拦截器作用的路径 -->
            <mvc:mapping path="/**"/>
            <!-- 定义在<mvc:interceptor>元素中，表示匹配指定路径的请求才进行拦截 -->
            <bean class="interceptor.Interceptor1"/>
        </mvc:interceptor>
        <mvc:interceptor>
            <!-- 配置拦截器作用的路径 -->
            <mvc:mapping path="/gotoTest"/>
            <!-- 定义在<mvc:interceptor>元素中，表示匹配指定路径的请求才进行拦截 -->
            <bean class="interceptor.Interceptor2"/>
        </mvc:interceptor>
    </mvc:interceptors>
</beans>
```

❻ 创建视图 JSP 文件

在 WEB-INF 目录下，创建一个 jsp 文件夹，并在该文件夹中创建一个 JSP 文件 test.jsp。test.jsp 的核心代码如下：

```
<body>
    视图
    <%System.out.println("视图渲染结束。"); %>
</body>
```

❼ 测试拦截器

首先，将应用 ch2_5 发布到 Tomcat 服务器，并启动 Tomcat 服务器。然后，通过地址 http://localhost:8080/ch2_5/gotoTest 测试拦截器。程序正确执行后，控制台的输出结果如图 2.5 所示。

```
Markers  Properties  Servers  Data Source Explorer  Snippets
Tomcat v9.0 Server at localhost [Apache Tomcat] C:\soft\javaee\eclipse\plug
preHandle方法在控制器的处理请求方法前执行
Interceptor1 preHandle方法执行
Interceptor2 preHandle方法执行
正在测试拦截器，执行控制器的处理请求方法中
Interceptor2 postHandle方法执行
Interceptor1 postHandle方法执行
postHandle方法在控制器的处理请求方法调用之后，解析视图之前执行
视图渲染结束。
Interceptor2 afterCompletion方法执行
Interceptor1 afterCompletion方法执行
afterCompletion方法在控制器的处理请求方法执行完成后执行，即视图渲染结束之后执行
```

图 2.5　多个拦截器的执行过程

视频讲解

2.7 文件上传

文件上传是一个 Spring MVC 应用中经常使用的功能，Spring MVC 通过配置一个 MultipartResolver 来上传文件。在 Spring MVC 的控制器中，可以通过 MultipartFile 来接收单个文件上传，通过 List<MultipartFile>来接收多个文件上传。

由于 Spring MVC 框架的文件上传是基于 commons-fileupload 组件的文件上传，因此，需要将 commons-fileupload 组件相关的 jar（commons-fileupload.jar 和 commons-io.jar）复制到 Spring MVC 应用的 WEB-INF/lib 目录下。下面讲解一下如何下载相关 jar 包。

Commons 是 Apache 开放源代码组织中的一个 Java 子项目，该项目包括文件上传、命令行处理、数据库连接池、XML 配置文件处理等模块。fileupload 是其中用来处理基于表单的文件上传的子项目，commons-fileupload 组件性能优良，并支持任意大小文件的上传。

commons-fileupload 组件可通过 http://commons.apache.org/proper/commons-fileupload/下载，本书采用的版本是 1.4。下载它的 Binaries 压缩包（commons-fileupload-1.4-bin.zip），解压后有个 JAR 文件 commons-fileupload-1.4.jar，该文件是 commons-fileupload 组件的类库。

commons-fileupload 组件依赖于 Apache 的另外一个项目：commons-io，该组件可以从 http://commons.apache.org/proper/commons-io/下载，本书采用的版本是 2.8.0。下载它的 Binaries 压缩包（commons-io-2.8.0-bin.zip），解压缩后的目录中有 5 个 JAR 文件，其中有一个 commons-io-2.8.0.jar 文件，该文件是 commons-io 的类库。

下面通过一个实例讲解如何上传多个文件。

【例 2-6】上传多个文件。

具体步骤如下。

❶ 创建 Web 应用并导入相关的 JAR 包

创建 Web 应用 ch2_6，导入如图 2.6 所示的 JAR 包。

```
lib
    commons-fileupload-1.4.jar
    commons-io-2.8.0.jar
    commons-logging-1.2.jar
    spring-aop-5.3.2.jar
    spring-beans-5.3.2.jar
    spring-context-5.3.2.jar
    spring-core-5.3.2.jar
    spring-expression-5.3.2.jar
    spring-web-5.3.2.jar
    spring-webmvc-5.3.2.jar
    taglibs-standard-impl-1.2.5.jar
    taglibs-standard-spec-1.2.5.jar
```

图 2.6 应用 ch2_6 的 JAR 包

❷ 创建多文件选择页面

在应用 ch2_6 的 WebContent 目录下，创建 JSP 页面 multiFiles.jsp。在该页面中使用表单（别忘记了 enctype 属性值为 multipart/form-data）上传多个文件，具体核心代码如下：

```
<form action="multifile" method="post" enctype="multipart/form-data">
    选择文件 1:<input type="file" name="myfile"> <br>
```

```
    文件描述 1:<input type="text" name="description"> <br>
    选择文件 2:<input type="file" name="myfile"> <br>
    文件描述 2:<input type="text" name="description"> <br>
    选择文件 3:<input type="file" name="myfile"> <br>
    文件描述 3:<input type="text" name="description"> <br>
    <input type="submit" value="提交">
</form>
```

❸ 创建 POJO 类

在应用 ch2_6 的 src 目录下，创建名为 pojo 的包，在该包中创建实体类 MultiFileDomain。上传多个文件时，需要 POJO 类 MultiFileDomain 封装文件信息。MultiFileDomain 类的核心代码如下：

```
public class MultiFileDomain {
    private List<String> description;
    private List<MultipartFile> myfile;
    //省略 setter 和 getter 方法
}
```

❹ 创建控制器类

在应用 ch2_6 的 src 目录下，创建名为 controller 的包，在该包中创建控制器类 MultiFilesController，核心代码如下：

```
@Controller
public class MultiFilesController {
    @RequestMapping("/multifile")
    public String multiFileUpload(@ModelAttribute MultiFileDomain
      multiFileDomain, HttpServletRequest request){
        String realpath = request.getServletContext().getRealPath
          ("uploadfiles");
        //上传到 eclipse-workspace/.metadata/.plugins/
          org.eclipse.wst.server.core/tmp0/wtpwebapps/ch2_6/uploadfiles
        File targetDir = new File(realpath);
        if(!targetDir.exists()){
            targetDir.mkdirs();
        }
        List<MultipartFile> files = multiFileDomain.getMyfile();
        for (int i = 0; i < files.size(); i++) {
            MultipartFile file = files.get(i);
            String fileName = file.getOriginalFilename();
            File targetFile = new File(realpath,fileName);
            //上传
            try {
            file.transferTo(targetFile);
            } catch (Exception e) {
                e.printStackTrace();
            }
        }
        return "showMulti";
    }
}
```

❺ 创建 Web 与 Spring MVC 配置文件

在应用 ch2_6 的 WEB-INF 目录下，创建 web.xml 文件。为防止中文乱码，需要在 web.xml 文件中添加字符编码过滤器，具体代码与例 2-3 相同，为节省篇幅，不再赘述。

在应用 ch2_6 的 src 目录下，创建名为 config 的包，并在该包中创建 Spring MVC 配置文件 springmvc.xml。在 Spring MVC 配置文件中，使用 Spring 的 CommonsMultipartResolver 类配置 MultipartResolver 用于文件上传，具体代码如下：

```
<?xml version="1.0" encoding="UTF-8"?>
<beans xmlns="http://www.springframework.org/schema/beans"
    xmlns:xsi="http://www.w3.org/2001/XMLSchema-instance"
    xmlns:p="http://www.springframework.org/schema/p"
    xmlns:context="http://www.springframework.org/schema/context"
    xsi:schemaLocation="
    http://www.springframework.org/schema/beans
    http://www.springframework.org/schema/beans/spring-beans.xsd
        http://www.springframework.org/schema/context
        http://www.springframework.org/schema/context/spring-context.xsd">
    <!-- 使用扫描机制，扫描控制器类 -->
    <context:component-scan base-package="controller"/>
    <!-- 配置视图解析器 -->
    <bean class=
      "org.springframework.web.servlet.view.InternalResourceViewResolver"
          id="internalResourceViewResolver">
        <!-- 前缀 -->
        <property name="prefix" value="/WEB-INF/jsp/" />
        <!-- 后缀 -->
        <property name="suffix" value=".jsp" />
    </bean>
    <!--使用 Spring 的 CommosMultipartResolver，配置 MultipartResolver 用于文件
      上传 -->
    <bean id="multipartResolver" class=
        "org.springframework.web.multipart.commons.CommonsMultipartResolver"
        p:defaultEncoding="UTF-8"
        p:maxUploadSize="5400000"
        p:uploadTempDir="fileUpload/temp">
        <!--workspace\.metadata\.plugins\org.eclipse.wst.server.core\tmp0\
            wtpwebapps\fileUpload-->
    </bean>
    <!--defaultEncoding="UTF-8"是请求的编码格式，默认为 iso-8859-1；maxUploadSize=
    "5400000"是允许上传文件的最大值，单位为字节；uploadTempDir="fileUpload/temp"
    为上传文件的临时路径 -->
</beans>
```

❻ 创建成功显示页面

在应用 ch2_6 的 WEB-INF 目录下，创建名为 jsp 的文件夹，并在该文件夹中创建多文件上传成功显示页面 showMulti.jsp，核心代码如下：

```
<body>
    <table>
        <tr>
            <td>详情</td><td>文件名</td>
```

```
        </tr>
        <!-- 同时取两个数组的元素 -->
        <c:forEach items="${multiFileDomain.description}" var="description"
          varStatus="loop">
            <tr>
                <td>${description}</td>
                <td>${multiFileDomain.myfile[loop.count-1].originalFilename}
                   </td>
            </tr>
        </c:forEach>
        <!-- fileDomain.getMyfile().getOriginalFilename() -->
    </table>
</body>
```

❼ 发布并运行应用

发布 ch2_6 应用到 Tomcat 服务器，并启动 Tomcat 服务器。然后，通过网址 http://localhost:8080/ch2_6/multiFiles.jsp 运行多文件选择页面，运行结果如图 2.7 所示。

图 2.7　多文件选择页面

在图 2.7 中选择文件，并输入文件描述，然后单击“提交”按钮上传多个文件，成功显示如图 2.8 所示的结果。

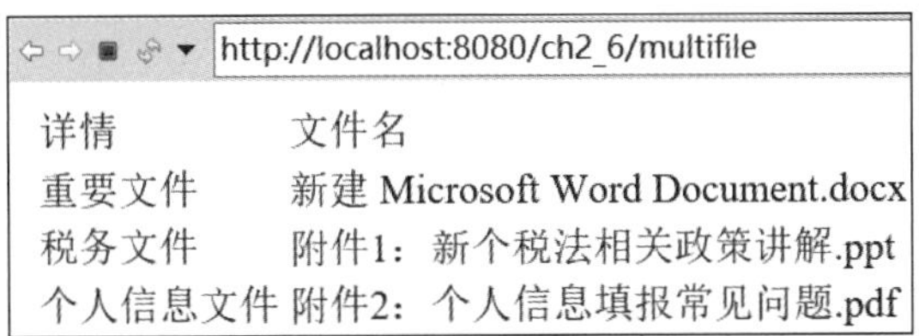

图 2.8　多文件成功上传结果

2.8　本章小结

本章简单介绍了 Spring MVC 框架基础，包括 Spring MVC 的工作流程、控制器、表单标签与数据绑定、JSON 数据交互、拦截器以及文件上传等内容。

习 题 2

1. 在开发 Spring MVC 应用时，如何配置 DispatcherServlet？
2. 简述 Spring MVC 的工作流程。

3．举例说明数据绑定的优点。

4．Spring MVC 有哪些表单标签？其中，可以绑定集合数据的标签有哪些？

5．@ModelAttribute 可实现哪些功能？

6．在 Spring MVC 中，JSON 类型的数据如何交互？请按照返回实体对象、ArrayList 集合、Map<String, Object>集合以及 List<Map<String, Object>>集合举例说明。

第 3 章 MyBatis

学习目的与要求

本章讲解 MyBatis 的环境构建、工作原理、SQL 映射文件以及 SSM 框架的整合开发。通过本章的学习，了解 MyBatis 的工作原理，掌握 MyBatis 的环境构建以及 SSM 框架的整合开发，了解 MyBatis 的核心配置文件的配置信息，掌握 MyBatis 的 SQL 映射文件的编写，熟悉级联查询的 MyBatis 实现，掌握 MyBatis 的动态 SQL 的编写。

主要内容

- ❖ MyBatis 的环境构建
- ❖ MyBatis 的工作原理
- ❖ SSM 框架整合开发
- ❖ 核心配置文件
- ❖ SQL 映射文件
- ❖ 级联查询
- ❖ 动态 SQL
- ❖ MyBatis 的缓存机制

MyBatis 是主流的 Java 持久层框架之一，它与 Hibernate 一样，也是一种 ORM（Object/Relational Mapping，即对象关系映射）框架。因其性能优异，且具有高度的灵活性、可优化性、易维护以及简单易学等特点，受到了广大互联网企业和编程爱好者的青睐。

3.1 MyBatis 简介

MyBatis 本是 apache 的一个开源项目 iBatis，2010 年这个项目由 apache software foundation 迁移到 google code，并改名为 MyBatis。

MyBatis 是一个基于 Java 的持久层框架，包括 SQL Maps 和 Data Access Objects（DAO），它消除了几乎所有的 JDBC 代码和参数的手工设置以及结果集的检索。MyBatis 使用简单的 XML 或注解用于配置和原始映射，将接口和 Java 的 POJOs（Plain Old Java Objects，普通的 Java 对象）映射成数据库中的记录。

目前，Java 的持久层框架产品有许多，常见的有 Hibernate 和 MyBatis。MyBatis 是一个半自动映射的框架，因为 MyBatis 需要手动匹配 POJO、SQL 和映射关系；而 Hibernate 是一个全表映射的框架，只需要提供 POJO 和映射关系即可。MyBatis 是一个小巧、方便、高效、简单、直接、半自动化的持久层框架；Hibernate 是一个强大、方便、高效、复杂、间接、全自动化的持久层框架。两个持久层框架各有优缺点，开发者应根据实际应用选择它们。

3.2 MyBatis 的环境构建

编写本书时，MyBatis 的最新版本是 3.5.6，因此笔者选择该版本作为本书的实践环境。也希望读者下载该版本，以便学习。

如果读者不使用 Maven 或 Gradle 下载 MyBatis，可通过网址 https://github.com/mybatis/mybatis-3/releases 下载。解压后得到如图 3.1 所示的目录。

lib
LICENSE
mybatis-3.5.6.jar
mybatis-3.5.6.pdf
NOTICE

图 3.1 MyBatis 解压后的目录

图 3.1 中，mybatis-3.5.6.jar 是 MyBatis 的核心包，mybatis-3.5.6.pdf 是 MyBatis 的使用手册，lib 文件夹下的 JAR 是 MyBatis 的依赖包。

使用 MyBatis 框架时，需要将它的核心包和依赖包引入应用程序中。如果是 Web 应用，只需将核心包和依赖包复制到/WEB-INF/lib 目录中即可。

3.3 MyBatis 的工作原理

在学习 MyBatis 程序之前，读者需要了解一下 MyBatis 的工作原理，以便理解程序。MyBatis 的工作原理如图 3.2 所示。

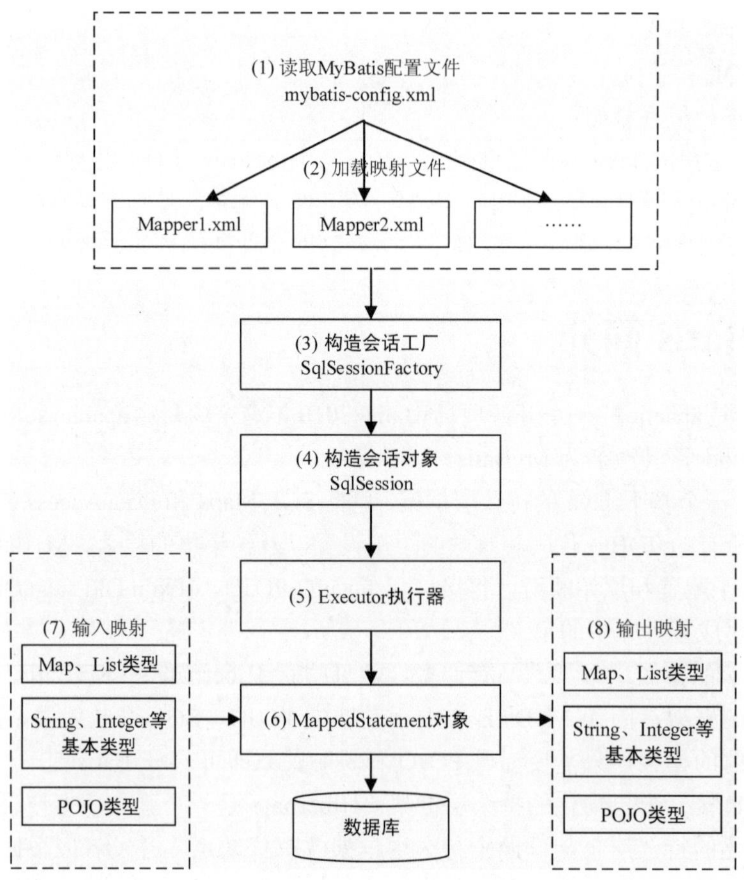

图 3.2 MyBatis 框架执行流程图

下面对图 3.2 中的每一步流程进行说明，具体如下。

（1）读取 MyBatis 配置文件 mybatis-config.xml。mybatis-config.xml 是 MyBatis 的全局配置文件，配置了 MyBatis 的运行环境等信息，如数据库连接信息。

（2）加载映射文件。映射文件即 SQL 映射文件，文件中配置了操作数据库的 SQL 语句，需要在 MyBatis 配置文件 mybatis-config.xml 中加载。mybatis-config.xml 文件可以加载多个映射文件。

（3）构造会话工厂。通过 MyBatis 的环境等配置信息，构建会话工厂 SqlSessionFactory。

（4）创建 SqlSession 对象。由会话工厂创建 SqlSession 对象，该对象中包含执行 SQL 语句的所有方法。

（5）MyBatis 底层定义了一个 Executor 接口来操作数据库，它将根据 SqlSession 传递的参数动态地生成需要执行的 SQL 语句，同时负责查询缓存的维护。

（6）在 Executor 接口的执行方法中，有一个 MappedStatement 类型的参数，该参数是对映射信息的封装，用于存储要映射的 SQL 语句的 id、参数等信息。

（7）输入参数映射。输入参数类型可以是 Map、List 等集合类型，也可以是基本数据类型和 POJO 类型。输入参数映射过程类似于 JDBC 对 preparedStatement 对象设置参数的过程。

（8）输出结果映射。输出结果类型可以是 Map、List 等集合类型，也可以是基本数据类型和 POJO 类型。输出结果映射过程类似于 JDBC 对结果集的解析过程。

通过上面的讲解，读者对 MyBatis 框架应该有一个初步了解，在后续的学习中，慢慢加深理解。

3.4 MyBatis 的核心配置

MyBatis 的核心配置文件配置了影响 MyBatis 行为的信息，这些信息通常只配置在一个文件中，并不轻易改动。另外，SSM 框架整合后，MyBatis 的核心配置信息将配置到 Spring 配置文件中。因此，在实际开发中很少编写或修改 MyBatis 的核心配置文件。本节的目的仅是了解 MyBatis 的核心配置文件的主要元素。

MyBatis 的核心配置文件模板代码如下：

```
<?xml version="1.0" encoding="UTF-8" ?>
<!DOCTYPE configuration
PUBLIC "-//mybatis.org//DTD Config 3.0//EN"
"http://mybatis.org/dtd/mybatis-3-config.dtd">
<configuration>
    <properties/><!-- 属性 -->
    <settings><!-- 设置 -->
        <setting name="" value=""/>
    </settings>
    <typeAliases/><!-- 类型命名（别名） -->
    <typeHandlers/><!-- 类型处理器 -->
    <objectFactory type=""/><!-- 对象工厂 -->
    <plugins><!-- 插件 -->
        <plugin interceptor=""></plugin>
    </plugins>
    <environments default=""><!-- 配置环境 -->
```

```
        <environment id=""><!-- 环境变量 -->
            <transactionManager type=""/><!-- 事务管理器 -->
            <dataSource type=""/><!-- 数据源 -->
        </environment>
    </environments>
    <databaseIdProvider type=""/><!-- 数据库厂商标识 -->
     <mappers><!-- 映射器，告诉 MyBatis 到哪里去找映射文件-->
       <mapper resource="com/mybatis/UserMapper.xml"/>
    </mappers>
</configuration>
```

MyBatis 的核心配置文件中的元素配置顺序不能颠倒，否则，在 MyBatis 启动阶段就将发生异常。

视频讲解

3.5　使用 Eclipse 开发 MyBatis 入门程序

本节使用第 1.6.2 节的 MySQL 数据库 springtest 的 user 数据表进行讲解。下面通过一个实例讲解如何使用 Eclipse 开发 MyBatis 入门程序。

【例 3-1】 使用 Eclipse 开发 MyBatis 入门程序。

具体实现步骤如下。

❶ 创建 Web 应用并导入相关 JAR 包

使用 Eclipse 创建一个名为 ch3_1 的 Web 应用，并将 MyBatis 的核心 JAR 包、MyBatis 的依赖 JAR 包以及 MySQL 的驱动连接 JAR 包复制到 WEB-INF/lib 目录中。

❷ 创建 Log4j 的日志配置文件

MyBatis 可使用 Log4j 输出日志信息，如果开发者需要查看控制台输出的 SQL 语句，那么需要在 classpath 路径下配置其日志文件。在应用 ch3_1 的 src 目录下创建 log4j.properties 文件，其内容如下：

```
# Global logging configuration
log4j.rootLogger=ERROR, stdout
# MyBatis logging configuration...
log4j.logger.com.mybatis.mapper=DEBUG
# Console output...
log4j.appender.stdout=org.apache.log4j.ConsoleAppender
log4j.appender.stdout.layout=org.apache.log4j.PatternLayout
log4j.appender.stdout.layout.ConversionPattern=%5p [%t] - %m%n
```

上述日志文件中配置了全局的日志配置、MyBatis 的日志配置和控制台输出，其中，MyBatis 的日志配置用于将 com.mybatis.mapper 包下所有类的日志记录级别设置为 DEBUG，该配置文件内容不需要开发者全部手写，可以从 MyBatis 使用手册中 Logging 小节复制，然后进行简单修改。

Log4j 是 Apache 的一个开源代码项目，通过使用 Log4j，可以控制日志信息输送的目的地是控制台、文件或 GUI 组件等，也可以控制每一条日志的输出格式；通过定义每一条日志信息的级别，能够更加详细地控制日志的生成过程。这些都可以通过一个配置文件来灵活地进行配置，而不需要修改应用的代码。有关 Log4j 的使用方法，读者可参考相关资料学习。

❸ 创建持久化类

在应用 ch3_1 的 src 目录下，创建一个名为 com.mybatis.po 的包，并在该包中创建持久化类 MyUser。类中声明的属性与数据表 user（创建表的代码请参见源代码的 user.sql）的字段一致。

MyUser 类的代码如下：

```
package com.mybatis.po;
/**
 *springtest 数据库中 user 表的持久化类
 */
public class MyUser {
    private Integer uid;//主键
    private String uname;
    private String usex;
    //此处省略 setter 和 getter 方法
    @Override
    public String toString() {//为了方便查看结果，重写了 toString 方法
        return "User [uid=" + uid +",uname=" + uname + ",usex=" + usex +"]";
    }
}
```

❹ 创建 MyBatis 的核心配置文件

在应用 ch3_1 的 src 目录下，创建 MyBatis 的核心配置文件 mybatis-config.xml。在该文件中，配置了数据库环境和映射文件的位置，具体内容如下：

```
<?xml version="1.0" encoding="UTF-8" ?>
<!DOCTYPE configuration
PUBLIC "-//mybatis.org//DTD Config 3.0//EN"
"http://mybatis.org/dtd/mybatis-3-config.dtd">
<configuration>
    <!-- 数据库连接信息 -->
    <properties>
        <property name="username" value="root" />
        <property name="password" value="root" />
        <property name="driver" value="com.mysql.jdbc.Driver" />
        <property name="url"
            value="jdbc:mysql://localhost:3306/springtest?characterEncoding=
                utf8" />
    </properties>
    <settings>
        <setting name="logImpl" value="LOG4J" />
    </settings>
    <!--为实体类 com.mybatis.po.MyUser 配置一个别名 MyUser -->
    <!--  <typeAliases>
        <typeAlias type="com.mybatis.po.MyUser" alias="MyUser" />
    </typeAliases>-->
    <!--为 com.mybatis.po 包下的所有实体类配置别名，MyBatis 默认的设置别名的方式是去
      除类所在的包后的简单的类名，例如 com.mybatis.po.MyUser 这个实体类的别名就会被
      设置为 MyUser -->
    <typeAliases>
        <package name="com.mybatis.po" />
```

```
    </typeAliases>
    <!-- SSM 整合后 environments 配置将废除 -->
    <environments default="development">
        <environment id="development">
            <!-- 使用 jdbc 事务管理 -->
            <transactionManager type="JDBC" />
            <!-- 数据库连接池 -->
            <dataSource type="POOLED">
                <property name="driver" value="${driver}" />
                <property name="url" value="${url}" />
                <property name="username" value="${username}" />
                <property name="password" value="${password}" />
            </dataSource>
        </environment>
    </environments>
    <!-- 加载映射文件 -->
    <mappers>
        <mapper resource="com/mybatis/mapper/UserMapper.xml" />
    </mappers>
</configuration>
```

上述映射文件和配置文件都不需要读者完全手动编写，都可以从 MyBatis 使用手册中复制，然后做简单修改即可。

❺ 创建 SQL 映射文件

在应用 ch3_1 的 src 目录下，创建一个名为 com.mybatis.mapper 的包，并在该包中创建 SQL 映射文件 UserMapper.xml。

SQL 映射文件 UserMapper.xml 内容如下：

```
<?xml version="1.0" encoding="UTF-8" ?>
<!DOCTYPE mapper
PUBLIC "-//mybatis.org//DTD Mapper 3.0//EN"
"http://mybatis.org/dtd/mybatis-3-mapper.dtd">
<mapper namespace="com.mybatis.mapper.UserMapper">
    <!-- 根据 uid 查询一个用户信息 -->
    <select id="selectUserById" parameterType="Integer"
        resultType="com.mybatis.po.MyUser">
        select * from user where uid = #{uid}
    </select>
    <!-- 查询所有用户信息 -->
    <select id="selectAllUser"  resultType="MyUser">
        select * from user
    </select>
    <!-- 添加一个用户，#{uname}为 MyUser 的属性值-->
    <insert id="addUser" parameterType="MyUser">
        insert into user (uname,usex) values(#{uname},#{usex})
    </insert>
    <!-- 修改一个用户 -->
    <update id="updateUser" parameterType="MyUser">
        update user set uname = #{uname},usex = #{usex} where uid = #{uid}
    </update>
    <!-- 删除一个用户 -->
```

```
        <delete id="deleteUser" parameterType="Integer">
            delete from user where uid = #{uid}
        </delete>
    </mapper>
```

上述映射文件中，<mapper>元素是配置文件的根元素，它包含了一个 namespace 属性，该属性值通常设置为“包名+SQL 映射文件名”，指定了唯一的命名空间。子元素<select>、<insert>、<update>以及<delete>中的信息是用于执行查询、添加、修改以及删除操作的配置。在定义的 SQL 语句中，“#{}”表示一个占位符，相当于“?”，而“#{uid}”表示该占位符待接收参数的名称为 uid。

❻ 创建测试类

在应用 ch3_1 的 src 目录下，创建一个名为 com.mybatis.test 的包，并在该包中创建 MyBatisTest 测试类。在测试类中，首先使用输入流读取配置文件，然后根据配置信息构建 SqlSessionFactory 对象。接下来通过 SqlSessionFactory 对象创建 SqlSession 对象，并使用 SqlSession 对象执行数据库操作。

MyBatisTest 测试类的核心代码如下：

```
public class MyBatisTest {
    public static void main(String[] args) {
        try {
            //读取配置文件 mybatis-config.xml
            InputStream config = Resources.getResourceAsStream("mybatis-
              config.xml");
            //根据配置文件构建 SqlSessionFactory
            SqlSessionFactory ssf = new SqlSessionFactoryBuilder().build
              (config);
            //通过 SqlSessionFactory 创建 SqlSession
            SqlSession ss = ssf.openSession();
            //SqlSession 执行映射文件中定义的 SQL，并返回映射结果
            /*com.mybatis.mapper.UserMapper.selectUserById 为
             UserMapper.xml 中的命名空间+select 的 id*/
            //查询一个用户
            MyUser mu = ss.selectOne("com.mybatis.mapper.UserMapper
              .selectUserById", 1);
            System.out.println(mu);
            //添加一个用户
            MyUser addmu = new MyUser();
            addmu.setUname("陈恒");
            addmu.setUsex("男");
            ss.insert("com.mybatis.mapper.UserMapper.addUser",addmu);
            //修改一个用户
            MyUser updatemu = new MyUser();
            updatemu.setUid(1);
            updatemu.setUname("张三");
            updatemu.setUsex("女");
            ss.update("com.mybatis.mapper.UserMapper.updateUser", updatemu);
            //删除一个用户
            ss.delete("com.mybatis.mapper.UserMapper.deleteUser", 2);
            //查询所有用户
```

```
            List<MyUser> listMu = ss.selectList("com.mybatis.mapper
              .UserMapper.selectAllUser");
            for (MyUser myUser : listMu) {
                System.out.println(myUser);
            }
            //提交事务
            ss.commit();
            //关闭 SqlSession
            ss.close();
        } catch (IOException e) {
            e.printStackTrace();
        }
    }
}
```

上述测试类的运行结果如图 3.3 所示。

```
Markers  Properties  Servers  Data Source Explorer  Snippets  Console
<terminated> MyBatisTest [Java Application] C:\soft\javaee\eclipse\plugins\org.eclipse.justj.openjdk.h
DEBUG [main] - ==>  Preparing: select * from user where uid = ?
DEBUG [main] - ==> Parameters: 1(Integer)
DEBUG [main] - <==      Total: 1
User [uid=1,uname=张三,usex=女]
DEBUG [main] - ==>  Preparing: insert into user (uname,usex) values(?,?)
DEBUG [main] - ==> Parameters: 陈恒(String), 男(String)
DEBUG [main] - <==    Updates: 1
DEBUG [main] - ==>  Preparing: update user set uname = ?,usex = ? where uid = ?
DEBUG [main] - ==> Parameters: 张三(String), 女(String), 2(Integer)
DEBUG [main] - <==    Updates: 1
DEBUG [main] - ==>  Preparing: delete from user where uid = ?
DEBUG [main] - ==> Parameters: 2(Integer)
DEBUG [main] - <==    Updates: 1
DEBUG [main] - ==>  Preparing: select * from user
DEBUG [main] - ==> Parameters: 
DEBUG [main] - <==      Total: 2
User [uid=1,uname=张三,usex=女]
User [uid=3,uname=陈恒,usex=男]
```

图 3.3 MyBatis 入门程序的运行结果

视频讲解

3.6 SSM 框架整合开发

从 3.5 节测试类的代码中可以看出，直接使用 MyBatis 框架的 SqlSession 访问数据库并不简便。MyBatis 框架的重点是 SQL 映射文件，因此为方便后续学习，本节就开始讲解 SSM 框架整合开发。在本书 MyBatis 的后续学习中，将使用整合后的框架进行演示。

3.6.1 相关 JAR 包

实现 SSM 框架整合开发需要导入相关 JAR 包，包括 MyBatis、Spring、Spring MVC、MySQL 连接器、MyBatis 与 Spring 桥接器、Log4j 以及 DBCP 等。

❶ MyBatis 框架所需的 JAR 包

MyBatis 框架所需的 JAR 包包括核心包和依赖包，详情见 3.2 节。

❷ Spring 框架所需的 JAR 包

Spring 框架所需的 JAR 包包括核心模块 JAR、AOP 开发使用的 JAR、JDBC 和事务的 JAR 包以及 Spring MVC 所需要的 JAR 包，具体如下：

```
commons-logging-1.2.jar
spring-aop-5.3.2.jar
spring-beans-5.3.2.jar
spring-context-5.3.2.jar
spring-core-5.3.2.jar
spring-expression-5.3.2.jar
spring-jdbc-5.3.2.jar
spring-tx-5.3.2.jar
spring-web-5.3.2.jar
spring-webmvc-5.3.2.jar
```

❸ MyBatis 与 Spring 整合的中间 JAR 包

编写本书时，该中间 JAR 包的最新版本为 mybatis-spring-2.0.6.jar。此版本可从网址 http://mvnrepository.com/artifact/org.mybatis/mybatis-spring 下载。

❹ 数据库驱动 JAR 包

本书所使用的 MySQL 数据库驱动 JAR 包为 mysql-connector-java-5.1.45-bin.jar。

❺ 数据源所需的 JAR 包

整合时使用的是 DBCP 数据源，需要准备 DBCP 和连接池的 JAR 包。编写本书时，最新版本的 DBCP 的 JAR 包为 commons-dbcp2-2.8.0.jar，可从网址 http://commons.apache.org/proper/commons-dbcp/download_dbcp.cgi 下载；最新版本的连接池的 JAR 包为 commons-pool2-2.9.0.jar，可从网址 http://commons.apache.org/proper/commons-pool/download_pool.cgi 下载。

3.6.2 MapperScannerConfigurer 方式

一般情况下，将数据源及 MyBatis 工厂配置在 Spring 的配置文件中，实现 MyBatis 与 Spring 的无缝整合。在 Spring 的配置文件中，首先使用 org.apache.commons.dbcp2.BasicDataSource 配置数据源；其次使用 org.springframework.jdbc.datasource.DataSourceTransactionManager 为数据源添加事务管理器；最后使用 org.mybatis.spring.SqlSessionFactoryBean 配置 MyBatis 工厂，同时指定数据源，并与 MyBatis 完美整合。

使用 Spring 管理 MyBatis 的数据操作接口的方式有多种。其中，最常用、最简捷的一种接口方式是基于 org.mybatis.spring.mapper.MapperScannerConfigurer 的整合，实现 Mapper 代理开发。MapperScannerConfigurer 将包（<property name="basePackage" value="xxx" />）中所有接口自动装配为 MyBatis 映射接口 Mapper 的实现类的实例（映射器），所有映射器都被自动注入 SqlSessionFactory 实例，同时扫描包中 SQL 映射文件，MyBatis 核心配置文件不再加载 SQL 映射文件（但要保证接口与 SQL 映射文件名相同）。配置文件的示例代码如下：

```
<!-- 配置数据源 -->
<bean id="dataSource" class="org.apache.commons.dbcp2.BasicDataSource">
    <property name="driverClassName" value="${jdbc.driver}" />
    <property name="url" value="${jdbc.url}" />
    <property name="username" value="${jdbc.username}" />
    <property name="password" value="${jdbc.password}" />
    <!-- 最大连接数 -->
```

```
        <property name="maxTotal" value="${jdbc.maxTotal}" />
        <!-- 最大空闲连接数 -->
        <property name="maxIdle" value="${jdbc.maxIdle}" />
        <!-- 初始化连接数 -->
        <property name="initialSize" value="${jdbc.initialSize}" />
    </bean>
    <!-- 添加事务支持 -->
    <bean id="txManager" class=
      "org.springframework.jdbc.datasource.DataSourceTransactionManager">
        <property name="dataSource" ref="dataSource" />
    </bean>
    <!-- 开启事务注解 -->
    <tx:annotation-driven transaction-manager="txManager" />
    <!-- 配置 MyBatis 工厂，同时指定数据源，并与 MyBatis 完美整合 -->
    <bean id="sqlSessionFactory" class="org.mybatis.spring.SqlSessionFactoryBean">
        <property name="dataSource" ref="dataSource" />
        <!-- configLocation 的属性值为 MyBatis 的核心配置文件 -->
        <property name="configLocation" value="classpath:config/mybatis-
          config.xml" />
    </bean>
    <!--Mapper 代理开发，MapperScannerConfigurer 将包中所有接口自动装配为 MyBatis
      映射接口 Mapper 的实现类的实例（映射器），所有映射器都被自动注入 SqlSessionFactory
      实例，同时扫描包中 SQL 映射文件，MyBatis 核心配置文件不再加载 SQL 映射文件 -->
    <bean class="org.mybatis.spring.mapper.MapperScannerConfigurer">
        <!-- mybatis-spring 组件的扫描器，basePackage 属性可以包含多个包名，多个包名之
           间可以用逗号或分号隔开 -->
        <property name="basePackage" value="dao" />
        <property name="sqlSessionFactoryBeanName" value="sqlSessionFactory" />
    </bean>
```

▶ 3.6.3 整合示例

下面通过 SSM 框架整合，实现例 3-1 的功能。

【例 3-2】SSM 框架整合开发。

具体实现步骤如下。

❶ 创建 Web 应用并导入相关 JAR 包

使用 Eclipse 创建一个名为 ch3_2 的 Web 应用，并参考 3.6.1 节，将相关 JAR 包复制到 WEB-INF/lib 目录中。

❷ 创建数据库连接信息属性文件及 Log4j 的日志配置文件

在应用 ch3_2 的 src 目录下，创建名为 config 的包，并在该包中创建数据库连接信息属性文件 jdbc.properties，具体内容如下：

```
jdbc.driver=com.mysql.jdbc.Driver
jdbc.url=jdbc:mysql://localhost:3306/springtest?characterEncoding=utf8
jdbc.username=root
jdbc.password=root
jdbc.maxTotal=30
jdbc.maxIdle=10
jdbc.initialSize=5
```

在应用 ch3_2 的 src 目录下，创建 Log4j 的日志配置文件 log4j.properties，其内容与 3.5 节“2.”相同，为节省篇幅，不再赘述。

❸ 创建持久化类

在应用 ch3_2 的 src 目录下，创建一个名为 com.mybatis.po 的包，并在该包中创建持久化类 MyUser。该类与 3.5 节“3.”相同，为节省篇幅，不再赘述。

❹ 创建 SQL 映射文件

在应用 ch3_2 的 src 目录下，创建一个名为 com.mybatis.mapper 的包，并在该包中创建 SQL 映射文件 UserMapper.xml。该文件与 3.5 节“5.”相同，为节省篇幅，不再赘述。

❺ 创建 MyBatis 的核心配置文件

在应用 ch3_2 的 config 包中，创建 MyBatis 的核心配置文件 mybatis-config.xml。在该文件中，配置实体类别名、日志输出等，具体内容如下：

```
<?xml version="1.0" encoding="UTF-8" ?>
<!DOCTYPE configuration
PUBLIC "-//mybatis.org//DTD Config 3.0//EN"
"http://mybatis.org/dtd/mybatis-3-config.dtd">
<configuration>
    <settings>
        <setting name="logImpl" value="LOG4J" />
    </settings>
    <typeAliases>
        <package name="com.mybatis.po" />
    </typeAliases>
</configuration>
```

❻ 创建 Mapper 接口

在应用 ch3_2 的 com.mybatis.mapper 包中，创建接口 UserMapper。使用@Repository 注解标注该接口是数据访问层。该接口中的方法与 SQL 映射文件 UserMapper.xml 的 id 一致。UserMapper 接口的核心代码如下：

```
@Repository
public interface UserMapper {
    public MyUser selectUserById(Integer id);
    public List<MyUser> selectAllUser();
    public int addUser(MyUser myUser);
    public int updateUser(MyUser myUser);
    public int deleteUser(Integer id);
}
```

❼ 创建控制器类

在应用 ch3_2 的 src 目录下，创建一个名为 controller 的包，并在该包中创建控制器类 TestController。在该控制器类中，调用 Mapper 接口中的方法操作数据库，核心代码如下：

```
@Controller
public class TestController {
    @Autowired
    private UserMapper userMapper;
    @RequestMapping("/test")
    public String test() {
```

```
            //查询一个用户
            MyUser mu = userMapper.selectUserById(1);
            System.out.println(mu);
            //添加一个用户
            MyUser addmu = new MyUser();
            addmu.setUname("陈恒");
            addmu.setUsex("男");
            userMapper.addUser(addmu);
            //修改一个用户
            MyUser updatemu = new MyUser();
            updatemu.setUid(1);
            updatemu.setUname("张三");
            updatemu.setUsex("女");
            userMapper.updateUser(updatemu);
            //删除一个用户
            userMapper.deleteUser(3);
            //查询所有用户
            List<MyUser> listMu = userMapper.selectAllUser();
            for (MyUser myUser : listMu) {
                System.out.println(myUser);
            }
            return "test";
        }
    }
```

❽ 创建测试页面

在/WEB-INF/目录下，创建一个名为 jsp 的文件夹，并在该文件夹中创建 test.jsp 文件。test.jsp 的代码略。

❾ 创建 Web、Spring、Spring MVC 的配置文件

在应用 ch3_2 的 config 包中，创建 Spring 配置文件 applicationContext.xml 和 Spring MVC 配置文件 springmvc.xml；在应用 ch3_2 的/WEB-INF/目录中，创建 Web 配置文件 web.xml。

在 Spring 配置文件 applicationContext.xml 中，首先使用<context:property-placeholder/>加载数据库连接信息属性文件；然后使用 org.apache.commons.dbcp2.BasicDataSource 配置数据源，并使用 org.springframework.jdbc.datasource.DataSourceTransactionManager 为数据源添加事务管理器；然后使用 org.mybatis.spring.SqlSessionFactoryBean 配置 MyBatis 工厂，同时指定数据源，并与 MyBatis 完美整合；最后使用 org.mybatis.spring.mapper.MapperScannerConfigurer 实现 Mapper 代理开发，将 basePackage 属性指定包中所有接口自动装配为 MyBatis 映射接口 Mapper 的实现类的实例（映射器），所有映射器都被自动注入 SqlSessionFactory 实例，同时扫描包中 SQL 映射文件，MyBatis 核心配置文件不再加载 SQL 映射文件。Spring 配置文件 applicationContext.xml 的具体内容如下：

```
<?xml version="1.0" encoding="UTF-8"?>
<beans xmlns="http://www.springframework.org/schema/beans"
xmlns:xsi="http://www.w3.org/2001/XMLSchema-instance"
xmlns:tx="http://www.springframework.org/schema/tx"
xmlns:context="http://www.springframework.org/schema/context"
    xsi:schemaLocation="http://www.springframework.org/schema/beans
        http://www.springframework.org/schema/beans/spring-beans.xsd
```

```
        http://www.springframework.org/schema/tx
    http://www.springframework.org/schema/tx/spring-tx.xsd
    http://www.springframework.org/schema/context
        http://www.springframework.org/schema/context/spring-context.xsd">
    <!-- 加载数据库配置文件 -->
<context:property-placeholder location="classpath:config/db.properties" />
    <!-- 配置数据源 -->
<bean id="dataSource" class="org.apache.commons.dbcp2.BasicDataSource">
    <property name="driverClassName" value="${jdbc.driver}" />
    <property name="url" value="${jdbc.url}" />
    <property name="username" value="${jdbc.username}" />
    <property name="password" value="${jdbc.password}" />
    <!-- 最大连接数 -->
    <property name="maxTotal" value="${jdbc.maxTotal}" />
    <!-- 最大空闲连接数 -->
    <property name="maxIdle" value="${jdbc.maxIdle}" />
    <!-- 初始化连接数 -->
    <property name="initialSize" value="${jdbc.initialSize}" />
    </bean>
        <!-- 添加事务支持 -->
    <bean id="txManager" class=
      "org.springframework.jdbc.datasource.DataSourceTransactionManager">
        <property name="dataSource" ref="dataSource" />
    </bean>
        <!-- 开启事务注解 -->
    <tx:annotation-driven transaction-manager="txManager" />
        <!-- 配置 MyBatis 工厂，同时指定数据源，并与 MyBatis 完美整合 -->
    <bean id="sqlSessionFactory" class=
      "org.mybatis.spring.SqlSessionFactoryBean">
        <property name="dataSource" ref="dataSource" />
        <property name="configLocation" value="classpath:config/mybatis-
          config.xml" />
    </bean>
    <bean class="org.mybatis.spring.mapper.MapperScannerConfigurer">
        <property name="basePackage" value="com.mybatis.mapper" />
        <property name="sqlSessionFactoryBeanName" value=
          "sqlSessionFactory" />
    </bean>
</beans>
```

在 Spring MVC 配置文件 springmvc.xml 中，使用<context:component-scan/>扫描控制器包，并使用 org.springframework.web.servlet.view.InternalResourceViewResolver 配置视图解析器。具体代码如下：

```
<?xml version="1.0" encoding="UTF-8"?>
<beans xmlns="http://www.springframework.org/schema/beans"
    xmlns:xsi="http://www.w3.org/2001/XMLSchema-instance"
    xmlns:context="http://www.springframework.org/schema/context"
    xsi:schemaLocation="
    http://www.springframework.org/schema/beans
    http://www.springframework.org/schema/beans/spring-beans.xsd
        http://www.springframework.org/schema/context
```

```
        http://www.springframework.org/schema/context/spring-context.xsd">
    <context:component-scan base-package="controller"/>
    <bean class=
      "org.springframework.web.servlet.view.InternalResourceViewResolver"
          id="internalResourceViewResolver">
        <property name="prefix" value="/WEB-INF/jsp/" />
        <property name="suffix" value=".jsp" />
    </bean>
</beans>
```

在 Web 配置文件 web.xml 中，首先通过<context-param>加载 Spring 配置文件 applicationContext.xml，并通过 org.springframework.web.context.ContextLoaderListener 启动 Spring 容器；然后配置 Spring MVC DispatcherServlet，并加载 Spring MVC 配置文件 springmvc.xml。Web 配置文件 web.xml 的代码如下：

```
<?xml version="1.0" encoding="UTF-8"?>
<web-app xmlns:xsi="http://www.w3.org/2001/XMLSchema-instance"
    xmlns="http://xmlns.jcp.org/xml/ns/javaee"
    xsi:schemaLocation="http://xmlns.jcp.org/xml/ns/javaee
http://xmlns.jcp.org/xml/ns/javaee/web-app_4_0.xsd"
    id="WebApp_ID" version="4.0">
    <!-- 实例化 ApplicationContext 容器 -->
    <context-param>
        <!-- 加载 applicationContext.xml 文件 -->
        <param-name>contextConfigLocation</param-name>
        <param-value>
            classpath:config/applicationContext.xml
        </param-value>
    </context-param>
    <!-- 指定以 ContextLoaderListener 方式启动 Spring 容器 -->
    <listener>
        <listener-class>org.springframework.web.context.ContextLoaderListener
            </listener-class>
    </listener>
    <!--配置 Spring MVC DispatcherServlet -->
    <servlet>
        <servlet-name>springmvc</servlet-name>
        <servlet-class>org.springframework.web.servlet.DispatcherServlet</
          servlet-class>
        <init-param>
            <param-name>contextConfigLocation</param-name>
            <!-- classpath 是指到 src 目录查找配置文件 -->
            <param-value>classpath:config/springmvc.xml</param-value>
        </init-param>
        <load-on-startup>1</load-on-startup>
    </servlet>
    <servlet-mapping>
        <servlet-name>springmvc</servlet-name>
        <url-pattern>/</url-pattern>
    </servlet-mapping>
</web-app>
```

⑩ 测试应用

发布应用 ch3_2 到 Web 服务器 Tomcat 后，通过网址 http://localhost:8080/ch3_2/test 测试应用。成功运行后，控制台信息输出结果如图 3.4 所示。

```
Markers Properties Servers Data Source Explorer Snippets Console
Tomcat v9.0 Server at localhost [Apache Tomcat] C:\soft\javaee\eclipse\plugins\org.eclipse.justj.openjdk.hotspot.jre.full.win
DEBUG [http-nio-8080-exec-6] - <==      Total: 1
User [uid=1,uname=张三,usex=女]
DEBUG [http-nio-8080-exec-6] - ==>  Preparing: insert into user (uname,usex) values(?,?)
DEBUG [http-nio-8080-exec-6] - ==> Parameters: 陈恒(String), 男(String)
DEBUG [http-nio-8080-exec-6] - <==    Updates: 1
DEBUG [http-nio-8080-exec-6] - ==>  Preparing: update user set uname = ?,usex = ? where uid = ?
DEBUG [http-nio-8080-exec-6] - ==> Parameters: 张三(String), 女(String), 1(Integer)
DEBUG [http-nio-8080-exec-6] - <==    Updates: 1
DEBUG [http-nio-8080-exec-6] - ==>  Preparing: delete from user where uid = ?
DEBUG [http-nio-8080-exec-6] - ==> Parameters: 3(Integer)
DEBUG [http-nio-8080-exec-6] - <==    Updates: 0
DEBUG [http-nio-8080-exec-6] - ==>  Preparing: select * from user
DEBUG [http-nio-8080-exec-6] - ==> Parameters: 
DEBUG [http-nio-8080-exec-6] - <==      Total: 4
User [uid=1,uname=张三,usex=女]
User [uid=4,uname=陈恒,usex=男]
User [uid=5,uname=陈恒,usex=男]
User [uid=6,uname=陈恒,usex=男]
```

图 3.4 应用 ch3_2 的控制台信息输出结果

3.6.4 SqlSessionDaoSupport 方式

从 3.6.3 节的例 3.2 可知，在 MyBatis 中，当编写好访问数据库的映射器接口后，MapperScannerConfigurer 就能自动根据这些接口生成 DAO 对象，然后使用@Autowired 把这些 DAO 对象注入业务逻辑层或控制层。因此，在这种情况下的 DAO 层中，几乎不用编写代码，而且也没有地方编写，因为只有接口。这固然方便，不过当需要在 DAO 层编写代码时，这种方式就无能为力。幸运的是，MyBatis-Spring 提供了继承 SqlSessionDaoSupport 类的方式访问数据库。

类 org.mybatis.spring.support.SqlSessionDaoSupport 继承了 org.springframework.dao.support.DaoSupport 类，是一个抽象类，是作为 DAO 的基类使用，需要一个 SqlSessionFactory。在继承 SqlSessionDaoSupport 类的子类中通过调用 SqlSessionDaoSupport 类的 getSqlSession()方法获取这个 SqlSessionFactory 提供的 SqlSessionTemplate 对象。而 SqlSessionTemplate 类实现了 SqlSession 接口，即可以进行数据库访问。所以，需要 Spring 框架给 SqlSessionDaoSupport 类的子类的对象（多个 DAO 对象）注入一个 SqlSessionFactory。

但自 mybatis-spring-1.2.0 以来，SqlSessionDaoSupport 的 setSqlSessionTemplate 和 setSqlSessionFactory 两个方法上的@Autowired 注解被删除，这就意味着继承于 SqlSessionDaoSupport 的 DAO 类，它们的对象不能被自动注入 SqlSessionFactory 或 SqlSessionTemplate 对象。如果在 Spring 的配置文件中一个一个地配置，显然太麻烦了。比较好的解决办法是在 DAO 类中覆盖这两个方法之一，并加上@Autowired 或@Resource 注解。那么如果在每个 DAO 类中都这么做的话，显然很低效。更合理的做法是，写一个继承于 SqlSessionDaoSupport 的 BaseDao，在 BaseDao 中完成这个工作，然后其他的 DAO 类再都继承 BaseDao。BaseDao 的示例代码如下：

```
package dao;
import javax.annotation.Resource;
```

```
import org.apache.ibatis.session.SqlSessionFactory;
import org.mybatis.spring.support.SqlSessionDaoSupport;
public class BaseDao extends SqlSessionDaoSupport {
    //依赖注入 sqlSession 工厂
    @Resource(name = "sqlSessionFactory")
    public void setSqlSessionFactory(SqlSessionFactory sqlSessionFactory) {
        super.setSqlSessionFactory(sqlSessionFactory);
    }
}
```

下面通过实例讲解继承 SqlSessionDaoSupport 类的方式访问数据库。

【例 3-3】在 3.6.3 节例 3-2 的基础上，实现继承 SqlSessionDaoSupport 类的方式访问数据库。为节省篇幅，相同的实现不再赘述。其他的具体实现如下。

❶ 创建 Web 应用并导入相关 JAR 包

使用 Eclipse 创建一个名为 ch3_3 的 Web 应用，并参考 3.6.1 节，将相关 JAR 包复制到 WEB-INF/lib 目录中。

❷ 复制数据库连接信息属性文件及 Log4j 的日志配置文件

在应用 ch3_3 的 src 目录下，创建名为 config 的包，将应用 ch3_2 的数据库连接信息属性文件 jdbc.properties 复制到该包中。

将应用 ch3_2 的 Log4j 日志配置文件 log4j.properties 复制到 ch3_3 的 src 目录中，并将其中的“log4j.logger.com.mybatis.mapper=DEBUG”修改为“log4j.logger.dao=DEBUG”。

❸ 创建持久化类

在应用 ch3_3 的 src 目录下，创建一个名为 po 的包，并在该包中创建持久化类 MyUser。该类与 3.5 节“3.”相同，为节省篇幅，不再赘述。

❹ 创建 SQL 映射文件

在应用 ch3_3 的 src 目录下，创建一个名为 dao 的包，并在该包中创建 SQL 映射文件 UserMapper.xml。该文件内容如下：

```
<?xml version="1.0" encoding="UTF-8" ?>
<!DOCTYPE mapper
PUBLIC "-//mybatis.org//DTD Mapper 3.0//EN"
"http://mybatis.org/dtd/mybatis-3-mapper.dtd">
<mapper namespace="dao.UserMapper">
        <!-- 根据 uid 查询一个用户信息 -->
    <select id="selectUserById" parameterType="Integer"  resultType="MyUser">
        select * from user where uid = #{uid}
    </select>
        <!-- 查询所有用户信息 -->
    <select id="selectAllUser"  resultType="MyUser">
        select * from user
    </select>
</mapper>
```

❺ 创建 MyBatis 的核心配置文件

在应用 ch3_3 的 config 包中，创建 MyBatis 的核心配置文件 mybatis-config.xml。在该文件中，配置实体类别名、日志输出、指定映射文件位置等，具体内容如下：

```
<?xml version="1.0" encoding="UTF-8" ?>
```

```
<!DOCTYPE configuration
PUBLIC "-//mybatis.org//DTD Config 3.0//EN"
"http://mybatis.org/dtd/mybatis-3-config.dtd">
<configuration>
    <settings>
        <setting name="logImpl" value="LOG4J" />
    </settings>
    <typeAliases>
        <package name="po" />
    </typeAliases>
    <!-- 告诉 MyBatis 到哪里去找映射文件 -->
    <mappers>
        <mapper resource="dao/UserMapper.xml"/>
    </mappers>
</configuration>
```

❻ 创建 DAO 接口和接口实现类

在应用 ch3_3 的 dao 包中，创建接口 UserMapper。UserMapper 接口代码如下：

```
package dao;
import java.util.List;
import po.MyUser;
public interface UserMapper {
    public MyUser selectUserById(int id);
    public List<MyUser> selectAllUser();
}
```

在应用 ch3_3 的 dao 包中，创建 BaseMapper 类，在该类中使用@Resource(name = "sqlSessionFactory")注解依赖注入 sqlSession 工厂。BaseMapper 类的核心代码如下：

```
public class BaseMapper extends SqlSessionDaoSupport  {
    //依赖注入 sqlSession 工厂
    @Resource(name = "sqlSessionFactory")
    public void setSqlSessionFactory(SqlSessionFactory sqlSessionFactory) {
        super.setSqlSessionFactory(sqlSessionFactory);
    }
}
```

在应用 ch3_3 的 dao 包中，创建接口 UserMapper 的实现类 UserMapperImpl，在该类中使用@Repository 注解标注该类的实例是数据访问对象。UserMapperImpl 类的核心代码如下：

```
@Repository
public class UserMapperImpl extends BaseMapper implements UserMapper {
    public MyUser selectUserById(int id) {
        //获取 SqlSessionFactory 提供的 SqlSessionTemplate 对象
        SqlSession session = getSqlSession();
        return session.selectOne("dao.UserMapper.selectUserById", id);
    }
    public List<MyUser> selectAllUser() {
        SqlSession session = getSqlSession();
        return session.selectList("dao.UserMapper.selectAllUser");
    }
}
```

❼ 创建控制类

在应用 ch3_3 的 src 目录下，创建一个名为 controller 的包，并在该包中创建控制器类 MyController。在该控制器类中，调用 UserMapper 接口中的方法操作数据库，核心代码如下：

```
@Controller
public class MyController {
        @Autowired
        private UserMapper userMapper;
        @RequestMapping("/test")
        public String test() {
        // 查询一个用户
        MyUser mu=userMapper.selectUserById(1);
        System.out.println(mu);
        // 查询所有用户
        List<MyUser> listMu=userMapper.selectAllUser();
        for (MyUser myUser : listMu) {
            System.out.println(myUser);
        }
        return "test";
    }
}
```

❽ 创建测试页面

在/WEB-INF/目录下，创建一个名为 jsp 的文件夹，并在该文件夹中创建 test.jsp 文件。test.jsp 的代码略。

❾ 创建 Web、Spring、Spring MVC 的配置文件

在应用 ch3_3 的 config 包中，创建 Spring 配置文件 applicationContext.xml 和 Spring MVC 配置文件 springmvc.xml；在应用 ch3_3 的/WEB-INF/目录中，创建 Web 配置文件 web.xml。

在 Spring 配置文件 applicationContext.xml 中，首先使用<context:property-placeholder/>加载数据库连接信息属性文件；然后使用 org.apache.commons.dbcp2.BasicDataSource 配置数据源，并使用 org.springframework.jdbc.datasource.DataSourceTransactionManager 为数据源添加事务管理器；最后使用 org.mybatis.spring.SqlSessionFactoryBean 配置 MyBatis 工厂，同时指定数据源，并与 MyBatis 完美整合。Spring 配置文件 applicationContext.xml 的具体内容如下：

```
<?xml version="1.0" encoding="UTF-8"?>
<beans xmlns="http://www.springframework.org/schema/beans"
xmlns:xsi="http://www.w3.org/2001/XMLSchema-instance"
xmlns:tx="http://www.springframework.org/schema/tx"
xmlns:context="http://www.springframework.org/schema/context"
    xsi:schemaLocation="http://www.springframework.org/schema/beans
        http://www.springframework.org/schema/beans/spring-beans.xsd
        http://www.springframework.org/schema/tx
        http://www.springframework.org/schema/tx/spring-tx.xsd
        http://www.springframework.org/schema/context
        http://www.springframework.org/schema/context/spring-context.xsd">
    <!-- 加载数据库配置文件 -->
<context:property-placeholder location="classpath:config/jdbc.properties" />
        <!-- 配置数据源 -->
<bean id="dataSource" class="org.apache.commons.dbcp2.BasicDataSource">
    <property name="driverClassName" value="${jdbc.driver}" />
```

```
        <property name="url" value="${jdbc.url}" />
        <property name="username" value="${jdbc.username}" />
        <property name="password" value="${jdbc.password}" />
        <!-- 最大连接数 -->
        <property name="maxTotal" value="${jdbc.maxTotal}" />
        <!-- 最大空闲连接数 -->
        <property name="maxIdle" value="${jdbc.maxIdle}" />
        <!-- 初始化连接数 -->
        <property name="initialSize" value="${jdbc.initialSize}" />
    </bean>
        <!-- 添加事务支持 -->
    <bean id="txManager" class=
          "org.springframework.jdbc.datasource.DataSourceTransactionManager">
            <property name="dataSource" ref="dataSource" />
    </bean>
        <!-- 开启事务注解 -->
    <tx:annotation-driven transaction-manager="txManager" />
    <bean id="sqlSessionFactory" class=
            "org.mybatis.spring.SqlSessionFactoryBean">
        <property name="dataSource" ref="dataSource" />
            <property name="configLocation" value="classpath:config/mybatis-
             config.xml"></property>
        </bean>
    </beans>
```

在 Spring MVC 配置文件 springmvc.xml 中，使用<context:component-scan/>扫描包，并使用 org.springframework.web.servlet.view.InternalResourceViewResolver 配置视图解析器。具体代码如下：

```
    <?xml version="1.0" encoding="UTF-8"?>
    <beans xmlns="http://www.springframework.org/schema/beans"
        xmlns:xsi="http://www.w3.org/2001/XMLSchema-instance"
        xmlns:context="http://www.springframework.org/schema/context"
        xsi:schemaLocation="
        http://www.springframework.org/schema/beans
        http://www.springframework.org/schema/beans/spring-beans.xsd
            http://www.springframework.org/schema/context
            http://www.springframework.org/schema/context/spring-context.xsd">
        <context:component-scan base-package="controller"/>
        <context:component-scan base-package="dao"/>
        <bean class=
          "org.springframework.web.servlet.view.InternalResourceViewResolver"
              id="internalResourceViewResolver">
            <property name="prefix" value="/WEB-INF/jsp/" />
            <property name="suffix" value=".jsp" />
        </bean>
    </beans>
```

在 Web 配置文件 web.xml 中，首先通过<context-param>加载 Spring 配置文件 applicationContext.xml，并通过 org.springframework.web.context.ContextLoaderListener 启动 Spring 容器；然后配置 Spring MVC DispatcherServlet，并加载 Spring MVC 配置文件 springmvc.xml。Web 配置文件 web.xml 的代码与 ch3_2 的相同，不再赘述。

⑩ 测试应用

发布应用 ch3_3 到 Web 服务器 Tomcat 后，通过网址 http://localhost:8080/ch3_3/test 测试应用。

视频讲解

3.7 使用 MyBatis Generator 插件自动生成映射文件

使用 MyBatis Generator 插件自动生成 MyBatis 的 DAO 接口、实体模型类、Mapper 映射文件，将生成的代码复制到项目工程中即可，把更多精力放在业务逻辑上。

MyBatis Generator 有三种常用方法自动生成代码：命令行、Eclipse 插件和 Maven 插件。本节使用比较简单的方法（命令行）自动生成相关代码。具体步骤如下。

❶ 准备相关 JAR 包

需要准备的 JAR 包：mysql-connector-java-5.1.45-bin.jar 和 mybatis-generator-core-1.4.0.jar（https://mvnrepository.com/artifact/org.mybatis.generator/mybatis-generator-core/1.4.0）。

❷ 创建文件目录

在某磁盘根目录下新建一个文件目录，例如 C:\generator，并将 mysql-connector-java-5.1.45-bin.jar 和 mybatis-generator-core-1.4.0.jar 文件复制到 generator 目录下。另外，在 generator 目录下，创建 src 子目录存放生成的相关代码文件。

❸ 创建配置文件

在第 2 步创建的文件目录（C:\generator）下创建配置文件，例如 C:\generator\generator.xml。文件目录如图 3.5 所示。

名称	修改日期
src	2020/2/15 7:41
generator.xml	2020/2/15 7:41
mybatis-generator-core-1.4.0.jar	2020/2/15 7:36
mysql-connector-java-5.1.45-bin.jar	2018/1/22 6:57

图 3.5　generator 目录

generator.xml 配置文件内容如下（具体含义见注释）：

```
<?xml version="1.0" encoding="UTF-8"?>
<!DOCTYPE generatorConfiguration PUBLIC "-//mybatis.org//DTD MyBatis Generator
  Configuration 1.0//EN" "http://mybatis.org/dtd/mybatis-generator-config_
  1_0.dtd">
<generatorConfiguration>
<!-- 数据库驱动包位置 -->
    <classPathEntry location="C:\generator\mysql-connector-java-5.1.45-
      bin.jar" />
    <context id="mysqlTables" targetRuntime="MyBatis3">
        <commentGenerator>
            <property name="suppressAllComments" value="true" />
        </commentGenerator>
        <!-- 数据库链接 URL、用户名、密码（前提是数据库 springtest 存在） -->
         <jdbcConnection
            driverClass="com.mysql.jdbc.Driver"
```

```
            connectionURL="jdbc:mysql://localhost:3306/
              springtest?characterEncoding=utf8"
            userId="root" password="root">
        </jdbcConnection>
        <javaTypeResolver>
           <property name="forceBigDecimals" value="false" />
        </javaTypeResolver>
        <!-- 生成模型（MyBatis 里面用到实体类）的包名和位置-->
        <javaModelGenerator targetPackage="com.po" targetProject="C:\generator\
         src">
           <property name="enableSubPackages" value="true" />
           <property name="trimStrings" value="true" />
        </javaModelGenerator>
        <!-- 生成的映射文件（MyBatis 的 SQL 语句 xml 文件）的包名和位置-->
        <sqlMapGenerator targetPackage="mybatis" targetProject="C:\generator\
          src">
           <property name="enableSubPackages" value="true" />
        </sqlMapGenerator>
        <!-- 生成 DAO 的包名和位置 -->
        <javaClientGenerator type="XMLMAPPER" targetPackage="com.dao"
         targetProject="C:\generator\src">
           <property name="enableSubPackages" value="true" />
        </javaClientGenerator>
        <!-- 要生成那些表（更改 tableName 和 domainObjectName 就可以，前提是数据库
           springtest 中的 user 表已创建）-->
        <table tableName="user" domainObjectName="User" enableCountByExample=
         "false" enableUpdateByExample="false" enableDeleteByExample="false"
          enableSelectByExample="false" selectByExampleQueryId="false" />
     </context>
  </generatorConfiguration>
```

❹ 使用命令生成代码

打开命令提示符，进入 C:\generator，输入命令：java -jar mybatis-generator-core-1.4.0.jar -configfile generator.xml -overwrite，如图 3.6 所示。该命令成功执行的前提是配置 Java 的系统环境变量 classpath。

```
C:\>cd generator

C:\generator>java -jar mybatis-generator-core-1.4.0.jar -configfile generator.xml -overwrite
MyBatis Generator finished successfully.

C:\generator>
```

图 3.6 使用命令行生成映射文件

3.8 映射器概述

映射器是 MyBatis 最复杂且最重要的组件，由一个接口加一个 XML 文件（SQL 映射文件）组成。MyBatis 的映射器也可以使用注解完成，但在实际应用中使用不多，原因主要来自这几个方面：其一，面对复杂的 SQL 会显得无力；其二，注解的可读性较差；其三，注解丢失了 XML 上下文相互引用的功能。因此，推荐使用 XML 文件开发 MySQL 的映射器。

SQL 映射文件的常用配置元素如表 3.1 所示。

表 3.1　SQL 映射文件的常用配置元素

元素名称	描　述	备　注
select	查询语句，最常用、最复杂的元素之一	可以自定义参数，返回结果集等
insert	插入语句	执行后返回一个整数，代表插入的行数
update	更新语句	执行后返回一个整数，代表更新的行数
delete	删除语句	执行后返回一个整数，代表删除的行数
sql	定义一部分 SQL，在多个位置被引用	例如，一张表列名，一次定义，可以在多个 SQL 语句中使用
resultMap	用来描述从数据库结果集中加载对象，是最复杂、最强大的元素之一	提供映射规则

视频讲解

3.9　<select>元素

在 SQL 映射文件中，<select>元素用于映射 SQL 的 select 语句，其示例代码如下：

```
<!-- 根据 uid 查询一个用户信息 -->
<select id="selectUserById" parameterType="Integer" resultType="MyUser">
    select * from user where uid = #{uid}
</select>
```

上述示例代码中，id 的值是唯一标识符（对应 Mapper 接口的某个方法），它接收一个 Integer 类型的参数，返回一个 MyUser 类型的对象，结果集自动映射到 MyUser 的属性。但需要注意的是，MyUser 的属性名称一定与查询结果的列名相同。

<select>元素除上述示例代码中的几个属性外，还有一些常用的属性，如表 3.2 所示。

表 3.2　<select>元素的常用属性

属性名称	描　述
id	它和 Mapper 的命名空间组合起来使用（对应 Mapper 接口的某个方法），是唯一标识符，供 MyBatis 调用
parameterType	表示传入 SQL 语句的参数类型的全限定名或别名，是可选属性，MyBatis 能推断出具体传入语句的参数
resultType	SQL 语句执行后返回的类型（全限定名或者别名）。如果是集合类型，返回的是集合元素的类型。返回时可以使用 resultType 或 resultMap 之一
resultMap	它是映射集的引用，与<resultMap>元素一起使用。返回时可以使用 resultType 或 resultMap 之一
flushCache	它的作用是在调用 SQL 语句后，判断是否要求 MyBatis 清空之前查询本地缓存和二级缓存。默认值为 false。如果设置为 true，则任何时候只要 SQL 语句被调用，都将清空本地缓存和二级缓存
useCache	启动二级缓存的开关。默认值为 true，表示将查询结果存入二级缓存中
timeout	用于设置超时参数，单位是秒。超时将抛出异常
fetchSize	设定获取记录的总条数
statementType	告诉 MyBatis 使用哪个 JDBC 的 Statement 工作，取值为 STATEMENT（Statement）、PREPARED（PreparedStatement）、CALLABLE（CallableStatement），默认值为 PREPARED
resultSetType	这是针对 JDBC 的 ResultSet 接口而言，其值可设置为 FORWARD_ONLY（只允许向前访问）、SCROLL_SENSITIVE（双向滚动，但不及时更新）、SCROLL_INSENSITIVE（双向滚动，及时更新）

▶ 3.9.1 使用 Map 接口传递参数

在实际开发中，查询 SQL 语句经常需要多个参数，例如多条件查询。传递多个参数时，<select>元素的 parameterType 属性值的类型是什么呢？在 MyBatis 中，允许 Map 接口通过键值对传递多个参数。

假设数据操作接口中有一个实现查询陈姓男性用户信息功能的方法：

```
public List<MyUser> testMapSelect(Map<String, Object> param);
```

此时，传递给 MyBatis 映射器的是一个 Map 对象，使用该 Map 对象在 SQL 中设置对应的参数，对应 SQL 映射文件代码如下：

```
<!-- 查询陈姓男性用户信息 -->
<select id="testMapSelect" resultType="MyUser" parameterType="map">
    select * from user
    where uname like concat('%',#{u_name},'%')
    and usex = #{u_sex}
</select>
```

上述 SQL 映射文件中，参数名 u_name 和 u_sex 是 Map 中的 key。

【例 3-4】在 3.6.3 节例 3-2 的基础上实现 Map 接口传递参数。为节省篇幅，相同的实现不再赘述。其他的具体实现如下。

❶ 添加接口方法

在应用 ch3_2 的 com.mybatis.mapper.UserMapper 接口中，添加接口方法（见上述），实现查询陈姓男性用户信息。

❷ 添加 SQL 映射

在应用 ch3_2 的 SQL 映射文件 UserMapper.xml 中，添加 SQL 映射（见上述），实现查询陈姓男性用户信息。

❸ 添加请求处理方法

在应用 ch3_2 的 TestController 控制器类中，添加测试方法 testMapSelect，具体代码如下：

```
@RequestMapping("/testMapSelect")
public String testMapSelect(Model model) {
    //查询所有陈姓男性用户
    Map<String, Object> map = new HashMap<>();
    map.put("u_name", "陈");
    map.put("u_sex", "男");
    List<MyUser> unameAndUsexList = userMapper.testMapSelect;
    model.addAttribute("unameAndUsexList", unameAndUsexList);
    return "showUnameAndUsexUser";
}
```

❹ 创建查询结果显示页面

在应用 ch3_2 的 WEB-INF/jsp 目录下创建查询结果显示页面 showUnameAndUsexUser.jsp。在该页面中使用 JSTL 标签，所以需要将 taglibs-standard-impl-1.2.5.jar 和 taglibs-standard-spec-1.2.5.jar 复制到 WEB-INF/lib 目录中。另外，在页面中使用 BootStrap 美化页面，所以需要将相关的 css 及 js 复制到 WebContent/static 目录中，同时在应用 ch3_2 的 springmvc.xml 文件中使用<mvc:resources location="/static/" mapping="/static/**"></mvc:resources>允许

WebContent/static 目录下所有静态资源可见。testMapSelect.jsp 的核心代码如下：

```
<body>
    <div class="container">
        <div class="panel panel-primary">
            <div class="panel-heading">
                <h3 class="panel-title">陈姓男性用户列表</h3>
            </div>
            <div class="panel-body">
                <div class="table table-responsive">
                    <table class="table table-bordered table-hover">
                        <tbody class="text-center">
                            <tr>
                                <th>用户 ID</th>
                                <th>姓名</th>
                                <th>性别</th>
                            </tr>
                            <c:forEach items="${unameAndUsexList}" var="user">
                                <tr>
                                    <td>${user.uid}</td>
                                    <td>${user.uname}</td>
                                    <td>${user.usex}</td>
                                </tr>
                            </c:forEach>
                        </tbody>
                    </table>
                </div>
            </div>
        </div>
    </div>
</body>
```

❺ 测试应用

发布应用 ch3_2 到 Web 服务器 Tomcat 后，通过网址 http://localhost:8080/ch3_2/testMapSelect 测试应用。成功运行后，如图 3.7 所示。

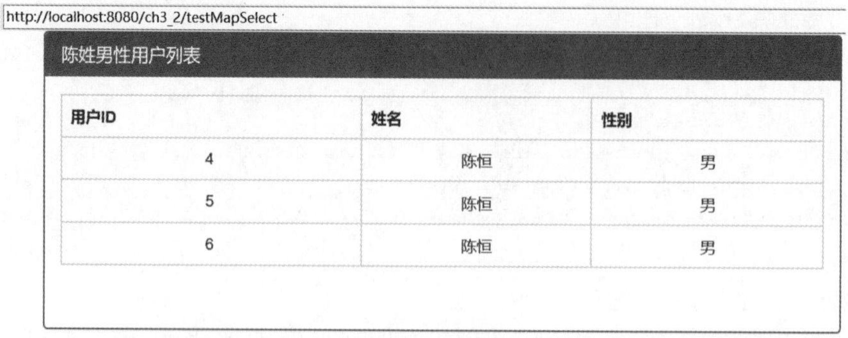

http://localhost:8080/ch3_2/testMapSelect

陈姓男性用户列表

用户ID	姓名	性别
4	陈恒	男
5	陈恒	男
6	陈恒	男

图 3.7　查询所有陈姓男性用户

Map 是一个键值对应的集合，使用者需要通过阅读它的键，了解其作用。另外，使用 Map 不能限定其传递的数据类型，所以业务性不强，可读性差。如果 SQL 语句很复杂，参数很多，使用 Map 很不方便。幸运的是，MyBatis 还提供使用 Java Bean 传递参数的方法。

▶ 3.9.2 使用 Java Bean 传递参数

在 MyBatis 中，需要将多个参数传递给映射器时，可以将它们封装在一个 Java Bean 中。下面通过具体实例讲解如何使用 Java Bean 传递参数。

【例 3-5】在 3.6.3 节例 3-2 的基础上实现 Java Bean 传递参数。为节省篇幅，相同的实现不再赘述。其他的具体实现如下。

❶ 添加接口方法

在应用 ch3_2 的 com.mybatis.mapper.UserMapper 接口中，添加接口方法 selectAllUserByJavaBean()，在该方法中使用 MyUser 类的对象将参数信息封装。接口方法 selectAllUserByJavaBean()的定义如下：

```
public List<MyUser> selectAllUserByJavaBean(MyUser user);
```

❷ 添加 SQL 映射

在应用 ch3_2 的 SQL 映射文件 UserMapper.xml 中，添加接口方法对应的 SQL 映射，具体代码如下：

```
<!--通过 Java Bean 传递参数查询陈姓男性用户信息，#{uname}的 uname 为参数 MyUser 的属性-->
<select id="selectAllUserByJavaBean" resultType="MyUser" parameterType=
      "MyUser">
    select * from user
    where uname like concat('%',#{uname},'%')
    and usex = #{usex}
</select>
```

❸ 添加请求处理方法

在应用 ch3_2 的控制器类 TestController 中添加请求处理方法 selectAllUserByJavaBean()，具体代码如下：

```
@RequestMapping("/selectAllUserByJavaBean")
public String selectAllUserByJavaBean(Model model) {
    //通过 MyUser 封装参数，查询所有陈姓男性用户
    MyUser mu = new MyUser();
    mu.setUname("陈");
    mu.setUsex("男");
    List<MyUser> unameAndUsexList = userMapper.selectAllUserByJavaBean(mu);
    model.addAttribute("unameAndUsexList", unameAndUsexList);
    return "showUnameAndUsexUser";
}
```

❹ 测试应用

重启 Web 服务器 Tomcat，通过网址 http://localhost:8080/ch3_2/selectAllUserByJavaBean 测试应用。

▶ 3.9.3 使用@Param 注解传递参数

不管是 Map 传递参数，还是 Java Bean 传递参数，它们都是将多个参数封装在一个对象中，实际上传递的还是一个参数。而使用@Param 注解可以将多个参数依次传递给 MyBatis 映射器。示例代码如下：

```
public List<MyUser> selectByParam(@Param("puname") String uname, @Param("pusex")
    String usex);
```

在上述示例代码中，puname 和 pusex 是传递给 MyBatis 映射器的参数名。

下面通过实例讲解如何使用@Param 注解传递参数。

【例 3-6】在 3.6.3 节例 3-2 的基础上实现@Param 注解传递参数。为节省篇幅，相同的实现不再赘述。其他的具体实现如下。

❶ 添加接口方法

在应用 ch3_2 的 UserMapper 接口中，添加数据操作接口方法 selectAllUserByParam()，在该方法中使用@Param 注解传递两个参数。接口方法 selectAllUserByParam()的定义如下：

```
public List<MyUser> selectAllUserByParam(@Param("puname") String uname,
  @Param("pusex") String usex);
```

❷ 添加 SQL 映射

在应用 ch3_2 的 SQL 映射文件 UserMapper.xml 中，添加接口方法 selectAllUserByParam()对应的 SQL 映射，具体代码如下：

```
<!-- 通过@Param 注解传递参数查询陈姓男性用户信息,这里不需要定义参数类型-->
<select id="selectAllUserByParam" resultType="MyUser">
    select * from user
    where uname like concat('%',#{puname},'%')
    and usex = #{pusex}
</select>
```

❸ 添加请求处理方法

在应用 ch3_2 的控制器类 TestController 中，添加请求处理方法 selectAllUserByParam()，具体代码如下：

```
@RequestMapping("/selectAllUserByParam")
public String selectAllUserByParam(Model model) {
    //通过@Param 注解传递参数，查询所有陈姓男性用户
    List<MyUser> unameAndUsexList = userMapper.selectAllUserByParam("陈",
      "男");
    model.addAttribute("unameAndUsexList", unameAndUsexList);
    return "showUnameAndUsexUser";
}
```

❹ 测试应用

重启 Web 服务器 Tomcat，通过网址 http://localhost:8080/ch3_2/selectAllUserByParam 测试应用。

在实际应用中是选择 Map、Java Bean，还是选择@Param 传递多个参数，应根据实际情况。如果参数较少，建议选择@Param；如果参数较多，建议选择 Java Bean。

▶ 3.9.4 <resultMap>元素

<resultMap>元素表示结果映射集，是 MyBatis 中最重要也是最强大的元素之一，主要用来定义映射规则、类型转化器以及进行级联的更新等。<resultMap>元素包含了一些子元素，结构如下所示：

```
<resultMap type="" id="">
    <constructor><!-- 类在实例化时,用来注入结果到构造方法 -->
        <idArg/><!-- ID 参数,结果为 ID -->
        <arg/><!-- 注入到构造方法的一个普通结果 -->
    </constructor>
    <id/><!-- 用于表示哪个列是主键 -->
    <result/><!-- 注入到字段或 POJO 属性的普通结果 -->
    <association property=""/><!-- 用于一对一关联 -->
    <collection property=""/><!-- 用于一对多、多对多关联 -->
    <discriminator javaType=""><!-- 使用结果值来决定使用哪个结果映射 -->
        <case value=""/>    <!-- 基于某些值的结果映射 -->
    </discriminator>
</resultMap>
```

<resultMap>元素的 type 属性表示需要的 POJO，id 属性是 resultMap 的唯一标识。子元素<constructor>用于配置构造方法（当 POJO 未定义无参数的构造方法时使用）。子元素<id>用于表示哪个列是主键。子元素<result>用于表示 POJO 和数据表普通列的映射关系。子元素<association>、<collection>和<discriminator>是用在级联的情况下。关于级联的问题比较复杂，将在 3.11 节学习。

一条 SQL 语句执行后，结果可以使用 Map 存储，也可以使用 POJO（Java Bean）存储。

▶ 3.9.5 使用 POJO 存储结果集

在 3.9.1 节至 3.9.3 节中，都是直接使用 Java Bean（MyUser）存储的结果集，这是因为 MyUser 的属性名与查询结果集的列名相同。如果查询结果集的列名与 Java Bean 的属性名不同，那么可以结合<resultMap>元素将 Java Bean 的属性与查询结果集的列名一一对应。

下面通过一个实例讲解如何使用<resultMap>元素将 Java Bean 的属性与查询结果集的列名一一对应。

【例 3-7】在 3.6.3 节例 3-2 的基础上，使用<resultMap>元素将 Java Bean 的属性与查询结果集的列名一一对应。为节省篇幅，相同的实现不再赘述。其他的具体实现如下。

❶ 创建 POJO 类

在应用 ch3_2 的 com.mybatis.po 包中，创建一个名为 MapUser 的 POJO（Plain Ordinary Java Object，普通的 Java 类）类，具体代码如下：

```
package com.mybatis.po;
public class MapUser {
    private Integer m_uid;
    private String m_uname;
    private String m_usex;
    //此处省略 setter 和 getter 方法
    @Override
    public String toString() {
        return "User [uid=" + m_uid +",uname=" + m_uname + ",usex=" + m_usex +"]";
    }
}
```

❷ 添加接口方法

在应用 ch3_2 的 UserMapper 接口中，添加数据操作接口方法 selectAllUserPOJO()，该方

法的返回值类型是 List<MapUser>。接口方法 selectAllUserPOJO()的定义如下：

```
public List<MapUser> selectAllUserPOJO();
```

❸ 添加 SQL 映射

在应用 ch3_2 的 SQL 映射文件 UserMapper.xml 中，首先，使用<resultMap>元素将 MapUser 类的属性与查询结果列名一一对应；然后，添加接口方法 selectAllUserPOJO()对应的 SQL 映射。具体代码如下：

```
<!-- 使用自定义结果集类型 -->
<resultMap type="com.mybatis.po.MapUser" id="myResult">
    <!-- property 是 MapUser 类中的属性-->
    <!-- column 是查询结果的列名，可以来自不同的表 -->
    <id property="m_uid" column="uid"/>
    <result property="m_uname" column="uname"/>
    <result property="m_usex" column="usex"/>
</resultMap>
<!-- 使用自定义结果集类型查询所有用户 -->
<select id="selectAllUserPOJO" resultMap="myResult">
    select * from user
</select>
```

❹ 添加请求处理方法

在应用 ch3_2 的控制器类 TestController 中，添加请求处理方法 selectAllUserPOJO()，具体代码如下：

```
@RequestMapping("/selectAllUserPOJO")
public String selectAllUserPOJO(Model model) {
    List<MapUser> unameAndUsexList = userMapper.selectAllUserPOJO();
    model.addAttribute("unameAndUsexList", unameAndUsexList);
    return "showUnameAndUsexUserPOJO";
}
```

❺ 创建显示查询结果的页面

在应用 ch3_2 的 WEB-INF/jsp 目录下，创建 showUnameAndUsexUserPOJO.jsp 文件显示查询结果，核心代码如下：

```
<c:forEach items="${unameAndUsexList}" var="user">
    <tr>
        <td>${user.m_uid}</td>
        <td>${user.m_uname}</td>
        <td>${user.m_usex}</td>
    </tr>
</c:forEach>
```

❻ 测试应用

重启 Web 服务器 Tomcat，通过网址 http://localhost:8080/ch3_2/selectAllUserPOJO 测试应用。

3.9.6 使用 Map 存储结果集

在 MyBatis 中，任何查询结果都可以使用 Map 存储。下面通过一个实例讲解如何使用 Map 存储查询结果。

【例 3-8】在 3.6.3 节例 3-2 的基础上，使用 Map 存储查询结果。为节省篇幅，相同的实现不再赘述。其他的具体实现如下。

❶ 添加接口方法

在应用 ch3_2 的 UserMapper 接口中，添加数据操作接口方法 selectAllUserMap()，该方法的返回值类型是 List<Map<String, Object>>。接口方法 selectAllUserMap()的定义如下：

```
public List<Map<String, Object>> selectAllUserMap();
```

❷ 添加 SQL 映射

在应用 ch3_2 的 SQL 映射文件 UserMapper.xml 中，添加接口方法 selectAllUserMap()对应的 SQL 映射，具体代码如下：

```
<!-- 使用 Map 存储查询结果，查询结果的列名作为 Map 的 key，列值作为 Map 的 value -->
<select id="selectAllUserMap" resultType="map">
    select * from user
</select>
```

❸ 添加请求处理方法

在应用 ch3_2 的控制器类 TestController 中，添加请求处理方法 selectAllUserMap()，具体代码如下：

```
@RequestMapping("/selectAllUserMap")
public String selectAllUserMap(Model model) {
    //使用 Map 存储查询结果
    List<Map<String, Object>> unameAndUsexList = userMapper.selectAllUserMap();
    model.addAttribute("unameAndUsexList", unameAndUsexList);
    //在 showUnameAndUsexUser.jsp 页面中遍历时，属性名与查询结果的列名（Map 的 key）
    //相同
    return "showUnameAndUsexUser";
}
```

❹ 测试应用

重启 Web 服务器 Tomcat，通过网址 http://localhost:8080/ch3_2/selectAllUserMap 测试应用。

3.10 <insert>、<update>、<delete>以及<sql>元素

视频讲解

3.10.1 <insert>元素

<insert>元素用于映射添加语句，MyBatis 执行完一条添加语句后，将返回一个整数表示其影响的行数。它的属性与<select>元素的属性大部分相同，本节讲解它的几个特有属性。

keyProperty：添加时将自动生成的主键值回填给 PO（Persistant Object）类的某个属性，通常会设置为主键对应的属性。如果是联合主键，可以在多个值之间用逗号隔开。

keyColumn：设置第几列是主键，当主键列不是表中的第一列时需要设置。如果是联合主键时，可以在多个值之间用逗号隔开。

useGeneratedKeys：该属性将使 MyBatis 使用 JDBC 的 getGeneratedKeys()方法获取由数据库内部产生的主键，如 MySQL、SQL Server 等自动递增的字段，其默认值为 false。

❶ 主键（自动递增）回填

MySQL、SQL Server 等数据库的表格可以采用自动递增的字段作为主键。有时可能需要使用这个刚刚产生的主键，用以关联其他业务。因为本书采用的数据库是 MySQL 数据库，所以可以直接使用 ch3_2 应用讲解自动递增主键回填的使用方法。

【例 3-9】在 3.6.3 节例 3-2 的基础上，实现自动递增主键回填。为节省篇幅，相同的实现不再赘述。其他的具体实现如下。

1）添加接口方法

在应用 ch3_2 的 UserMapper 接口中，添加数据操作接口方法 addUserBack()，该方法的返回值类型是 int。接口方法 addUserBack()的定义如下：

```
public int addUserBack(MyUser mu);
```

2）添加 SQL 映射

在应用 ch3_2 的 SQL 映射文件 UserMapper.xml 中，添加接口方法 addUserBack()对应的 SQL 映射，具体代码如下：

```
<!-- 添加一个用户，成功后将主键值回填给 uid（po 类的属性）-->
<insert id="addUserBack" parameterType="MyUser" keyProperty="uid"
 useGeneratedKeys="true">
    insert into user (uname,usex) values(#{uname},#{usex})
</insert>
```

3）添加请求处理方法

在应用 ch3_2 的控制器类 TestController 中，添加请求处理方法 addUserBack()，具体代码如下：

```
@RequestMapping("/addUserBack")
public StringaddUserBack(Model model) {
    //添加一个用户
    MyUser addmu = new MyUser();
    addmu.setUname("陈恒主键回填");
    addmu.setUsex("男");
    userMapper.addUserBack(addmu);
    model.addAttribute("addmu", addmu);
    return "showAddUser";
}
```

4）创建显示被添加的用户信息页面

在应用 ch3_2 的/WEB-INF/jsp 目录下，创建 showAddUser.jsp 文件显示添加的用户信息，核心代码如下：

```
<div class="container">
    <div class="panel panel-primary">
        <div class="panel-heading">
            <h3 class="panel-title">添加的用户信息</h3>
        </div>
        <div class="panel-body">
            <div class="table table-responsive">
                <table class="table table-bordered table-hover">
```

```
                    <tbody class="text-center">
                        <tr>
                            <th>用户 ID（回填的主键）</th>
                            <th>姓名</th>
                            <th>性别</th>
                        </tr>
                        <tr>
                            <td>${addmu.uid}</td>
                            <td>${addmu.uname}</td>
                            <td>${addmu.usex}</td>
                        </tr>
                    </tbody>
                </table>
            </div>
        </div>
    </div>
</div>
```

5）测试应用

重启 Web 服务器 Tomcat，通过网址 http://localhost:8080/ch3_2/addUserBack 测试应用。运行结果如图 3.8 所示。

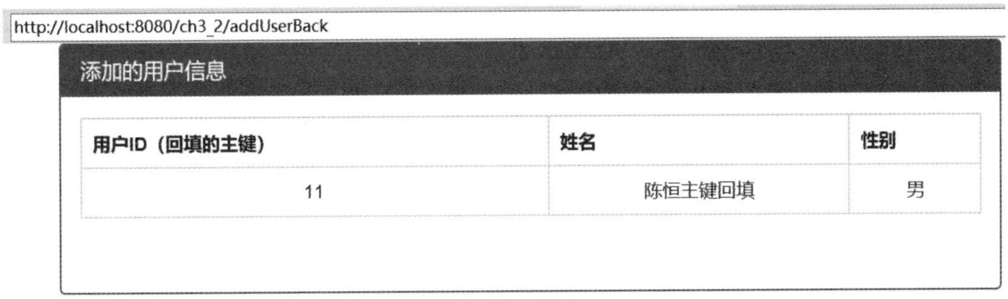

http://localhost:8080/ch3_2/addUserBack

添加的用户信息

用户ID（回填的主键）	姓名	性别
11	陈恒主键回填	男

图 3.8　回填主键值

❷ 自定义主键

如果实际工程中使用的数据库不支持主键自动递增（如 Oracle），或者取消了主键自动递增的规则，可以使用 MyBatis 的<selectKey>元素来自定义生成主键。具体配置示例代码如下：

```
<insert id="insertUser" parameterType="MyUser">
    <!-- 先使用 selectKey 元素定义主键，然后再定义 SQL 语句 -->
    <selectKey keyProperty="uid" resultType="Integer" order="BEFORE">
        select decode(max(uid), null, 1 , max(uid)+1) as newUid from user
    </selectKey>
    insert into user (uid,uname,usex) values(#uid,#{uname},#{usex})
</insert>
```

在执行上述示例代码时，<selectKey>元素首先被执行，该元素通过自定义的语句设置数据表的主键，然后执行添加语句。

<selectKey>元素的 keyProperty 属性指定了新生主键值返回给 PO 类（MyUser）的哪个属性。order 属性可以设置为 BEFORE 或 AFTER，BEFORE 表示先执行<selectKey>元素然后执行插入语句；AFTER 表示先执行插入语句再执行<selectKey>元素。

▶ 3.10.2 <update>与<delete>元素

<update>和<delete>元素比较简单，它们的属性和<insert>元素的属性基本一样，执行后也返回一个整数，表示影响数据库的记录行数。配置示例代码如下：

```
<!-- 修改一个用户 -->
<update id="updateUser" parameterType="MyUser">
    update user set uname = #{uname},usex = #{usex} where uid = #{uid}
</update>
<!-- 删除一个用户 -->
<delete id="deleteUser" parameterType="Integer">
    delete from user where uid = #{uid}
</delete>
```

▶ 3.10.3 <sql>元素

<sql>元素的作用是定义 SQL 语句的一部分（代码片段），方便后续的 SQL 语句引用它，例如反复使用的列名。在 MyBatis 中只需使用<sql>元素编写一次便能在其他元素中引用它。配置示例代码如下：

```
<sql id="comColumns">id,uname,usex</sql>
<select id="selectUser" resultType="MyUser">
    select <include refid="comColumns"/> from user
</select>
```

在上述代码中，使用<include>元素的 refid 属性引用了自定义的代码片段。

视频讲解

3.11 级联查询

级联关系是一个数据库实体的概念。级联关系有 3 种，分别是一对一级联、一对多级联以及多对多级联。级联的优点是获取关联数据十分方便，但是级联过多会增加数据库系统的复杂度，同时降低系统的性能。在实际开发中，要根据实际情况判断是否需要使用级联。数据库的级联操作包括更新、删除和查询。更新和删除的级联关系很简单，由数据库内在机制即可完成。本节仅讲述级联查询的相关实现。

如果表 A 中有一个外键引用了表 B 的主键，表 A 就是子表，表 B 就是父表。当查询表 A 的数据时，通过表 A 的外键，也将表 B 的相关记录返回，这就是级联查询。例如，查询一个人的信息的同时，根据外键（身份证号）也将他的身份证信息返回。

▶ 3.11.1 一对一级联查询

一对一级联关系在现实生活中是十分常见的。例如一个大学生只有一张一卡通，一张一卡通只属于一个学生；再如人与身份证的关系。

MyBatis 如何处理一对一级联查询呢？在 MyBatis 中，通过<resultMap>元素的子元素<association>处理这种一对一级联关系。在<association>元素中，通常使用以下属性。

- property：指定映射到实体类的对象属性。
- column：指定表中对应的字段（即查询返回的列名）。

- javaType：指定映射到实体对象属性的类型。
- select：指定引入嵌套查询的子 SQL 语句，该属性用于关联映射中的嵌套查询。

下面以个人与身份证之间的关系为例，讲解一对一级联查询的处理过程，读者只需参考该实例即可学会一对一级联查询的 MyBatis 实现。

【例 3-10】在 3.6.3 节例 3-2 的基础上，进行一对一级联查询的 MyBatis 实现。为节省篇幅，相同的实现不再赘述。其他的具体实现如下。

❶ 创建数据表

本实例需要在数据库 springtest 中创建两张数据表，一张是身份证表 idcard，一张是个人信息表 person。这两张表具有一对一的级联关系，它们的创建代码如下：

```
CREATE TABLE 'idcard' (
    'id' int(11) NOT NULL AUTO_INCREMENT,
    'code' varchar(18) COLLATE utf8_unicode_ci DEFAULT NULL,
    PRIMARY KEY ('id')
);
CREATE TABLE 'person' (
    'id' int(11) NOT NULL AUTO_INCREMENT,
    'name' varchar(20) COLLATE utf8_unicode_ci DEFAULT NULL,
    'age' int(11) DEFAULT NULL,
    'idcard_id' int(11) DEFAULT NULL,
    PRIMARY KEY ('id'),
    KEY 'idcard_id' ('idcard_id'),
    CONSTRAINT 'idcard_id' FOREIGN KEY ('idcard_id') REFERENCES 'idcard' ('id')
);
```

❷ 创建实体类

在应用 ch3_2 的 com.mybatis.po 包中创建数据表对应的持久化类 Idcard 和 Person。

Idcard 的代码如下：

```
package com.mybatis.po;
public class Idcard {
    private Integer id;
    private String code;
    //省略 setter 和 getter 方法
    @Override
    public String toString() {
        return "Idcard [id=" + id + ",code="+ code + "]";
    }
}
```

Person 的代码如下：

```
package com.mybatis.po;
public class Person {
    private Integer id;
    private String name;
    private Integer age;
    //个人身份证关联
    private Idcard card;
    //省略 setter 和 getter 方法
```

```
    @Override
    public String toString() {
        return "Person [id=" + id + ",name=" + name + ",age=" + age +",card="
          + card +"]" ;
    }
}
```

❸ 创建 SQL 映射文件

在应用 ch3_2 的 com.mybatis.mapper 包中创建两张表对应的映射文件 IdCardMapper.xml 和 PersonMapper.xml。在 PersonMapper.xml 文件中以 3 种方式实现“根据个人 id 查询个人信息”的功能，详情请看代码备注。

IdCardMapper.xml 的代码如下：

```
<?xml version="1.0" encoding="UTF-8" ?>
<!DOCTYPE mapper
PUBLIC "-//mybatis.org//DTD Mapper 3.0//EN"
"http://mybatis.org/dtd/mybatis-3-mapper.dtd">
<mapper namespace="com.mybatis.mapper.IdCardMapper">
    <select id="selectCodeById" parameterType="Integer" resultType="Idcard">
        select * from idcard where id=#{id}
    </select>
</mapper>
```

PersonMapper.xml 的代码如下：

```
<?xml version="1.0" encoding="UTF-8" ?>
<!DOCTYPE mapper
PUBLIC "-//mybatis.org//DTD Mapper 3.0//EN"
"http://mybatis.org/dtd/mybatis-3-mapper.dtd">
<mapper namespace="com.mybatis.mapper.PersonMapper">
    <!-- 一对一 根据 id 查询个人信息：级联查询的第一种方法（嵌套查询，执行两个 SQL 语句）-->
    <resultMap type="Person" id="cardAndPerson1">
        <id property="id" column="id"/>
        <result property="name" column="name"/>
        <result property="age" column="age"/>
        <!-- 一对一关联查询 -->
        <association property="card" column="idcard_id" javaType="Idcard"
        select="com.mybatis.mapper.IdCardMapper.selectCodeById"/>
    </resultMap>
    <select id="selectPersonById1" parameterType="Integer" resultMap=
      "cardAndPerson1">
        select * from person where id=#{id}
    </select>
    <!-- 一对一 根据 id 查询个人信息：级联查询的第二种方法（嵌套结果，执行一个 SQL 语句）-->
    <resultMap type="Person" id="cardAndPerson2">
        <id property="id" column="id"/>
        <result property="name" column="name"/>
        <result property="age" column="age"/>
        <!-- 一对一关联查询 -->
        <association property="card" javaType="Idcard">
            <id property="id" column="idcard_id"/>
            <result property="code" column="code"/>
```

```
        </association>
    </resultMap>
    <select id="selectPersonById2" parameterType="Integer" resultMap=
      "cardAndPerson2">
        select p.*,ic.code
        from person p, idcard ic
        where p.idcard_id = ic.id and p.id=#{id}
    </select>
    <!-- 一对一 根据 id 查询个人信息：连接查询（使用 POJO 存储结果） -->
    <select id="selectPersonById3" parameterType="Integer" resultType=
     "SelectPersonById">
        select p.*,ic.code
        from person p, idcard ic
        where p.idcard_id = ic.id and p.id=#{id}
    </select>
</mapper>
```

❹ 创建 POJO 类

在应用 ch3_2 的 com.mybatis.po 包中创建 POJO 类 SelectPersonById（第 3 步使用的 POJO 类）。

SelectPersonById 的代码如下：

```
package com.mybatis.po;
public class SelectPersonById {
    private Integer id;
    private String name;
    private Integer age;
    private String code;
    //省略 setter 和 getter 方法
    @Override
    public String toString() {
        return "Person [id=" + id + ",name=" + name + ",age=" + age
        + ",code=" + code + "]";
    }
}
```

❺ 创建 Mapper 接口

在应用 ch3_2 的 com.mybatis.mapper 包中创建第 3 步映射文件对应的数据操作接口 IdCardMapper 和 PersonMapper。

IdCardMapper 的核心代码如下：

```
@Repository
public interface IdCardMapper{
    public Idcard selectCodeById(Integer i);
}
```

PersonMapper 的核心代码如下：

```
@Repository
public interface PersonMapper{
    public Person selectPersonById1(Integer id);
    public Person selectPersonById2(Integer id);
```

```
    public SelectPersonById selectPersonById3(Integer id);
}
```

❻ 创建控制器类

在应用 ch3_2 的 controller 包中创建 OneToOneController 控制器类，在该类中调用第 5 步的接口方法。核心代码如下：

```
@Controller
public class OneToOneController{
    @Autowired
    private PersonMapper personMapper;
    @RequestMapping("/oneToOneTest")
    public String oneToOneTest() {
        System.out.println("级联查询的第一种方法（嵌套查询，执行两个 SQL 语句）");
        Person p1 = personMapper.selectPersonById1(1);
        System.out.println(p1);
        System.out.println("=======================");
        System.out.println("级联查询的第二种方法（嵌套结果，执行一个 SQL 语句）");
        Person p2 = personMapper.selectPersonById2(1);
        System.out.println(p2);
        System.out.println("=======================");
        System.out.println("连接查询（使用 POJO 存储结果）");
        SelectPersonById p3 = personMapper.selectPersonById3(1);
        System.out.println(p3);
        return "test";
    }
}
```

❼ 测试应用

发布应用到 Web 服务器 Tomcat，通过网址 http://localhost:8080/ch3_2/oneToOneTest 测试应用。测试时，需要事先为数据表手动添加数据。运行结果如图 3.9 所示。

```
Markers  Properties  Servers  Data Source Explorer  Snippets  Console
Tomcat v9.0 Server at localhost [Apache Tomcat] C:\soft\javaee\eclipse\plugins\org.eclip
级联查询的第一种方法（嵌套查询，执行两个SQL语句）
DEBUG [http-nio-8080-exec-3] - ==>  Preparing: select * from person w
DEBUG [http-nio-8080-exec-3] - ==> Parameters: 1(Integer)
DEBUG [http-nio-8080-exec-3] - ====>  Preparing: select * from idcarc
DEBUG [http-nio-8080-exec-3] - ====> Parameters: 1(Integer)
DEBUG [http-nio-8080-exec-3] - <====      Total: 1
DEBUG [http-nio-8080-exec-3] - <==      Total: 1
Person [id=1,name=陈恒,age=88,card=Idcard [id=1,code=123456789]]
=======================
级联查询的第二种方法（嵌套结果，执行一个SQL语句）
DEBUG [http-nio-8080-exec-3] - ==>  Preparing: select p.*,ic.code frc
DEBUG [http-nio-8080-exec-3] - ==> Parameters: 1(Integer)
DEBUG [http-nio-8080-exec-3] - <==      Total: 1
Person [id=1,name=陈恒,age=88,card=Idcard [id=1,code=123456789]]
=======================
连接查询（使用POJO存储结果）
DEBUG [http-nio-8080-exec-3] - ==>  Preparing: select p.*,ic.code frc
DEBUG [http-nio-8080-exec-3] - ==> Parameters: 1(Integer)
DEBUG [http-nio-8080-exec-3] - <==      Total: 1
Person [id=1,name=陈恒,age=88,code=123456789]
```

图 3.9　一对一级联查询结果

▶ 3.11.2 一对多级联查询

在 3.11.1 节学习了 MyBatis 如何处理一对一级联查询，那么 MyBatis 又如何处理一对多级联查询呢？在实际生活中有许多一对多关系，例如一个用户可以有多个订单，而一个订单只属于一个用户。

下面以用户和订单之间的关系为例，讲解一对多级联查询（实现“根据用户 id 查询用户及其关联的订单信息”的功能）的处理过程，读者只需参考该实例即可学会一对多级联查询的 MyBatis 实现。

【例 3-11】在 3.6.3 节例 3-2 的基础上，进行一对多级联查询的 MyBatis 实现。为节省篇幅，相同的实现不再赘述。其他的具体实现如下。

❶ 创建数据表

本实例需要两张数据表：一张是用户表 user；一张是订单表 orders。这两张表具有一对多级联关系。user 表在前面已创建，orders 的创建代码如下：

```
CREATE TABLE 'orders' (
    'id' int(11) NOT NULL AUTO_INCREMENT,
    'ordersn' varchar(10) COLLATE utf8_unicode_ci DEFAULT NULL,
    'user_id' int(11) DEFAULT NULL,
    PRIMARY KEY ('id'),
    KEY 'user_id' ('user_id'),
    CONSTRAINT 'user_id' FOREIGN KEY ('user_id') REFERENCES 'user' ('uid')
);
```

❷ 创建持久化类

在应用 ch3_2 的 com.mybatis.po 包中，创建数据表 orders 对应的持久化类 Orders、数据表 user 对应的持久化类 MyUserOrder。

Orders 类的代码如下：

```
package com.mybatis.po;
public class Orders {
    private Integer id;
    private  String ordersn;
    //省略 setter 和 getter 方法
    @Override
    public String toString() {
        return "Orders [id=" + id + ",ordersn=" + ordersn + "]";
    }
}
```

MyUserOrder 类的代码如下：

```
package com.mybatis.po;
import java.util.List;
public class MyUserOrder{
    private Integer uid;//主键
    private String uname;
    private String usex;
    private List<Orders> ordersList;
```

```
//省略 setter 和 getter 方法
    @Override
    public String toString() {
    return "User [uid=" + uid +", uname=" + uname + ", usex=" + usex + ", ordersList="
     + ordersList +"]";
    }
}
```

③ 创建并修改映射文件

在应用 ch3_2 的 com.mybatis.mapper 中，创建 orders 表对应的映射文件 OrdersMapper.xml。在映射文件 UserMapper.xml 中，添加实现一对多级联查询（根据用户 id 查询用户及其关联的订单信息）的 SQL 映射。

在 UserMapper.xml 文件中添加的 SQL 映射如下：

```
<!-- 一对多 根据 uid 查询用户及其关联的订单信息：级联查询的第一种方法（嵌套查询） -->
<resultMap type="MyUserOrder" id="userAndOrders1">
    <id property="uid" column="uid"/>
    <result property="uname" column="uname"/>
    <result property="usex" column="usex"/>
    <!-- 一对多关联查询，ofType 表示集合中的元素类型，将 uid 传递给
      selectOrdersById-->
    <collection property="ordersList" ofType="Orders" column="uid" select=
      "com.mybatis.mapper.OrdersMapper.selectOrdersById"/>
</resultMap>
<select id="selectUserOrdersById1" parameterType="Integer" resultMap=
      "userAndOrders1">
    select * from user where uid = #{id}
</select>
<!-- 一对多 根据 uid 查询用户及其关联的订单信息：级联查询的第二种方法（嵌套结果） -->
<resultMap type="MyUserOrder" id="userAndOrders2">
    <id property="uid" column="uid"/>
    <result property="uname" column="uname"/>
    <result property="usex" column="usex"/>
    <!-- 一对多关联查询，ofType 表示集合中的元素类型 -->
    <collection property="ordersList" ofType="Orders" >
        <id property="id" column="id"/>
        <result property="ordersn" column="ordersn"/>
    </collection>
</resultMap>
<select id="selectUserOrdersById2" parameterType="Integer" resultMap=
      "userAndOrders2">
    select u.*,o.id,o.ordersn from user u, orders o where u.uid = o.user_
      id and u.uid=#{id}
</select>
<!-- 一对多 根据 uid 查询用户及其关联的订单信息：连接查询（使用 map 存储结果） -->
<select id="selectUserOrdersById3" parameterType="Integer" resultType=
      "map">
    select u.*,o.id,o.ordersn from user u, orders o where u.uid = o.user_
      id and u.uid=#{id}
</select>
```

OrdersMapper.xml 的代码如下：

```
<?xml version="1.0" encoding="UTF-8" ?>
<!DOCTYPE mapper
PUBLIC "-//mybatis.org//DTD Mapper 3.0//EN"
"http://mybatis.org/dtd/mybatis-3-mapper.dtd">
<mapper namespace=" com.mybatis.mapper.OrdersMapper">
    <!-- 根据用户 uid 查询订单信息 -->
    <select id="selectOrdersById" parameterType="Integer" resultType=
      "Orders">
        select * from orders where user_id=#{id}
    </select>
</mapper>
```

❹ 创建并修改 Mapper 接口

在应用 ch3_2 的 com.mybatis.mapper 包中，创建第 3 步映射文件对应的数据操作接口 OrdersMapper，并在 UserMapper 接口中添加接口方法。

OrdersMapper 的核心代码如下：

```
@Repository
public interface OrdersMapper{
    public List<Orders> selectOrdersById(Integer uid);
}
```

在 UserMapper 接口中添加如下接口方法：

```
public MyUserOrder selectUserOrdersById1(Integer uid);
public MyUserOrder selectUserOrdersById2(Integer uid);
public List<Map<String, Object>> selectUserOrdersById3(Integer uid);
```

❺ 创建控制器类

在应用 ch3_2 的 controller 包中，创建控制器类 OneToMoreController，在该类中调用第 4 步的接口方法。核心代码如下：

```
@Controller
public class OneToMoreController {
    @Autowired
    private UserMapper userMapper;
    @RequestMapping("/oneToMoreTest")
    public String oneToMoreTest() {
        //查询一个用户及订单信息
        System.out.println("级联查询的第一种方法（嵌套查询，执行两个 SQL 语句）");
        MyUserOrder auser1 = userMapper.selectUserOrdersById1(1);
        System.out.println(auser1);
        System.out.println("====================================");
        System.out.println("级联查询的第二种方法（嵌套结果，执行一个 SQL 语句）");
        MyUserOrder auser2 = userMapper.selectUserOrdersById2(1);
        System.out.println(auser2);
        System.out.println("====================================");
        System.out.println("连接查询（使用 map 存储结果）");
        List<Map<String, Object>> auser3 =
          userMapper.selectUserOrdersById3(1);
        System.out.println(auser3);
```

```
        return "test";
    }
}
```

❻ 测试应用

重启 Web 服务器 Tomcat，通过网址 http://localhost:8080/ch3_2/oneToMoreTest 测试应用。测试时，应该事先为数据表手动添加数据。运行结果如图 3.10 所示。

```
Markers  Properties  Servers  Data Source Explorer  Snippets  Console
Tomcat v9.0 Server at localhost [Apache Tomcat] C:\soft\javaee\eclipse\plugins\org.eclipse.justj.openjdk.hotspot.jre.full.win32.x86_64_15.0.
级联查询的第一种方法（嵌套查询，执行两个SQL语句）
DEBUG [http-nio-8080-exec-4] - ==>  Preparing: select * from user where uid = ?
DEBUG [http-nio-8080-exec-4] - ==> Parameters: 1(Integer)
DEBUG [http-nio-8080-exec-4] - ====>  Preparing: select * from orders where user_id=?
DEBUG [http-nio-8080-exec-4] - ====> Parameters: 1(Integer)
DEBUG [http-nio-8080-exec-4] - <====      Total: 2
DEBUG [http-nio-8080-exec-4] - <==      Total: 1
User [uid=1,uname=张三,usex=女,ordersList=[Orders [id=1,ordersn=123456789], Orders [id=2,ordersn=987654321]]]
===================================
级联查询的第二种方法（嵌套结果，执行一个SQL语句）
DEBUG [http-nio-8080-exec-4] - ==>  Preparing: select u.*,o.id,o.ordersn from user u, orders o where u.uid =
DEBUG [http-nio-8080-exec-4] - ==> Parameters: 1(Integer)
DEBUG [http-nio-8080-exec-4] - <==      Total: 2
User [uid=1,uname=张三,usex=女,ordersList=[Orders [id=1,ordersn=123456789], Orders [id=2,ordersn=987654321]]]
===================================
连接查询（使用map存储结果）
DEBUG [http-nio-8080-exec-4] - ==>  Preparing: select u.*,o.id,o.ordersn from user u, orders o where u.uid =
DEBUG [http-nio-8080-exec-4] - ==> Parameters: 1(Integer)
DEBUG [http-nio-8080-exec-4] - <==      Total: 2
[{uid=1, uname=张三, ordersn=123456789, usex=女, id=1}, {uid=1, uname=张三, ordersn=987654321, usex=女, id=2}]
```

图 3.10　一对多级联查询结果

3.11.3　多对多级联查询

其实，MyBatis 没有实现多对多级联，这是因为多对多级联可以通过两个一对多级联进行替换。例如，一个订单可以有多种商品，一种商品可以对应多个订单，订单与商品就是多对多级联关系。使用一个中间表——订单记录表，就可以将多对多级联转换成两个一对多关系。下面以订单和商品（实现“查询所有订单以及每个订单对应的商品信息”的功能）为例，讲解多对多级联查询。

【例 3-12】在 3.11.2 节例 3-11 的基础上，进行多对多级联查询的 MyBatis 实现。为节省篇幅，相同的实现不再赘述。其他的具体实现如下。

❶ 创建数据表

订单表前文已创建，这里需要再创建商品表 product 和订单记录表 orders_detail。创建代码如下：

```
CREATE TABLE 'product' (
    'id' int(11) NOT NULL AUTO_INCREMENT,
    'name' varchar(50) COLLATE utf8_unicode_ci DEFAULT NULL,
    'price' double DEFAULT NULL,
    PRIMARY KEY ('id')
);
CREATE TABLE 'orders_detail' (
    'id' int(11) NOT NULL AUTO_INCREMENT,
    'orders_id' int(11) DEFAULT NULL,
    'product_id' int(11) DEFAULT NULL,
    PRIMARY KEY ('id'),
    KEY 'orders_id' ('orders_id'),
```

```
    KEY 'product_id' ('product_id'),
    CONSTRAINT 'orders_id' FOREIGN KEY ('orders_id') REFERENCES 'orders' ('id'),
    CONSTRAINT 'product_id' FOREIGN KEY ('product_id') REFERENCES 'product' ('id')
);
```

❷ 创建持久化类

在应用 ch3_2 的 com.mybatis.po 包中，创建数据表 product 对应的持久化类 Product，而中间表 orders_detail 不需要持久化类，但需要在订单表 orders 对应的持久化类 Orders 中添加关联属性。

Product 的代码如下：

```
package com.mybatis.po;
import java.util.List;
public class Product {
    private Integer id;
    private String name;
    private Double price;
    //多对多中的一个一对多
    private List<Orders> orders;
    //省略 setter 和 getter 方法
    @Override
    public String toString() {
        return "Product [id=" + id + ",name=" + name + ",price=" + price +"]";
    }
}
```

修改后的 Orders 代码如下：

```
package com.mybatis.po;
import java.util.List;
public class Orders {
    private Integer id;
    private  String ordersn;
    //多对多中的另一个一对多
    private List<Product> products;
    //省略 setter 和 getter 方法
    @Override
    public String toString() {
        return "Orders [id=" + id + ",ordersn=" + ordersn + ",products=" +
          products + "]";
    }
}
```

❸ 创建映射文件

本实例只需在 com.mybatis.mapper 的 OrdersMapper.xml 文件中追加以下配置，即可实现多对多关联查询。

```
<!-- 多对多关联查询所有订单以及每个订单对应的商品信息(嵌套结果) -->
<resultMap type="Orders" id="allOrdersAndProducts">
    <id property="id" column="id"/>
    <result property="ordersn" column="ordersn"/>
    <!-- 多对多关联 -->
```

```
        <collection property="products" ofType="Product">
            <id property="id" column="pid"/>
            <result property="name" column="name"/>
            <result property="price" column="price"/>
        </collection>
    </resultMap>
    <select id="selectallOrdersAndProducts" resultMap="allOrdersAndProducts">
        select o.*,p.id as pid,p.name,p.price
        from orders o,orders_detail od,product p
        where od.orders_id = o.id
        and od.product_id = p.id
    </select>
```

❹ 添加 Mapper 接口方法

在 OrdersMapper 接口中添加以下接口方法：

```
public List<Orders> selectallOrdersAndProducts();
```

❺ 创建控制器类

在应用 ch3_2 的 controller 包中，创建类 MoreToMoreController，在该类中调用第 4 步的接口方法。MoreToMoreController 的核心代码如下：

```
@Controller
public class MoreToMoreController {
    @Autowired
    private OrdersMapper ordersMapper;
    @RequestMapping("/moreToMoreTest")
    public String test() {
        List<Orders> os = ordersMapper.selectallOrdersAndProducts();
        for (Orders orders : os) {
            System.out.println(orders);
        }
        return "test";
    }
}
```

❻ 测试应用

重启 Web 服务器 Tomcat，通过网址 http://localhost:8080/ch3_2/moreToMoreTest 测试应用。测试时，需要事先为数据表手动添加数据。运行结果如图 3.11 所示。

```
Markers  Properties  Servers  Data Source Explorer  Snippets  Console ☒
Tomcat v9.0 Server at localhost [Apache Tomcat] C:\soft\javaee\eclipse\plugins\org.eclipse.justj.openjdk.hotspot.jre.full.win32.x86_64_15.0.1.v20
DEBUG [http-nio-8080-exec-7] - ==>  Preparing: select o.*,p.id as pid,p.name,p.price from orders o,orders_detail
DEBUG [http-nio-8080-exec-7] - ==> Parameters: 
DEBUG [http-nio-8080-exec-7] - <==      Total: 2
Orders [id=1,ordersn=123456789,products=[Product [id=1,name=苹果,price=10.0], Product [id=2,name=桔子,price=8.0]]]
```

图 3.11　多对多级联查询结果

3.12　动态 SQL

视频讲解

开发人员通常根据需求手动拼接 SQL 语句，这是一个极其麻烦的工作，而 MyBatis 提供了对 SQL 语句动态组装的功能，恰能解决这一问题。MyBatis 的动态 SQL 元素和使用 JSTL

或其他类似基于 XML 的文本处理器相似，常用元素有<if>、<choose>、<when>、<otherwise>、<trim>、<where>、<set>、<foreach>和<bind>等。

3.12.1 <if>元素

动态 SQL 通常要做的事情是有条件地包含 where 子句的一部分。所以在 MyBatis 中，<if>元素是最常用的元素。它类似于 Java 中的 if 语句。下面通过一个实例讲解<if>元素的使用过程。

【例 3-13】在 3.6.3 节例 3-2 的基础上，讲解<if>元素的使用过程。为节省篇幅，相同的实现不再赘述。其他的具体实现如下。

❶ 添加 Mapper 接口方法

在应用 ch3_2 的 UserMapper 接口中，添加数据操作接口方法 selectAllUserByIf()，在该方法中使用 MyUser 类的对象将参数信息封装。接口方法 selectAllUserByIf()的定义如下：

```
public List<MyUser> selectAllUserByIf(MyUser user);
```

❷ 添加 SQL 映射

在应用 ch3_2 的 SQL 映射文件 UserMapper.xml 中，添加接口方法 selectAllUserByIf()对应的 SQL 映射，具体代码如下：

```
<!-- 使用 if 元素，根据条件动态查询用户信息 -->
<select id="selectAllUserByIf"  resultType="MyUser" parameterType="MyUser">
    select * from user where 1=1
    <if test="uname !=null and uname!=''">
        and uname like concat('%',#{uname},'%')
    </if>
    <if test="usex !=null and usex!=''">
        and usex = #{usex}
    </if>
</select>
```

❸ 添加请求处理方法

在应用 ch3_2 的控制器类 TestController 中，添加请求处理方法 selectAllUserByIf()，具体代码如下：

```
@RequestMapping("/selectAllUserByIf")
public StringselectAllUserByIf(Model model) {
    MyUser mu = new MyUser();
    mu.setUname("陈");
    mu.setUsex("男");
    List<MyUser> unameAndUsexList = userMapper.selectAllUserByIf(mu);
    model.addAttribute("unameAndUsexList", unameAndUsexList);
    return "showUnameAndUsexUser";
}
```

❹ 测试应用

重启 Web 服务器 Tomcat，通过网址 http://localhost:8080/ch3_2/selectAllUserByIf 测试应用。

3.12.2 <choose>、<when>、<otherwise>元素

有时不需要用到所有的条件语句，而只需从中择其一二。针对这种情况，MyBatis 提供

了<choose>元素，它有点像 Java 中的 switch 语句。下面通过一个实例讲解<choose>元素的使用过程。

【例 3-14】在 3.6.3 节例 3-2 的基础上，讲解<choose>元素的使用过程。为节省篇幅，相同的实现不再赘述。其他的具体实现如下。

❶ 添加 Mapper 接口方法

在应用 ch3_2 的 UserMapper 接口中，添加数据操作接口方法 selectUserByChoose()，在该方法中使用 MyUser 类的对象将参数信息封装。接口方法 selectUserByChoose()的定义如下：

```
public List<MyUser> selectUserByChoose(MyUser user);
```

❷ 添加 SQL 映射

在应用 ch3_2 的 SQL 映射文件 UserMapper.xml 中，添加接口方法 selectUserByChoose()对应的 SQL 映射，具体代码如下：

```
<!-- 使用 choose、when、otherwise 元素，根据条件动态查询用户信息 -->
<select id="selectUserByChoose"  resultType="MyUser" parameterType="MyUser">
    select * from user where 1=1
    <choose>
    <when test="uname !=null and uname!=''">
        and uname like concat('%',#{uname},'%')
    </when>
    <when test="usex !=null and usex!=''">
        and usex = #{usex}
    </when>
    <otherwise>
        and uid > 3
    </otherwise>
    </choose>
</select>
```

❸ 添加请求处理方法

在应用 ch3_2 的控制器类 TestController 中，添加请求处理方法 selectUserByChoose()，具体代码如下：

```
@RequestMapping("/selectUserByChoose")
public String selectUserByChoose(Model model) {
    MyUser mu = new MyUser();
    mu.setUname("");
    mu.setUsex("");
    List<MyUser> unameAndUsexList = userMapper.selectUserByChoose(mu);
    model.addAttribute("unameAndUsexList", unameAndUsexList);
    return "showUnameAndUsexUser";
}
```

❹ 测试应用

重启 Web 服务器 Tomcat，通过网址 http://localhost:8080/ch3_2/selectUserByChoose 测试应用。

3.12.3 <trim>元素

<trim>元素的主要功能是可以在自己包含的内容前加上某些扩展名，也可以在其后加上

某些扩展名，与之对应的属性是 prefix 和 suffix；可以把包含内容的首部的某些内容覆盖，即忽略，也可以把尾部的某些内容覆盖，对应的属性是 prefixOverrides 和 suffixOverrides。正因为<trim>元素有这样的功能，所以也可以非常简单地利用<trim>来代替<where>元素的功能。下面通过一个实例讲解<trim>元素的使用过程。

【例 3-15】在 3.6.3 节例 3-2 的基础上，讲解<trim>元素的使用过程。为节省篇幅，相同的实现不再赘述。其他的具体实现如下。

❶ 添加 Mapper 接口方法

在应用 ch3_2 的 UserMapper 接口中，添加数据操作接口方法 selectUserByTrim()，在该方法中使用 MyUser 类的对象将参数信息封装。接口方法 selectUserByTrim()的定义如下：

```
public List<MyUser> selectUserByTrim(MyUser user);
```

❷ 添加 SQL 映射

在应用 ch3_2 的 SQL 映射文件 UserMapper.xml 中，添加接口方法 selectUserByTrim()对应的 SQL 映射，具体代码如下：

```
<!-- 使用 trim 元素，根据条件动态查询用户信息 -->
<select id="selectUserByTrim"  resultType="MyUser" parameterType="MyUser">
    select * from user
    <trim prefix="where" prefixOverrides="and|or">
        <if test="uname !=null and uname!=''">
            and uname like concat('%',#{uname},'%')
        </if>
        <if test="usex !=null and usex!=''">
            and usex = #{usex}
        </if>
    </trim>
</select>
```

❸ 添加请求处理方法

在应用 ch3_2 的控制器类 TestController 中，添加请求处理方法 selectUserByTrim()，具体代码如下：

```
@RequestMapping("/selectUserByTrim")
public String selectUserByTrim(Model model) {
    MyUser mu = new MyUser();
    mu.setUname("陈");
    mu.setUsex("男");
    List<MyUser> unameAndUsexList = userMapper.selectUserByTrim(mu);
    model.addAttribute("unameAndUsexList", unameAndUsexList);
    return "showUnameAndUsexUser";
}
```

❹ 测试应用

重启 Web 服务器 Tomcat，通过网址 http://localhost:8080/ch3_2/selectUserByTrim 测试应用。

3.12.4 <where>元素

<where>元素的作用是输出一个 where 语句，优点是不考虑<where>元素的条件输出，MyBatis 将智能处理。如果所有的条件都不满足，那么 MyBatis 将会查出所有记录；如果输

出是 and 开头，MyBatis 将把第一个 and 忽略；如果输出是 or 开头，MyBatis 也将把第一个 or 忽略。此外，在<where>元素中不考虑空格的问题，MyBatis 将智能加上。下面通过一个实例讲解<where>元素的使用过程。

【例 3-16】在 3.6.3 节例 3-2 的基础上，讲解<where>元素的使用过程。为节省篇幅，相同的实现不再赘述。其他的具体实现如下。

❶ 添加 Mapper 接口方法

在应用 ch3_2 的 UserMapper 接口中，添加数据操作接口方法 selectUserByWhere()，在该方法中使用 MyUser 类的对象将参数信息封装。接口方法 selectUserByWhere()的定义如下：

```
public List<MyUser> selectUserByWhere(MyUser user);
```

❷ 添加 SQL 映射

在应用 ch3_2 的 SQL 映射文件 UserMapper.xml 中，添加接口方法 selectUserByWhere() 对应的 SQL 映射，具体代码如下：

```
<!-- 使用 where 元素，根据条件动态查询用户信息 -->
<select id="selectUserByWhere" resultType="MyUser" parameterType="MyUser">
    select * from user
    <where>
        <if test="uname !=null and uname!=''">
            and uname like concat('%',#{uname},'%')
        </if>
        <if test="usex !=null and usex!=''">
            and usex = #{usex}
        </if>
    </where>
</select>
```

❸ 添加请求处理方法

在应用 ch3_2 的控制器类 TestController 中，添加请求处理方法 selectUserByWhere()，具体代码如下：

```
@RequestMapping("/selectUserByWhere")
public StringselectUserByWhere(Model model) {
    MyUser mu = new MyUser();
    mu.setUname("陈");
    mu.setUsex("男");
    List<MyUser> unameAndUsexList = userMapper.selectUserByWhere(mu);
    model.addAttribute("unameAndUsexList", unameAndUsexList);
    return "showUnameAndUsexUser";
}
```

❹ 测试应用

重启 Web 服务器 Tomcat，通过网址 http://localhost:8080/ch3_2/selectUserByWhere 测试应用。

3.12.5 <set>元素

在 update 语句中，可以使用<set>元素动态更新列。下面通过一个实例讲解<set>元素的使用过程。

【例 3-17】在 3.6.3 节例 3-2 的基础上，讲解<set>元素的使用过程。为节省篇幅，相同的实现不再赘述。其他的具体实现如下。

❶ 添加 Mapper 接口方法

在应用 ch3_2 的 UserMapper 接口中，添加数据操作接口方法 updateUserBySet()，在该方法中使用 MyUser 类的对象将参数信息封装。接口方法 updateUserBySet()的定义如下：

```
public int updateUserBySet(MyUser user);
```

❷ 添加 SQL 映射

在应用 ch3_2 的 SQL 映射文件 UserMapper.xml 中，添加接口方法 updateUserBySet()对应的 SQL 映射，具体代码如下：

```
<!-- 使用 set 元素，动态修改一个用户 -->
<update id="updateUserBySet" parameterType="MyUser">
    update user
    <set>
        <if test="uname != null">uname=#{uname},</if>
        <if test="usex != null">usex=#{usex}</if>
    </set>
    where uid = #{uid}
</update>
```

❸ 添加请求处理方法

在应用 ch3_2 的控制器类 TestController 中，添加请求处理方法 updateUserBySet()，具体代码如下：

```
@RequestMapping("/updateUserBySet")
public String updateUserBySet(Model model) {
    MyUser setmu = new MyUser();
    setmu.setUid(3);
    setmu.setUname("张九");
    userMapper.updateUserBySet(setmu);
    //查询出来看看 id 为 3 的用户是否被修改
    List<Map<String, Object>> unameAndUsexList = userMapper.selectAllUserMap();
    model.addAttribute("unameAndUsexList", unameAndUsexList);
    return "showUnameAndUsexUser";
}
```

❹ 测试应用

重启 Web 服务器 Tomcat，通过网址 http://localhost:8080/ch3_2/updateUserBySet 测试应用。

3.12.6 <foreach>元素

<foreach>元素主要用于构建 in 条件，它可以在 SQL 语句中迭代一个集合。<foreach>元素的属性主要有 item、index、collection、open、separator、close。item 表示集合中每一个元素进行迭代时的别名；index 指定一个名字，用于表示在迭代过程中，每次迭代到的位置；open 表示该语句以什么开始；separator 表示在每次进行迭代之间以什么符号作为分隔符；close 表示该语句以什么结束。在使用<foreach>时，最关键的也是最容易出错的是 collection 属性，该属性是必选的，但在不同情况下，该属性的值是不一样的，主要有以下 3 种情况：

- 如果传入的是单参数且参数类型是一个 List 时，collection 属性值为 list。
- 如果传入的是单参数且参数类型是一个 array 数组时，collection 属性值为 array。
- 如果传入的参数是多个时，需要把它们封装成一个 Map，当然单参数也可以封装成 Map。Map 的 key 是参数名，所以 collection 属性值是传入的 List 或 array 对象在自己封装的 Map 中的 key。

下面通过一个实例讲解<foreach>元素的使用过程。

【例 3-18】 在 3.6.3 节例 3-2 的基础上，讲解<foreach>元素的使用过程。为节省篇幅，相同的实现不再赘述。其他的具体实现如下。

❶ 添加 Mapper 接口方法

在应用 ch3_2 的 UserMapper 接口中，添加数据操作接口方法 selectUserByForeach()，在该方法中使用 List 作为参数。接口方法 selectUserByForeach()的定义如下：

```
public List<MyUser> selectUserByForeach(List<Integer> listId);
```

❷ 添加 SQL 映射

在应用 ch3_2 的 SQL 映射文件 UserMapper.xml 中，添加接口方法 selectUserByForeach()对应的 SQL 映射，具体代码如下：

```
<!-- 使用 foreach 元素，查询用户信息 -->
<select id="selectUserByForeach" resultType="MyUser"  parameterType="List">
    select * from user where uid in
    <foreach item="item" index="index" collection="list"
    open="(" separator="," close=")">
        #{item}
    </foreach>
</select>
```

❸ 添加请求处理方法

在应用 ch3_2 的控制器类 TestController 中，添加请求处理方法 selectUserByForeach()，具体代码如下：

```
@RequestMapping("/selectUserByForeach")
public String selectUserByForeach(Model model) {
    List<Integer> listId = new ArrayList<Integer>();
    listId.add(4);
    listId.add(5);
    List<MyUser> unameAndUsexList = userMapper.selectUserByForeach(listId);
    model.addAttribute("unameAndUsexList", unameAndUsexList);
    return "showUnameAndUsexUser";
}
```

❹ 测试应用

重启 Web 服务器 Tomcat，通过网址 http://localhost:8080/ch3_2/selectUserByForeach 测试应用。

3.12.7 <bind>元素

在模糊查询时，如果使用“${}”拼接字符串，则无法防止 SQL 注入问题；如果使用字符串拼接函数或连接符号，不同数据库的拼接函数或连接符号不同，如 MySQL 的 concat 函

数、Oracle 的连接符号“||”。这样，SQL 映射文件就需要根据不同的数据库提供不同的实现，显然比较麻烦，且不利于代码的移植。幸运的是，MyBatis 提供了<bind>元素来解决这一问题。

下面通过一个实例讲解<bind>元素的使用过程。

【例 3-19】在 3.6.3 节例 3-2 的基础上，讲解<bind>元素的使用过程。为节省篇幅，相同的实现不再赘述。其他的具体实现如下。

❶ 添加 Mapper 接口方法

在应用 ch3_2 的 UserMapper 接口中，添加数据操作接口方法 selectUserByBind()。接口方法 selectUserByBind()的定义如下：

```
public List<MyUser> selectUserByBind(MyUser user);
```

❷ 添加 SQL 映射

在应用 ch3_2 的 SQL 映射文件 UserMapper.xml 中，添加接口方法 selectUserByBind()对应的 SQL 映射，具体代码如下：

```
<!-- 使用bind元素进行模糊查询 -->
<select id="selectUserByBind" resultType="MyUser" parameterType="MyUser">
    <!-- bind中uname是com.po.MyUser的属性名 -->
    <bind name="paran_uname" value="'%' + uname + '%'"/>
    select * from user where uname like #{paran_uname}
</select>
```

❸ 添加请求处理方法

在应用 ch3_2 的控制器类 TestController 中，添加请求处理方法 selectUserByBind()，具体代码如下：

```
@RequestMapping("/selectUserByBind")
public String selectUserByBind(Model model) {
    MyUser bindmu = new MyUser();
    bindmu.setUname("陈");
    List<MyUser> unameAndUsexList = userMapper.selectUserByBind(bindmu);
    model.addAttribute("unameAndUsexList", unameAndUsexList);
    return "showUnameAndUsexUser";
}
```

❹ 测试应用

重启 Web 服务器 Tomcat，通过网址 http://localhost:8080/ch3_2/selectUserByBind 测试应用。

3.13 MyBatis 的缓存机制

视频讲解

我们知道内存的读取速度远大于硬盘的读取速度。当需要重复地获取相同数据时，一次一次地请求数据库或者远程服务，导致大量的时间消耗在数据库查询或者远程方法调用上，最终导致程序性能降低。这就是数据缓存要解决的问题。

MyBatis 提供数据查询缓存，用于减轻数据库压力，提高数据库性能。MyBatis 提供一级缓存和二级缓存。

▶ 3.13.1 一级缓存（SqlSession 级别的缓存）

在操作数据库时，需要构造 SqlSession 对象，在对象中有一个数据结构（HashMap）用于存储缓存数据，不同的 SqlSession 之间的缓存区域是互相不影响的。

❶ 一级缓存配置

MyBatis 的一级缓存不需要任何配置，在每一个 SqlSession 中都有一个一级缓存区，作用范围是 SqlSession。

MyBatis 一级缓存中，当第一次发起查询 ID 为 1 的用户信息时，先去找缓存中是否有 ID 为 1 的用户信息，如果没有则从数据库查询用户信息，将用户信息存储到一级缓存中；如果 sqlSession 去执行插入、更新、删除操作（执行 commit 操作），将会清空 SqlSession 中的一级缓存，这样做的目的是让缓存存储的是最新的信息，避免脏读；当第二次发起查询 ID 为 1 的用户信息时，先去找缓存中是否有 ID 为 1 的用户信息，如果缓存中有则直接从缓存中获取用户信息。这里涉及一个缓存命中率（Cache Hit Ratio），指的是在缓存中查询到的次数与总共在缓存中查询的次数的比值。

❷ 一级缓存实验

【例 3-20】在 3.5 节例 3-1 的基础上，讲解一级缓存实验。为节省篇幅，相同的实现不再赘述。其他的具体实现如下。

1）修改测试类

在应用 ch3_1 的测试类 MyBatisTest 中，多次发起查询用户 ID 为 1 的用户信息，修改后的代码如下：

```
public class MyBatisTest {
    public static void main(String[] args) {
        try {
            //读取配置文件mybatis-config.xml
            InputStream config =
              Resources.getResourceAsStream("mybatis-config.xml");
            //根据配置文件构建SqlSessionFactory
            SqlSessionFactory ssf =
              new SqlSessionFactoryBuilder().build(config);
            //通过SqlSessionFactory创建SqlSession
            SqlSession ss = ssf.openSession();
            //查询一个用户
            MyUser mu = ss.selectOne
             ("com.mybatis.mapper.UserMapper.selectUserById", 1);
            System.out.println(mu);
            //测试一级缓存
            mu = ss.selectOne
              ("com.mybatis.mapper.UserMapper.selectUserById", 1);
            System.out.println(mu);
            //添加一个用户
            MyUser addmu = new MyUser();
            addmu.setUname("陈恒");
            addmu.setUsex("男");
            ss.insert("com.mybatis.mapper.UserMapper.addUser",addmu);
            //修改一个用户
```

```
            MyUser updatemu = new MyUser();
            updatemu.setUid(2);
            updatemu.setUname("张三");
            updatemu.setUsex("女");
            ss.update("com.mybatis.mapper.UserMapper.updateUser", updatemu);
            //测试一级缓存
            mu = ss.selectOne
              ("com.mybatis.mapper.UserMapper.selectUserById", 1);
            System.out.println(mu);
            //提交事务
            ss.commit();
            //关闭 SqlSession
            ss.close();
        } catch (IOException e) {
            // TODO Auto-generated catch block
            e.printStackTrace();
        }
    }
}
```

2）测试缓存，运行程序

运行修改后的测试类 MyBatisTest，运行结果如图 3.12 所示。

```
Markers  Properties  Servers  Data Source Explorer  Snippets  Console
<terminated> MyBatisTest [Java Application] C:\soft\javaee\eclipse\plugins\org.eclipse.justj.openjdk.h
DEBUG [main] - ==>  Preparing: select * from user where uid = ?
DEBUG [main] - ==> Parameters: 1(Integer)
DEBUG [main] - <==      Total: 1
User [uid=1,uname=张三,usex=女]
User [uid=1,uname=张三,usex=女]
DEBUG [main] - ==>  Preparing: insert into user (uname,usex) values(?,?)
DEBUG [main] - ==> Parameters: 陈恒(String), 男(String)
DEBUG [main] - <==    Updates: 1
DEBUG [main] - ==>  Preparing: update user set uname = ?,usex = ? where uid = ?
DEBUG [main] - ==> Parameters: 张三(String), 女(String), 2(Integer)
DEBUG [main] - <==    Updates: 0
DEBUG [main] - ==>  Preparing: select * from user where uid = ?
DEBUG [main] - ==> Parameters: 1(Integer)
DEBUG [main] - <==      Total: 1
User [uid=1,uname=张三,usex=女]
```

图 3.12　一级缓存测试结果

从图 3.12 运行的结果可知，前两次发起查询 ID 为 1 的用户信息时，只执行了一次 SQL 语句（即只查询一次数据库），第二次直接从缓存返回数据。但经过添加和修改操作后，第三次发起查询 ID 为 1 的用户信息时，又重新查询了数据库（清空 SqlSession 中的一级缓存）。

3.13.2　二级缓存（Mapper 级别的缓存）

MyBatis 的二级缓存需要手动开启才能启动，与一级缓存的最大区别在于二级缓存的作用范围比一级缓存大，二级缓存是多个 SqlSession 可以共享一个 Mapper 的二级缓存区域，二级缓存作用的范围是 Mapper 中的同一个命名空间（namespace）的 statement。在 MyBatis 的核心配置文件中，默认开启二级缓存。在默认开启二级缓存的情况下，如果每一个 namespace 都开启了二级缓存，则都对应一个二级缓存区，同一个 namespace 共用一个二级缓存区。

❶ 二级缓存配置

在 MyBatis 默认开启二级缓存的情况下，当 SqlSession 1 查询 ID 为 1 的用户信息时，查询到用户信息会将查询数据存储到二级缓存中；当 SqlSession 2 执行相同 Mapper 下的 statement 时，执行 commit 提交，清空该 Mapper 下的二级缓存区域的数据；当 SqlSession 3 查询 ID 为 1 的用户信息时，先去缓存中找是否存在数据，如果存在则直接从缓存中取出数据，不存在就去数据库查询读取。二级缓存配置具体如下。

1）开启二级缓存

第一个需要配置的地方是核心配置文件（此步可以省略，因为默认是开启的，配置的目的是方便维护）。配置示例代码如下：

```
<settings>
    <!-- 开启二级缓存 -->
    <setting name="cacheEnabled" value="true"/>
</settings>
```

2）开启 namespace 下的二级缓存

在需要开启二级缓存的 statement 的命名空间（namespace）中配置标签<cache></cache>。配置示例代码如下：

```
<mapper namespace="dao.UserMapper">
    <!-- 开启 namespace 下的二级缓存 -->
    <cache></cache>
</mapper>
```

<cache>标签有以下 6 个参数。

type：指定缓存（cache）接口的实现类型，当需要和 ehcache 整合时更改该参数值即可。

flushInterval：刷新间隔。可被设置为任意的正整数，单位为毫秒。默认不设置。

size：引用数目。可被设置为任意正整数，缓存的对象数目等于运行环境的可用内存资源数目。默认是 1024。

readOnly：只读，true 或 false。只读的缓存会给所有的调用者返回缓存对象的相同实例。默认是 false。

eviction：缓存收回策略。取值为 LRU（最近最少使用的）、FIFO（先进先出）、SOFT（软引用）、WEAK（弱引用）。默认是 LRU。

在 Mapper 的 select 中可设置 useCache="false"禁用缓存，默认是开启的；在 insert、update、delete 中可设置 flushCache="true"清空缓存（刷新缓存），默认是清空缓存。

3）POJO 类实现序列化

使用二级缓存时，持久化类需要序列化，即 POJO 类实现 Serializable 接口。示例代码如下：

```
public class MyUser implements Serializable{}
```

❷ 二级缓存实验

【例 3-21】在 3.6.4 节例 3-3 的基础上，讲解二级缓存实验。为节省篇幅，相同的实现不再赘述。其他的具体实现如下。

1）开启 namespace 下的二级缓存

在应用 ch3_3 的 Mapper 映射文件 UserMapper.xml 中，添加标签<cache></cache>。

2）POJO 类实现序列化

对应用 ch3_3 的持久化类 MyUser 实现序列化接口 Serializable。

3）修改控制器类

将应用 ch3_3 的控制器类 MyController 修改如下：

```
@Controller
public class MyController {
    @Autowired
    private UserMapper userMapper;
    @RequestMapping("/test")
    public String test() {
        // 查询一个用户
        MyUser mu = userMapper.selectUserById(1);
        System.out.println(mu);
        //测试二级缓存
        mu = userMapper.selectUserById(1);
        System.out.println(mu);
        return "test";
    }
}
```

4）测试缓存，运行程序

发布应用 ch3_3 到 Web 服务器 Tomcat 后，通过网址 http://localhost:8080/ch3_3/test 测试应用。运行结果如图 3.13 所示。

```
Markers  Properties  Servers  Data Source Explorer  Snippets  Console
Tomcat v9.0 Server at localhost [Apache Tomcat] C:\soft\javaee\eclipse\plugins\org.eclipse.justj.openjc
DEBUG [http-nio-8080-exec-6] - Cache Hit Ratio [dao.UserMapper]: 0.0
DEBUG [http-nio-8080-exec-6] - ==>  Preparing: select * from user where uid = ?
DEBUG [http-nio-8080-exec-6] - ==> Parameters: 1(Integer)
DEBUG [http-nio-8080-exec-6] - <==      Total: 1
User [uid=1,uname=张三,usex=女]
DEBUG [http-nio-8080-exec-6] - Cache Hit Ratio [dao.UserMapper]: 0.5
User [uid=1,uname=张三,usex=女]
```

图 3.13　二级缓存测试结果

从图 3.13 可以看出，第一次查询 ID 为 1 的用户信息时，缓存命中率为 0，说明先访问缓存，读取缓存中是否有 ID 为 1 的用户数据，如果发现缓存中没有，就去数据库查询用户信息；第二次查询 ID 为 1 的用户信息时，如果发现缓存中有对应数据，就直接从缓存中读取。

二级缓存一般应用在访问多的查询请求且对查询结果的实时性要求不高的场合，此时可采用 MyBatis 二级缓存技术降低数据库访问量，提高访问速度。例如耗时比较高的统计分析的 SQL。

3.14　本章小结

本章重点讲述了 MyBatis 的 SQL 映射文件的编写以及 SSM 框架整合开发的流程。通过本章的学习，读者不仅掌握了 SSM 框架整合开发的流程，还应该熟悉 MyBatis 的基本应用。

习 题 3

1．MyBatis Generator 有哪几种方法自动生成代码？

2．简述 SSM 框架集成的步骤。

3．MyBatis 实现查询时，返回的结果集有几种常见的存储方式？请举例说明。

4．在 MyBatis 中针对不同的数据库软件，<insert>元素如何将主键回填？

5．在 MyBatis 中，如何给 SQL 语句传递参数？

6．在动态 SQL 元素中，类似分支语句的元素有哪些？如何使用它们？

第 4 章 名片管理系统的设计与实现（SSM + JSP）

视频讲解

学习目的与要求

本章通过名片管理系统的设计与实现，讲述如何使用 SSM 框架实现一个 Web 应用。通过本章的学习，掌握 SSM 框架应用开发的流程、方法以及技术。

主要内容

- 系统设计
- 数据库设计
- 系统管理
- 组件设计
- 系统实现

本章系统使用 SSM 框架实现各个模块，Web 引擎为 Tomcat 9.0，数据库采用的是 MySQL 5.x，集成开发环境为 Eclipse。

4.1 系统设计

4.1.1 系统功能需求

名片管理系统是针对注册用户使用的系统。系统提供的功能如下：

（1）非注册用户可以注册为注册用户。

（2）成功注册的用户可以登录系统。

（3）成功登录的用户可以添加、修改、删除以及浏览自己客户的名片信息。

（4）成功登录的用户可以修改密码。

4.1.2 系统模块划分

用户登录成功后，进入管理主页面（main.jsp），可以对自己的客户名片进行管理。系统模块划分如图 4.1 所示。

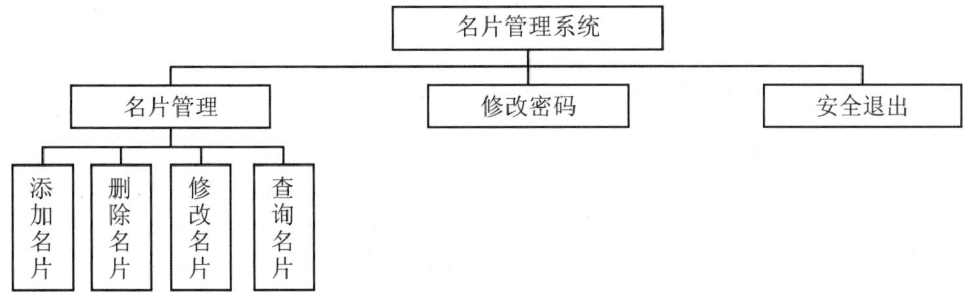

图 4.1　名片管理系统

4.2 数据库设计

系统采用加载纯 Java 数据库驱动程序的方式连接 MySQL 5.x 数据库。在 MySQL 5.x 的数据库 ch4 中，共创建两张与系统相关的数据表：usertable 和 cardtable。

4.2.1 数据库概念结构设计

根据系统设计与分析，可以设计出如下数据结构。

❶ 用户

用户包括 ID、用户名以及密码，注册用户名唯一。

❷ 名片

名片包括 ID、名称、电话、邮箱、单位、职务、地址、Logo 以及所属用户。其中，ID 唯一，“所属用户”与“1. 用户”的用户 ID 关联。

根据以上数据结构，结合数据库设计特点，可画出如图 4.2 所示的数据库概念结构图。

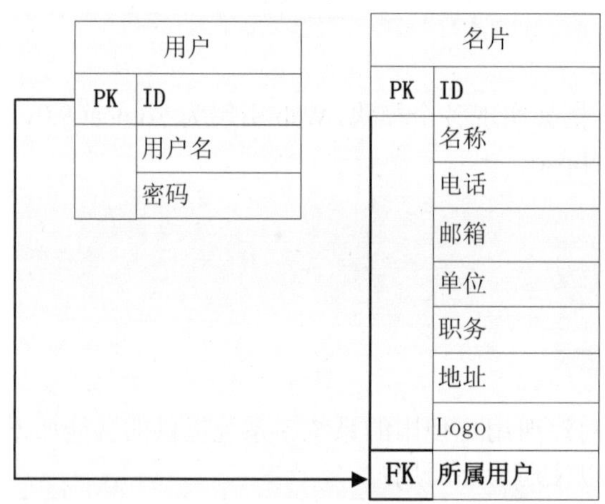

图 4.2 数据库概念结构图

其中，ID 为正整数，值是从 1 开始递增的序列。

4.2.2 数据库逻辑结构设计

将数据库概念结构图转换为 MySQL 数据库所支持的实际数据模型，即数据库的逻辑结构。

用户信息表（usertable）的设计如表 4.1 所示。

表 4.1 用户信息表

字 段	含 义	类 型	长 度	是否为空
id	编号（PK）	int	11	no
uname	用户名	varchar	50	no
upwd	密码	varchar	32	no

名片信息表（cardtable）的设计如表 4.2 所示。

表 4.2　名片信息表

字　段	含　义	类　型	长　度	是否为空
id	编号（PK）	int	11	no
name	名称	varchar	50	no
telephone	电话	varchar	20	no
email	邮箱	varchar	50	
company	单位	varchar	50	
post	职务	varchar	50	
address	地址	varchar	50	
logoName	图片	varchar	30	
user_id	所属用户	int	11	no

4.3　系统管理

4.3.1　所需 JAR 包

使用 Eclipse 创建一个名为 ch4 的 Web 应用，并将所依赖的 JAR 包（包括 MyBatis、Spring、Spring MVC、Spring JDBC、MySQL 连接器、MyBatis 与 Spring 桥接器、Log4j、Fileupload、Jackson、DBCP 以及 JSTL 等）复制到/WEB-INF/lib 目录中，具体参见源代码。

4.3.2　JSP 页面管理

为方便管理，在/WebContent/static 目录下存放与系统相关的静态资源，如 BootStrap 相关的 CSS 与 JS；在/WEB-INF/jsp 目录下存放与系统相关的 JSP 页面。由于篇幅受限，本章仅附上部分 JSP 和 Java 文件的核心代码，具体代码请读者参见本书提供的源代码 ch4。

❶ 首页面

在/WebContent/目录下创建应用的首页面 index.jsp，首页面重定向到 user/toLogin 请求，打开登录页面。index.jsp 的核心代码如下：

```
<body>
    <%response.sendRedirect("user/toLogin");%>
</body>
```

❷ 异常信息显示页面

本系统使用 Spring 框架的统一异常处理机制，处理未登录异常和程序错误异常。为显示异常信息，需要在/WEB-INF/jsp 目录下创建一个名为 error.jsp 的页面。error.jsp 的核心代码如下：

```
<body>
    <c:if test="${mymessage=='noLogin'}">
        <h2>没登录，您没有权限访问，请<a href="user/toLogin">登录</a>! </h2>
    </c:if>
    <c:if test="${mymessage=='noError'}">
        <h2>服务器内部错误或资源不存在! </h2>
```

```
    </c:if>
</body>
```

▶ 4.3.3 包管理

❶ config 包

该包存放的配置文件是系统的配置，包括 Spring 配置、Spring MVC 配置以及 MyBatis 的核心配置。

❷ controller 包

该包存放的类是系统的控制器类和异常处理类，包括名片管理相关的控制器类、用户相关的控制器类、验证码控制器类以及全局异常处理类。

❸ dao 包

该包存放的 Java 程序是@Repository 注解的数据操作接口以及 SQL 映射文件，包括名片和用户相关的数据访问接口和 SQL 映射文件。

❹ model 包

该包存放的类是两个领域模型类，与表单对应：Card 封装名片信息；MyUser 封装用户信息。

❺ po 包

该包存放的类是两个持久化类，与两个数据表对应。

❻ service 包

该包存放的类是业务处理类，是控制器和 dao 的桥梁。包下有 Service 接口和 Service 实现类。

❼ util 包

该包存放的类是工具类，包括 MyUtil 类（文件重命名）和 MD5Util 类（MD5 加密）。

▶ 4.3.4 配置管理

名片管理系统共有 4 个配置，分别是 Web 的配置 web.xml、Spring 的配置 applicationContext.xml、Spring MVC 的配置 springmvc.xml 和 MyBatis 的核心配置 mybatis-config.xml。具体代码请读者参见本书提供的源代码 ch4。

在 web.xml 文件中，实例化 ApplicationContext 容器、配置 Spring MVC DispatcherServlet 以及部署字符编码过滤器；在 applicationContext.xml 文件中，配置数据源、为数据源添加事务管理器、配置 MyBatis 工厂以及 Mapper 代理开发；在 springmvc.xml 文件中，扫描注解的包、配置视图解析器、静态资源可见以及上传文件的相关设置；在 mybatis-config.xml 文件中，配置实体类别名以及日志实现 logImpl。

4.4 组件设计

名片管理系统的组件包括工具类、统一异常处理类和验证码类。

▶ 4.4.1 工具类

名片管理系统的工具类包括 MyUtil 和 MD5Util。在 MyUtil 类中定义一个文件重命名方法 getNewFileName；在 MD5Util 类中，定义 MD5 加密方法。具体代码请读者参见本书提供的源代码 ch4。

▶ 4.4.2　统一异常处理

名片管理系统采用@ControllerAdvice 注解（控制器增强）实现异常的统一处理，统一处理了 NoLoginException 和 Exception 异常，核心代码如下：

```
/**
 * 统一异常处理
 */
@ControllerAdvice
public class GlobalExceptionHandleController {
    @ExceptionHandler(value=Exception.class)
    public String exceptionHandler(Exception e, Model model) {
        String message = "";
        if (e instanceof NoLoginException) {
            message = "noLogin";
        } else {//未知异常
            message =  "noError";
        }
        model.addAttribute("mymessage",message);
        return "error";
    }
}
```

未登录异常类 NoLoginException 的代码如下：

```
package controller;
public class NoLoginException extends Exception{
    private static final long serialVersionUID = 1L;
    public NoLoginException() {
        super();
    }
    public NoLoginException(String message) {
        super(message);
    }
}
```

▶ 4.4.3　验证码

本系统验证码的使用步骤如下。

❶ 创建产生验证码的控制器类

在 controller 包中，创建产生验证码的控制器类 ValidateCodeController，具体代码请读者参见本书提供的源代码 ch4。

❷ 使用验证码

在需要使用验证码的 JSP 页面中，调用产生验证码的控制器显示验证码，示例代码片段如下：

```
<td><img src="validateCode" id="mycode"></td>
```

4.5 名片管理

▶ 4.5.1 领域模型与持久化类

在本系统中，领域模型简单地作为视图对象，它的作用是将某个指定页面的所有数据封装起来，与表单对应。数据传递方向为 View → Controller → Service → Dao。与名片管理相关的领域模型是 Card（位于 model 包），具体代码请读者参见本书提供的源代码 ch4。

在本系统中，持久层是关系型数据库，所以，持久化类的每个属性对应数据表中的每个字段。数据传递方向为 Dao → Service → Controller → View。与名片管理相关的持久化类是 CardTable（位于 po 包），具体代码请读者参见本书提供的源代码 ch4。

▶ 4.5.2 Controller 实现

在本系统中，与名片管理相关的功能包括添加、修改、删除、查询等，由控制器类 CardController 负责处理。由系统功能需求可知，用户必须成功登录才能管理自己的名片，所以，CardController 处理添加、修改、删除、查询名片等功能前，需要进行登录权限验证。在 CardController 中，使用@ModelAttribute 注解的方法进行登录权限验证。CardController 的核心代码如下：

```
@Controller
@RequestMapping("/card")
public class CardController {
    @Autowired
    private CardService cardService;
    /**
     * 权限控制
     */
    @ModelAttribute
    public void checkLogin(HttpSession session) throws NoLoginException{
        if(session.getAttribute("userLogin") == null) {
            throw new NoLoginException();
        }
    }
    /**
     * 查询、修改查询、删除查询
     */
    @RequestMapping("/selectAllCardsByPage")
    public String selectAllCardsByPage(Model model, int currentPage,
      HttpSession session) {
        return cardService.selectAllCardsByPage(model, currentPage,  session);
    }
```

```
/**
 * 打开添加页面
 */
@RequestMapping("/toAddCard")
public String toAddCard(@ModelAttribute Card card) {
    return "addCard";
}
/**
 * 实现添加及修改功能
 */
@RequestMapping("/addCard")
public String addCard(@ModelAttribute Card card, HttpServletRequest
  request, String act, HttpSession session) throws IllegalStateException,
  IOException {
    return cardService.addCard(card, request, act, session);
}
/**
 * 打开详情及修改页面
 */
@RequestMapping("/detail")
public String detail(Model model, int id, String act) {
    return cardService.detail(model, id, act);
}
/**
 * 删除
 */
@RequestMapping("/delete")
@ResponseBody
public String delete(int id) {
    return cardService.delete(id);
}
/**
 * 安全退出
 */
@RequestMapping("/loginOut")
public String loginOut(Model model, HttpSession session) {
    return cardService.loginOut(model, session);
}
/**
 * 打开修改密码页面
 */
@RequestMapping("/toUpdatePwd")
public String toUpdatePwd(Model model, HttpSession session) {
    return cardService.toUpdatePwd(model, session);
}
/**
```

```
     * 修改密码
     */
    @RequestMapping("/updatePwd")
    public String updatePwd(@ModelAttribute MyUser myuser) {
        return cardService.updatePwd(myuser);
    }
}
```

4.5.3 Service 实现

与名片管理相关的 Service 接口和实现类分别为 CardService 和 CardServiceImpl。控制器获取一个请求后，需要调用 Service 层的业务处理方法，在 Service 层需要调用 Dao 层。所以，Service 层是控制器层和 Dao 层的桥梁。CardService 接口代码略。

CardServiceImpl 实现类的核心代码如下：

```
@Service
public class CardServiceImpl implements CardService{
    @Autowired
    private CardMapper cardMapper;
    /**
     * 查询、修改查询、删除查询、分页查询
     */
    @Override
    public String selectAllCardsByPage(Model model, int currentPage,
      HttpSession session) {
        MyUserTable mut = (MyUserTable)session.getAttribute("userLogin");
        List<Map<String, Object>> allUser = cardMapper.selectAllCards
          (mut.getId());
        //共多少个用户
        int totalCount = allUser.size();
        //计算共多少页
        int pageSize = 5;
        int totalPage = (int)Math.ceil(totalCount*1.0/pageSize);
        List<Map<String, Object>> cardsByPage = cardMapper.selectAllCardsByPage
          ((currentPage-1)*pageSize, pageSize, mut.getId());
        model.addAttribute("allCards", cardsByPage);
        model.addAttribute("totalPage", totalPage);
        model.addAttribute("currentPage", currentPage);
        return "main";
    }
    /**
     * 添加与修改名片
     */
    @Override
    public String addCard(Card card, HttpServletRequest  request, String act,
      HttpSession session) throws IllegalStateException, IOException {
```

```
        MultipartFile myfile = card.getLogo();
        //如果选择了上传文件，将文件上传到指定的目录 static/images
        if(!myfile.isEmpty()) {
            //上传文件路径（生产环境）
            String path = request.getServletContext().getRealPath("/static/
              images/");
            //获得上传文件原名
            String fileName = myfile.getOriginalFilename();
            //对文件重命名
            String fileNewName = MyUtil.getNewFileName(fileName);
            File filePath = new File(path + File.separator + fileNewName);
            //如果文件目录不存在，创建目录
            if(!filePath.getParentFile().exists()) {
                filePath.getParentFile().mkdirs();
            }
            //将上传文件保存到一个目标文件中
            myfile.transferTo(filePath);
            //将重命名后的图片名存到 card 对象中，添加时使用
            card.setLogoName(fileNewName);
        }
        if("add".equals(act)) {
            MyUserTable mut = (MyUserTable)session.getAttribute("userLogin");
            card.setUser_id(mut.getId());
            int n = cardMapper.addCard(card);
            if(n > 0)//成功
                return "redirect:/card/selectAllCardsByPage?currentPage=1";
            //失败
            return "addCard";
        }else {//修改
            int n = cardMapper.updateCard(card);
            if(n > 0)//成功
                return "redirect:/card/selectAllCardsByPage?currentPage=1";
            //失败
            return "updateCard";
        }
    }
    /**
     * 打开详情与修改页面
     */
    @Override
    public String detail(Model model, int id, String act) {
        CardTable ct = cardMapper.selectACard(id);
        model.addAttribute("card", ct);
        if("detail".equals(act)) {
            return "cardDetail";
        }else {
```

```
                return "updateCard";
            }
        }
        /**
         * 删除
         */
        @Override
        public String delete(int id) {
            cardMapper.deleteACard(id);
            return "/card/selectAllCardsByPage?currentPage=1";
        }
        /**
         * 安全退出
         */
        @Override
        public String loginOut(Model model, HttpSession session) {
            session.invalidate();
            model.addAttribute("myUser", new MyUser());
            return "login";
        }
        /**
         * 打开修改密码页面
         */
        @Override
        public String toUpdatePwd(Model model, HttpSession session) {
            MyUserTable mut = (MyUserTable)session.getAttribute("userLogin");
            model.addAttribute("myuser", mut);
            return "updatePwd";
        }
        /**
         * 修改密码
         */
        @Override
        public String updatePwd(MyUser myuser) {
            //将明文变成密文
            myuser.setUpwd(MD5Util.MD5(myuser.getUpwd()));
            cardMapper.updatePwd(myuser);
            return "login";
        }
}
```

4.5.4 Dao 实现

Dao 层是数据访问层，即@Repository 注解的数据操作接口（接口中的方法与 SQL 映射文件中元素的 id 对应），与名片管理相关的数据访问层为 CardMapper。CardMapper 接口代码略。

▶ 4.5.5　SQL 映射文件

SQL 映射文件的 namespace 属性与数据操作接口对应。与名片管理功能相关的 SQL 映射文件是 CardMapper.xml（位于 dao 包中），具体代码如下：

```
<?xml version="1.0" encoding="UTF-8" ?>
<!DOCTYPE mapper
PUBLIC "-//mybatis.org//DTD Mapper 3.0//EN"
"http://mybatis.org/dtd/mybatis-3-mapper.dtd">
<mapper namespace="dao.CardMapper">
    <!-- 查询所有名片 -->
    <select id="selectAllCards"  resultType="map">
        select * from cardtable where user_id = #{uid}
    </select>
    <!-- 分页查询名片 -->
    <select id="selectAllCardsByPage"  resultType="map">
        select * from cardtable where user_id = #{uid}  limit #{startIndex},
        #{perPageSize}
    </select>
    <!-- 添加名片 -->
    <insert id="addCard" parameterType="Card">
        insert into cardtable (id, name, telephone, email, company, post,
          address, logoName, user_id)
         values (null, #{name}, #{telephone}, #{email}, #{company}, #{post},
           #{address}, #{logoName}, #{user_id})
    </insert>
    <!-- 修改名片 -->
    <update id="updateCard" parameterType="Card">
        update cardtable set
            name = #{name},
            telephone =  #{telephone},
            email = #{email},
            company = #{company},
            post = #{post},
            address = #{address},
            logoName = #{logoName}
        where id = #{id}
    </update>
    <!-- 查询一个名片，修改及详情使用 -->
    <select id="selectACard" parameterType="integer" resultType="CardTable">
        select * from cardtable where id = #{id}
    </select>
    <!-- 删除一个名片 -->
    <delete id="deleteACard" parameterType="integer">
        delete from cardtable where id = #{id}
    </delete>
```

```
    <!-- 修改密码 -->
    <update id="updatePwd" parameterType="myuser">
        update usertable set upwd = #{upwd} where id = #{id}
    </update>
</mapper>
```

▶ 4.5.6 添加名片

首先，用户登录成功后，进入名片管理系统的主页面；然后，用户在名片管理主页面单击“添加名片”超链接打开添加名片页面；最后，用户输入客户名片的姓名、电话号码、E-mail、单位、职务、地址、照片后，单击“添加”按钮实现添加。如果成功，则跳转到名片管理主页面；如果失败，则回到添加名片页面。

addCard.jsp 页面是实现添加名片信息的输入界面，如图 4.3 所示。addCard.jsp 的代码请读者参见本书提供的源代码 ch4。

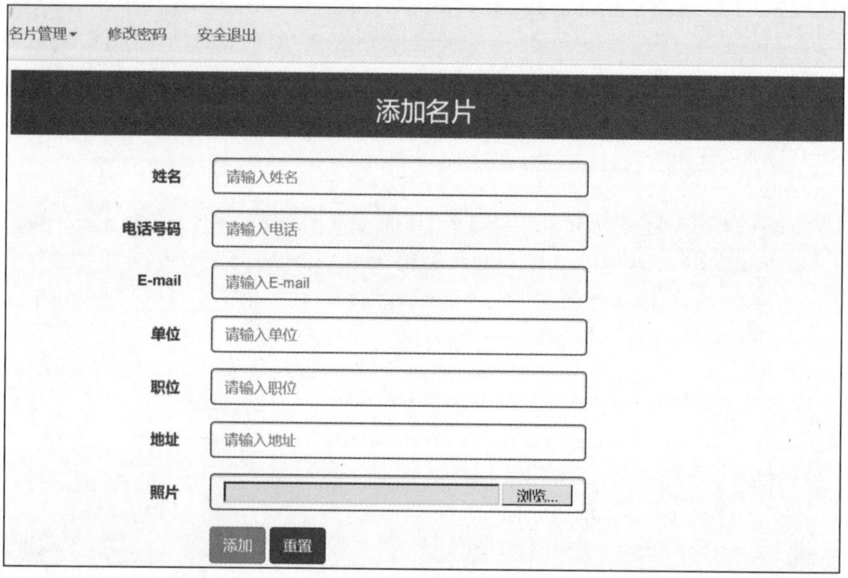

图 4.3　添加名片页面

单击图 4.3 中的“添加”按钮，将添加请求通过“card/addCard?act=add”提交给控制器类 CardController（4.5.2 节）的 addCard 方法进行添加功能处理。添加成功跳转到名片管理主页面；添加失败回到添加名片页面。

▶ 4.5.7 名片管理主页面

用户登录成功后，进入名片管理系统的主页面（main.jsp），运行效果如图 4.4 所示。

在名片管理主页面中，单击“详情”超链接，打开名片详细信息页面 cardDetail.jsp。“详情”超链接的目标地址是个 url 请求，该请求路径为“card/detail?id=${card.id}&act=detail”。根据请求路径找到对应控制器类 CardController 的 detail 方法实现查询一个名片功能。根据动作类型（“修改”以及“详情”），将查询结果转发到不同视图。名片详细信息页面 cardDetail.jsp 的运行效果如图 4.5 所示。

名片管理▾　修改密码　安全退出

名片列表

名片ID	姓名	单位	职位	操作
9	名片1	好单位	董事长	详情　修改　删除
10	名片2	好单位2	厂长	详情　修改　删除
11	名片3	好单位3	总经理	详情　修改　删除
12	名片4	好单位4	董秘	详情　修改　删除
13	名片5	好单位5	经理助理	详情　修改　删除

第1页　共2页　下一页

添加名片

图 4.4　名片管理主页面

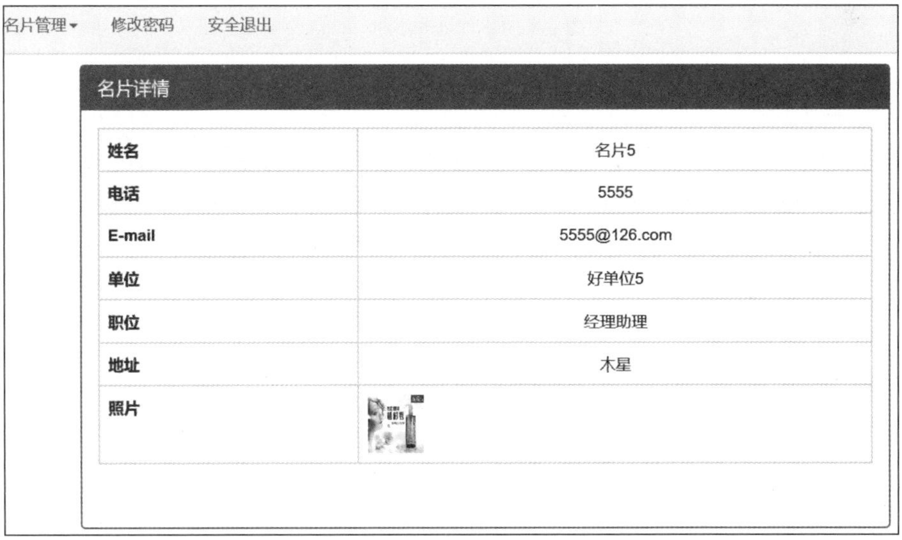

名片管理▾　修改密码　安全退出

名片详情

姓名	名片5
电话	5555
E-mail	5555@126.com
单位	好单位5
职位	经理助理
地址	木星
照片	

图 4.5　名片详情

▶ 4.5.8　修改名片

在名片管理主页面中，单击“修改”超链接，打开修改名片信息页面 updateCard.jsp。“修改”超链接的目标地址是 url 请求“card/detail?id=${card.id}&act=update”。找到对应控制器类 CardController 的方法 detail，在该方法中，根据动作类型，将查询结果转发给 updateCard.jsp 页面显示。

输入要修改的信息后，单击“修改”按钮，将名片信息提交给控制器类，找到对应控制器类 CardController 的方法 addCard，在 addCard 方法中根据动作类型，执行修改的业务处理。修改成功，进入名片管理主页面；修改失败，回到 updateCard.jsp 页面。

updateCard.jsp 页面的运行效果如图 4.6 所示。

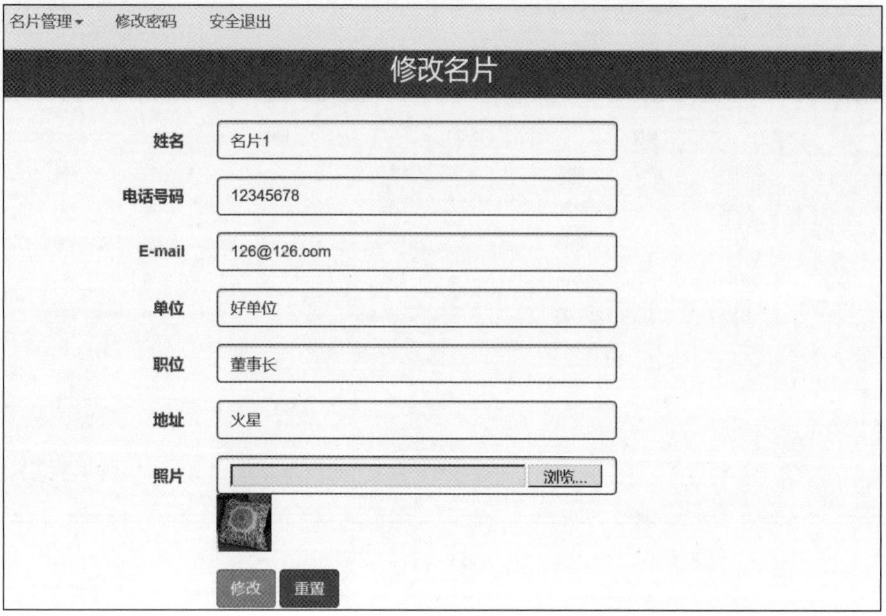

图 4.6 updateCard.jsp 页面

4.5.9 删除名片

在名片管理主页面中，单击“删除”超链接，将要删除名片的 ID 通过 Ajax 提交给控制器类。找到对应控制器类 CardController 的方法 delete，在该方法中，执行删除的业务处理。删除成功后，进入管理主页面。

4.6 用户相关

4.6.1 领域模型与持久化类

与用户相关的领域模型是 MyUser（位于 model 包），与用户相关的持久化类是 MyUserTable（位于 po 包），具体代码请读者参见本书提供的源代码 ch4。

4.6.2 Controller 实现

在本系统中，与用户相关的功能包括用户注册、用户登录以及用户检查等，由控制器类 UserController 负责处理。UserController 的核心代码如下：

```
@Controller
@RequestMapping("/user")
public class UserController {
    @Autowired
    private UserService userService;
    @RequestMapping("/toLogin")
    public String toLogin(@ModelAttribute MyUser myUser) {
        return "login";
```

```
    }
    @RequestMapping("/toRegister")
    public String toRegister(@ModelAttribute MyUser myUser) {
        return "register";
    }
    @RequestMapping("/checkUname")
    @ResponseBody
    public String checkUname(@RequestBody MyUser myUser) {
        return userService.checkUname(myUser);
    }
    @RequestMapping("/register")
    public String register(@ModelAttribute MyUser myUser, Model model) {
        return userService.register(myUser);
    }
    @RequestMapping("/login")
    public String login(@ModelAttribute MyUser myUser, Model model, HttpSession
      session) {
        return userService.login(myUser, model, session);
    }
}
```

▶ 4.6.3　Service 实现

与用户相关的 Service 接口和实现类分别为 UserService 和 UserServiceImpl。控制器获取一个请求后，需要调用 Service 层的业务处理方法，在 Service 层中调用 Dao 层。所以，Service 层是控制器层和 Dao 层的桥梁。UserService 接口的代码略。

UserServiceImpl 实现类的核心代码如下：

```
@Service
public class UserServiceImpl implements UserService{
    @Autowired
    private UserMapper userMapper;
    /***
     * 检查用户名是否可用
     */
    @Override
    public String checkUname(MyUser myUser) {
        List<MyUserTable> userList = userMapper.selectByUname(myUser);
        if(userList.size() > 0)
            return "no";
        return "ok";
    }
    /**
     * 实现注册功能
     */
    @Override
```

```
    public String register(MyUser myUser) {
        //将明文变成密文
        myUser.setUpwd(MD5Util.MD5(myUser.getUpwd()));
        if(userMapper.register(myUser) > 0)
            return "login";
        return "register";
    }
    /**
     * 实现登录功能
     */
    @Override
    public String login(MyUser myUser, Model model, HttpSession session) {
        //ValidateCodeController中的rand
        String code = (String)session.getAttribute("rand");
        if(!code.equalsIgnoreCase(myUser.getCode())) {
            model.addAttribute("errorMessage", "验证码错误！");
            return "login";
        }else {
            //将明文变成密文
            myUser.setUpwd(MD5Util.MD5(myUser.getUpwd()));
            List<MyUserTable> list = userMapper.login(myUser);
            if(list.size() > 0){
                session.setAttribute("userLogin", list.get(0));
                return "redirect:/card/selectAllCardsByPage?currentPage=1";
            }else {
                model.addAttribute("errorMessage", "用户名或密码错误！");
                return "login";
            }
        }
    }
}
```

4.6.4 Dao 实现

Dao 层是数据访问层，即@Repository 注解的数据操作接口（接口中的方法与 SQL 映射文件中元素的 id 对应），与用户相关的数据访问层为 UserMapper。UserMapper 的代码略。

4.6.5 SQL 映射文件

SQL 映射文件的 namespace 属性与数据操作接口对应。与用户相关的 SQL 映射文件是 UserMapper.xml（位于 dao 包中），具体代码如下：

```
<?xml version="1.0" encoding="UTF-8" ?>
<!DOCTYPE mapper PUBLIC "-//mybatis.org//DTD Mapper 3.0//EN"
"http://mybatis.org/dtd/mybatis-3-mapper.dtd">
<mapper namespace="dao.UserMapper">
    <select id="selectByUname"  resultType="MyUserTable" parameterType=
```

```
        "MyUser">
            select * from usertable where uname = #{uname}
        </select>
        <insert id="register" parameterType="MyUser">
            insert into usertable (id,uname,upwd) values(null,#{uname},#{upwd})
        </insert>
        <select id="login" parameterType="MyUser" resultType="MyUserTable">
            select * from usertable where uname=#{uname} and upwd=#{upwd}
        </select>
    </mapper>
```

4.6.6　注册

在登录页面 login.jsp，单击“注册”链接，打开注册页面 register.jsp，效果如图 4.7 所示。

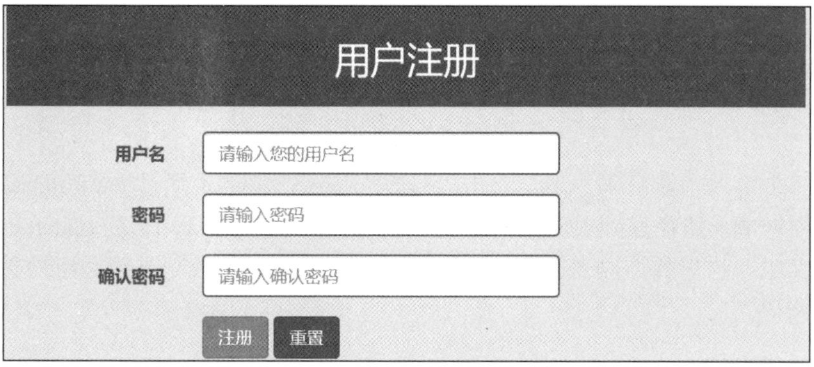

图 4.7　注册页面

在图 4.7 所示的注册页面中，输入“用户名”后，系统将通过 Ajax 提交 user/checkUname 请求检测“用户名”是否可用。输入合法的用户信息后，单击“注册”按钮，实现注册功能。

4.6.7　登录

在浏览器中，通过网址 http://localhost:8080/ch4 打开登录页面 login.jsp，效果如图 4.8 所示。

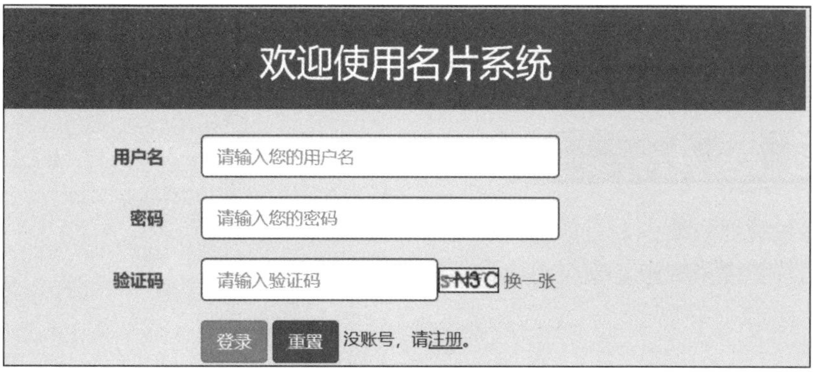

图 4.8　登录页面

用户输入用户名、密码和验证码后，系统将对用户名、密码和验证码进行验证。如果用户名、密码和验证码同时正确，则登录成功，将用户信息保存到 session 对象，并进入系统管理主页面（main.jsp）；如果输入有误，则提示错误。

4.6.8 修改密码

在名片管理主页面中，单击“修改密码”菜单，打开密码修改页面 updatePwd.jsp。密码修改页面效果如图 4.9 所示。

图 4.9 密码修改页面

在图 4.9 中输入“新密码”后，单击“修改”按钮，将请求通过 card/updatePwd 提交给控制器类。根据请求路径找到对应控制器类 CardController（4.5.2 节）的 updatePwd 方法，处理密码修改请求。这里找控制器类 CardController 处理密码修改，是因为用户必须登录成功后才能修改密码。

4.6.9 安全退出

在名片管理主页面中，单击“安全退出”菜单，将返回登录页面。“安全退出”超链接的目标地址是一个请求 card/loginOut，找到控制器类 CardController（4.5.2 节）的对应处理方法 loginOut。这里找控制器类 CardController 处理安全退出，是因为用户必须登录成功后才能安全退出。

4.7 本章小结

本章讲述了名片管理系统的设计与实现。通过本章的学习，读者不仅应该掌握 SSM 框架整合开发的流程、方法和技术，还应该熟悉名片管理的业务需求、设计以及实现。

习 题 4

1．在名片管理系统中，是如何控制登录权限的？

2．在名片管理系统中，安全退出功能的程序做了什么工作？

第 5 章 Spring Boot 入门

本章首先介绍什么是 Spring Boot，然后介绍 Spring Boot 应用的开发环境。通过本章的学习，掌握如何构建 Spring Boot 应用的开发环境以及 Spring Boot 应用。

主要内容

❖ Spring Boot 概述

❖ Spring Boot 应用的开发环境

Spring 框架非常优秀，但问题在于“配置过多”，造成开发效率低、部署流程复杂以及集成难度大等问题。为解决上述问题，Spring Boot 应运而生。作者在编写本书时，Spring Boot 的最新正式版是 2.4.1。读者测试本书示例代码时，建议使用 2.4.1 或更高版本。

5.1 Spring Boot 概述

5.1.1 什么是 Spring Boot

Spring Boot 是由 Pivotal 团队提供的全新框架，其设计目的是简化新 Spring 应用的初始搭建以及开发过程。使用 Spring Boot 框架可以做到专注于 Spring 应用的开发，无须过多关注样板化的配置。

在 Spring Boot 框架中，使用“约定优于配置（Convention Over Configuration，COC）”的理念。针对企业应用开发，提供了符合各种场景的 spring-boot-starter 自动配置依赖模块，这些模块都是基于“开箱即用”的原则，进而使企业应用开发更加快捷和高效。可以说，Spring Boot 是开发者和 Spring 框架的中间层，目的是帮助开发者管理应用的配置，提供应用开发中常见配置的默认处理（即约定优于配置），简化 Spring 应用的开发和运维，降低开发人员对框架的关注度，使开发人员把更多精力放在业务逻辑代码上。通过“约定优于配置”的原则，Spring Boot 致力于在蓬勃发展的快速应用开发领域中成为领导者。

5.1.2 Spring Boot 的优点

Spring Boot 之所以能够应运而生，是因为它具有如下优点。

（1）使编码变得简单：推荐使用注解。

（2）使配置变得快捷：自动配置、快速构建项目、快速集成第三方技术。

（3）使部署变得简便：内嵌 Tomcat、Jetty 等 Web 容器。

（4）使监控变得容易：自带项目监控。

5.1.3 Spring Boot 的主要特性

❶ 约定优于配置

Spring Boot 遵循"约定优于配置"的原则，只需很少的配置，大多数情况下直接使用默认配置即可。

❷ 独立运行的 Spring 应用

Spring Boot 可以以 jar 包的形式独立运行。使用 java -jar 命令或者在项目的主程序中执行 main 方法运行 Spring Boot 应用（项目）。

❸ 内嵌 Web 容器

内嵌 Servlet 容器，Spring Boot 可以选择内嵌 Tomcat、Jetty 等 Web 容器，无须以 war 包形式部署应用。

❹ 提供 starter 简化 Maven 配置

Spring Boot 提供了一系列的 starter pom 简化 Maven 的依赖加载，基本上可以做到自动化配置，高度封装，开箱即用。

❺ 自动配置 Spring

Spring Boot 根据项目依赖（在类路径中的 jar 包、类）自动配置 Spring 框架，极大地减少了项目的配置。

❻ 提供准生产的应用监控

Spring Boot 提供基于 HTTP、SSH、TELNET 对运行的项目进行跟踪监控。

❼ 无代码生成和 XML 配置

Spring Boot 不是借助于代码生成来实现的，而是通过条件注解来实现的；提倡使用 Java 配置和注解配置相结合的配置方式，方便快捷。

5.2 第一个 Spring Boot 应用

视频讲解

因为 Spring Boot 使用 Maven 配置 spring-boot-starter，所以在讲解 Spring Boot 应用之前，先了解 Maven 的相关基础知识。

5.2.1 Maven 简介

Apache Maven 是一个软件项目管理工具。基于项目对象模型（Project Object Model，POM）的理念，通过一段核心描述信息来管理项目构建、报告和文档信息。在 Java 项目中，Maven 主要完成两件工作：①统一开发规范与工具；②统一管理 jar 包。

Maven 统一管理项目开发所需要的 jar 包，但这些 jar 包将不再包含在项目内（即不在 lib 目录下），而是存放于仓库中。仓库主要包括以下两种。

❶ 中央仓库

存放开发过程中的所有 jar 包，例如 JUnit，都可以通过互联网从中央仓库中下载，仓库网址为 http://mvnrepository.com。

❷ 本地仓库

本地仓库是指本地计算机中的仓库。官方下载 Maven 的本地仓库，配置在"%MAVEN_HOME%\conf\settings.xml"文件中，找到 localRepository 即可。

Maven 项目首先会从本地仓库中获取所需要的 jar 包，当无法获取指定的 jar 包时，本地仓库将从远程仓库（中央仓库）中下载 jar 包，并放入本地仓库以备将来使用。

▶ 5.2.2 Maven 的 pom.xml

Maven 是基于项目对象模型的理念管理项目的，所以 Maven 的项目都有一个 pom.xml 配置文件来管理项目的依赖以及项目的编译等功能。

在 Maven 项目中，重点关注以下元素。

❶ properties 元素

在<properties></properties>之间可以定义变量，以便在<dependency></dependency>中引用，示例代码如下：

```
<properties>
    <!-- spring 版本号 -->
    <spring.version>5.3.2.RELEASE</spring.version>
</properties>
<dependencies>
    <dependency>
        <groupId>org.springframework</groupId>
        <artifactId>spring-core</artifactId>
        <version>${spring.version}</version>
    </dependency>
</dependencies>
```

❷ dependencies 元素

<dependencies></dependencies>元素包含多个项目依赖需要使用的<dependency></dependency>元素。

❸ dependency 元素

<dependency></dependency>元素内部通过<groupId></groupId>、<artifactId> </artifactId>、<version></version>三个子元素确定唯一的依赖，也可以称为三个坐标。示例代码如下：

```
<dependency>
    <!--groupId 组织的唯一标识 -->
    <groupId>org.springframework</groupId>
    <!--artifactId 项目的唯一标识 -->
    <artifactId>spring-core</artifactId>
    <!--version 项目的版本号 -->
    <version>${spring.version}</version>
</dependency>
```

❹ scope 子元素

在<dependency></dependency>元素中，有时使用<scope></scope>子元素管理依赖的部署。<scope></scope>子元素可以使用以下 5 个值。

1）compile（编译范围）

compile 是缺省值，即默认范围。依赖如果没有提供范围，那么该依赖的范围就是编译范围。编译范围的依赖在所有的 classpath 中可用，同时也会被打包发布。

2）provided（已提供范围）

provided 表示已提供范围，只有当 JDK 或者容器已提供该依赖时才可以使用。已提供范围的依赖不是传递性的，也不会被打包发布。

3）runtime（运行时范围）

runtime 在运行和测试系统时需要，但在编译时不需要。

4）test（测试范围）

test 在一般的编译和运行时都不需要，它们只有在测试编译和测试运行阶段可用。test 不会随项目发布。

5）system（系统范围）

system 范围与 provided 范围类似，但需要显式提供包含依赖的 JAR 包，Maven 不会在 Repository 中查找它。

▶ 5.2.3　使用 STS 快速构建 Spring Boot 应用

可以使用 Spring Tool Suite（STS）便捷地构建 Spring Boot 应用。STS 是一个定制版的 Eclipse，专为 Spring 开发定制，方便创建、调试、运行、维护 Spring 应用。通过该工具，可以很轻易地生成一个 Spring 工程，例如 Web 工程，最令人兴奋的是工程里的配置文件都将自动生成，开发者再也不用关注配置文件的格式及各种配置了。可以通过官网 https://spring.io/tools 下载 Spring Tools for Eclipse，本书采用的版本是 spring-tool-suite-4-4.9.0.RELEASE-e4.18.0-win32.win32.x86_64.self-extracting.jar。该版本与 Eclipse 一样免安装，解压即可使用。另外，STS 自带 Java SE，所以也不需要安装 JDK。

下面详细讲解如何使用 STS 集成开发工具快速构建一个 Spring Boot 应用，具体步骤如下。

❶ 新建 Spring Starter Project

通过选择菜单 File | New | Spring Starter Project，打开如图 5.1 所示的 New Spring Starter Project 对话框。

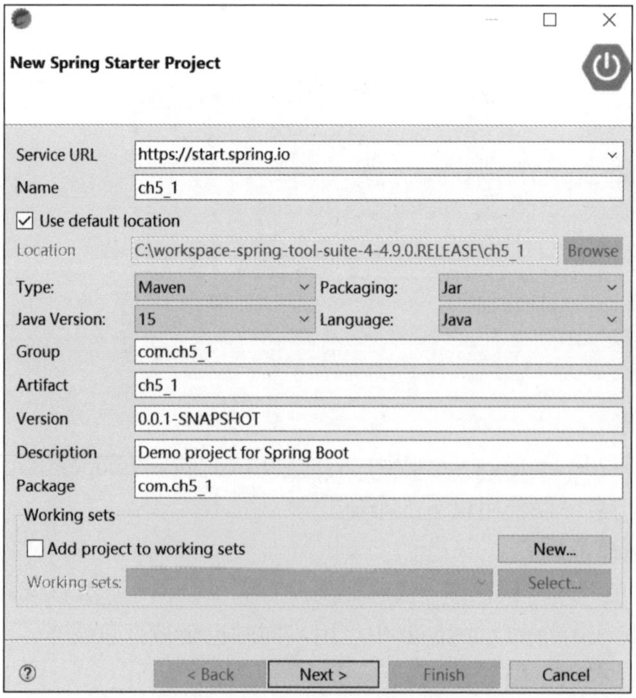

图 5.1　New Spring Starter Project

❷ 选择项目依赖

在图5.1中输入项目信息后，单击Next按钮，打开如图5.2所示的New Spring Starter Project Dependencies对话框，并在图中选择项目依赖，如Spring Web。

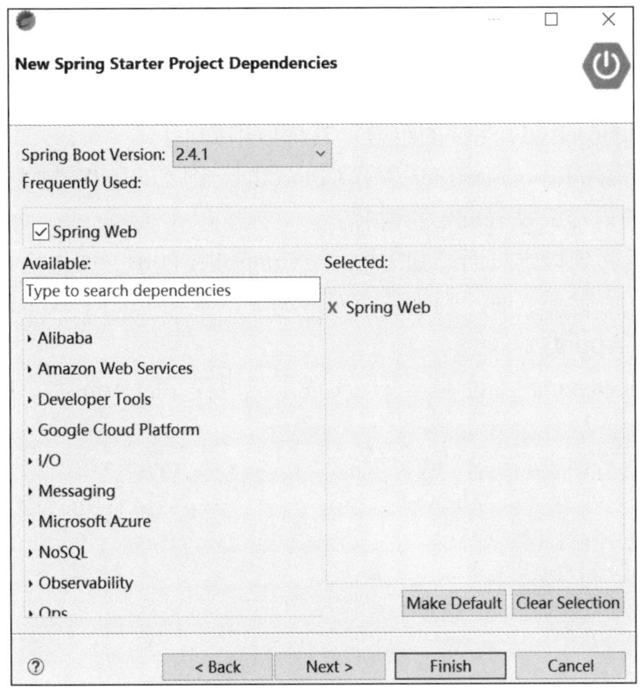

图5.2　New Spring Starter Project Dependencies

单击图5.2中的Finish按钮，即可完成Web应用的创建。ch5_1的项目结构如图5.3所示。此时，可以在项目ch5_1中编写自己的Web应用程序了。

图5.3　ch5_1的项目结构

❸ 编写测试代码

在应用ch5_1的src/main/java目录下，创建包com.ch5_1.test，并在该包中创建TestController类，代码如下：

```
package com.ch5_1.test;
import org.springframework.web.bind.annotation.RequestMapping;
import org.springframework.web.bind.annotation.RestController;
@RestController
```

```
public class TestController {
    @RequestMapping("/hello")
    public String hello() {
        return "您好，Spring Boot!";
    }
}
```

上述代码中使用的@RestController 注解是一个组合注解，相当于 Spring MVC 中的@Controller 和@ResponseBody 注解的组合，具体应用如下：

（1）如果只是使用@RestController 注解 Controller，则 Controller 中的方法无法返回 JSP、html 等视图，返回的内容就是 return 的内容。

（2）如果需要返回到指定页面，则需要用@Controller 注解。如果需要返回 JSON、XML 或自定义 mediaType 内容到页面，则需要在对应的方法上加上@ResponseBody 注解。

❹ 应用程序的 App 类

在应用 ch5_1 的 com.ch5_1 包中，自动生成了应用程序的 App 类，具体代码如下：

```
package com.ch5_1;
import org.springframework.boot.SpringApplication;
import org.springframework.boot.autoconfigure.SpringBootApplication;
@SpringBootApplication
public class Ch51Application {
    public static void main(String[] args) {
        SpringApplication.run(Ch51Application.class, args);
    }
}
```

上述代码中使用@SpringBootApplication 注解指定该程序是一个 Spring Boot 应用，该注解也是一个组合注解，相当于@SpringBootConfiguration、@EnableAutoConfiguration 和@ComponentScan 注解的组合，具体细节在第 6 章讲解。SpringApplication 类调用 run 方法启动 Spring Boot 应用。

❺ 运行 main 方法启动 Spring Boot 应用

运行 Ch51Application 类的 main 方法后，控制台信息如图 5.4 所示。

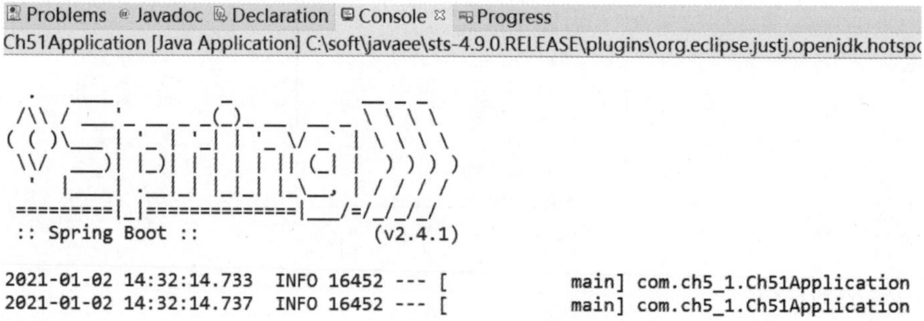

```
Problems  Javadoc  Declaration  Console  Progress
Ch51Application [Java Application] C:\soft\javaee\sts-4.9.0.RELEASE\plugins\org.eclipse.justj.openjdk.hotsp

  .   ____          _            __ _ _
 /\\ / ___'_ __ _ _(_)_ __  __ _ \ \ \ \
( ( )\___ | '_ | '_| | '_ \/ _` | \ \ \ \
 \\/  ___)| |_)| | | | | || (_| |  ) ) ) )
  '  |____| .__|_| |_|_| |_\__, | / / / /
 =========|_|==============|___/=/_/_/_/
 :: Spring Boot ::                (v2.4.1)

2021-01-02 14:32:14.733  INFO 16452 --- [           main] com.ch5_1.Ch51Application
2021-01-02 14:32:14.737  INFO 16452 --- [           main] com.ch5_1.Ch51Application
```

图 5.4　启动 Spring Boot 应用后的控制台信息

从控制台信息可以看到 Tomcat 的启动过程、Spring MVC 的加载过程。注意：Spring Boot 内嵌 Tomcat 容器，因此 Spring Boot 应用不需要开发者配置与启动 Tomcat。

❻ 测试 Spring Boot 应用

启动 Spring Boot 应用后，默认访问网址为 http://localhost:8080/，将项目路径直接设为根路径，这是 Spring Boot 的默认设置。因此，可以通过 http://localhost:8080/hello 测试应用（hello 与测试类 TestController 中的@RequestMapping("/hello")对应），测试效果如图 5.5 所示。

http://localhost:8080/hello

您好，Spring Boot!

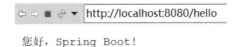

图 5.5 访问 Spring Boot 应用

▶ 5.2.4 使用 IntelliJ IDEA 快速构建 Spring Boot 应用

如果你的计算机上已经安装了 IntelliJ IDEA，那么可以使用 IntelliJ IDEA 便捷地构建 Spring Boot 应用，需要事先安装 JDK 并配置环境变量。具体步骤如下。

❶ 配置环境变量

安装 JDK 后，需要配置环境变量 Java_Home 和 path。配置环境变量 Java_Home 示例如图 5.6 所示。在 path 环境变量中，新建%Java_Home%\bin，示例如图 5.7 所示。

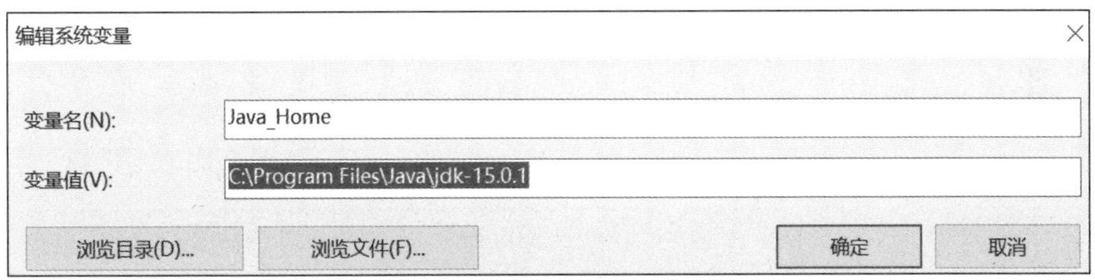

图 5.6 配置环境变量 Java_Home

图 5.7 新建 path 变量值

❷ 新建 Spring Starter Project

打开 IntelliJ IDEA，通过选择菜单 File | New | Project，打开如图 5.8 所示的 New Project 对话框。

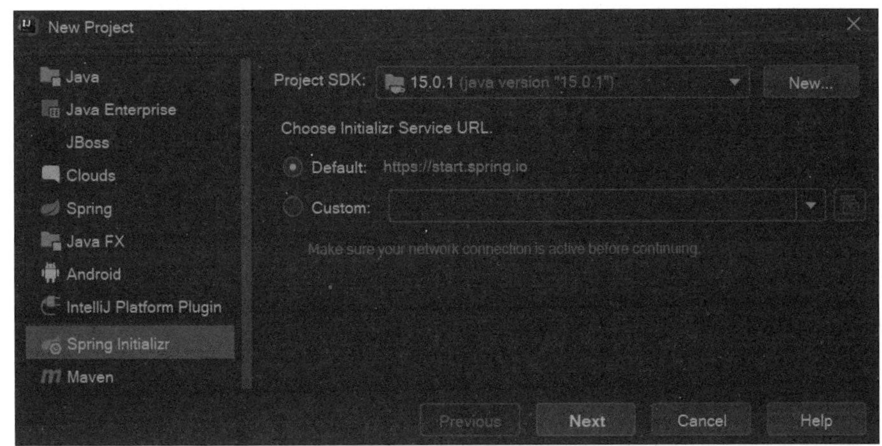

图 5.8 New Project

在图 5.8 左侧选择 Spring Initializr 项，单击 Next 按钮，打开如图 5.9 所示的 Project Metadata 对话框。

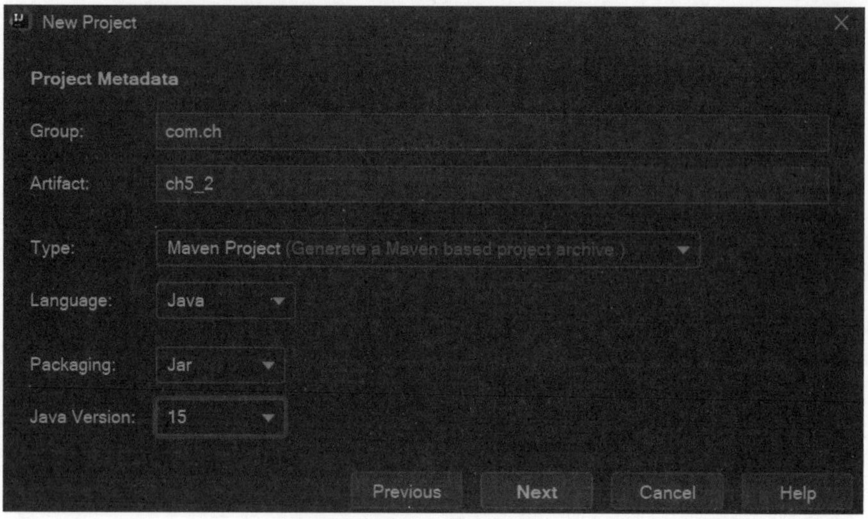

图 5.9　Project Metadata

在图 5.9 中，输入项目的 Metadata，单击 Next 按钮，打开如图 5.10 所示的对话框。

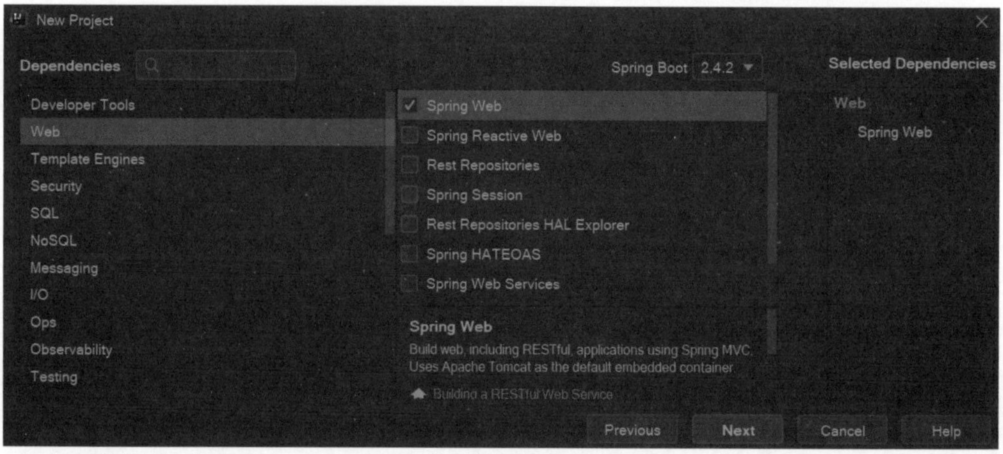

图 5.10　选择项目依赖

在图 5.10 中，选择项目依赖 Web | Spring Web，然后单击 Next 和 Finish 按钮，即可完成快速构建 Spring Boot 应用。测试程序与 5.2.3 节一样，不再赘述。

为方便机房教学，可以使用 STS 编写程序。需要说明的是，本书部分章节使用 STS 编写，但读者可以将本书第 5 章及以后章节提供的源程序直接导入 IntelliJ IDEA 中运行。另外，本书也会使用 IntelliJ IDEA 编写部分章节的程序。

5.3　本章小结

本章首先简单介绍了 Spring Boot 应运而生的缘由，然后讲述了如何使用 Spring Tool Suite（STS）和 IntelliJ IDEA 快速构建 Spring Boot 应用。开发者要构建 Spring Boot 应用，可根据实际工程需要选择合适的 IDE。

习 题 5

1．Spring、Spring MVC、Spring Boot 三者有什么联系？为什么还要学习 Spring Boot？
2．在 STS 中如何快速构建 Spring Boot 的 Web 应用？
3．在 IntelliJ IDEA 中如何快速构建 Spring Boot 的 Web 应用？

第 6 章 Spring Boot 核心

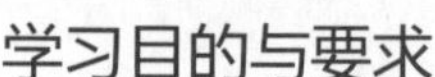

本章将详细介绍 Spring Boot 的核心注解、基本配置、自动配置原理以及条件注解。通过本章的学习，掌握 Spring Boot 的核心注解与基本配置，理解 Spring Boot 的自动配置原理与条件注解。

主要内容

- ❖ Spring Boot 的基本配置
- ❖ 读取应用配置
- ❖ Spring Boot 的自动配置原理
- ❖ Spring Boot 的条件注解

在 Spring Boot 产生之前，Spring 项目会存在多个配置文件，例如 web.xml、application.xml，应用程序自身也需要多个配置文件，同时需要编写程序读取这些配置文件。现在 Spring Boot 简化了 Spring 项目配置的管理和读取，仅需要一个 application.properties 文件，并提供了多种读取配置文件的方式。本章将学习 Spring Boot 的基本配置与运行原理。

6.1 Spring Boot 的基本配置

视频讲解

6.1.1 启动类和核心注解@SpringBootApplication

Spring Boot 应用通常都有一个名为*Application 的程序入口类，该入口类需要使用 Spring Boot 的核心注解@SpringBootApplication 标注为应用的启动类。另外，该入口类有一个标准的 Java 应用程序的 main 方法，在 main 方法中通过“SpringApplication.run(*Application.class, args);”启动 Spring Boot 应用。

Spring Boot 的核心注解@SpringBootApplication 是一个组合注解，主要组合了@SpringBootConfiguration、@EnableAutoConfiguration 和@ComponentScan 注解。源代码可以从 spring-boot-autoconfigure-2.4.1.jar 依赖包中查看 org/springframework/boot/autoconfigure/SpringBootApplication.java。

❶ @SpringBootConfiguration 注解

@SpringBootConfiguration 是 Spring Boot 应用的配置注解，该注解也是一个组合注解，源代码可以从 spring-boot-2.4.1.jar 依赖包中查看 org/springframework/boot/SpringBootConfiguration.java。在 Spring Boot 应用中推荐使用@SpringBootConfiguration 注解替代@Configuration 注解。

❷ @EnableAutoConfiguration 注解

@EnableAutoConfiguration 注解可以让 Spring Boot 根据当前应用项目所依赖的 jar 自动配置项目的相关配置。例如，在 Spring Boot 项目的 pom.xml 文件中添加了 spring-boot-starter-web

依赖，Spring Boot 项目会自动添加 Tomcat 和 Spring MVC 的依赖，同时对 Tomcat 和 Spring MVC 进行自动配置。打开 pom.xml 文件，选择 Dependency Hierarchy 页面查看 spring-boot-starter-web 的自动配置，如图 6.1 所示。

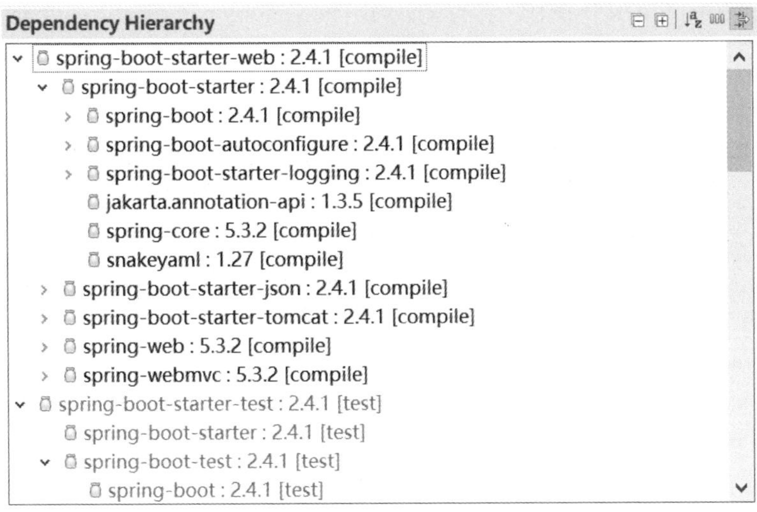

图 6.1　spring-boot-starter-web 的自动配置

❸ @ComponentScan 注解

该注解的功能是让 Spring Boot 自动扫描@SpringBootApplication 所在类的同级包以及它的子包中的配置，所以建议将@SpringBootApplication 注解的入口类放置在项目包下（Group Id+Artifact Id 组合的包名），这样可以保证 Spring Boot 自动扫描项目所有包中的配置。

▶ 6.1.2　关闭某个特定的自动配置

从 6.1.1 节可知，使用@EnableAutoConfiguration 注解可以让 Spring Boot 根据当前应用项目所依赖的 jar 自动配置项目的相关配置。如果开发者不需要 Spring Boot 的某一项自动配置，该如何实现呢？通过查看@SpringBootApplication 的源代码可知，应该使用@SpringBootApplication 注解的 exclude 参数关闭特定的自动配置。以关闭 neo4j 自动配置为例，代码如下：

```
@SpringBootApplication(exclude={Neo4jDataAutoConfiguration.class})
```

▶ 6.1.3　定制 banner

Spring Boot 项目启动时，在控制台可以看到如图 6.2 所示的默认启动图案。

```
Problems  Javadoc  Declaration  Console  Progress
Ch51Application [Java Application] C:\soft\javaee\sts-4.9.0.RELEASE\plugins\org.eclipse.justj.openjdk.hotsp

  .   ____          _            __ _ _
 /\\ / ___'_ __ _ _(_)_ __  __ _ \ \ \ \
( ( )\___ | '_ | '_| | '_ \/ _` | \ \ \ \
 \\/  ___)| |_)| | | | | || (_| |  ) ) ) )
  '  |____| .__|_| |_|_| |_\__, | / / / /
 =========|_|==============|___/=/_/_/_/
 :: Spring Boot ::                (v2.4.1)

2021-01-02 14:32:14.733  INFO 16452 --- [           main] com.ch5_1.Ch51Application
2021-01-02 14:32:14.737  INFO 16452 --- [           main] com.ch5_1.Ch51Application
```

图 6.2　Spring Boot 项目的默认启动图案

如果开发者希望指定自己的启动信息，又该如何配置呢？首先，在 src/main/resources 目录下新建 banner.txt 文件，并在文件中添加任意字符串内容，如"#Hello, Spring Boot!"。然后，重新启动 Spring Boot 项目，将发现控制台启动信息已经发生改变。如果开发者想把启动字符串信息换成字符串图案，具体操作为：首先，打开网站 http://patorjk.com/software/taag，输入自定义字符串，单击网页下方的 Select & Copy 按钮，如图 6.3 所示；然后，将自定义 banner 字符串图案复制到 src/main/resources 目录下的 banner.txt 文件中，重新启动 Spring Boot 项目即可。

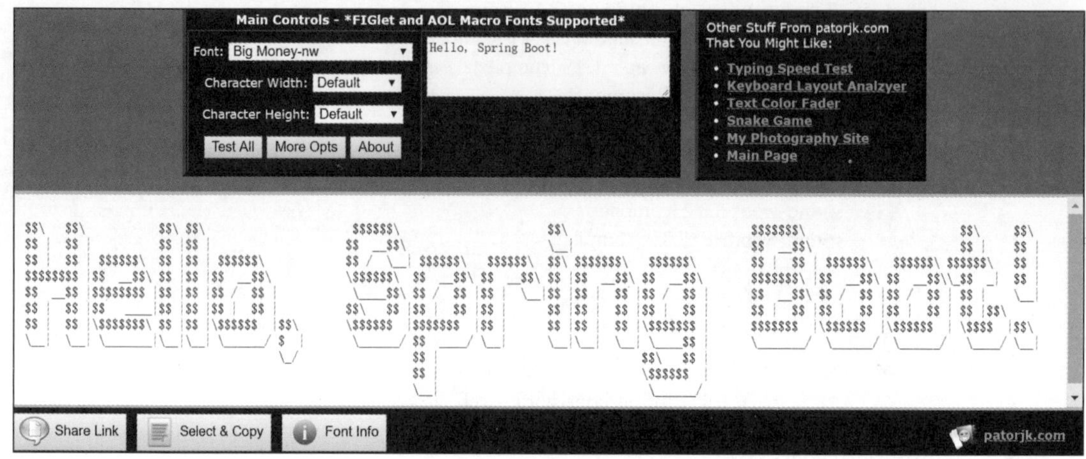

图 6.3　自定义 banner 图案

▶ 6.1.4　关闭 banner

开发者如果需要关闭 banner，可以在 src/main/resources 目录下的 application.properties 文件中添加如下配置：

```
spring.main.banner-mode = off
```

▶ 6.1.5　Spring Boot 的全局配置文件

Spring Boot 的全局配置文件（application.properties 或 application.xml）位于 Spring Boot 应用的 src/main/resources 目录下。

❶ 设置端口号

全局配置文件主要用于修改项目的默认配置，例如修改内嵌的 Tomcat 的默认端口。例如，在 Spring Boot 应用 ch6_1 的 src/main/resources 目录下找到名为 application.properties 的全局配置文件，添加如下配置内容：

```
server.port=8888
```

可以将内嵌的 Tomcat 的默认端口改为 8888。启动应用时，端口修改为 8888，如图 6.4 所示。

❷ 设置 Web 应用的上下文路径

如果开发者想设置一个 Web 应用程序的上下文路径，可以在 application.properties 文件中配置如下内容：

```
server.servlet.context-path=/XXX
```

```
Starting Ch61Application using Java 15.0.1 on LAPTOP-1GNNC309 with PID 21156
No active profile set, falling back to default profiles: default
Tomcat initialized with port(s): 8888 (http)
Starting service [Tomcat]
Starting Servlet engine: [Apache Tomcat/9.0.41]
Initializing Spring embedded WebApplicationContext
Root WebApplicationContext: initialization completed in 3465 ms
Exception encountered during context initialization - cancelling refresh att
Stopping service [Tomcat]
```

图 6.4　修改 Tomcat 的默认端口

这时应该通过 http://localhost:8080/XXX/testStarters 访问如下控制器类中的请求处理方法：

```
@RequestMapping("/testStarters")
public String index() {
}
```

❸ 配置文档

在 Spring Boot 的全局配置文件中，可以配置与修改多个参数。读者想了解参数的详细说明和描述，可以查看官方文档说明：https://docs.spring.io/spring-boot/docs/2.4.1/reference/htmlsingle/#common-application-properties。

6.1.6　Spring Boot 的 Starters

Spring Boot 提供了很多简化企业级开发的“开箱即用”的 Starters。Spring Boot 项目只要使用了所需要的 Starters，Spring Boot 即可自动关联项目开发所需要的相关依赖。例如，在 ch6_1 的 pom.xml 文件中添加如下依赖配置：

```
<dependency>
    <groupId>org.springframework.boot</groupId>
    <artifactId>spring-boot-starter-web</artifactId>
</dependency>
```

Spring Boot 将自动关联 Web 开发的相关依赖，例如 tomcat、spring-webmvc 等，进而对 Web 开发给予支持，并对相关技术的配置实现自动配置。

通过访问 https://docs.spring.io/spring-boot/docs/2.4.1/reference/htmlsingle/#using-boot-starter 官网，可以查看 Spring Boot 官方提供的 Starters，如表 6.1 所示。

表 6.1　Spring Boot 官方提供的 Starters

名　称	描　述
spring-boot-starter	核心 starter，包括自动配置、日志记录和 YAML 文件的支持
spring-boot-starter-actuator	支持准生产特性的 starter，用来监控和管理应用
spring-boot-starter-activemq	为 JMS 使用 Apache ActiveMQ 进行消息传递的 starter。ActiveMQ 是 Apache 出品的最流行、能力强的开源消息总线
spring-boot-starter-amqp	使用 spring-rabbit 支持 AMQP 协议（Advanced Message Queuing Protocol）的 starter
spring-boot-starter-aop	支持面向切面编程（AOP）的 starter，包括 spring-aop 和 AspectJ
spring-boot-starter-artemis	使用 Apache Artemis 支持 JMS 消息传递的 starter

续表

名　称	描　述
spring-boot-starter-batch	支持 Spring Batch 的 starter，包括 HSQLDB（HyperSQL DataBase）数据库
spring-boot-starter-cache	支持 Spring Cache 的 starter
spring-boot-starter-cloud-connectors	支持 Spring Cloud Connectors 的 starter，简化了 Cloud Foundry、Heroku 等云平台中的服务连接
spring-boot-starter-data-cassandra	使用 Spring Data Cassandra 支持 Cassandra 分布式数据库的 starter
spring-boot-starter-data-couchbase	使用 Spring Data Couchbase 支持 Couchbase 文件存储数据库的 starter
spring-boot-starter-data-elasticsearch	使用 Spring Data Elasticsearch 支持 Elasticsearch 搜索和分析引擎的 starter
spring-boot-starter-data-jdbc	支持 Spring Data JDBC 的 starter
spring-boot-starter-data-jpa	支持 JPA（Java Persistence API）的 starter，包括 spring-data-jpa、spring-orm 和 Hibernate
spring-boot-starter-data-ldap	支持 Spring Data LDAP（轻量级目录访问协议，Lightweight Directory Access Protocol）的 starter
spring-boot-starter-data-mongodb	使用 Spring Data MongoDB 支持 MongoDB 的 starter
spring-boot-starter-data-neo4j	使用 Spring Data Neo4j 支持 Neo4j 图数据库的 starter
spring-boot-starter-data-redis	使用 Spring Data Redis 支持 Redis 键值存储数据库的 starter，包括 Lettuce 客户端
spring-boot-starter-data-rest	使用 Spring Data REST 支持通过 REST 公开 Spring Data 数据仓库的 starter
spring-boot-starter-data-solr	使用 Spring Data Solr 支持 Apache Solr 搜索平台的 starter
spring-boot-starter-freemarker	支持 FreeMarker 模板引擎构建 MVC Web 应用的 starter
spring-boot-starter-groovy-templates	支持 Groovy 模板引擎构建 MVC Web 应用的 starter
spring-boot-starter-hateoas	使用 Spring MVC、Spring HATEOAS 构建基于超媒体的 RESTful Web 应用程序的 starter
spring-boot-starter-integration	支持通用的 spring-integration 模块的 starter
spring-boot-starter-jdbc	支持 JDBC 的 starter，包括 HikariCP 连接池
spring-boot-starter-jersey	使用 JAX-RS 和 Jersey 支持 RESTful Web 应用程序的 starter，替代 spring-boot-starter-web
spring-boot-starter-jetty	使用 Jetty 作为嵌入式 servlet 容器替代 Tomcat 的 starter
spring-boot-starter-jooq	使用 jOOQ 支持访问 SQL 数据库的 starter，替代 spring-boot-starter-jpa 或 spring-boot-starter-jdbc
spring-boot-starter-json	读写 json 的 starter
spring-boot-starter-jta-atomikos	使用 Atomikos 支持 JTA 分布式事务处理的 starter
spring-boot-starter-jta-bitronix	使用 Bitronix 支持 JTA 分布式事务处理的 starter
spring-boot-starter-log4j2	支持使用 Log4j2 日志框架的 starter，替代 spring-boot-starter-logging
spring-boot-starter-logging	支持 Spring Boot 的默认日志框架 Logback 的 starter
spring-boot-starter-mail	支持 javax.mail 的 starter
spring-boot-starter-mustache	支持 Mustache 模板引擎构建 Web 应用的 starter
spring-boot-starter-oauth2-client	使用 Spring Security 的 OAuth2/OpenID Connect 支持客户端功能的 starter
spring-boot-starter-oauth2-resource-server	使用 Spring Security 的 OAuth2 支持资源服务器功能的 starter
spring-boot-starter-quartz	支持 Quartz 调度器的 starter
spring-boot-starter-reactor-netty	使用 Reactor Netty 作为嵌入式响应 HTTP 服务器的 starter

续表

名　　称	描　　述
spring-boot-starter-security	支持 spring-security 的 starter
spring-boot-starter-test	支持常规的测试依赖的 starter，包括 Junit、Hamcrest、Mockito 以及 spring-test 模块
spring-boot-starter-thymeleaf	支持 Thymeleaf 模板引擎构建 MVC Web 应用的 starter
spring-boot-starter-tomcat	支持 Spring Boot 的默认 Servlet 容器 Tomcat 的 starter
spring-boot-starter-undertow	使用 Undertow 作为嵌入式 servlet 容器替代 Tomcat 的 starter
spring-boot-starter-validation	支持 Java Bean 验证的 starter，包括 Hibernate 验证器
spring-boot-starter-web	支持 Web 应用开发的 starter，包括 Tomcat（默认嵌入式容器）和 spring-webmvc
spring-boot-starter-web-services	支持 Spring Web Services 的 starter
spring-boot-starter-webflux	支持 WebFlux（一个非阻塞异步框架）开发的 starter
spring-boot-starter-websocket	支持 WebSocket 开发的 starter

除了 Spring Boot 官方提供的 Starters 外，还可以通过访问 https://github.com/spring-projects/spring-boot/blob/master/spring-boot-project/spring-boot-starters/README.adoc 网站，查看第三方为 Spring Boot 贡献的 Starters。

6.2　读取应用配置

视频讲解

Spring Boot 提供了三种方式读取项目的 application.properties 配置文件的内容，分别为 Environment 类、@Value 注解以及@ConfigurationProperties 注解。

6.2.1　Environment

Environment 是一个通用的读取应用程序运行时的环境变量的类，可以通过 key-value 方式读取 application.properties、命令行输入参数、系统属性、操作系统环境变量等。下面通过一个实例演示如何使用 Environment 类读取 application.properties 配置文件的内容。

【例 6-1】使用 Environment 类读取 application.properties 配置文件的内容。

具体实现步骤如下。

❶ 创建 Spring Boot 项目 ch6_1

使用 STS 快速创建 Spring Web 应用 ch6_1（参考 5.2.3 节）。

❷ 添加配置文件内容

在 src/main/resources 目录下，找到全局配置文件 application.properties，并添加如下内容：

```
test.msg=read config
```

❸ 创建控制器类 EnvReaderConfigController

在 src/main/java 目录下，创建名为 com.ch6_1.controller 的包（是 com.ch6_1 包（主类所在的包）的子包，保障注解全部被扫描），并在该包下创建控制器类 EnvReaderConfigController。在控制器类 EnvReaderConfigController 中，使用@Autowired 注解依赖注入 Environment 类的对象，核心代码如下：

```
@RestController
public class EnvReaderConfigController{
    @Autowired
    private Environment env;
    @RequestMapping("/testEnv")
    public String testEnv() {
        return "方法一： " + env.getProperty("test.msg");
        //test.msg 为配置文件 application.properties 中的 key
    }
}
```

❹ 启动 Spring Boot 应用

运行 Ch61Application 类的 main 方法，启动 Spring Boot 应用。

❺ 测试应用

启动 Spring Boot 应用后，默认访问网址为 http://localhost:8080/，将项目路径直接设为根路径，这是 Spring Boot 的默认设置。因此，可以通过 http://localhost:8080/testEnv 测试应用（testEnv 与控制器类 ReaderConfigController 中的@RequestMapping("/testEnv")对应）。

6.2.2 @Value

使用@Value 注解读取配置文件内容示例代码如下：

```
@Value("${test.msg}")//test.msg 为配置文件 application.properties 中的 key
private String msg;//通过@Value 注解将配置文件中 key 对应的 value 赋值给变量 msg
```

下面通过实例讲解如何使用@Value 注解读取配置文件内容。

【例 6-2】使用@Value 注解读取配置文件内容。

具体实现步骤如下。

❶ 创建控制器类 ValueReaderConfigController

在 ch6_1 应用的 com.ch6_1.controller 包中，创建名为 ValueReaderConfigController 的控制器类，在该控制器类中使用@Value 注解读取配置文件内容。核心代码如下：

```
@RestController
public class ValueReaderConfigController {
    @Value("${test.msg}")
    private String msg;
    @RequestMapping("/testValue")
    public String testValue() {
        return "方法二： " + msg ;
    }
}
```

❷ 启动并测试应用

首先，运行 Ch61Application 类的 main 方法，启动 Spring Boot 应用。然后，通过 http://localhost:8080/testValue 测试应用。

6.2.3 @ConfigurationProperties

使用@ConfigurationProperties 首先建立配置文件与对象的映射关系，然后在控制器方法中使用@Autowired 注解将对象注入。

下面通过实例讲解如何使用@ConfigurationProperties 读取配置文件内容。

【例 6-3】 使用@ConfigurationProperties 读取配置文件内容。

具体实现步骤如下。

❶ 添加配置文件内容

在 ch6_1 应用的 src/main/resources 目录下，找到全局配置文件 application.properties，并添加如下内容：

```
# nest Simple properties
obj.sname=chenheng
obj.sage=88
#List properties
obj.hobby[0]=running
obj.hobby[1]=basketball
#Map Properties
obj.city.cid=dl
obj.city.cname=dalian
```

❷ 建立配置文件与对象的映射关系

在 ch6_1 项目的 src/main/java 目录下创建名为 com.ch6_1.model 的包，并在包中创建实体类 StudentProperties，在该类中使用@ConfigurationProperties 注解建立配置文件与对象的映射关系。核心代码如下：

```
@Component//使用 Component 注解，声明一个组件，被控制器依赖注入
@ConfigurationProperties(prefix = "obj")//obj 为配置文件中 key 的前缀
public class StudentProperties {
    private String sname;
    private int sage;
    private List<String> hobby;
    private Map<String, String> city;
    //省略 set 方法和 get 方法
    @Override
    public String toString() {
        return "StudentProperties [sname=" + sname
            + ", sage=" + sage
            + ", hobby0=" + hobby.get(0)
            + ", hobby1=" + hobby.get(1)
            + ", city=" + city + "]";
    }
}
```

❸ 创建控制器类 ConfigurationPropertiesController

在 ch6_1 项目的 com.ch6_1.controller 包中，创建名为 ConfigurationPropertiesController 的控制器类，在该控制器类中使用@Autowired 注解依赖注入 StudentProperties 对象。核心代码如下：

```
@RestController
public class ConfigurationPropertiesController {
    @Autowired
    StudentProperties studentProperties;
    @RequestMapping("/testConfigurationProperties")
```

```
    public String testConfigurationProperties() {
        return studentProperties.toString();
    }
}
```

❹ 启动并测试应用

首先，运行 Ch61Application 类的 main 方法，启动 Spring Boot 应用。然后，通过 http://localhost:8080/testConfigurationProperties 测试应用。

▶ 6.2.4 @PropertySource

开发者希望读取项目的其他配置文件，而不是全局配置文件 application.properties，该如何实现呢？可以使用@PropertySource 注解找到项目的其他配置文件，然后结合 6.2.1～6.2.3 节中任意一种方式读取。

下面通过实例讲解如何使用@PropertySource + @Value 读取其他配置文件内容。

【例 6-4】使用@PropertySource + @Value 读取其他配置文件内容。

具体实现步骤如下。

❶ 创建配置文件

在 ch6_1 的 src/main/resources 目录下创建配置文件 ok.properties 和 test.properties，并在 ok.properties 文件中添加如下内容：

```
your.msg=hello.
```

在 test.properties 文件中添加如下内容：

```
my.msg=test PropertySource
```

❷ 创建控制器类 PropertySourceValueReaderOtherController

在 ch6_1 项目的 com.ch6_1.controller 包中，创建名为 PropertySourceValueReaderOtherController 的控制器类。在该控制器类中，首先使用@PropertySource 注解找到其他配置文件，然后使用@Value 注解读取配置文件内容。核心代码如下：

```
@RestController
@PropertySource({"test.properties","ok.properties"})
public class PropertySourceValueReaderOtherController {
    @Value("${my.msg}")
    private String mymsg;
    @Value("${your.msg}")
    private String yourmsg;
    @RequestMapping("/testProperty")
    public String testProperty() {
        return "其他配置文件 test.properties: " + mymsg + "<br>"
            + "其他配置文件 ok.properties: " + yourmsg;
    }
}
```

❸ 启动并测试应用

首先，运行 Ch61Application 类的 main 方法，启动 Spring Boot 应用。然后，通过 http://localhost:8080/testProperty 测试应用。

6.3　日志配置

视频讲解

默认情况下，Spring Boot 应用使用 LogBack 实现日志，使用 apache Commons Logging 作为日志接口，因此使用日志的代码通常如下：

```
@RestController
public class LogTestController {
    private Log log = LogFactory.getLog(LogTestController.class);
    @RequestMapping("/testLog")
    public String testLog() {
        log.info("测试日志");
        return "测试日志" ;
    }
}
```

通过网址 http://localhost:8080/testLog 运行上述控制器类代码，可以在控制台输出“测试日志”信息。

日志级别有 ERROR、WARN、INFO、DEBUG 和 TRACE。Spring Boot 默认的日志级别为 INFO，日志信息可以打印到控制台。但开发者可以自己设定 Spring Boot 项目的日志输出级别，例如在 application.properties 配置文件中加入以下配置：

```
#设定日志的默认级别为 info
logging.level.root=info
#设定 org 包下的日志级别为 warn
logging.level.org=warn
#设定 com.ch.ch4_1 包下的日志级别为 debug
logging.level.com.ch.ch4_1=debug
```

Spring Boot 项目默认并没有输出日志到文件，但开发者可以在 application.properties 配置文件中指定日志输出到文件，配置示例如下：

```
logging.file=my.log
```

日志输出到 my.log 文件，该日志文件位于 Spring Boot 项目运行的当前目录（项目工程目录下）。也可以指定日志文件目录，配置示例如下：

```
logging.file=C:/log/my.log
```

这样将在 C:/log 目录下生成一个名为 my.log 的日志文件。不管日志文件位于何处，当日志文件大小达到 10MB 时，将自动生成一个新日志文件。

Spring Boot 使用内置的 LogBack 支持对控制台日志输出和文件输出进行格式控制，例如开发者可以在 application.properties 配置文件中添加如下配置：

```
logging.pattern.console=%level %date{yyyy-MM-dd HH:mm:ss:SSS} %logger{50}.%M %L:%m%n
logging.pattern.file=%level %date{ISO8601} %logger{50}.%M %L:%m%n
```

logging.pattern.console：指定控制台日志格式。

logging.pattern.file：指定日志文件格式。

%level：指定输出日志级别。

%date：指定日志发生的时间。ISO8601 表示标准日期，相当于 yyyy-MM-dd HH:mm:ss:SSS。

%logger{n}：指定输出 Looger 的名字，包名+类名，{n}限定了输出长度。

%M：指定日志发生时的方法名。

%L：指定日志调用时所在代码行，适用于开发调试，线上运行时不建议使用此参数，因为获取代码行对性能有消耗。

%m：表示日志消息。

%n：表示日志换行。

视频讲解

6.4 Spring Boot 的自动配置原理

从 6.1.1 节可知，Spring Boot 使用核心注解@SpringBootApplication 将一个带有 main 方法的类标注为应用的启动类。@SpringBootApplication 注解最主要的功能之一是为 Spring Boot 开启了一个@EnableAutoConfiguration 注解的自动配置功能。

@EnableAutoConfiguration 注解主要利用一个类名为 AutoConfigurationImportSelector 的选择器向 Spring 容器自动配置一些组件。@EnableAutoConfiguration 注解的源代码可以从 spring-boot-autoconfigure-2.4.1.jar（org.springframework.boot.autoconfigure）依赖包中查看，核心代码如下：

```
@Import(AutoConfigurationImportSelector.class)
public @interface EnableAutoConfiguration {
    String ENABLED_OVERRIDE_PROPERTY = "spring.boot.enableautoconfiguration";
    Class<?>[] exclude() default {};
    String[] excludeName() default {};
}
```

AutoConfigurationImportSelector（源代码位于 org.springframework.boot.autoconfigure 包）类中有一个名为 selectImports 的方法，该方法规定了向 Spring 容器自动配置的组件。

selectImports 方法的代码如下：

```
@Override
public String[] selectImports(AnnotationMetadata annotationMetadata) {
    //判断@EnableAutoConfiguration 注解有没有开启，默认开启
    if (!isEnabled(annotationMetadata)) {
        return NO_IMPORTS;
    }
    //获得自动配置
    AutoConfigurationEntry autoConfigurationEntry = getAutoConfigurationEntry
        (annotationMetadata);
    return StringUtils.toStringArray(autoConfigurationEntry.getConfigurations());
}
```

在方法 selectImports 中，调用 getAutoConfigurationEntry 方法获得自动配置。进入该方法，查看到的源代码如下：

```
protected AutoConfigurationEntry getAutoConfigurationEntry(AnnotationMetadata
    annotationMetadata) {
    if (!isEnabled(annotationMetadata)) {
        return EMPTY_ENTRY;
```

```
    }
    AnnotationAttributes attributes = getAttributes(annotationMetadata);
    //获取 META-INF/spring.factoies 的配置数据
    List<String>configurations=getCandidateConfigurations(annotationMetadata,
        attributes);
    //去重
    configurations = removeDuplicates(configurations);
    //去除一些多余的类
    Set<String> exclusions = getExclusions(annotationMetadata, attributes);
    checkExcludedClasses(configurations, exclusions);
    configurations.removeAll(exclusions);
    //过滤掉一些条件没有满足的配置
    configurations = getConfigurationClassFilter().filter(configurations);
    fireAutoConfigurationImportEvents(configurations, exclusions);
    return new AutoConfigurationEntry(configurations, exclusions);
}
```

在方法 getAutoConfigurationEntry 中，调用 getCandidateConfigurations 方法获取 META-INF/spring.factories 的配置数据。进入该方法，查看到的源代码如下：

```
protected List<String> getCandidateConfigurations(AnnotationMetadata metadata,
        AnnotationAttributes attributes) {
    List<String> configurations = SpringFactoriesLoader.loadFactoryNames(
        getSpringFactoriesLoaderFactoryClass(), getBeanClassLoader());
    …
}
```

在方法 getCandidateConfigurations 中，调用了 loadFactoryNames 方法。进入该方法，查看到的源代码如下：

```
public static List<String> loadFactoryNames(Class<?> factoryClass, @Nullable
    ClassLoader classLoader) {
    …
return loadSpringFactories(classLoaderToUse).getOrDefault(factoryTypeName,
    Collections.emptyList());
}
```

在方法 loadFactoryNames 中，调用了 loadSpringFactories 方法。进入该方法，查看到的源代码如下：

```
private static Map<String, List<String>> loadSpringFactories(ClassLoader
    classLoader) {
    …
    Enumeration<URL> urls = classLoader.getResources(FACTORIES_RESOURCE_
    LOCATION);
    …
}
```

在方法 loadSpringFactories 中，可以看到加载一个常量：FACTORIES_RESOURCE_LOCATION，该常量的源代码如下：

```
public static final String FACTORIES_RESOURCE_LOCATION =
    "META-INF/spring. factories";
```

从上述源代码可以看出，最终 Spring Boot 是通过加载所有(in multiple JAR files)META-INF/spring.factories 配置文件进行自动配置的。所以，@SpringBootApplication 注解通过使用@EnableAutoConfiguration 注解自动配置的原理是：从 classpath 中搜索所有 META-INF/spring.factories 配置文件，并将其中 org.springframework.boot.autoconfigure.EnableAutoConfiguration 对应的配置项通过 Java 反射机制进行实例化，然后汇总并加载到 Spring 的 IoC 容器。

在 Spring Boot 项目的 Maven Dependencies 的 spring-boot-autoconfigure-2.4.1.jar 目录下，可以找到 META-INF/spring.factories 配置文件，该文件中定义了许多自动配置。

6.5 Spring Boot 的条件注解

打开 spring.factories 配置文件中任意一个 AutoConfiguration，一般都可以找到条件注解。例如，打开 org.springframework.boot.autoconfigure.aop.AopAutoConfiguration 的源代码，可以看到@ConditionalOnClass 和@ConditionalOnProperty 等条件注解。

通过 org.springframework.boot.autoconfigure.aop.AopAutoConfiguration 的源代码可以看出，Spring Boot 的自动配置是使用 Spring 的@Conditional 注解实现的。因此，本节将介绍相关的条件注解，并讲述如何自定义 Spring 的条件注解。

▶ 6.5.1 条件注解

所谓 Spring 的条件注解，就是应用程序的配置类满足某些特定条件才会自动启用此配置类的配置项。Spring Boot 的条件注解位于 spring-boot-autoconfigure-2.4.1.jar 的 org.springframework.boot.autoconfigure.condition 包下，具体如表 6.2 所示。

表 6.2 Spring Boot 的部分条件注解

注解名	条件实现类	条　件
@ConditionalOnBean	OnBeanCondition	Spring 容器中存在指定的实例 Bean
@ConditionalOnClass	OnClassCondition	类加载器（类路径）中存在对应的类
@ConditionalOnCloudPlatform	OnCloudPlatformCondition	是否在云平台
@ConditionalOnExpression	OnExpressionCondition	判断 SpEL 表达式是否成立
@ConditionalOnJava	OnJavaCondition	指定 Java 版本是否符合要求
@ConditionalOnJndi	OnJndiCondition	在 JNDI（Java 命名和目录接口）存在的条件下查找指定的位置
@ConditionalOnMissingBean	OnBeanCondition	Spring 容器中不存在指定的实例 Bean
@ConditionalOnMissingClass	OnClassCondition	类加载器（类路径）中不存在对应的类
@ConditionalOnNotWebApplication	OnWebApplicationCondition	当前应用程序不是 Web 程序
@ConditionalOnProperty	OnPropertyCondition	应用环境中属性是否存在指定的值
@ConditionalOnResource	OnResourceCondition	是否存在指定的资源文件
@ConditionalOnSingleCandidate	OnBeanCondition	Spring 容器中是否存在且只存在一个对应的实例 Bean
@ConditionalOnWebApplication	OnWebApplicationCondition	当前应用程序是 Web 程序

表 6.2 中的条件注解都是组合了@Conditional 元注解，只是针对不同的条件去实现。

【例 6-5】 通过@ConditionalOnWebApplication 注解的源码分析，讲解条件注解的实现方法。

@ConditionalOnWebApplication 是一个标记注解，源代码如下：

```
...
@Target({ElementType.TYPE, ElementType.METHOD })
@Retention(RetentionPolicy.RUNTIME)
@Documented
@Conditional(OnWebApplicationCondition.class)
public @interface ConditionalOnWebApplication {
}
```

下面通过代码注释的形式，分析@ConditionalOnWebApplication 注解的实现类 OnWebApplicationCondition 的源代码，具体分析如下：

```
class OnWebApplicationCondition extends FilteringSpringBootCondition {
    ...
    /** ConditionOutcome 记录了匹配结果*/
    @Override
    public ConditionOutcome getMatchOutcome(ConditionContext context,
            AnnotatedTypeMetadata metadata) {
        //1. 检查是否被@ConditionalOnWebApplication 注解
        boolean required = metadata
            .isAnnotated(ConditionalOnWebApplication.class.getName());
        //2. 判断是否是 WebApplication
        ConditionOutcome outcome = isWebApplication(context, metadata,
            required);
        /**3. 如果被@ConditionalOnWebApplication 注解，但不是 WebApplication
           环境,则返回不匹配*/
        if (required && !outcome.isMatch()) {
            return ConditionOutcome.noMatch(outcome.getConditionMessage());
        }
        /**4. 如果没有被@ConditionalOnWebApplication 注解，并且是 WebApplication
           环境,则返回不匹配*/
        if (!required && outcome.isMatch()) {
            return ConditionOutcome.noMatch(outcome.getConditionMessage());
        }
        /**5. 如果被@ConditionalOnWebApplication 注解，并且是 WebApplication
           环境,则返回匹配*/
        return ConditionOutcome.match(outcome.getConditionMessage());
    }
    /**判断是否是 Web 环境*/
    private ConditionOutcome isWebApplication(ConditionContext context,
            AnnotatedTypeMetadata metadata, boolean required) {
        switch (deduceType(metadata)) {
        //1. 基于 servlet 的 Web 应用程序
        case SERVLET:
            return isServletWebApplication(context);
        //2. 基于 reactive（响应）的 Web 应用程序
        case REACTIVE:
            return isReactiveWebApplication(context);
        //3. 任意的 Web 应用程序
```

```
        default:
            return isAnyWebApplication(context, required);
        }
    }
    /**任意的 Web 应用程序*/
    private ConditionOutcome isAnyWebApplication(ConditionContext context,
            boolean required) {
        …
    }
    /**基于 servlet 的 Web 应用程序判断*/
    private ConditionOutcome isServletWebApplication(ConditionContext context){
        ConditionMessage.Builder message = ConditionMessage.forCondition("");
        // 1. 判断 GenericWebApplicationContext 是否在类路径中,如果不存在,
        // 则返回不匹配
        if (!ClassNameFilter.isPresent(SERVLET_WEB_APPLICATION_CLASS,
                context.getClassLoader())) {
            return ConditionOutcome.noMatch(
                message.didNotFind("servlet web application classes").atAll());
        }
        // 2. 容器里是否有名为 session 的 scope,如果是,则返回匹配
        if (context.getBeanFactory() != null) {
            String[] scopes = context.getBeanFactory().getRegisteredScopeNames();
            if (ObjectUtils.containsElement(scopes, "session")) {
                return ConditionOutcome.match(message.foundExactly("'session'
                  scope"));
            }
        }
        //3. Environment 是否为 ConfigurableWebEnvironment,如果是,则返回匹配
        if (context.getEnvironment() instanceof ConfigurableWebEnvironment) {
            return ConditionOutcome
                .match(message.foundExactly("ConfigurableWebEnvironment"));
        }
        // 4. 当前 ResourceLoader 是否为 WebApplicationContext,如果是,则返回匹配
        if (context.getResourceLoader() instanceof WebApplicationContext) {
            return ConditionOutcome.match(message.foundExactly
                ("WebApplicationContext"));
        }
        // 5. 其他情况,返回不匹配
        return ConditionOutcome.noMatch(message.because("not a servlet web
            application"));
    }
    /**基于 reactive（响应）的 Web 应用程序*/
    private ConditionOutcome isReactiveWebApplication(ConditionContext
            context) {
        ConditionMessage.Builder message = ConditionMessage.forCondition("");
        // 1. 判断 HandlerResult 是否在类路径中,如果否,则返回不匹配
        if (!ClassNameFilter.isPresent(REACTIVE_WEB_APPLICATION_CLASS,
                context.getClassLoader())) {
            return ConditionOutcome.noMatch(
                message.didNotFind("reactive web application classes").atAll());
        }
        // 2. Environment 是否为 ConfigurableReactiveWebEnvironment,如果是,
        // 则返回匹配
```

```
        if (context.getEnvironment() instanceof
                ConfigurableReactiveWebEnvironment) {
            return ConditionOutcome.match(message.foundExactly
                    ("ConfigurableReactiveWebEnvironment"));
        }
        // 3. 当前 ResourceLoader 是否为 ReactiveWebApplicationContext，如果是，
        // 则返回匹配
        if (context.getResourceLoader() instanceof
          ReactiveWebApplicationContext) {
            return ConditionOutcome
                .match(message.foundExactly
                ("ReactiveWebApplicationContext"));
        }
        // 4. 其他情况，返回不匹配
        return ConditionOutcome
            .noMatch(message.because("not a reactive web application"));
    }
    …
}
```

从上述源代码可以看出，实现类 OnWebApplicationCondition 的 getMatchOutcome 方法返回条件匹配结果。其中，最重要的一步是判断是否是 Web 环境（ConditionOutcome outcome = isWebApplication(context, metadata, required);）。

6.5.2　实例分析

在 6.5.1 节了解了 Spring Boot 的条件注解后，下面通过实例分析 Spring Boot 内置的一个简单的自动配置：HTTP 编码配置。

【例 6-6】分析 HTTP 编码配置。

在 Spring MVC 常规项目中，可以在 web.xml 中配置一个 filter 进行 HTTP 编码设置，例如：

```
<filter>
    <filter-name>characterEncodingFilter</filter-name>
    <filter-class>org.springframework.web.filter.CharacterEncodingFilter</
        filter-class>
    <init-param>
        <param-name>encoding</param-name>
        <param-value>UTF-8</param-value>
    </init-param>
    <init-param>
        <param-name>forceEncoding</param-name>
        <param-value>true</param-value>
    </init-param>
</filter>
<filter-mapping>
    <filter-name>characterEncodingFilter</filter-name>
    <url-pattern>/*</url-pattern>
</filter-mapping>
```

从上述 web.xml 文件中过滤器的配置可知，Spring Boot 自动配置 HTTP 编码需要满足的条件是：配置 CharacterEncodingFilter 的 Bean，并设置 encoding 和 forceEncoding 这两个参数。

通过查看源代码可知，org.springframework.boot.autoconfigure.web.servlet.HttpEncoding-AutoConfiguration 类根据条件注解配置了 CharacterEncodingFilter 的 Bean，并设置了 encoding 和 forceEncoding 这两个参数。代码分析见代码中的注释部分，具体分析如下：

```
package org.springframework.boot.autoconfigure.web.servlet;
…
@Configuration(proxyBeanMethods = false)//配置类
@EnableConfigurationProperties(ServerProperties.class)//开启属性注入，使用
  @Autowired 注入
@ConditionalOnWebApplication(type = ConditionalOnWebApplication.Type.SERVLET)
  /**基于 Servlet 的 Web 应用程序*/
@ConditionalOnClass(CharacterEncodingFilter.class)
  //当 CharacterEncodingFilter 在类路径下
@ConditionalOnProperty(prefix = "server.servlet.encoding", value = "enabled",
  matchIfMissing = true)/**当设置 server.servlet.encoding=enabled 时，如果没有
  该设置默认为 true，即符合条件*/
public class HttpEncodingAutoConfiguration {
    private final Encoding properties;
    public HttpEncodingAutoConfiguration(ServerProperties properties) {
        this.properties = properties.getServlet().getEncoding();
    }
    @Bean//使用 Java 配置的方式配置 CharacterEncodingFilter 的 Bean
    @ConditionalOnMissingBean/**当 Spring 容器中没有 CharacterEncodingFilter 的
      Bean 时，新建该 Bean*/
    public CharacterEncodingFilter characterEncodingFilter() {
        CharacterEncodingFilter filter = new OrderedCharacterEncodingFilter();
        //设置 encoding 参数
        filter.setEncoding(this.properties.getCharset().name());
        //设置 forceEncoding 参数
        filter.setForceRequestEncoding(this.properties.shouldForce
          (Type.REQUEST));
        filter.setForceResponseEncoding(this.properties.shouldForce
          (Type.RESPONSE));
        return filter;
    }
    …
}
```

▶ 6.5.3 自定义条件

视频讲解

Spring 的@Conditional 注解根据满足某特定条件创建一个特定的 Bean。例如，当某 jar 包在类路径下时，自动配置一个或多个 Bean。即根据特定条件控制 Bean 的创建行为，这样就可以利用此特性进行一些自动配置。那么，开发者如何自己构造条件呢？在 Spring 框架中，可以通过实现 Condition 接口，并重写 matches 方法来构造条件。下面通过实例讲解条件的构造过程。

【例 6-7】如果类路径 classpath（src/main/resources）下存在文件 test.properties，则输出“test.properties 文件存在。”，否则输出“test.properties 文件不存在！”

具体实现步骤如下。

❶ 构造条件

在 Spring Boot 应用 ch6_1 的 src/main/java 目录下创建包 com.ch6_1.conditional，并在该包中分别创建条件实现类 MyCondition（存在文件 test.properties）和 YourCondition（不存在文件 test.properties）。

MyCondition 的核心代码如下：

```
public class MyCondition implements Condition{
    @Override
    public boolean matches(ConditionContext context, AnnotatedTypeMetadata
     metadata) {
        return context.getResourceLoader().getResource("classpath:
         test.properties").exists();
    }
}
```

YourCondition 的核心代码如下：

```
public class YourCondition implements Condition{
    @Override
    public boolean matches(ConditionContext context, AnnotatedTypeMetadata
     metadata) {
        return !context.getResourceLoader().getResource("classpath:test.
         properties").exists();
    }
}
```

❷ 创建不同条件下 Bean 的类

在包 com.ch6_1.conditional 中，创建接口 MessagePrint，并分别创建该接口的实现类 MyMessagePrint 和 YourMessagePrint。

MessagePrint 的代码如下：

```
package com.ch6_1.conditional;
public interface MessagePrint {
    public String showMessage();
}
```

MyMessagePrint 的代码如下：

```
package com.ch6_1.conditional;
public class MyMessagePrint implements MessagePrint{
    @Override
    public String showMessage() {
        return "test.properties 文件存在。";
    }
}
```

YourMessagePrint 的代码如下：

```
package com.ch6_1.conditional;
public class YourMessagePrint implements MessagePrint{
    @Override
    public String showMessage() {
```

```
            return "test.properties 文件不存在！";
        }
    }
```

❸ 创建配置类

在包 com.ch6_1.conditional 中，创建配置类 ConditionConfig，并在该配置类中使用@Bean 和@Conditional 实例化符合条件的 Bean。

ConditionConfig 的核心代码如下：

```
@Configuration
public class ConditionConfig {
    @Bean
    @Conditional(MyCondition.class)
    public MessagePrint myMessage() {
        return new MyMessagePrint();
    }
    @Bean
    @Conditional(YourCondition.class)
    public MessagePrint yourMessage() {
        return new YourMessagePrint();
    }
}
```

❹ 创建测试类

在包 com.ch6_1.conditional 中，创建测试类 TestMain，具体代码如下：

```
package com.ch6_1.conditional;
import org.springframework.context.annotation.AnnotationConfigApplicationContext;
public class TestMain {
    private static AnnotationConfigApplicationContext context;
    public static void main(String[] args) {
        context=newAnnotationConfigApplicationContext(ConditionConfig.class);
        MessagePrint mp = context.getBean(MessagePrint.class);
        System.out.println(mp.showMessage());
    }
}
```

❺ 运行

当 Spring Boot 应用 ch6_1 的 src/main/resources 目录下存在 test.properties 文件时，运行测试类，控制台显示“test.properties 文件存在。”；当 Spring Boot 应用 ch6_1 的 src/main/resources 目录下不存在 test.properties 文件时，运行测试类，控制台显示“test.properties 文件不存在！”。

6.5.4 自定义 Starters

视频讲解

从 6.1.6 节可知，第三方为 Spring Boot 贡献了许多 Starters。那么，我们作为开发者是否也可以贡献自己的 Starters？学习 Spring Boot 的自动配置机制后，答案是肯定的。下面通过实例讲解如何自定义 Starters。

【例 6-8】自定义一个 Starters（spring_boot_mystarters）。要求：当类路径中存在 MyService 类时，自动配置该类的 Bean，并可以将相应 Bean 的属性在 application.properties 中配置。

具体实现步骤如下。

❶ 新建 Spring Boot 项目 spring_boot_mystarters

首先，通过选择菜单 File | New | Spring Starter Project 打开 New Spring Starter Project 对话框。然后，在对话框中输入项目名称“spring_boot_mystarters”。最后，单击 Next 与 Finish 按钮。

❷ 修改 pom 文件

修改 Spring Boot 项目 spring_boot_mystarters 的 pom 文件，增加 Spring Boot 自身的自动配置作为依赖，代码如下：

```
<dependency>
    <groupId>org.springframework.boot</groupId>
    <artifactId>spring-boot-autoconfigure</artifactId>
</dependency>
```

❸ 创建属性配置类 MyProperties

在项目 spring_boot_mystarters 的 com.ch.spring_boot_mystarters 包中，创建属性配置类 MyProperties。在使用 spring_boot_mystarters 的 Spring Boot 项目的 application.properties 中，可以使用 my.msg=设置属性；若不设置，则 my.msg 为默认值。属性配置类 MyProperties 的代码如下：

```
package com.ch.spring_boot_mystarters;
import org.springframework.boot.context.properties.ConfigurationProperties;
//在 application.properties 中通过 my.msg=设置属性
@ConfigurationProperties(prefix="my")
public class MyProperties {
    private String msg = "默认值";
    public String getMsg() {
        return msg;
    }
    public void setMsg(String msg) {
        this.msg = msg;
    }
}
```

❹ 创建判断依据类 MyService

在项目 spring_boot_mystarters 的 com.ch.spring_boot_mystarters 包中，创建判断依据类 MyService。本例自定义的 Starters 将根据该类的存在与否来创建该类的 Bean，该类可以是第三方类库的类。判断依据类 MyService 的代码如下：

```
package com.ch.spring_boot_mystarters;
public class MyService {
    private String msg;
    public String sayMsg() {
        return "my " + msg;
    }
    public String getMsg() {
        return msg;
    }
    public void setMsg(String msg) {
        this.msg = msg;
    }
}
```

❺ 创建自动配置类 MyAutoConfiguration

在项目 spring_boot_mystarters 的 com.ch.spring_boot_mystarters 包中，创建自动配置类 MyAutoConfiguration。在该类中使用@EnableConfigurationProperties 注解开启属性配置类 MyProperties 提供参数；使用@ConditionalOnClass 注解判断类加载器（类路径）中是否存在 MyService 类；使用@ConditionalOnMissingBean 注解判断容器中是否存在 MyService 的 Bean，如果不存在，自动配置这个 Bean。自动配置类 MyAutoConfiguration 的核心代码如下：

```
@Configuration
//开启属性配置类 MyProperties 提供参数
@EnableConfigurationProperties(MyProperties.class)
//类加载器（类路径）中是否存在对应的类
@ConditionalOnClass(MyService.class)
//应用环境中属性是否存在指定的值
@ConditionalOnProperty(prefix = "my", value = "enabled", matchIfMissing = true)
public class MyAutoConfiguration {
    @Autowired
    private MyProperties myProperties;
    @Bean
    //当容器中不存在 MyService 的 Bean 时，自动配置这个 Bean
    @ConditionalOnMissingBean(MyService.class)
    public MyService myService() {
        MyService myService = new MyService();
        myService.setMsg(myProperties.getMsg());
        return myService;
    }
}
```

❻ 注册配置

在项目 spring_boot_mystarters 的 src/main/resources 目录下新建文件夹 META-INF，并在该文件夹下创建名为 spring.factories 的文件。在 spring.factories 文件中添加如下内容注册自动配置类 MyAutoConfiguration：

```
org.springframework.boot.autoconfigure.EnableAutoConfiguration=\
com.ch.spring_boot_mystarters.MyAutoConfiguration
```

上述文件内容中，若有多个自动配置，则使用“,”分开，此处“\”是为了换行后仍然能读到属性值。

至此，经过上述 6 个步骤后，自定义 Starters（spring_boot_mystarters）已经完成。可以将 spring_boot_mystarters 安装到 Maven 的本地库，或者将 jar 包发布到 Maven 的私服上。

【例 6-9】创建 Spring Boot 的 Web 应用 ch6_2，并在 ch6_2 中使用 spring_boot_mystarters。具体实现步骤如下。

❶ 创建 Spring 的 Web 应用 ch6_2

使用 STS 快速创建 Spring 的 Web 应用 ch6_2。

❷ 添加 spring_boot_mystarters 的依赖

在 Web 应用 ch6_2 的 pom.xml 文件中添加 spring_boot_mystarters 的依赖，代码如下：

```
<dependency>
    <groupId>com.ch</groupId>
```

```
        <artifactId>spring_boot_mystarters</artifactId>
        <version>0.0.1-SNAPSHOT</version>
    </dependency>
```

添加依赖后，可以在 Maven 的依赖里查看到 spring_boot_mystarters 依赖。

❸ 修改程序入口类 Ch62Application，测试 spring_boot_mystarters

类 Ch62Application 修改后的核心代码如下：

```
@RestController
@SpringBootApplication
public class Ch62Application {
    @Autowired MyService myService;
    public static void main(String[] args) {
        SpringApplication.run(Ch62Application.class, args);
    }
    @RequestMapping("/testStarters")
    public String index() {
        return myService.sayMsg();
    }
}
```

运行 Ch62Application 应用程序，启动 Web 应用。通过访问 http://localhost:8080/testStarters 测试 spring_boot_mystarters，运行效果如图 6.5 所示。

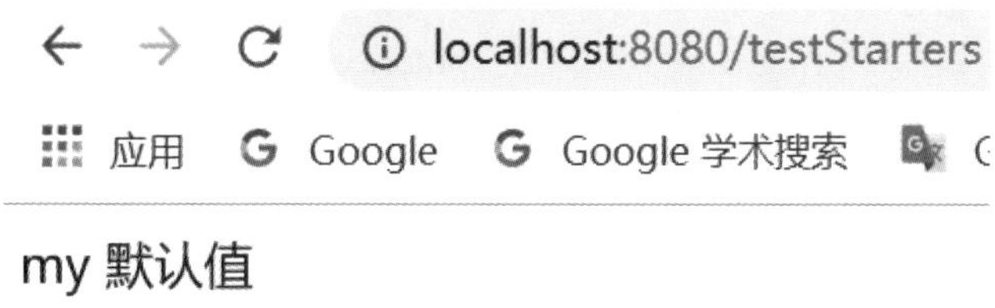

图 6.5　访问 http://localhost:8080/testStarters

这时，在 Web 应用 ch6_2 的 application.properties 文件中配置 msg 的内容：

```
my.msg=starter pom
```

然后，运行 Ch62Application 应用程序，重新启动 Web 应用。再次访问 http://localhost:8080/testStarters，运行效果如图 6.6 所示。

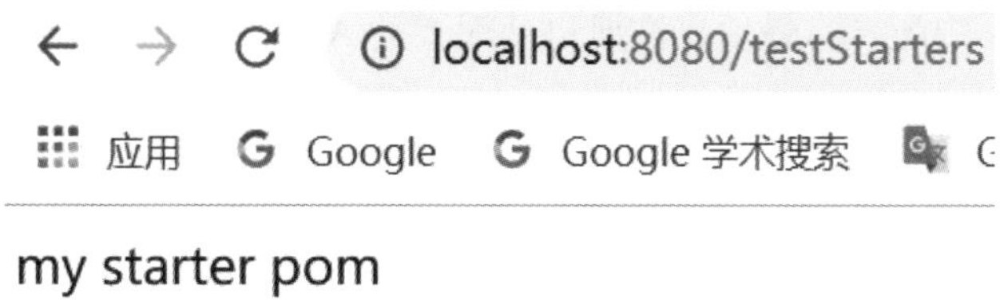

图 6.6　配置 msg 后的效果

另外，可以在 Web 应用 ch6_2 的 application.properties 文件中配置 debug 属性（debug=true），查看自动配置报告。重新启动 Web 应用，可以在控制台中查看到如图 6.7 所示的自定义的自动配置。

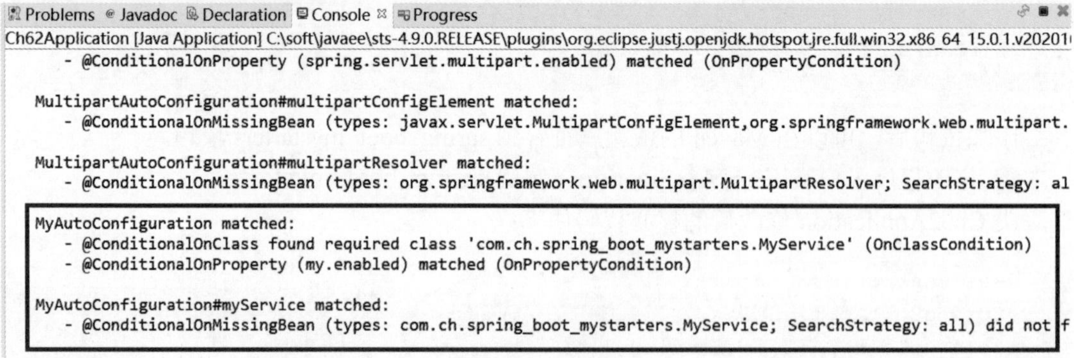

```
Problems  Javadoc  Declaration  Console  Progress
Ch62Application [Java Application] C:\soft\javaee\sts-4.9.0.RELEASE\plugins\org.eclipse.justj.openjdk.hotspot.jre.full.win32.x86_64_15.0.1.v20201
      - @ConditionalOnProperty (spring.servlet.multipart.enabled) matched (OnPropertyCondition)

   MultipartAutoConfiguration#multipartConfigElement matched:
      - @ConditionalOnMissingBean (types: javax.servlet.MultipartConfigElement,org.springframework.web.multipart.

   MultipartAutoConfiguration#multipartResolver matched:
      - @ConditionalOnMissingBean (types: org.springframework.web.multipart.MultipartResolver; SearchStrategy: al

   MyAutoConfiguration matched:
      - @ConditionalOnClass found required class 'com.ch.spring_boot_mystarters.MyService' (OnClassCondition)
      - @ConditionalOnProperty (my.enabled) matched (OnPropertyCondition)

   MyAutoConfiguration#myService matched:
      - @ConditionalOnMissingBean (types: com.ch.spring_boot_mystarters.MyService; SearchStrategy: all) did not f
```

图 6.7　查看自动配置报告

6.6　本章小结

本章重点讲解了 Spring Boot 的基本配置、自动配置原理以及条件注解。通过本章的学习，开发者可以利用 Spring Boot 的自动配置与条件注解贡献自己的 Starters。

习 题 6

1．如何读取 Spring Boot 项目的应用配置？请举例说明。

2．参考例 6-7，编写 Spring Boot 应用程序 practice6_2。要求如下：以不同的操作系统作为条件，若在 Windows 系统下运行程序，则输出列表命令为 dir；若在 Linux 操作系统下运行程序，则输出列表命令为 ls。

3．参考例 6-8 与例 6-9，自定义一个 Starter（spring_boot_addstarters）和 Spring Boot 的 Web 应用 practice6_3。在 practice6_3 中，使用 spring_boot_addstarters 计算两个整数的和，通过访问 http://localhost:8080/testAddStarters 返回两个整数的和。在 spring_boot_addstarters 中，首先，创建属性配置类 AddProperties（有 Integer 类型的 number1 与 number2 两个属性），在该属性配置类中使用@ConfigurationProperties(prefix="add")注解设置属性前缀为 add；然后，创建判断依据类 AddService（有 Integer 类型的 number1 与 number2 两个属性），在 AddService 类中提供 add 方法（计算 number1 与 number2 的和）；再然后，创建自动配置类 AddAutoConfiguration，当类路径中存在 AddService 类时，自动配置该类的 Bean，并可以将相应 Bean 的属性在 application.properties 中配置；最后，注册自动配置类 AddAutoConfiguration。

第 7 章 Spring Boot 的 Web 开发

学习目的与要求

本章首先介绍 Spring Boot 的 Web 开发支持，然后介绍 Thymeleaf 视图模板引擎技术，最后介绍 Spring Boot 的 Web 开发技术（JSON 数据交互、文件上传与下载、异常统一处理以及对 JSP 的支持）。通过本章的学习，掌握 Spring Boot 的 Web 开发技术。

主要内容

- Thymeleaf 模板引擎
- Spring Boot 处理 JSON 数据
- Spring Boot 的文件上传与下载
- Spring Boot 的异常处理

Web 开发是一种基于 B/S 架构（即浏览器/服务器）的应用软件开发技术，分为前端（用户接口）和后端（业务逻辑和数据），前端的可视化及用户交互由浏览器实现，即通过浏览器作为客户端，实现客户与服务器远程的数据交互。Spring Boot 的 Web 开发内容主要包括内嵌 Servlet 容器和 Spring MVC。

7.1 Spring Boot 的 Web 开发支持

Spring Boot 提供了 spring-boot-starter-web 依赖模块，该依赖模块包含了 Spring Boot 预定义的 Web 开发常用依赖包，为 Web 开发者提供了内嵌的 Servlet 容器（Tomcat）以及 Spring MVC 的依赖。如果开发者希望开发 Spring Boot 的 Web 应用程序，可以在 Spring Boot 项目的 pom.xml 文件中添加如下依赖配置：

```
<dependency>
    <groupId>org.springframework.boot</groupId>
    <artifactId>spring-boot-starter-web</artifactId>
</dependency>
```

Spring Boot 将自动关联 Web 开发的相关依赖，如 tomcat、spring-webmvc 等，进而对 Web 开发提供支持，并对相关技术实现自动配置。

另外，开发者也可以使用 Spring Tool Suite 集成开发工具快速创建 Spring Starter Project，在 New Spring Starter Project Dependencies 窗口中添加 Spring Boot 的 Web 依赖。

7.2 Thymeleaf 模板引擎

在 Spring Boot 的 Web 应用中，建议开发者使用 HTML 完成动态页面。Spring Boot 提供了许多模板引擎，主要包括 FreeMarker、Groovy、Thymeleaf、Velocity 和 Mustache。因为

Thymeleaf 提供了完美的 Spring MVC 支持，所以在 Spring Boot 的 Web 应用中推荐使用 Thymeleaf 作为模板引擎。

Thymeleaf是一个 Java 类库，是一个 xml/xhtml/html5 的模板引擎，能够处理 HTML、XML、JavaScript 以及 CSS，可以作为 MVC Web 应用的 View 层显示数据。

7.2.1 Spring Boot 的 Thymeleaf 支持

在 Spring Boot 1.X 版本中，spring-boot-starter-thymeleaf 依赖包含了 spring-boot-starter-web 模块。但是，在 Spring 5 中，WebFlux 的出现对于 Web 应用的解决方案将不再唯一。所以，spring-boot-starter-thymeleaf 依赖不再包含 spring-boot-starter-web 模块，需要开发人员自己选择 spring-boot-starter-web 模块依赖。下面通过一个实例，讲解如何创建基于 Thymeleaf 模板引擎的 Spring Boot Web 应用 ch7_1。

【例 7-1】 创建基于 Thymeleaf 模板引擎的 Spring Boot Web 应用 ch7_1。

具体实现步骤如下。

❶ 创建 Spring Starter Project

选择菜单 File | New | Spring Starter Project，打开 New Spring Starter Project 对话框，在该对话框中选择和输入相关信息后，单击 Next 按钮，打开 New Spring Starter Project Dependencies 对话框。

❷ 选择依赖

在 New Spring Starter Project Dependencies 对话框中，选择 Thymeleaf 和 Spring Web 依赖，如图 7.1 所示。

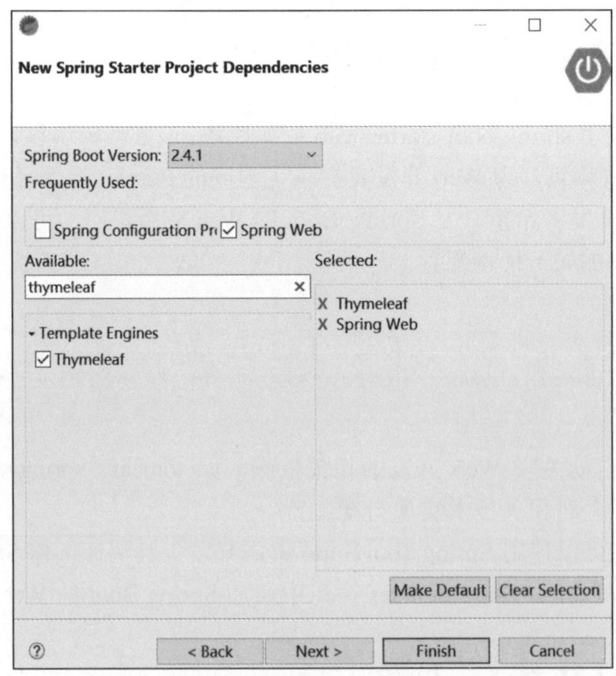

图 7.1 选择 Thymeleaf 和 Spring Web 依赖

❸ 打开项目目录

单击图 7.1 中的 Finish 按钮，创建如图 7.2 所示的基于 Thymeleaf 模板引擎的 Spring Boot

Web 应用 ch7_1。

```
ch7_1 [boot]
  > src/main/java
  v src/main/resources
      static
      templates
      application.properties
  > src/test/java
  > JRE System Library [JavaSE-15]
  > Maven Dependencies
  > src
    target
    HELP.md
    mvnw
    mvnw.cmd
    pom.xml
```

图 7.2　基于 Thymeleaf 模板引擎的 Spring Boot Web 应用 ch7_1

Thymeleaf 模板将 JS 脚本、CSS 样式、图片等静态文件默认放置在 src/main/resources/static 目录下，将视图页面放在 src/main/resources/templates 目录下。

❹ 创建控制器类

创建一个名为 com.ch.ch7_1.controller 的包，并在该包中创建控制器类 TestThymeleafController，核心代码如下：

```
@Controller
public class TestThymeleafController {
    @RequestMapping("/")
    public String test(){
        //根据 Thymeleaf 模板，默认将返回 src/main/resources/templates/index.html
        return "index";
    }
}
```

❺ 新建 index.html 页面

在 src/main/resources/templates 目录下新建 index.html 页面，代码如下：

```
<body>
测试 Spring Boot 的 Thymeleaf 支持
</body>
```

❻ 设置 Web 应用的上下文路径

在 src/main/resources/application.properties 文件中配置如下内容：

```
server.servlet.context-path=/ch7_1
```

❼ 运行测试

首先，运行 Ch71Application 主类。然后，访问 http://localhost:8080/ch7_1 测试程序。

▶ 7.2.2　Thymeleaf 基础语法

视频讲解

❶ 引入 Thymeleaf

首先，将 View 层页面文件的 html 标签修改如下：

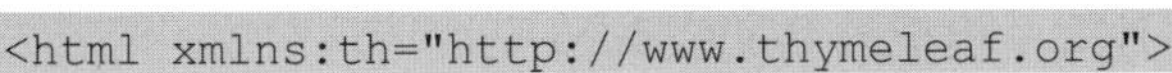

```
<html xmlns:th="http://www.thymeleaf.org">
```

然后，在 View 层页面文件的其他标签里使用 th:*动态处理页面。示例代码如下：

```
<img th:src="'images/' + ${aBook.picture}"/>
```

其中，${aBook.picture}获得数据对象 aBook 的 picture 属性。

❷ 输出内容

使用 th:text 和 th:utext（不对文本转义，正常输出）将文本内容输出到所在标签的 body 中。假如在国际化资源文件 messages_en_US.properties 中有消息文本“test.myText=<strong>Test International Message</strong>”，那么在页面中可以使用如下两种方式获得消息文本：

```
<p th:text="#{test.myText}"></p>
<!-- 对文本转义，即输出<strong>Test International Message</strong> -->
<p th:utext="#{test.myText}"></p>
<!-- 对文本不转义，即输出加粗的“Test International Message” -->
```

❸ 基本表达式

1）变量表达式：${...}

用于访问容器上下文环境中的变量，示例代码如下：

```
<span th:text="${information}">
```

2）选择变量表达式：*{...}

选择变量表达式计算的是选定的对象（th:object 属性绑定的对象），示例代码如下：

```
<div th:object="${session.user}">
    name: <span th: text="*{firstName}"></span><br>
    <!-- firstName 为 user 对象的属性-->
    surname: <span th: text="*{lastName}"></span><br>
    nationality: <span th: text="*{nationality}"></span><br>
</div>
```

3）信息表达式：#{...}

一般用于显示页面静态文本。将可能需要根据需求而整体变动的静态文本放在 properties 文件中以便维护（如国际化）。通常与 th:text 属性一起使用，示例代码如下：

```
<p th:text="#{test.myText}"></p>
```

❹ 引入 URL

Thymeleaf 模板通过@{...}表达式引入 URL，示例代码如下：

```
<!-- 默认访问 src/main/resources/static 下的 css 文件夹-->
<link rel="stylesheet" th:href="@{css/bootstrap.min.css}" />
<!--访问相对路径-->
<a th:href="@{/}">去看看</a>
<!--访问绝对路径-->
<a th:href="@{http://www.tup.tsinghua.edu.cn/index.html(param1='传参')}">去
清华大学出版社</a>
<!-- 默认访问 src/main/resources/static 下的'images 文件夹-->
<img th:src="'images/' + ${aBook.picture}"/>
```

❺ 访问 WebContext 对象中的属性

Thymeleaf 模板通过一些专门的表达式从模板的 WebContext 获取请求参数、请求、会话和应用程序中的属性，具体如下。

${xxx}：返回存储在 Thymeleaf 模板上下文中的变量 xxx 或请求 request 作用域中的属性 xxx。

${param.xxx}：返回一个名为 xxx 的请求参数（可能是多个值）。

${session.xxx}：返回一个名为 xxx 的 HttpSession 作用域中的属性。

${application.xxx}：返回一个名为 xxx 的全局 ServletContext 上下文作用域中的属性。

与 EL 表达式一样，使用${xxx}获得变量值，使用${对象变量名.属性名}获取 JavaBean 属性值。但需要注意的是，${}表达式只能在 th 标签内部有效。

❻ 运算符

在 Thymeleaf 模板的表达式中可以使用+、–、*、/、%等各种算术运算符，也可以使用>、<、<=、>=、==、!=等各种逻辑运算符。示例代码如下：

```
<tr th:class="(${row}== 'even')? 'even' : 'odd'">...</tr>
```

❼ 条件判断

1）if 和 unless

标签只有在 th:if 条件成立时才显示；th:unless 与 th:if 相反，只有条件不成立时，才显示标签内容。示例代码如下：

```
<a href="success.html" th:if="${user != null}">成功</a>
<a href="success.html" th:unless="${user = null}">成功</a>
```

2）switch 语句

Thymeleaf 模板也支持多路选择 switch 语句结构，默认属性 default 可用“*”表示。示例代码如下：

```
<div th:switch="${user.role}">
    <p th:case="'admin'">User is an administrator</p>
    <p th:case="'teacher'">User is a teacher</p>
    <p th:case="*">User is a student </p>
</div>
```

❽ 循环

1）基本循环

Thymeleaf 模板使用 th:each="obj,iterStat:${objList}"标签进行迭代循环，迭代对象可以是 java.util.List、java.util.Map 或数组等。示例代码如下：

```
<!-- 循环取出集合数据 -->
<div class="col-md-4 col-sm-6" th:each="book:${books}">
    <a href="">
        <img th:src="'images/'+${book.picture}" alt="图书封面" style="height:
         180px; width: 40%;"/>
    </a>
    <div class="caption">
        <h4 th:text="${book.bname}"></h4>
        <p th:text="${book.author}"></p>
        <p th:text="${book.isbn}"></p>
        <p th:text="${book.price}"></p>
```

```
        <p th:text="${book.publishing}"></p>
    </div>
</div>
```

2）循环状态的使用

在 th:each 标签中可以使用循环状态变量，该变量有如下属性。

index：当前迭代对象的 index（从 0 开始计数）。

count：当前迭代对象的 index（从 1 开始计数）。

size：迭代对象的大小。

current：当前迭代变量。

even/odd：布尔值，当前循环是否是偶数/奇数（从 0 开始计数）。

first：布尔值，当前循环是否是第一个。

last：布尔值，当前循环是否是最后一个。

使用循环状态变量的示例代码如下：

```
<!-- 循环取出集合数据 -->
<div class="col-md-4 col-sm-6" th:each="book,bookStat:${books}">
    <a href="">
        <img th:src="'images/'+${book.picture}" alt="图书封面" style="height:
         180px; width: 40%;"/>
    </a>
    <div class="caption">
        <!--循环状态 bookStat-->
        <h3 th:text="${bookStat.count}"></h3>
        <h4 th:text="${book.bname}"></h4>
        <p th:text="${book.author}"></p>
        <p th:text="${book.isbn}"></p>
        <p th:text="${book.price}"></p>
        <p th:text="${book.publishing}"></p>
    </div>
</div>
```

⑨ 内置对象

在实际 Web 项目开发中，经常传递列表、日期等数据。所以，Thymeleaf 模板提供了很多内置对象，可以通过#直接访问。这些内置对象一般都以“s”结尾，如 dates、lists、numbers、strings 等。Thymeleaf 模板通过${#...}表达式访问内置对象。常见的内置对象如下。

#dates：日期格式化的内置对象，操作的方法是 java.util.Date 类的方法。

#calendars：类似于#dates，但操作的方法是 java.util.Calendar 类的方法。

#numbers：数字格式化的内置对象。

#strings：字符串格式化的内置对象，操作的方法参照 java.lang.String。

#objects：参照 java.lang.Object。

#bools：判断 boolean 类型的内置对象。

#arrays：数组操作的内置对象。

#lists：列表操作的内置对象，参照 java.util.List。

#sets：Set 操作的内置对象，参照 java.util.Set。

#maps：Map 操作的内置对象，参照 java.util.Map。

#aggregates：创建数组或集合的聚合的内置对象。

#messages：在变量表达式内部获取外部消息的内置对象。

假如有如下控制器方法：

```
@RequestMapping("/testObject")
public String testObject(Model model) {
    //系统时间 new Date()
    model.addAttribute("nowTime", new Date());
    //系统日历对象
    model.addAttribute("nowCalendar", Calendar.getInstance());
    //创建 BigDecimal 对象
    BigDecimal money = new BigDecimal(2019.613);
    model.addAttribute("myMoney", money);
    //字符串
    String tsts = "Test strings";
    model.addAttribute("str", tsts);
    //boolean 类型
    boolean b = false;
    model.addAttribute("bool", b);
    //数组（这里不能使用 int 定义数组）
    Integer aint[] = {1,2,3,4,5};
    model.addAttribute("mya", aint);
    //List 列表 1
    List<String> nameList1 = new ArrayList<String>();
    nameList1.add("陈恒 1");
    nameList1.add("陈恒 3");
    nameList1.add("陈恒 2");
    model.addAttribute("myList1", nameList1);
    //Set 集合
    Set<String> st = new HashSet<String>();
    st.add("set1");
    st.add("set2");
    model.addAttribute("mySet", st);
    //Map 集合
    Map<String, Object> map = new HashMap<String, Object>();
    map.put("key1", "value1");
    map.put("key2", "value2");
    model.addAttribute("myMap", map);
    //List 列表 2
    List<String> nameList2 = new ArrayList<String>();
    nameList2.add("陈恒 6");
    nameList2.add("陈恒 5");
    nameList2.add("陈恒 4");
    model.addAttribute("myList2", nameList2);
    return "showObject";
}
```

那么，可以在 src/main/resources/templates/showObject.html 视图页面文件中，使用内置对象操作数据。showObject.html 的核心代码如下：

```
<body>
    格式化控制器传递过来的系统时间 nowTime
```

```
        <span th:text="${#dates.format(nowTime, 'yyyy/MM/dd')}"></span>
        <br>
        创建一个日期对象
        <span th:text="${#dates.create(2019,6,13)}"></span>
        <br>
        格式化控制器传递过来的系统日历 nowCalendar:
        <span th:text="${#calendars.format(nowCalendar, 'yyyy-MM-dd')}"></span>
        <br>
        格式化控制器传递过来的 BigDecimal 对象 myMoney:
        <span th:text="${#numbers.formatInteger(myMoney,3)}"></span>
        <br>
        计算控制器传递过来的字符串 str 的长度:
        <span th:text="${#strings.length(str)}"></span>
        <br>
        返回对象，当控制器传递过来的 BigDecimal 对象 myMoney 为空时，返回默认值 9999:
        <span th:text="${#objects.nullSafe(myMoney, 9999)}"></span>
        <br>
        判断 boolean 数据是否是 false:
        <span th:text="${#bools.isFalse(bool)}"></span>
        <br>
        判断数组 mya 中是否包含元素 5:
        <span th:text="${#arrays.contains(mya, 5)}"></span>
        <br>
        排序列表 myList1 的数据:
        <span th:text="${#lists.sort(myList1)}"></span>
        <br>
        判断集合 mySet 中是否包含元素 set2:
        <span th:text="${#sets.contains(mySet, 'set2')}"></span>
        <br>
        判断 myMap 中是否包含 key1 关键字:
        <span th:text="${#maps.containsKey(myMap, 'key1')}"></span>
        <br>
        将数组 mya 中的元素求和:
        <span th:text="${#aggregates.sum(mya)}"></span>
        <br>
        将数组 mya 中的元素求平均:
        <span th:text="${#aggregates.avg(mya)}"></span>
        <br>
        如果未找到消息，则返回默认消息（如"??msgKey_zh_CN??"）:
        <span th:text="${#messages.msg('msgKey')}"></span>
    </body>
```

视频讲解

▶ 7.2.3 Thymeleaf 的常用属性

通过 7.2.2 节的学习，得知在 html 页面的标签中添加 th:xxx 关键字实现 Thymeleaf 模板套用，且 Thymeleaf 的标签属性与 html 页面的标签属性基本类似。常用属性如下。

❶ th:action

th:action 定义后台控制器路径，类似于<form>标签的 action 属性。示例代码如下：

```
<form th:action="@{/login}">...</form>
```

❷ th:each

集合对象遍历，功能类似于 JSTL 标签<c:forEach>。示例代码如下：

```
<div class="col-md-4 col-sm-6" th:each="gtype:${gtypes}">
    <div class="caption">
        <p th:text="${gtype.id}"></p>
        <p th:text="${gtype.typename}"></p>
    </div>
</div>
```

❸ th:field

th:field 常用于表单参数绑定，通常与 th:object 一起使用。示例代码如下：

```
<form th:action="@{/login}" th:object="${user}">
    <input type="text" value="" th:field="*{username}"></input>
    <input type="text" value="" th:field="*{role}"></input>
</form>
```

❹ th:href

th:href 定义超链接，类似于<a>标签的 href 属性。value 形式为@{/logout}，示例代码如下：

```
<a th:href="@{/gogo}"></a>
```

❺ th:id

th:id 用于 div 的 id 声明，类似于 html 标签中的 id 属性。示例代码如下：

```
<div th:id ="stu+(${rowStat.index}+1)"></div>
```

❻ th:if

th:if 用于条件判断。如果为否则标签不显示，示例代码如下：

```
<div th:if="${rowStat.index} == 0">... do something ...</div>
```

❼ th:fragment

th:fragment 声明定义该属性的 div 为模板片段，常用于头文件、页尾文件的引入。常与 th:include、th:replace 一起使用。

假如在 ch7_1 的 src/main/resources/templates 目录下声明模板片段文件 footer.html，核心代码如下：

```
<body>
    <!-- 声明片段 content -->
    <div th:fragment="content" >
        主体内容
    </div>
    <!-- 声明片段 copy -->
    <div th:fragment="copy" >
        ©清华大学出版社
    </div>
</body>
```

那么，可以在 ch7_1 的 src/main/resources/templates/index.html 文件中引入模板片段，核心代码如下：

```
<body>
    测试 Spring Boot 的 Thymeleaf 支持<br>
    引入主体内容模板片段：
    <div th:include="footer::content"></div>
    引入版权所有模板片段：
    <div th:replace="footer::copy" ></div>
</body>
```

❽ th:object

th:object 用于表单数据对象绑定，将表单绑定到后台 controller 的一个 JavaBean 参数。th:object 常与 th:field 一起使用，进行表单数据绑定。下面通过实例讲解表单提交及数据绑定的实现过程。

【例 7-2】表单提交及数据绑定的实现过程。

具体实现步骤如下。

1）创建实体类

在 Web 应用 ch7_1 的 src/main/java 目录下，创建 com.ch.ch7_1.model 包，并在该包中创建实体类 LoginBean，代码如下：

```
package com.ch.ch7_1.model;
public class LoginBean {
    String uname;
    String urole;
    //省略 set 方法和 get 方法
}
```

2）创建控制器类

在 Web 应用 ch7_1 的 com.ch.ch7_1.controller 包中，创建控制器类 LoginController，核心代码如下：

```
@Controller
public class LoginController {
    @RequestMapping("/toLogin")
    public String toLogin(Model model) {
        /*loginBean 与 login.html 页面中的 th:object="${loginBean}"相同,类似于
          Spring MVC 的表单绑定*/
        model.addAttribute("loginBean", new LoginBean());
        return "login";
    }
    @RequestMapping("/login")
     public String greetingSubmit(@ModelAttribute LoginBean loginBean) {
        /*@ModelAttribute LoginBean loginBean 接收 login.html 页面中的表单数据,
          并将 loginBean 对象保存到 model 中返回给 result.html 页面显示*/
        System.out.println("测试提交的数据：" + loginBean.getUname());
        return "result";
    }
}
```

3）创建页面表示层

在 Web 应用 ch7_1 的 src/main/resources/templates 目录下，创建页面 login.html 和 result.html。

页面 login.html 的核心代码如下：

```
<body>
    <h1>Form</h1>
    <form action="#" th:action="@{/login}" th:object="${loginBean}" method=
      "post">
    <!--th:field="*{uname}"的 uname 与实体类的属性相同，即绑定 loginBean 对象 -->
        <p>Uname: <input type="text" th:field="*{uname}" th:placeholder=
          "请输入用户名" /></p>
        <p>Urole: <input type="text" th:field="*{urole}" th:placeholder=
          "请输入角色" /></p>
        <p><input type="submit" value="Submit" /> <input type="reset" value=
          "Reset" /></p>
    </form>
</body>
```

页面 result.html 的核心代码如下：

```
<body>
    <h1>Result</h1>
    <p th:text="'Uname: ' + ${loginBean.uname}" />
    <p th:text="'Urole: ' + ${loginBean.urole}" />
    <a href="toLogin">继续提交</a>
</body>
```

4）运行

首先，运行 Ch71Application 主类。然后，访问 http://localhost:8080/ch7_1/toLogin，运行结果如图 7.3 所示。

在图 7.3 的文本框中输入信息后，单击 Submit 按钮，打开如图 7.4 所示的页面。

Form

Uname: 请输入用户名

Urole: 请输入角色

Submit Reset

图 7.3 页面 login.html 的运行结果

Result

Uname: 陈恒

Urole: 教师

继续提交

图 7.4 页面 result.html 的运行结果

⑨ th:src

th:src 用于外部资源引入，类似于<script>标签的 src 属性。示例代码如下：

```
<img th:src="'images/' + ${aBook.picture}" />
```

⑩ th:text

th:text 用于文本显示，将文本内容显示到所在标签的 body 中。示例代码如下：

```
<td th:text="${username}"></td>
```

⑪ th:value

th:value 用于标签赋值，类似于标签的 value 属性。示例代码如下：

```
<option th:value="Adult">Adult</option>
<input type="hidden" th:value="${msg}" />
```

⑫ th:style

th:style 用于修改标签 style，示例代码如下：

```
<span th:style="'display:' + @{(${myVar} ? 'none' : 'inline-block')}"> myVar
 是一个变量</span>
```

⑬ th:onclick

th:onclick 用于修改点击事件，示例代码如下：

```
<button th:onclick="'getCollect()'"></button>
```

视频讲解

7.2.4 Spring Boot 与 Thymeleaf 实现页面信息国际化

在 Spring Boot 的 Web 应用中实现页面信息国际化非常简单，下面通过实例讲解国际化的实现过程。

【例 7-3】国际化的实现过程。

具体实现步骤如下。

❶ 编写国际化资源属性文件

1）编写管理员模块的国际化信息

在 ch7_1 的 src/main/resources 目录下创建 i18n/admin 文件夹，并在该文件夹下创建 adminMessages.properties、adminMessages_en_US.properties 和 adminMessages_zh_CN.properties 资源属性文件。adminMessages.properties 表示默认加载的信息；adminMessages_en_US.properties 表示英文信息（en 代表语言代码，US 代表国家或地区）；adminMessages_zh_CN.properties 表示中文信息。

adminMessages.properties 的内容如下：

```
test.admin=\u6D4B\u8BD5\u540E\u53F0
admin=\u540E\u53F0\u9875\u9762
```

adminMessages_en_US.properties 的内容如下：

```
test.admin=test admin
admin=admin
```

adminMessages_zh_CN.properties 的内容如下：

```
test.admin=\u6D4B\u8BD5\u540E\u53F0
admin=\u540E\u53F0\u9875\u9762
```

2）编写用户模块的国际化信息

在 ch7_1 的 src/main/resources 目录下创建 i18n/before 文件夹，并在该文件夹下创建 beforeMessages.properties、beforeMessages_en_US.properties 和 beforeMessages_zh_CN.properties 资源属性文件。

beforeMessages.properties 的内容如下：

```
test.before=\u6D4B\u8BD5\u524D\u53F0
before=\u524D\u53F0\u9875\u9762
```

beforeMessages_en_US.properties 的内容如下：

```
test.before=test before
before=before
```

beforeMessages_zh_CN.properties 的内容如下：

```
test.before=\u6D4B\u8BD5\u524D\u53F0
before=\u524D\u53F0\u9875\u9762
```

3）编写公共模块的国际化信息

在 ch7_1 的 src/main/resources 目录下创建 i18n/common 文件夹，并在该文件夹下创建 commonMessages.properties、commonMessages_en_US.properties 和 commonMessages_zh_CN.properties 资源属性文件。

commonMessages.properties 的内容如下：

```
chinese.key=\u4E2D\u6587\u7248
english.key=\u82F1\u6587\u7248
return=\u8FD4\u56DE\u9996\u9875
```

commonMessages_en_US.properties 的内容如下：

```
chinese.key=chinese
english.key=english
return=return
```

commonMessages_zh_CN.properties 的内容如下：

```
chinese.key=\u4E2D\u6587\u7248
english.key=\u82F1\u6587\u7248
return=\u8FD4\u56DE\u9996\u9875
```

❷ 添加配置文件内容，引入资源属性文件

在 ch7_1 应用的配置文件 application.properties 中添加如下内容，引入资源属性文件。

```
spring.messages.basename=i18n/admin/adminMessages,i18n/before/beforeMessages,
i18n/common/commonMessages
```

❸ 重写 localeResolver 方法配置语言区域选择

在 ch7_1 应用的 com.ch.ch7_1 包中，创建配置类 LocaleConfig，该配置类实现 WebMvcConfigurer 接口，并配置语言区域选择。LocaleConfig 的核心代码如下：

```
@Configuration
@EnableAutoConfiguration
public class LocaleConfig implements WebMvcConfigurer {
    /**
     *根据用户本次会话过程中的语义设定语言区域
     *（如用户进入首页时选择的语言种类）
     * @return
     */
    @Bean
    public LocaleResolver localeResolver() {
        SessionLocaleResolver slr = new SessionLocaleResolver();
        //默认语言
        slr.setDefaultLocale(Locale.CHINA);
        return slr;
    }
    /**
     * 使用 SessionLocaleResolver 存储语言区域时，
```

```
     * 必须配置 localeChangeInterceptor 拦截器
     * @return
     */
    @Bean
    public LocaleChangeInterceptor localeChangeInterceptor() {
        LocaleChangeInterceptor lci = new LocaleChangeInterceptor();
        //选择语言的参数名
        lci.setParamName("locale");
        return lci;
    }
    /**
    * 注册拦截器
    */
    @Override
    public void addInterceptors(InterceptorRegistry registry) {
        registry.addInterceptor(localeChangeInterceptor());
    }
}
```

❹ 创建控制器类 I18nTestController

在 ch7_1 应用的 com.ch.ch7_1.controller 包中，创建控制器类 I18nTestController。具体代码如下：

```
package com.ch.ch7_1.controller;
import org.springframework.stereotype.Controller;
import org.springframework.web.bind.annotation.RequestMapping;
@Controller
@RequestMapping("/i18n")
public class I18nTestController {
    @RequestMapping("/first")
    public String testI18n(){
        return "/i18n/first";
    }
    @RequestMapping("/admin")
    public String admin(){
        return "/i18n/admin";
    }
    @RequestMapping("/before")
    public String before(){
        return "/i18n/before";
    }
}
```

❺ 创建视图页面，并获得国际化信息

在 ch7_1 应用的 src/main/resources/templates 目录下，创建文件夹 i18n，并在该文件夹中创建 admin.html、before.html 和 first.html 视图页面，并在这些视图页面中使用 th:text="#{xxx}" 获得国际化信息。

admin.html 的核心代码如下：

```
<body>
    <span th:text="#{admin}"></span><br>
```

```
    <a th:href="@{/i18n/first}" th:text="#{return}"></a>
</body>
```

before.html 的核心代码如下：

```
<body>
    <span th:text="#{before}"></span><br>
    <a th:href="@{/i18n/first}" th:text="#{return}"></a>
</body>
```

first.html 的核心代码如下：

```
<body>
    <a th:href="@{/i18n/first(locale='zh_CN')}" th:text="#{chinese.key}"></a>
    <a th:href="@{/i18n/first(locale='en_US')}" th:text="#{english.key}"></a>
    <br>
    <a th:href="@{/i18n/admin}" th:text="#{test.admin}"></a><br>
    <a th:href="@{/i18n/before}"  th:text="#{test.before}"></a><br>
</body>
```

❻ 运行

首先，运行 Ch71Application 主类。然后，访问 http://localhost:8080/ch7_1/i18n/first。运行效果如图 7.5 所示。

单击图 7.5 中的“英文版”，打开如图 7.6 所示的页面。

中文版 英文版
测试后台
测试前台

图 7.5 程序入口页面

Chinese English
test admin
test before

图 7.6 英文版效果

▶ 7.2.5 Spring Boot 与 Thymeleaf 的表单验证

视频讲解

本节使用 Hibernate Validator 对表单进行验证，注意它和 Hibernate 无关，只是使用它进行数据验证。因为 spring-boot-starter-web 不再依赖 hibernate-validator 的 jar 包，所以在 Spring Boot 的 Web 应用中，使用 Hibernate Validator 对表单进行验证时，需要加载 Hibernate Validator 所依赖的 jar 包。示例代码如下：

```
<dependency>
    <groupId>org.hibernate.validator</groupId>
    <artifactId>hibernate-validator</artifactId>
</dependency>
```

使用 Hibernate Validator 验证表单时，需要利用它的标注类型在实体模型的属性上嵌入约束。

❶ 空检查

@Null：验证对象是否为 null。

@NotNull：验证对象是否不为 null，无法查验长度为 0 的字符串。

@NotBlank：检查约束字符串是不是 null，还有被 trim 后的长度是否大于 0，只针对字符串，且会去掉前后空格。

@NotEmpty：检查约束元素是否为 null 或者是 empty。

示例如下：

```
@NotBlank(message="{goods.gname.required}")//goods.gname.required 为属性文件
    的错误代码
private String gname;
```

❷ booelan 检查

@AssertTrue：验证 boolean 属性是否为 true。

@AssertFalse：验证 boolean 属性是否为 false。

示例如下：

```
@AssertTrue
private boolean isLogin;
```

❸ 长度检查

@Size(min=, max=)：验证对象（Array，Collection，Map，String）长度是否在给定的范围之内。

@Length(min=, max=)：验证字符串长度是否在给定的范围之内。

示例如下：

```
@Length(min=1,max=100)
private String gdescription;
```

❹ 日期检查

@Past：验证 Date 和 Calendar 对象是否在当前时间之前。

@Future：验证 Date 和 Calendar 对象是否在当前时间之后。

@Pattern：验证 String 对象是否符合正则表达式的规则。

示例如下：

```
@Past(message="{gdate.invalid}")
private Date gdate;
```

❺ 数值检查

@Min：验证 Number 和 String 对象是否大于或等于指定的值。

@Max：验证 Number 和 String 对象是否小于或等于指定的值。

@DecimalMax：被标注的值必须不大于约束中指定的最大值，这个约束的参数是一个通过 BigDecimal 定义的最大值的字符串表示，小数存在精度。

@DecimalMin：被标注的值必须不小于约束中指定的最小值，这个约束的参数是一个通过 BigDecimal 定义的最小值的字符串表示，小数存在精度。

@Digits：验证 Number 和 String 的构成是否合法。

@Digits(integer=,fraction=)：验证字符串是否符合指定格式的数字，integer 指定整数精度，fraction 指定小数精度。

@Range(min=, max=)：检查数字是否介于 min 和 max 之间。

@Valid：对关联对象进行校验，如果关联对象是个集合或者数组，那么对其中的元素进行校验；如果是一个 map，则对其中的值部分进行校验。

@CreditCardNumber：信用卡验证。

@Email：验证是否是邮件地址，如果为 null，不进行验证，通过验证。

示例如下：

```
@Range(min=0,max=100,message="{gprice.invalid}")
private double gprice;
```

下面通过实例讲解使用 Hibernate Validator 验证表单的过程。

【例 7-4】使用 Hibernate Validator 验证表单。

具体实现步骤如下。

❶ 加载 Hibernate Validator 依赖

在 ch7_1 应用的 pom.xml 文件中添加 Hibernate Validator 所依赖的 jar 包。

❷ 创建表单实体模型

在 ch7_1 应用的 com.ch.ch7_1.model 包中，创建表单实体模型类 Goods。在该类使用 Hibernate Validator 的标注类型进行表单验证，核心代码如下：

```
public class Goods {
    @NotBlank(message="商品名必须输入")
    @Length(min=1, max=5, message="商品名长度为 1~5")
    private String gname;
    @Range(min=0,max=100,message="商品价格为 0~100")
    private double gprice;
    //省略 set 方法和 get 方法
}
```

❸ 创建控制器

在 ch7_1 应用的 com.ch.ch7_1.controller 包中，创建控制器类 TestValidatorController。在该类中有两个处理方法，一个是界面初始化处理方法 testValidator，一个是添加请求处理方法 add。在 add 方法中，使用@Validated 注解使验证生效。核心代码如下：

```
@Controller
public class TestValidatorController {
    @RequestMapping("/testValidator")
    public String testValidator(@ModelAttribute("goodsInfo") Goods goods){
        goods.setGname("商品名初始化");
        goods.setGprice(0.0);
        return "testValidator";
    }
    @RequestMapping(value="/add")
    public String add(@ModelAttribute("goodsInfo") @Validated Goods goods,
      BindingResult rs){
        //@ModelAttribute("goodsInfo")与 th:object="${goodsInfo}"相对应
        if(rs.hasErrors()){//验证失败
            return "testValidator";
        }
        //验证成功，可以到任意地方，在这里直接到 testValidator 界面
        return "testValidator";
    }
}
```

❹ 创建视图页面

在 ch7_1 应用的 src/main/resources/templates 目录下，创建视图页面 testValidator.html。在

视图页面中，直接读取到 ModelAttribute 里面注入的数据，然后通过 th:errors="*{xxx}"获得验证错误信息。核心代码如下：

```
<body>
    <h2>通过 th:object 访问对象的方式</h2>
    <div th:object="${goodsInfo}">
        <p th:text="*{gname}"></p>
        <p th:text="*{gprice}"></p>
    </div>
    <h1>表单提交</h1>
    <!-- 表单提交用户信息，注意表单参数的设置，直接是*{} -->
    <form  th:action="@{/add}" th:object="${goodsInfo}" method="post">
        <div><span>商品名</span><input type="text" th:field="*{gname}"/><span
        th:errors="*{gname}"></span></div>
        <div><span>商品价格</span><input type="text" th:field="*{gprice}"/><span
        th:errors="*{gprice}"></span></div>
        <input type="submit" />
    </form>
</body>
```

❺ 运行

首先，运行 Ch71Application 主类。然后，访问 http://localhost:8080/ch7_1/testValidator。表单验证失败效果如图 7.7 所示。

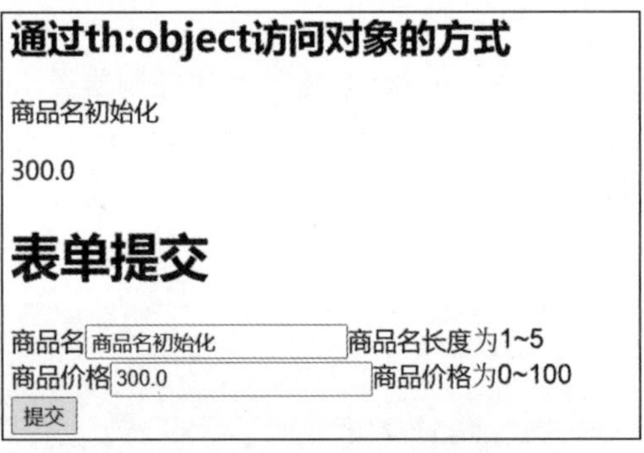

图 7.7　表单验证失败效果

7.2.6　基于 Thymeleaf 与 BootStrap 的 Web 开发实例

视频讲解

在本书的后续 Web 应用开发中，尽量使用 BootStrap、JavaScript 框架 jQuery 等前端开发工具包。BootStrap 和 jQuery 的相关知识，请读者自行学习。下面通过一个实例，讲解如何创建基于 Thymeleaf 模板引擎的 Spring Boot Web 应用 ch7_2。

【例 7-5】创建基于 Thymeleaf 模板引擎的 Spring Boot Web 应用 ch7_2。

具体实现步骤如下。

❶ 创建基于 Thymeleaf 模板引擎的 Spring Boot Web 应用 ch7_2

选择菜单 File | New | Spring Starter Project，打开 New Spring Starter Project 对话框，在该对话框中选择和输入相关信息后，单击 Next 按钮，打开 New Spring Starter Project Dependencies 对话框，选择 spring-boot-starter-thymeleaf 和 spring-boot-starter-web 依赖。单击 Finish 按钮，完成创建基于 Thymeleaf 模板引擎的 Spring Boot Web 应用 ch7_2。

❷ 设置 Web 应用 ch7_2 的上下文路径

在 ch7_2 的 application.properties 文件中配置如下内容：

```
server.servlet.context-path=/ch7_2
```

❸ 创建实体类 Book

创建名为 com.ch.ch7_2.model 的包，并在该包中创建名为 Book 的实体类，此实体类用于模板页面展示数据。具体代码如下：

```
package com.ch.ch7_2.model;
public class Book {
    String isbn;
    Double price;
    String bname;
    String publishing;
    String author;
    String picture;
    public Book(String isbn, Double price, String bname, String publishing,
      String author, String picture) {
        super();
        this.isbn = isbn;
        this.price = price;
        this.bname = bname;
        this.publishing = publishing;
        this.author = author;
        this.picture = picture;
    }
    //省略 set 方法和 get 方法
}
```

❹ 创建控制器类 ThymeleafController

创建名为 com.ch.ch7_2.controller 的包，并在该包中创建名为 ThymeleafController 的控制器类。在该控制器类中，实例化 Book 类的多个对象，并保存到集合 ArrayList<Book>中。核心代码如下：

```
@Controller
public class ThymeleafController {
    @RequestMapping("/")
    public String index(Model model) {
        Book teacherGeng = new Book("9787302464259", 59.5, "Java 2 实用教程（第
          5 版）", "清华大学出版社", "耿祥义", "073423-02.jpg");
        List<Book> chenHeng = new ArrayList<Book>();
        Book b1 = new Book("9787302529118", 69.8, "Java Web 开发从入门到实战（微
          课版）", "清华大学出版社", "陈恒","082526-01.jpg");
        chenHeng.add(b1);
        Book b2 = new Book("9787302502968", 69.8, "Java EE 框架整合开发入门到
          实战—Spring+Spring MVC+MyBatis（微课版）", "清华大学出版社", "陈恒",
          "079720-01.jpg");
        chenHeng.add(b2);
        model.addAttribute("aBook", teacherGeng);
        model.addAttribute("books", chenHeng);
```

```
        return "index";
    }
}
```

❺ 整理脚本样式静态文件

JS 脚本、CSS 样式、图片等静态文件默认放置在 src/main/resources/static 目录下，ch7_2 应用引入了 BootStrap 和 jQuery。

❻ View 视图页面

Thymeleaf 模板默认将视图页面放在 src/main/resources/templates 目录下。因此，在 src/main/resources/templates 目录下新建 html 页面文件 index.html。在该页面中，使用 Thymeleaf 模板显示控制器类 TestThymeleafController 中的 model 对象数据。具体代码如下：

```
<!DOCTYPE html>
<html xmlns:th="http://www.thymeleaf.org">
<head>
<meta charset="UTF-8">
<title>Insert title here</title>
<link rel="stylesheet" th:href="@{css/bootstrap.min.css}" />
<link rel="stylesheet" th:href="@{css/bootstrap-theme.min.css}" />
</head>
<body>
    <!-- 面板 -->
    <div class="panel panel-primary">
        <!-- 面板头信息 -->
        <div class="panel-heading">
            <!-- 面板标题 -->
            <h3 class="panel-title">第一个基于 Thymeleaf 模板引擎的 Spring Boot
              Web 应用</h3>
        </div>
    </div>
    <!-- 容器 -->
    <div class="container">
        <div>
        <h4>图书列表</h4>
        </div>
        <div class="row">
            <!-- col-md 针对桌面显示器 col-sm 针对平板 -->
            <div class="col-md-4 col-sm-6">
                <a href="">
                    <img th:src="'images/' + ${aBook.picture}" alt="图书封面"
                    style="height: 180px; width: 40%;"/>
                </a>
                <!-- caption 容器中放置其他基本信息，例如标题、文本描述等 -->
                <div class="caption">
                    <h4 th:text="${aBook.bname}"></h4>
                    <p th:text="${aBook.author}"></p>
                    <p th:text="${aBook.isbn}"></p>
                    <p th:text="${aBook.price}"></p>
                    <p th:text="${aBook.publishing}"></p>
                </div>
            </div>
```

```
                <!-- 循环取出集合数据 -->
                <div class="col-md-4 col-sm-6" th:each="book:${books}">
                    <a href="">
                        <img th:src="'images/' + ${book.picture}" alt="图书封面"
                         style="height: 180px; width: 40%;"/>
                    </a>
                    <div class="caption">
                        <h4 th:text="${book.bname}"></h4>
                        <p th:text="${book.author}"></p>
                        <p th:text="${book.isbn}"></p>
                        <p th:text="${book.price}"></p>
                        <p th:text="${book.publishing}"></p>
                    </div>
                </div>
            </div>
        </div>
    </body>
    </html>
```

❼ 运行

首先，运行 Ch72Application 主类。然后，访问 http://localhost:8080/ch7_2/。运行效果如图 7.8 所示。

图 7.8　例 7-5 运行结果

7.3　Spring Boot 处理 JSON 数据

视频讲解

在 Spring Boot 的 Web 应用中，内置了 JSON 数据的解析功能，默认使用 Jackson 自动完成解析（不需要加载 Jackson 依赖包），当控制器返回一个 Java 对象或集合数据时，Spring Boot 自动将其转换成 JSON 数据，使用起来很方便、简洁。

Spring Boot 处理 JSON 数据时，需要用到两个重要的 JSON 格式转换注解，分别是 @RequestBody 和@ResponseBody。

- @RequestBody：用于将请求体中的数据绑定到方法的形参中，该注解应用在方法的形参上。
- @ResponseBody：用于直接返回 JSON 对象，该注解应用在方法上。

下面通过一个实例讲解 Spring Boot 处理 JSON 数据的过程，该实例针对返回实体对象、ArrayList 集合、Map<String, Object>集合以及 List<Map<String, Object>>集合分别处理。

【例 7-6】 Spring Boot 处理 JSON 数据的过程。

具体实现步骤如下。

❶ 创建实体类

在 ch7_2 应用的 com.ch.ch7_2.model 包中，创建实体类 Person。具体代码如下：

```
package com.ch.ch7_2.model;
public class Person {
    private String pname;
    private String password;
    private Integer page;
    //省略 set 方法和 get 方法
}
```

❷ 创建视图页面

在 ch7_2 应用的 src/main/resources/templates 目录下，创建视图页面 input.html。在 input.html 页面中引入 jQuery 框架，并使用它的 ajax 方法进行异步请求。具体代码如下：

```
<!DOCTYPE html>
<html xmlns:th="http://www.thymeleaf.org">
<head>
<meta charset="UTF-8">
<title>Insert title here</title>
<link rel="stylesheet" th:href="@{css/bootstrap.min.css}" />
<!-- 默认访问 src/main/resources/static 下的 css 文件夹-->
<link rel="stylesheet" th:href="@{css/bootstrap-theme.min.css}" />
<!-- 引入 jQuery -->
<script type="text/javascript" th:src="@{js/jquery.min.js}"></script>
<script type="text/javascript">
    function testJson() {
        //获取输入的值 pname 为 id
        var pname = $("#pname").val();
        var password = $("#password").val();
        var page = $("#page").val();
        alert(password);
        $.ajax({
            //发送请求的 URL 字符串
            url : "testJson",
            //定义回调响应的数据格式为 JSON 字符串，该属性可以省略
            dataType : "json",
            //请求类型
            type : "post",
            //定义发送请求的数据格式为 JSON 字符串
            contentType : "application/json",
            //data 表示发送的数据
            data : JSON.stringify({pname:pname,password:password,page:page}),
            //成功响应的结果
            success : function(data){
                if(data != null){
                    //返回一个 Person 对象
```

```
                //alert("输入的用户名:"+data.pname+",密码:"+data.password+
                  //", 年龄: " +  data.page);
                //ArrayList<Person>对象
                /**for(var i = 0; i < data.length; i++){
                    alert(data[i].pname);
                }**/
                //返回一个 Map<String, Object>对象
                //alert(data.pname);//pname 为 key
                //返回一个 List<Map<String, Object>>对象
                for(var i = 0; i < data.length; i++){
                    alert(data[i].pname);
                }
            }
        },
        //请求出错
        error:function(){
            alert("数据发送失败");
        }
    });
  }
</script>
</head>
<body>
    <div class="panel panel-primary">
        <div class="panel-heading">
            <h3 class="panel-title">处理 JSON 数据</h3>
        </div>
    </div>
    <div class="container">
        <div>
        <h4>添加用户</h4>
        </div>
        <div class="row">
            <div class="col-md-6 col-sm-6">
                <form class="form-horizontal" action="">
                    <div class="form-group">
                        <div class="input-group col-md-6">
                            <span class="input-group-addon">
                                <i class="glyphicon glyphicon-pencil"></i>
                            </span>
                            <input class="form-control" type="text"
                             id="pname" th:placeholder="请输入用户名"/>
                        </div>
                    </div>
                    <div class="form-group">
                        <div class="input-group col-md-6">
                            <span class="input-group-addon">
                                <i class="glyphicon glyphicon-pencil"></i>
                            </span>
                            <input class="form-control" type="text"
                             id="password" th:placeholder="请输入密码"/>
                        </div>
                    </div>
```

```
                    <div class="form-group">
                        <div class="input-group col-md-6">
                            <span class="input-group-addon">
                                <i class="glyphicon glyphicon-pencil"></i>
                            </span>
                            <input class="form-control" type="text"
                             id="page" th:placeholder="请输入年龄"/>
                        </div>
                    </div>
                    <div class="form-group">
                        <div class="col-md-6">
                            <div class="btn-group btn-group-justified">
                                <div class="btn-group">
                                    <button type="button" onclick="testJson
                                      ()" class="btn btn-success">
                                        <span class="glyphicon glyphicon-
                                          share"></span>
                                         测试
                                    </button>
                                </div>
                            </div>
                        </div>
                    </div>
                </form>
            </div>
        </div>
    </div>
</body>
</html>
```

❸ 创建控制器

在 ch7_2 应用的 com.ch.ch7_2.controller 包中，创建控制器类 TestJsonController。在该类中有两个处理方法，一个是界面导航方法 input，一个是接收页面请求的方法。核心代码如下：

```
@Controller
public class TestJsonController {
    /**
     * 进入视图页面
     */
    @RequestMapping("/input")
    public String input() {
        return "input";
    }
    /**
     * 接收页面请求的 JSON 数据
     */
    @RequestMapping("/testJson")
    @ResponseBody
    /*@RestController 注解相当于@ResponseBody＋@Controller 合在一起的作用。
    ①如果只是使用@RestController注解Controller,则Controller中的方法无法返回JSP
      页面或者 HTML，返回的内容就是 return 的内容。
    ②如果需要返回指定页面，则需要使用@Controller 注解。如果需要返回 JSON、XML 或自定
      义 mediaType 内容到页面，则需要在对应的方法上加上@ResponseBody 注解。
```

```
     */
    public List<Map<String, Object>> testJson(@RequestBody Person user) {
        //打印接收的 JSON 格式数据
        System.out.println("pname=" + user.getPname() +
                ",password="+user.getPassword()+",page="+user.getPage());
        //返回 Person 对象
        //return user;
        /**ArrayList<Person> allp = new ArrayList<Person>();
        Person p1 = new Person();
        p1.setPname("陈恒 1");
        p1.setPassword("123456");
        p1.setPage(80);
        allp.add(p1);
        Person p2 = new Person();
        p2.setPname("陈恒 2");
        p2.setPassword("78910");
        p2.setPage(90);
        allp.add(p2);
        //返回 ArrayList<Person>对象
        return allp;
        **/
        Map<String, Object> map = new HashMap<String, Object>();
        map.put("pname", "陈恒 2");
        map.put("password", "123456");
        map.put("page", 25);
        //返回一个 Map<String, Object>对象
        //return map;
        //返回一个 List<Map<String, Object>>对象
        List<Map<String,Object>>allp=newArrayList<Map<String,Object>>();
        allp.add(map);
        Map<String, Object> map1 = new HashMap<String, Object>();
        map1.put("pname", "陈恒 3");
        map1.put("password", "54321");
        map1.put("page", 55);
        allp.add(map1);
        return allp;
    }
}
```

❹ 运行

首先，运行 Ch72Application 主类。然后，访问 http://localhost:8080/ch7_2/input。运行效果如图 7.9 所示。

处理JSON数据

添加用户

请输入用户名

请输入密码

请输入年龄

测试

图 7.9　input.html 运行效果

视频讲解

7.4 Spring Boot 文件上传与下载

文件上传与下载是 Web 应用开发中常用的功能之一。本节将讲解在 Spring Boot 的 Web 应用开发中，如何实现文件的上传与下载。

在实际的 Web 应用开发中，为了成功上传文件，必须将表单的 method 设置为 post，并将 enctype 设置为 multipart/form-data。只有这样设置，浏览器才能将所选文件的二进制数据发送给服务器。

从 Servlet 3.0 开始，就提供了处理文件上传的方法，但这种文件上传需要在 Java Servlet 中完成，而 Spring MVC 提供了更简单的封装。Spring MVC 是通过 Apache Commons FileUpload 技术实现一个 MultipartResolver 的实现类 CommonsMultipartResolver 完成文件上传的。因此，Spring MVC 的文件上传需要依赖 Apache Commons FileUpload 组件。

Spring MVC 将上传文件自动绑定到 MultipartFile 对象中，MultipartFile 提供了获取上传文件内容、文件名等方法，并通过 transferTo 方法将文件上传到服务器的磁盘中。MultipartFile 的常用方法如下。

- byte[] getBytes()：获取文件数据。
- String getContentType()：获取文件 MIME 类型，如 image/jpeg 等。
- InputStream getInputStream()：获取文件流。
- String getName()：获取表单中文件组件的名字。
- String getOriginalFilename()：获取上传文件的原名。
- long getSize()：获取文件的字节大小，单位为 byte。
- boolean isEmpty()：是否有（选择）上传文件。
- void transferTo(File dest)：将上传文件保存到一个目标文件中。

Spring Boot 的 spring-boot-starter-web 已经集成了 Spring MVC，所以使用 Spring Boot 实现文件上传更加便捷，只需要引入 Apache Commons FileUpload 组件依赖即可。

下面通过一个实例讲解 Spring Boot 文件上传与下载的实现过程。

【例 7-7】Spring Boot 文件上传与下载。

具体实现步骤如下。

❶ 引入 Apache Commons FileUpload 组件依赖

在 Web 应用 ch7_2 的 pom.xml 文件中，添加 Apache Commons FileUpload 组件依赖。具体代码如下：

```
<dependency>
    <groupId>commons-fileupload</groupId>
    <artifactId>commons-fileupload</artifactId>
    <!-- 由于 commons-fileupload 组件不属于 Spring Boot，所以需要加上版本 -->
    <version>1.4</version>
</dependency>
```

❷ 设置上传文件大小限制

在 Web 应用 ch7_2 的配置文件 application.properties 中，添加如下配置来限制上传文件大小：

```
#上传文件时，默认单个上传文件大小是 1MB，max-file-size 设置单个上传文件大小
spring.servlet.multipart.max-file-size=50MB
```

```
#默认总文件大小是 10MB，max-request-size 设置总上传文件大小
spring.servlet.multipart.max-request-size=500MB
```

❸ 创建选择文件视图页面

在 ch7_2 应用的 src/main/resources/templates 目录下，创建选择文件视图页面 uploadFile.html。该页面中有一个 enctype 属性值为 multipart/form-data 的 form 表单，具体代码如下：

```
<!DOCTYPE html>
<html xmlns:th="http://www.thymeleaf.org">
<head>
<meta charset="UTF-8">
<title>Insert title here</title>
<link rel="stylesheet" th:href="@{css/bootstrap.min.css}" />
<!-- 默认访问 src/main/resources/static 下的 css 文件夹-->
<link rel="stylesheet" th:href="@{css/bootstrap-theme.min.css}" />
</head>
<body>
<div class="panel panel-primary">
        <div class="panel-heading">
            <h3 class="panel-title">文件上传示例</h3>
        </div>
    </div>
    <div class="container">
        <div class="row">
            <div class="col-md-6 col-sm-6">
                <form class="form-horizontal" action="upload" method="post"
                 enctype="multipart/form-data">
                    <div class="form-group">
                        <div class="input-group col-md-6">
                            <span class="input-group-addon">
                                <i class="glyphicon glyphicon-pencil"></i>
                            </span>
                            <input class="form-control" type="text"
                             name="description" th:placeholder="文件描述"/>
                        </div>
                    </div>
                    <div class="form-group">
                        <div class="input-group col-md-6">
                            <span class="input-group-addon">
                                <i class="glyphicon glyphicon-search"></i>
                            </span>
                            <input class="form-control" type="file"
                             name="myfile" th:placeholder="请选择文件"/>
                        </div>
                    </div>
                    <div class="form-group">
                        <div class="col-md-6">
                            <div class="btn-group btn-group-justified">
                                <div class="btn-group">
                                    <button type="submit" class="btn btn-
                                         success">
                                        <span class="glyphicon glyphicon-
```

```
                                    share"></span>
                                 上传文件
                            </button>
                        </div>
                    </div>
                </div>
            </div>
        </form>
    </div>
  </div>
</div>
</body>
</html>
```

④ 创建控制器

在 ch7_2 应用的 com.ch.ch7_2.controller 包中，创建控制器类 TestFileUpload。在该类中有 4 个处理方法，一个是界面导航方法 uploadFile，一个是实现文件上传的 upload 方法，一个是显示将要被下载文件的 showDownLoad 方法，一个是实现下载功能的 download 方法。核心代码如下：

```
@Controller
public class TestFileUpload {
    @RequestMapping("/uploadFile")
    public String uploadFile() {
        return "uploadFile";
    }
    /**
     * 上传文件自动绑定到 MultipartFile 对象中，
     * 在这里使用处理方法的形参接收请求参数。
     */
    @RequestMapping("/upload")
    public String upload(
            HttpServletRequest request,
            @RequestParam("description") String description,
            @RequestParam("myfile") MultipartFile myfile)
            throws IllegalStateException, IOException {
        System.out.println("文件描述：" + description);
        //如果选择了上传文件，将文件上传到指定的目录 uploadFiles 中
        if(!myfile.isEmpty()) {
            //上传文件路径
            String path = request.getServletContext().getRealPath
              ("/uploadFiles/");
            //获得上传文件原名
            String fileName = myfile.getOriginalFilename();
            File filePath = new File(path + File.separator + fileName);
            //如果文件目录不存在，创建目录
            if(!filePath.getParentFile().exists()) {
                filePath.getParentFile().mkdirs();
            }
            //将上传文件保存到一个目标文件中
            myfile.transferTo(filePath);
        }
```

```
        //转发到一个请求处理方法，查询将要下载的文件
        return "forward:/showDownLoad";
    }
    /**
     * 显示要下载的文件
     */
    @RequestMapping("/showDownLoad")
    public String showDownLoad(HttpServletRequest request, Model model) {
        String path = request.getServletContext().getRealPath("/uploadFiles/");
        File fileDir = new File(path);
        //从指定目录获得文件列表
        File filesList[] = fileDir.listFiles();
        model.addAttribute("filesList", filesList);
        return "showFile";
    }
    /**
     * 实现下载功能
     */
    @RequestMapping("/download")
    public ResponseEntity<byte[]> download(
            HttpServletRequest request,
            @RequestParam("filename") String filename,
            @RequestHeader("User-Agent") String userAgent) throws IOException {
        //下载文件路径
        String path = request.getServletContext().getRealPath("/uploadFiles/");
        //构建将要下载的文件对象
        File downFile = new File(path + File.separator + filename);
        //ok 表示 HTTP 中的状态是 200
        BodyBuilder builder =  ResponseEntity.ok();
        //内容长度
        builder.contentLength(downFile.length());
        //application/octet-stream：二进制流数据（最常见的文件下载）
        builder.contentType(MediaType.APPLICATION_OCTET_STREAM);
        //使用 URLEncoder.encode 对文件名进行编码
        filename = URLEncoder.encode(filename,"UTF-8");
        /**
         * 设置实际的响应文件名，告诉浏览器文件要用于“下载”和“保存”。
         * 不同的浏览器，处理方式不同，根据浏览器的实际情况区别对待。
         */
        if(userAgent.indexOf("MSIE") > 0) {
            //IE 浏览器，只需要用 UTF-8 字符集进行 URL 编码
            builder.header("Content-Disposition", "attachment; filename=" +
              filename);
        }else {
            /**非 IE 浏览器，如 FireFox、Chrome 等浏览器，则需要说明编码的字符集
             * filename 后面有个*号，在 UTF-8 后面有两个单引号
             */
            builder.header("Content-Disposition", "attachment; filename*=
              UTF-8'' " + filename);
        }
        return builder.body(FileUtils.readFileToByteArray(downFile));
    }
}
```

❺ 创建文件下载视图页面

在 ch7_2 应用的 src/main/resources/templates 目录下，创建文件下载视图页面 showFile.html。核心代码如下：

```
<body>
    <div class="panel panel-primary">
        <div class="panel-heading">
            <h3 class="panel-title">文件下载示例</h3>
        </div>
    </div>
    <div class="container">
        <div class="panel panel-primary">
            <div class="panel-heading">
                <h3 class="panel-title">文件列表</h3>
            </div>
            <div class="panel-body">
                <div class="table table-responsive">
                    <table class="table table-bordered table-hover">
                        <tbody class="text-center">
                            <tr th:each="file,fileStat:${filesList}">
                                <td>
                                    <span th:text="${fileStat.count}"></span>
                                </td>
                                <td>
                                <!--file.name 相当于调用 getName()方法获得文件
                                  名称 -->
                                    <a th:href="@{download(filename=${file.
                                      name})}">
                                        <span th:text="${file.name}"></span>
                                    </a>
                                </td>
                            </tr>
                        </tbody>
                    </table>
                </div>
            </div>
        </div>
    </div>
</body>
```

❻ 运行

首先，运行 Ch72Application 主类。然后，访问 http://localhost:8080/ch7_2/uploadFile 测试文件上传与下载。

7.5 Spring Boot 的异常统一处理

视频讲解

在 Spring Boot 应用的开发中，不管是对底层数据库操作，还是业务层操作，还是控制层操作，都不可避免遇到各种可预知的、不可预知的异常需要处理。如果每个过程都单独处理异常，那么系统的代码耦合度高，工作量大且不好统一，以后维护的工作量也很大。

如果能将所有类型的异常处理从各层中解耦出来，就可以既保证相关处理过程的功能较单一，也实现异常信息的统一处理和维护。幸运的是，Spring 框架支持这样的实现。本节将以自定义 error 页面、@ExceptionHandler 注解以及@ControllerAdvice 三种方式讲解 Spring Boot 应用的异常统一处理。

7.5.1　自定义 error 页面

在 Spring Boot Web 应用的 src/main/resources/templates 目录下添加 error.html 页面，访问发生错误或异常时，Spring Boot 将自动找到该页面作为错误页面。Spring Boot 为错误页面提供了以下属性。

- timestamp：错误发生时间。
- status：HTTP 状态码。
- error：错误原因。
- exception：异常的类名。
- message：异常消息（如果这个错误是由异常引起的）。
- errors：BindingResult 异常里的各种错误（如果这个错误是由异常引起的）。
- trace：异常跟踪信息（如果这个错误是由异常引起的）。
- path：错误发生时请求的 URL 路径。

下面通过一个实例讲解在 Spring Boot 应用的开发中，如何使用自定义 error 页面。

【例 7-8】自定义 error 页面。

具体实现步骤如下。

❶ 创建基于 Thymeleaf 模板引擎的 Spring Boot Web 应用 ch7_3

参照 7.2.6 节的例 7-5，创建基于 Thymeleaf 模板引擎的 Spring Boot Web 应用 ch7_3。

❷ 设置 Web 应用 ch7_3 的上下文路径

在 ch7_3 的 application.properties 文件中配置如下内容：

```
server.servlet.context-path=/ch7_3
```

❸ 创建自定义异常类 MyException

创建名为 com.ch.ch7_3.exception 的包，并在该包中创建名为 MyException 的异常类。具体代码如下：

```
package com.ch.ch7_3.exception;
public class MyException extends Exception {
    private static final long serialVersionUID = 1L;
    public MyException() {
        super();
    }
    public MyException(String message) {
        super(message);
    }
}
```

❹ 创建控制器类 TestHandleExceptionController

创建名为 com.ch.ch7_3.controller 的包，并在该包中创建名为 TestHandleExceptionController 的控制器类。在该控制器类中，有 4 个请求处理方法，一个导航到 index.html，另外 3 个分别抛出不同的异常（并没有处理异常）。核心代码如下：

```
@Controller
public class TestHandleExceptionController {
    @RequestMapping("/")
    public String index() {
        return "index";
    }
    @RequestMapping("/db")
    public void db() throws SQLException {
        throw new SQLException("数据库异常");
    }
    @RequestMapping("/my")
    public void my() throws MyException {
        throw new MyException("自定义异常");
    }
    @RequestMapping("/no")
    public void no() throws Exception {
        throw new Exception("未知异常");
    }
}
```

❺ 整理脚本样式静态文件

JS 脚本、CSS 样式、图片等静态文件默认放置在 src/main/resources/static 目录下，ch7_3 应用引入了与 ch7_2 一样的 BootStrap 和 jQuery。

❻ View 视图页面

Thymeleaf 模板默认将视图页面放在 src/main/resources/templates 目录下。因此，在 src/main/resources/templates 目录下新建 HTML 页面文件 index.html 和 error.html。

在 index.html 页面中有 4 个超链接请求，3 个请求在控制器中有对应处理，另一个请求是 404 错误。核心代码如下：

```
<body>
    <div class="panel panel-primary">
        <div class="panel-heading">
            <h3 class="panel-title">异常处理示例</h3>
        </div>
    </div>
    <div class="container">
        <div class="row">
            <div class="col-md-4 col-sm-6">
                <a th:href="@{db}">处理数据库异常</a><br>
                <a th:href="@{my}">处理自定义异常</a><br>
                <a th:href="@{no}">处理未知错误</a>
                <hr>
                <a th:href="@{nofound}">404 错误</a>
            </div>
        </div>
    </div>
</body>
</html>
```

在 error.html 页面中，使用 Spring Boot 为错误页面提供的属性显示错误消息。核心代码如下：

```
<body>
    <div class="panel-l container clearfix">
        <div class="error">
            <p class="title"><span class="code" th:text="${status}"></span>
             非常抱歉，没有找到您要查看的页面</p>
            <div class="common-hint-word">
                <div th:text="${#dates.format(timestamp,'yyyy-MM-dd
                  HH:mm:ss')}"></div>
                <div th:text="${message}"></div>
                <div th:text="${error}"></div>
            </div>
        </div>
    </div>
</body>
```

❼ 运行

首先，运行 Ch73Application 主类。然后，访问 http://localhost:8080/ch7_3/打开 index.html 页面。运行效果如图 7.10 所示。

单击图 7.10 中的超链接时，Spring Boot 应用将根据链接请求，到控制器中找对应的处理。例如，单击图 7.10 中的“处理数据库异常”链接时，将执行控制器中的 public void db() throws SQLException 方法，而该方法仅仅抛出了 SQLException 异常，并没有处理异常。当 Spring Boot 发现有异常抛出并没有处理时，将自动在 src/main/resources/templates 目录下找到 error.html 页面显示异常信息，效果如图 7.11 所示。

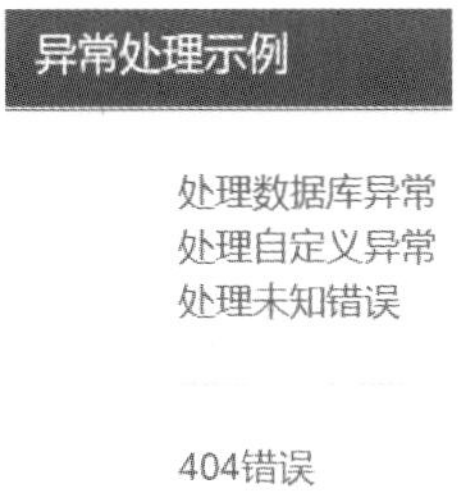

异常处理示例

处理数据库异常
处理自定义异常
处理未知错误

404错误

图 7.10　index.html 页面

500非常抱歉，没有找到您要查看的页面

2021-01-03 05:35:17
Internal Server Error

图 7.11　error.html 页面

从例 7-8 运行的结果可以看出，使用自定义 error 页面并没有真正处理异常，只是将异常或错误信息显示给客户端。因为在服务器控制台上同样抛出了异常，如图 7.12 所示。

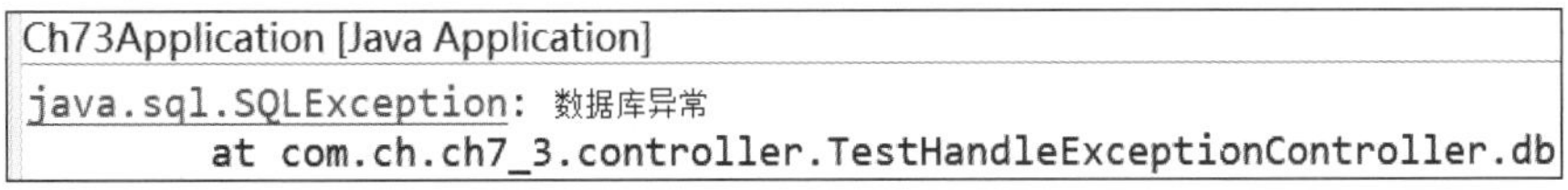

```
Ch73Application [Java Application]
java.sql.SQLException: 数据库异常
        at com.ch.ch7_3.controller.TestHandleExceptionController.db
```

图 7.12　异常信息

▶ 7.5.2　@ExceptionHandler 注解

在 7.5.1 节中使用自定义 error 页面并没有真正处理异常，本节可以使用@ExceptionHandler 注解处理异常。如果在 Controller 中有一个使用@ExceptionHandler 注解修饰的方法，那么当 Controller 的任何方法抛出异常时，都由该方法处理异常。

下面通过实例讲解如何使用@ExceptionHandler 注解处理异常。

【例 7-9】使用@ExceptionHandler 注解处理异常。

具体实现步骤如下。

❶ 在控制器类中添加使用@ExceptionHandler 注解修饰的方法

在例 7-8 的控制器类 TestHandleExceptionController 中，添加一个使用@ExceptionHandler 注解修饰的方法，具体代码如下：

```
@ExceptionHandler(value=Exception.class)
public String handlerException(Exception e) {
    //数据库异常
    if (e instanceof SQLException) {
        return "sqlError";
    } else if (e instanceof MyException) {//自定义异常
        return "myError";
    } else {//未知异常
        return "error";
    }
}
```

❷ 创建 sqlError、myError 和 noError 页面

在 ch7_3 的 src/main/resources/templates 目录下，创建 sqlError、myError 和 noError 页面。当发生 SQLException 异常时，Spring Boot 处理后，显示 sqlError 页面；当发生 MyException 异常时，Spring Boot 处理后，显示 myError 页面；当发生未知异常时，Spring Boot 处理后，显示 noError 页面。具体代码略。

❸ 运行

再次运行 Ch73Application 主类后，访问 http://localhost:8080/ch7_3/，打开 index.html 页面，单击“处理数据库异常”链接时，执行控制器中的 public void db() throws SQLException 方法，该方法抛出了 SQLException，这时 Spring Boot 会自动执行使用@ExceptionHandler 注解修饰的方法 public String handlerException(Exception e)进行异常处理并打开 sqlError.html 页面，同时观察控制台是否抛出异常信息。注意，单击“404 错误”链接时，还是由自定义 error 页面显示错误信息，这是因为没有执行控制器中抛出异常的方法，进而不会执行使用@ExceptionHandler 注解修饰的方法。

从例 7-9 可以看出，在控制器中添加使用@ExceptionHandler 注解修饰的方法才能处理异常。而一个 Spring Boot 应用中往往存在多个控制器，不太适合在每个控制器中添加使用@ExceptionHandler 注解修饰的方法进行异常处理。可以将使用@ExceptionHandler 注解修饰的方法放到一个父类中，然后所有需要处理异常的控制器继承该类即可。例如，可以将例 7-9 中使用@ExceptionHandler 注解修饰的方法移到一个父类 BaseController 中，然后让控制器类 TestHandleExceptionController 继承该父类即可处理异常。

7.5.3 @ControllerAdvice 注解

使用 7.5.2 节中的父类 Controller 进行异常处理也有其自身的缺点，就是代码耦合性太高。可以使用@ControllerAdvice 注解降低这种父子耦合关系。

@ControllerAdvice 注解，顾名思义，是一个增强的 Controller。使用该 Controller，可以实现三个方面的功能：全局异常处理、全局数据绑定以及全局数据预处理。本节将学习如何使用@ControllerAdvice 注解进行全局异常处理。

使用@ControllerAdvice 注解的类是当前 Spring Boot 应用中所有类的统一异常处理类，该类中使用@ExceptionHandler 注解的方法统一处理异常，不需要在每个 Controller 中逐一定义异常处理方法，这是因为@ExceptionHandler 注解的方法对所有注解了@RequestMapping 的控制器方法统一处理异常。

下面通过实例讲解如何使用@ControllerAdvice 注解进行全局异常处理。

【例 7-10】 使用@ControllerAdvice 注解进行全局异常处理。

具体实现步骤如下。

❶ 创建使用@ControllerAdvice 注解的类

在 ch7_3 的 com.ch.ch7_3.controller 包中，创建名为 GlobalExceptionHandlerController 的类。使用@ControllerAdvice 注解修饰该类，并将例 7-9 中使用@ExceptionHandler 注解修饰的方法移到该类中。核心代码如下：

```
@ControllerAdvice
public class GlobalExceptionHandlerController {
    @ExceptionHandler(value=Exception.class)
    public String handlerException(Exception e) {
        //数据库异常
        if (e instanceof SQLException) {
            return "sqlError";
        } else if (e instanceof MyException) {//自定义异常
            return "myError";
        } else {//未知异常
            return "noError";
        }
    }
}
```

❷ 运行

再次运行 Ch73Application 主类后，访问 http://localhost:8080/ch7_3/，打开 index.html 页面测试即可。

7.6　Spring Boot 对 JSP 的支持

视频讲解

尽管 Spring Boot 建议使用 HTML 完成动态页面，但也有部分 Java Web 应用使用 JSP 完成动态页面。遗憾的是 Spring Boot 官方不推荐使用 JSP 技术，但考虑到是常用的技术，本节将介绍 Spring Boot 如何集成 JSP 技术。

下面通过实例讲解 Spring Boot 如何集成 JSP 技术。

【例 7-11】 Spring Boot 集成 JSP 技术。

具体实现步骤如下。

❶ 创建 Spring Boot Web 应用 ch7_4

选择菜单 File | New | Spring Starter Project，打开 New Spring Starter Project 对话框，在该对话框中选择和输入相关信息后，单击 Next 按钮，打开 New Spring Starter Project Dependencies 对话框，选择 spring-boot-starter-web 依赖，单击 Finish 按钮，完成创建 Spring Boot Web 应用 ch7_4。

❷ 修改 pom.xml 文件，添加 Servlet、Tomcat 和 JSTL 依赖

因为在 JSP 页面中使用 EL 和 JSTL 标签显示数据，所以在 pom.xml 文件中，除了添加 Servlet 和 Tomcat 依赖外，还需要添加 JSTL 依赖。具体代码如下：

```
<!-- 添加 Servlet 依赖 -->
<dependency>
    <groupId>javax.servlet</groupId>
    <artifactId>javax.servlet-api</artifactId>
    <scope>provided</scope>
    </dependency>
    <!-- 添加 Tomcat 依赖 -->
    <dependency>
        <groupId>org.springframework.boot</groupId>
        <artifactId>spring-boot-starter-tomcat</artifactId>
        <scope>provided</scope>
    </dependency>
    <!-- Jasper 是 Tomcat 使用的引擎，使用 tomcat-embed-jasper 可以将 Web 应用在内嵌
      的 Tomcat 下运行 -->
    <dependency>
        <groupId>org.apache.tomcat.embed</groupId>
        <artifactId>tomcat-embed-jasper</artifactId>
        <scope>provided</scope>
    </dependency>
    <!-- 添加 JSTL 依赖 -->
    <dependency>
        <groupId>javax.servlet</groupId>
        <artifactId>jstl</artifactId>
    </dependency>
```

❸ 设置 Web 应用 ch7_4 的上下文路径及页面配置信息

在 ch7_4 的 application.properties 文件中配置如下内容：

```
server.servlet.context-path=/ch7_4
#设置页面前缀目录
spring.mvc.view.prefix=/WEB-INF/jsp/
#设置页面后缀
spring.mvc.view.suffix=.jsp
```

❹ 创建实体类 Book

创建名为 com.ch.ch7_4.model 的包，并在该包中创建名为 Book 的实体类。此实体类用于模板页面展示数据，代码与例 7-5 中的 Book 一样，不再赘述。

❺ 创建控制器类 ThymeleafController

创建名为 com.ch.ch7_4.controller 的包，并在该包中创建名为 ThymeleafController 的控制器类。在该控制器类中，实例化 Book 类的多个对象，并保存到集合 ArrayList<Book>中。代码与例 7-5 中的 ThymeleafController 一样，不再赘述。

❻ 整理脚本样式静态文件

JS 脚本、CSS 样式、图片等静态文件默认放置在 src/main/resources/static 目录下，ch7_4 应用引入的 BootStrap 和 jQuery 与例 7-5 中的一样，不再赘述。

❼ View 视图页面

从 application.properties 配置文件中可知，将 JSP 文件路径指定到/WEB-INF/jsp/目录。因此，需要在 src/main 目录下创建目录 webapp/WEB-INF/jsp/，并在该目录下创建 JSP 文件 index.jsp。核心代码如下：

```
<body>
    <div class="panel panel-primary">
        <div class="panel-heading">
            <h3 class="panel-title">第一个基于 JSP 技术的 Spring Boot Web 应用</h3>
        </div>
    </div>
    <div class="container">
        <div>
        <h4>图书列表</h4>
    </div>
        <div class="row">
            <div class="col-md-4 col-sm-6">
                <!-- 使用 EL 表达式 -->
                <a href="">
        <img src="images/${aBook.picture}" alt="图书封面" style="height:
          180px; width: 40%;"/>
                </a>
                <div class="caption">
                    <h4>${aBook.bname}</h4>
                    <p>${aBook.author}</p>
                    <p>${aBook.isbn}</p>
                    <p>${aBook.price}</p>
                    <p>${aBook.publishing}</p>
                </div>
            </div>
            <!-- 使用 JSTL 标签 forEach 循环取出集合数据 -->
            <c:forEach var="book" items="${books}">
                <div class="col-md-4 col-sm-6">
                <a href="">
            <img src="images/${book.picture}" alt="图书封面" style="height:
              180px; width: 40%;"/>
                </a>
                <div class="caption">
                    <h4>${book.bname}</h4>
                    <p>${book.author}</p>
                    <p>${book.isbn}</p>
                    <p>${book.price}</p>
                    <p>${book.publishing}</p>
                </div>
            </div>
            </c:forEach>
        </div>
    </div>
</body>
```

❽ 运行

首先，运行 Ch74Application 主类。然后，访问 http://localhost:8080/ch7_4/。运行效果如图 7.13 所示。

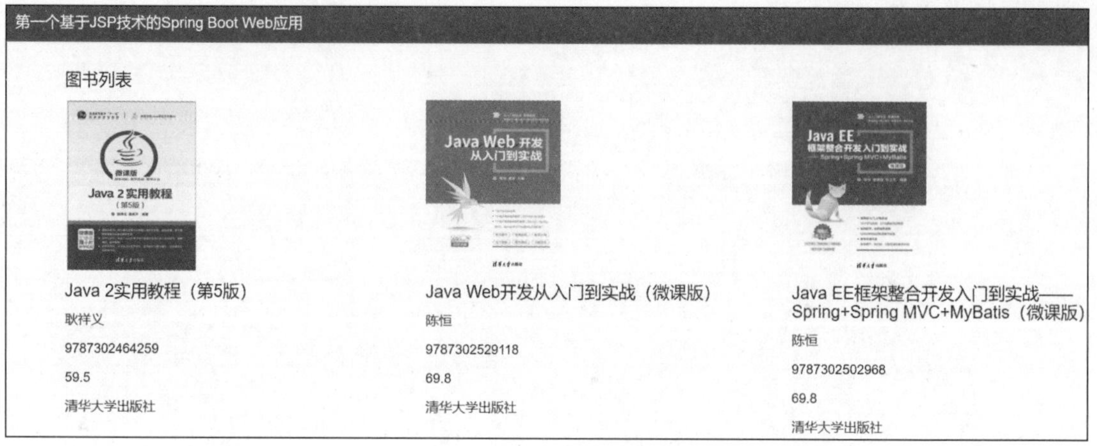

图 7.13 例 7-11 运行结果

7.7 本章小结

本章首先介绍了 Spring Boot 的 Web 开发支持，然后详细讲述了 Spring Boot 推荐使用的 Thymeleaf 模板引擎，包括 Thymeleaf 的基础语法、常用属性以及国际化。同时，本章还介绍了 Spring Boot 对 JSON 数据的处理、文件上传下载、异常统一处理和对 JSP 的支持等 Web 应用开发的常用功能。

习 题 7

使用 Hibernate Validator 验证如图 7.14 所示的表单信息，具体要求如下：

（1）用户名必须输入，并且长度为 5~20。

（2）年龄为 18~60。

（3）工作日期在系统时间之前。

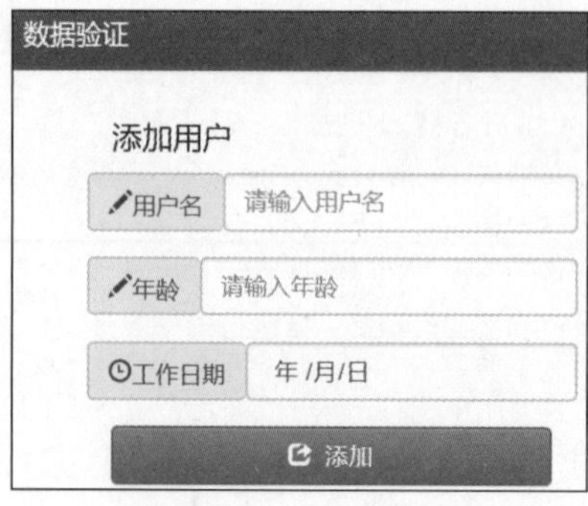

图 7.14 输入页面

第 8 章 Spring Boot 的数据访问 

学习目的与要求

本章将详细介绍 Spring Boot 访问数据库的解决方案。通过本章的学习，掌握 Spring Boot 访问数据库的解决方案。

主要内容

- ❖ Spring Data JPA
- ❖ Spring Boot 整合 MyBatis
- ❖ Spring Boot 整合 REST
- ❖ Spring Boot 整合 MongoDB
- ❖ Spring Boot 整合 Redis
- ❖ 数据缓存 Cache

Spring Data 是 Spring 访问数据库的一揽子解决方案，是一个伞形项目，包含大量关系型数据库及非关系型数据库的数据访问解决方案。本章将介绍 Spring Data JPA、Spring Data REST、Spring Data MongoDB、Spring Data Redis 等 Spring Data 子项目。

8.1 Spring Data JPA

Spring Data JPA 是 Spring Data 的子项目，在讲解 Spring Data JPA 之前，先了解一下 Hibernate。这是因为 Spring Data JPA 是由 Hibernate 默认实现的。

Hibernate 是一个开源的对象关系映射框架，它对 JDBC 进行了非常轻量级的对象封装，它将 POJO（Plain Ordinary Java Object）简单的 Java 对象与数据库表建立映射关系，是一个全自动的 ORM（Object Relational Mapping）框架。Hibernate 可以自动生成 SQL 语句、自动执行，使得 Java 开发人员可以随心所欲地使用对象编程思维来操纵数据库。

JPA（Java Persistence API）是官方提出的 Java 持久化规范。JPA 通过注解或 XML 描述对象-关系（表）的映射关系，并将内存中的实体对象持久化到数据库。

Spring Data JPA 通过提供基于 JPA 的 Repository 极大地简化了 JPA 的写法，在几乎不写实现的情况下，实现数据库的访问和操作。使用 Spring Data JPA 建立数据访问层十分方便，只要定义一个继承 JpaRepository 接口的接口即可。

继承了 JpaRepository 接口的自定义数据访问接口，具有 JpaRepository 接口的所有数据访问操作方法。JpaRepository 接口的核心源代码如下：

```
@NoRepositoryBean
public interface JpaRepository<T, ID> extends PagingAndSortingRepository<T, ID>,
QueryByExampleExecutor<T> {
    List<T> findAll();
    List<T> findAll(Sort sort);
```

```
    List<T> findAllById(Iterable<ID> ids);
    <S extends T> List<S> saveAll(Iterable<S> entities);
    void flush();
    <S extends T> S saveAndFlush(S entity);
    void deleteInBatch(Iterable<T> entities);
    void deleteAllInBatch();
    T getOne(ID id);
    <S extends T> List<S> findAll(Example<S> example);
    <S extends T> List<S> findAll(Example<S> example, Sort sort);
}
```

JpaRepository 接口提供的常用方法如下。

List<T> findAll()：查询所有实体对象数据，返回一个 List 集合。

List<T> findAll(Sort sort)：按照指定的排序规则查询所有实体对象数据，返回一个 List 集合。

List<T> findAllById(Iterable<ID> ids)：根据所提供的实体对象 id（多个），将对应的实体全部查询出来，并返回一个 List 集合。

<S extends T> List<S> saveAll(Iterable<S> entities)：将提供的集合中的实体对象数据保存到数据库。

void flush()：将缓存的对象数据操作更新到数据库。

<S extends T> S saveAndFlush(S entity)：保存对象的同时立即更新到数据库。

void deleteInBatch(Iterable<T> entities)：批量删除提供的实体对象。

void deleteAllInBatch()：批量删除所有的实体对象。

T getOne(ID id)：根据 id 获得对应的实体对象。

<S extends T> List<S> findAll(Example<S> example)：根据提供的 example 实例查询实体对象数据。

<S extends T> List<S> findAll(Example<S> example, Sort sort)：根据提供的 example 实例，并按照指定规则，查询实体对象数据。

▶ 8.1.1 Spring Boot 的支持

在 Spring Boot 应用中，如果需要使用 Spring Data JPA 访问数据库，那么可以通过 STS 创建 Spring Boot 应用时选择 Spring Data JPA 模块依赖，如图 8.1 所示。

❶ JDBC 的自动配置

Spring Data JPA 模块的依赖关系如图 8.2 所示。

从图 8.2 的依赖关系可知，spring-boot-starter-data-jpa 依赖于 spring-boot-starter-jdbc，而 Spring Boot 对 spring-boot-starter-jdbc 做了自动配置。JDBC 自动配置源码位于 org.springframework.boot.autoconfigure.jdbc 包下。从该包的 DataSourceProperties 类可以看出，可以使用 spring.datasource 为扩展名的属性在 application.properties 配置文件中配置 datasource。

❷ JPA 的自动配置

Spring Boot 对 JPA 的自动配置位于 org.springframework.boot.autoconfigure.orm.jpa 包下。从该包的 HibernateJpaAutoConfiguration 类可以看出，Spring Boot 对 JPA 的默认实现是 Hibernate；从该包的 JpaProperties 类可以看出，可以使用 spring.jpa 为扩展名的属性在 application.properties 配置文件中配置 JPA。

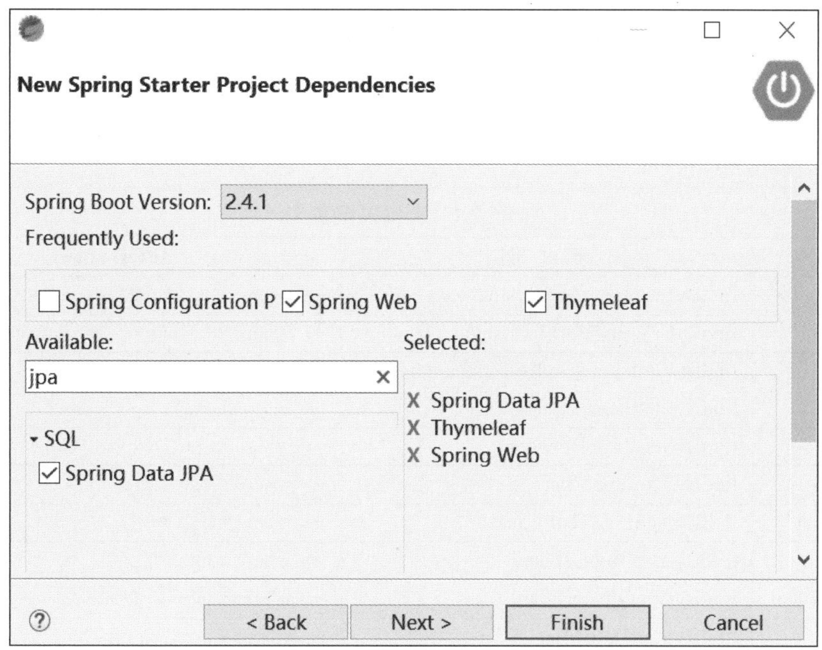

图 8.1 选择 Spring Data JPA 模块

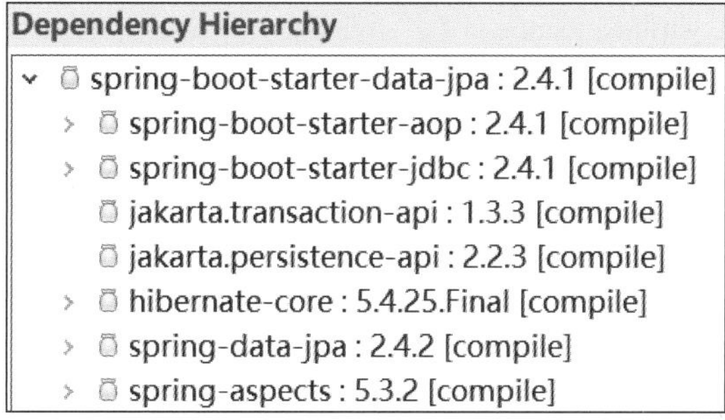

图 8.2 Spring Data JPA 模块的依赖关系

❸ Spring Data JPA 的自动配置

Spring Boot 对 Spring Data JPA 的自动配置位于 org.springframework.boot.autoconfigure.data.jpa 包下。从该包的 JpaRepositoriesAutoConfiguration 类可以看出，JpaRepositoriesAutoConfiguration 依赖于 HibernateJpaAutoConfiguration 配置；从该包的 JpaRepositoriesRegistrar 类可以看出，Spring Boot 自动开启了对 Spring Data JPA 的支持，即开发人员无须在配置类中显式声明@Enable-JpaRepositories。

❹ Spring Boot 应用的 Spring Data JPA

从上述分析可知，在 Spring Boot 应用中使用 Spring Data JPA 访问数据库时，除了添加 spring-boot-starter-data-jpa 依赖外，只需定义 DataSource、持久化实体类和数据访问层，并在需要使用数据访问的地方（如 Service 层）依赖注入数据访问层即可。

▶ 8.1.2 简单条件查询

视频讲解

从前面的学习可知，只需定义一个继承 JpaRepository 接口的接口即可使用 Spring Data JPA 建立数据访问层。因此，自定义的数据访问接口完全继承了 JpaRepository 的接口方法。

但更重要的是，在自定义的数据访问接口中可以根据查询关键字定义查询方法，这些查询方法需要符合它的命名规则，一般是根据持久化实体类的属性名来确定的。

❶ 查询关键字

目前，Spring Data JPA 支持的查询关键字如表 8.1 所示。

表 8.1 查询关键字

关键字	示 例	JPQL 代码段
And	findByLastnameAndFirstname	… where x.lastname = ?1 and x.firstname = ?2
Or	findByLastnameOrFirstname	… where x.lastname = ?1 or x.firstname = ?2
Is,Equals	findByFirstname,findByFirstnameIs, findByFirstnameEquals	… where x.firstname = ?1
Between	findByStartDateBetween	… where x.startDate between ?1 and ?2
LessThan	findByAgeLessThan	… where x.age < ?1
LessThanEqual	findByAgeLessThanEqual	… where x.age <= ?1
GreaterThan	findByAgeGreaterThan	… where x.age > ?1
GreaterThanEqual	findByAgeGreaterThanEqual	… where x.age >= ?1
After	findByStartDateAfter	… where x.startDate > ?1
Before	findByStartDateBefore	… where x.startDate < ?1
IsNull	findByAgeIsNull	… where x.age is null
IsNotNull,NotNull	findByAge(Is)NotNull	… where x.age not null
Like	findByFirstnameLike	… where x.firstname like ?1
NotLike	findByFirstnameNotLike	… where x.firstname not like ?1
StartingWith	findByFirstnameStartingWith	… where x.firstname like ?1 参数后加%，即以参数开头的模糊查询
EndingWith	findByFirstnameEndingWith	… where x.firstname like ?1 参数前加%，即以参数结尾的模糊查询
Containing	findByFirstnameContaining	… where x.firstname like ?1 参数两边加%，即包含参数的模糊查询
OrderBy	findByAgeOrderByLastnameDesc	… where x.age = ?1 order by x.lastname desc
Not	findByLastnameNot	… where x.lastname <> ?1
In	findByAgeIn(Collection<Age> ages)	… where x.age in ?1
NotIn	findByAgeNotIn(Collection<Age> ages)	… where x.age not in ?1
True	findByActiveTrue()	… where x.active = true
False	findByActiveFalse()	… where x.active = false
IgnoreCase	findByFirstnameIgnoreCase	… where UPPER(x.firstname) = UPPER(?1)

❷ 限制查询结果数量

在 Spring Data JPA 中，使用 Top 和 First 关键字限制查询结果数量。示例如下：

```
public interface UserRepository extends JpaRepository<MyUser, Integer>{
    /**
     * 获得符合查询条件的前 10 条
     */
    public List<MyUser> findTop10ByUnameLike(String uname);
    /**
     * 获得符合查询条件的前 15 条
```

```
    */
    public List<MyUser> findFirst15ByUnameLike(String uname);
}
```

❸ 简单条件查询示例

下面通过实例讲解在 Spring Boot Web 应用中如何使用 Spring Data JPA 进行简单条件查询。

【例 8-1】 使用 Spring Data JPA 进行简单条件查询。

具体实现步骤如下。

1）创建数据库

本书采用的关系型数据库是 MySQL 5.x，为了演示本例，首先通过命令“CREATE DATABASE springbootjpa;”创建名为 springbootjpa 的数据库。

2）创建基于 Thymeleaf 和 Spring Data JPA 依赖的 Spring Boot Web 应用 ch8_1

参考图 8.1 创建基于 Thymeleaf 和 Spring Data JPA 依赖的 Spring Boot Web 应用 ch8_1。

3）修改 pom.xml 文件，添加 MySQL 依赖

在 pom.xml 文件中添加如下依赖：

```
<dependency>
    <groupId>mysql</groupId>
    <artifactId>mysql-connector-java</artifactId>
    <version>5.1.45</version>
</dependency>
```

4）设置 Web 应用 ch8_1 的上下文路径及数据源配置信息

在 ch8_1 的 application.properties 文件中配置如下内容：

```
server.servlet.context-path=/ch8_1
###
##数据源信息配置
###
#数据库地址
spring.datasource.url=jdbc:mysql://localhost:3306/springbootjpa?characterEncoding=utf8
#数据库 MySQL 为 8.x 时，url 为
#jdbc:mysql://localhost:3306/springbootjpa?useSSL=false&serverTimezone=Asia/
 Beijing&characterEncoding=utf-8
#数据库用户名
spring.datasource.username=root
#数据库密码
spring.datasource.password=root
#数据库驱动
spring.datasource.driver-class-name=com.mysql.jdbc.Driver
#数据库 MySQL 为 8.x 时，驱动类为 com.mysql.cj.jdbc.Driver
####
#JPA 持久化配置
####
#指定数据库类型
spring.jpa.database=MYSQL
#指定是否在日志中显示 SQL 语句
spring.jpa.show-sql=true
#指定自动创建、更新数据库表等配置，update 表示如果数据库中存在持久化类对应的表就不创建，
```

```
#不存在就创建
spring.jpa.hibernate.ddl-auto=update
#让控制器输出的 JSON 字符串格式更美观
spring.jackson.serialization.indent-output=true
```

5）创建持久化实体类 MyUser

创建名为 com.ch.ch8_1.entity 的包，并在该包中创建名为 MyUser 的持久化实体类。核心代码如下：

```
@Entity
@Table(name = "user_table")
public class MyUser implements Serializable{
    private static final long serialVersionUID = 1L;
    @Id
    @GeneratedValue(strategy = GenerationType.IDENTITY)
    private int id;//主键
    /**使用@Column 注解，可以配置列相关属性(列名、长度等)，
     *可以省略，默认为属性名小写，如果属性名是词组，将在中间加上“_”。
     */
    private String uname;
    private String usex;
    private int age;
    //省略 get 方法和 set 方法
}
```

在持久化类中，@Entity 注解表明该实体类是一个与数据库表映射的实体类。@Table 表示实体类与哪个数据库表映射，如果没有通过 name 属性指定表名，默认为小写的类名。如果类名为词组，将在中间加上“_”（如 MyUser 类对应的表名为 my_user）。@Id 注解的属性表示该属性映射为数据库表的主键。@GeneratedValue 注解默认使用主键生成方式为自增，如果是 MySQL、SQL Server 等关系型数据库，可映射成一个递增的主键；如果是 Oracle 等关系型数据库 hibernate，将自动生成一个名为 HIBERNATE_SEQUENCE 的序列。

6）创建数据访问层

创建名为 com.ch.ch8_1.repository 的包，并在该包中创建名为 UserRepository 的接口，该接口继承 JpaRepository 接口。核心代码如下：

```
/**
 * 这里不需要使用@Repository 注解数据访问层，
 * 因为 Spring Boot 自动配置了 JpaRepository
 */
public interface UserRepository extends JpaRepository<MyUser, Integer>{
    public MyUser findByUname(String uname);
    public List<MyUser> findByUnameLike(String uname);
}
```

由于 UserRepository 接口继承了 JpaRepository 接口，因此 UserRepository 接口中除了上述自定义的两个接口方法外（方法名命名规范参照表 8.1），还拥有 JpaRepository 的接口方法。

7）创建业务层

创建名为 com.ch.ch8_1.service 的包，并在该包中创建 UserService 接口和接口的实现类 UserServiceImpl。UserService 接口代码略。

UserServiceImpl 实现类的核心代码如下：

```
@Service
public class UserServiceImpl implements UserService{
    @Autowired//依赖注入数据访问层
    private UserRepository userRepository;
    @Override
    public void saveAll() {
        MyUser mu1 = new MyUser();
        mu1.setUname("陈恒 1");
        mu1.setUsex("男");
        mu1.setAge(88);
        MyUser mu2 = new MyUser();
        mu2.setUname("陈恒 2");
        mu2.setUsex("女");
        mu2.setAge(18);
        MyUser mu3 = new MyUser();
        mu3.setUname("陈恒 3");
        mu3.setUsex("男");
        mu3.setAge(99);
        List<MyUser> users = new ArrayList<MyUser>();
        users.add(mu1);
        users.add(mu2);
        users.add(mu3);
        //调用父接口中的方法 saveAll
        userRepository.saveAll(users);
    }
    @Override
    public List<MyUser> findAll() {
        //调用父接口中的方法 findAll
        return userRepository.findAll();
    }
    @Override
    public MyUser findByUname(String uname) {
        return userRepository.findByUname(uname);
    }
    @Override
    public List<MyUser> findByUnameLike(String uname) {
        return userRepository.findByUnameLike("%" + uname + "%");
    }
    @Override
    public MyUser getOne(int id) {
        //调用父接口中的方法 getOne
        return userRepository.getOne(id);
    }
}
```

8）创建控制器类 UserTestController

创建名为 com.ch.ch8_1.controller 的包，并在该包中创建名为 UserTestController 的控制器类。核心代码如下：

```
@Controller
public class UserTestController {
```

```
    @Autowired
    private UserService userService;
    @RequestMapping("/save")
    @ResponseBody
    public String save() {
        userService.saveAll();
        return "保存用户成功！";
    }
    @RequestMapping("/findByUname")
    public String findByUname(String uname, Model model) {
        model.addAttribute("title", "根据用户名查询一个用户");
        model.addAttribute("auser", userService.findByUname(uname));
        return "showAuser";
    }
    @RequestMapping("/getOne")
    public String getOne(int id, Model model) {
        model.addAttribute("title", "根据用户 id 查询一个用户");
        model.addAttribute("auser",userService.getOne(id));
        return "showAuser";
    }
    @RequestMapping("/findAll")
    public String findAll(Model model){
        model.addAttribute("title", "查询所有用户");
        model.addAttribute("allUsers",userService.findAll());
        return "showAll";
    }
    @RequestMapping("/findByUnameLike")
    public String findByUnameLike(String uname, Model model){
        model.addAttribute("title", "根据用户名模糊查询所有用户");
        model.addAttribute("allUsers",userService.findByUnameLike(uname));
        return "showAll";
    }
}
```

9）整理脚本样式静态文件

JS 脚本、CSS 样式、图片等静态文件默认放置在 src/main/resources/static 目录下，ch8_1 应用引入的 BootStrap 和 jQuery 与例 7-5 中的一样，不再赘述。

10）创建 View 视图页面

在 src/main/resources/templates 目录下，创建视图页面 showAll.html 和 showAuser.html。

showAll.html 的核心代码如下：

```
<body>
    <div class="panel panel-primary">
        <div class="panel-heading">
            <h3 class="panel-title">Spring Data JPA 简单查询</h3>
        </div>
    </div>
    <div class="container">
        <div class="panel panel-primary">
            <div class="panel-heading">
                <h3 class="panel-title"><span th:text="${title}"></span></h3>
```

```
            </div>
            <div class="panel-body">
                <div class="table table-responsive">
                    <table class="table table-bordered table-hover">
                        <tbody class="text-center">
                            <tr th:each="user:${allUsers}">
                                <td><span th:text="${user.id}"></span></td>
                                <td><span th:text="${user.uname}"></span></td>
                                <td><span th:text="${user.usex}"></span></td>
                                <td><span th:text="${user.age}"></span></td>
                            </tr>
                        </tbody>
                    </table>
                </div>
            </div>
        </div>
    </div>
</body>
```

showAuser.html 的核心代码如下：

```
<body>
    <div class="panel panel-primary">
        <div class="panel-heading">
            <h3 class="panel-title">Spring Data JPA 简单查询</h3>
        </div>
    </div>
    <div class="container">
        <div class="panel panel-primary">
            <div class="panel-heading">
                <h3 class="panel-title"><span th:text="${title}"></span></h3>
            </div>
            <div class="panel-body">
                <div class="table table-responsive">
                    <table class="table table-bordered table-hover">
                        <tbody class="text-center">
                            <tr>
                                <td><span th:text="${auser.id}"></span></td>
                                <td><span th:text="${auser.uname}"></span></td>
                                <td><span th:text="${auser.usex}"></span></td>
                                <td><span th:text="${auser.age}"></span></td>
                            </tr>
                        </tbody>
                    </table>
                </div>
            </div>
        </div>
    </div>
</body>
```

11）运行

首先，运行 Ch81Application 主类。然后，访问 http://localhost:8080/ch8_1/save/保存用户数据。http://localhost:8080/ch8_1/save/成功运行后，在 MySQL 的 springbootjpa 数据库中创建一张名为 user_table 的数据库表，并插入三条记录。

通过访问“http://localhost:8080/ch8_1/findAll”查询所有用户，运行效果如图 8.3 所示。

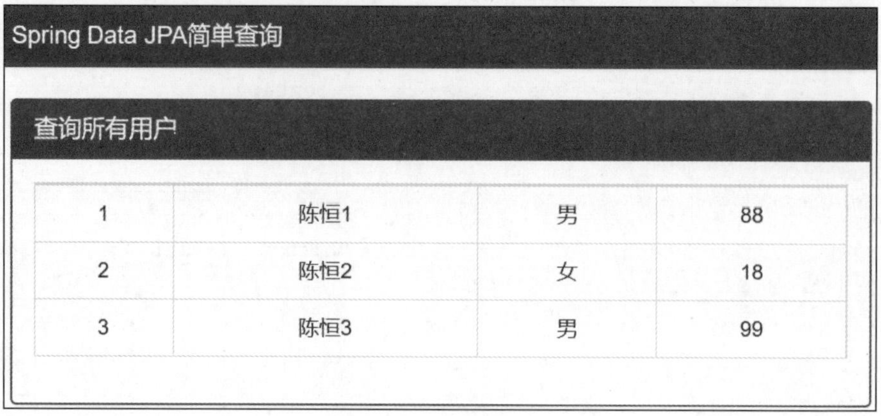

Spring Data JPA简单查询

查询所有用户

1	陈恒1	男	88
2	陈恒2	女	18
3	陈恒3	男	99

图 8.3　查询所有用户

通过访问“http://localhost:8080/ch8_1/findByUnameLike?uname=陈”模糊查询所有陈姓用户，运行效果如图 8.4 所示。

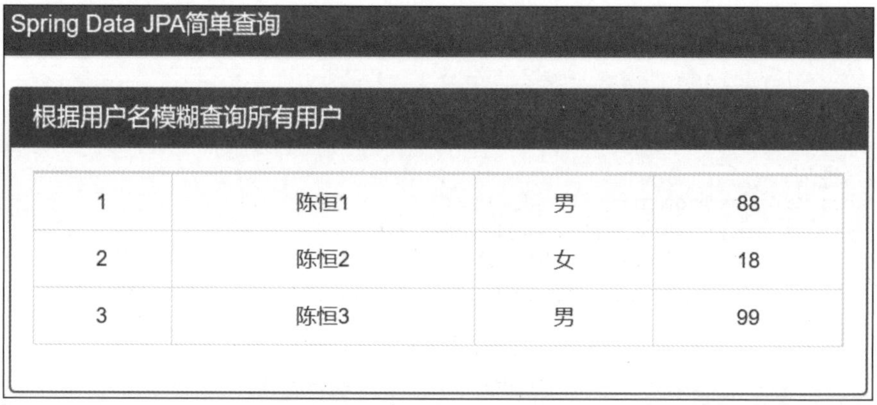

Spring Data JPA简单查询

根据用户名模糊查询所有用户

1	陈恒1	男	88
2	陈恒2	女	18
3	陈恒3	男	99

图 8.4　模糊查询所有陈姓用户

通过访问“http://localhost:8080/ch8_1/findByUname?uname=陈恒 2”查询一个名为“陈恒 2”的用户信息，运行效果如图 8.5 所示。

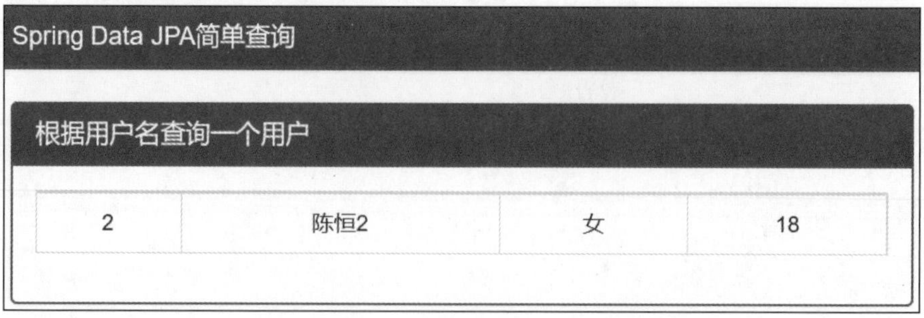

Spring Data JPA简单查询

根据用户名查询一个用户

2	陈恒2	女	18

图 8.5　查询一个名为“陈恒 2”的用户信息

通过访问“http://localhost:8080/ch8_1/getOne?id=1”查询一个 id 为 1 的用户信息，运行效果如图 8.6 所示。

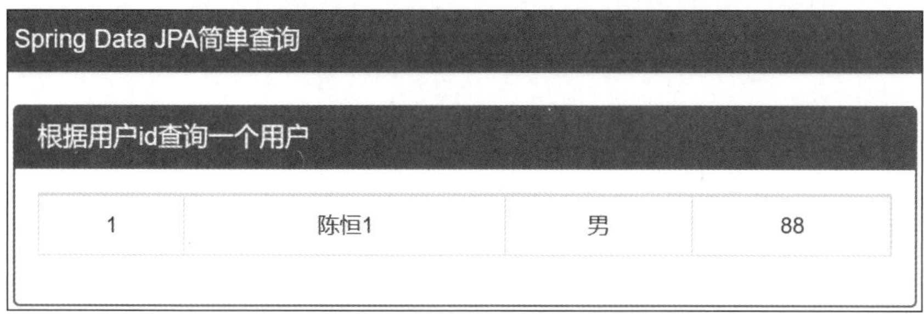

图 8.6 查询一个 id 为 1 的用户信息

8.1.3 关联查询

在 Spring Data JPA 中有一对一、一对多、多对多等关系映射。本节将针对这些关系映射进行讲解。

❶ @OneToOne

一对一关系，在现实生活中是十分常见的。例如一个大学生只有一张一卡通，一张一卡通只属于一个大学生。再如人与身份证的关系也是一对一的关系。

在 Spring Data JPA 中，可用两种方式描述一对一关系映射：一种是通过外键的方式（一个实体通过外键关联到另一个实体的主键）；一种是通过一张关联表来保存两个实体一对一的关系。下面通过外键的方式讲解一对一关系映射。

【例 8-2】使用 Spring Data JPA 实现人与身份证的一对一关系映射。

首先，为例 8-2 创建基于 Spring Data JPA 依赖的 Spring Boot Web 应用 ch8_2。ch8_2 应用的数据库、pom.xml 以及 application.properties 与 ch8_1 应用基本一样，不再赘述。

其他具体实现步骤如下。

1）创建持久化实体类

创建名为 com.ch.ch8_2.entity 的包，并在该包中创建名为 Person 和 IdCard 的持久化实体类。

Person 的核心代码如下：

```
@Entity
@Table(name="person_table")
/**解决 No serializer found for class org.hibernate.proxy.pojo.bytebuddy.
ByteBuddyInterceptor 异常*/
@JsonIgnoreProperties(value = {"hibernateLazyInitializer"})
public class Person implements Serializable{
    private static final long serialVersionUID = 1L;
    @Id
    @GeneratedValue(strategy = GenerationType.IDENTITY)
    private int id;//自动递增的主键
    private String pname;
    private String psex;
    private int page;
    @OneToOne(
            optional = true,
            fetch = FetchType.LAZY,
            targetEntity = IdCard.class,
```

```
                cascade = CascadeType.ALL
                )
        /**
        *指明 Person 对应表的 id_Card_id 列作为外键与 IdCard 对应表的 id 列进行关联
        * unique= true 指明 id_Card_id 列的值不可重复
        */
        @JoinColumn(
                name = "id_Card_id",
                referencedColumnName = "id",
                unique= true
                )
        @JsonIgnore
        //如果 A 对象持有 B 的引用，B 对象持有 A 的引用，这样就形成了循环引用，
        //如果直接使用 JSON 转换会报错，使用@JsonIgnore 解决该错误
        private IdCard idCard;
        //省略 get 方法和 set 方法
}
```

上述实体类 Person 中，@OneToOne 注解有 5 个属性：optional、fetch、targetEntity、cascade 和 mappedBy。

optional = true：表示 idCard 属性可以为 null，也就是允许没有身份证，如未成年人没有身份证。

FetchType.LAZY：懒加载，加载一个实体时，定义懒加载的属性不会马上从数据库中加载。FetchType.EAGER：急加载，加载一个实体时，定义急加载的属性会立即从数据库中加载。

targetEntity 属性：class 类型属性。定义关系类的类型，默认是该成员属性对应的类类型，所以通常不需要提供定义。

cascade 属性：CascadeType[]类型。该属性定义类和类之间的级联关系。定义的级联关系将被容器视为对当前类对象及其关联类对象采取相同的操作，而且这种关系是递归调用的。cascade 的值只能从 CascadeType.PERSIST（级联新建）、CascadeType.REMOVE（级联删除）、CascadeType.REFRESH（级联刷新）、CascadeType.MERGE（级联更新）中选择一个或多个。还有一个选择是使用 CascadeType.ALL，表示选择全部四项。

mappedBy 标签：一定是定义在关系的被维护端，它指向关系的维护端；只有@OneToOne、@OneToMany、@ManyToMany 有 mappedBy 属性，ManyToOne 不存在该属性。拥有 mappedBy 注解的实体类为关系的被维护端。

IdCard 的核心代码如下：

```
@Entity
@Table(name = "idcard_table")
@JsonIgnoreProperties(value = { "hibernateLazyInitializer"})
public class IdCard implements Serializable{
    private static final long serialVersionUID = 1L;
    @Id
    @GeneratedValue(strategy = GenerationType.IDENTITY)
    private int id;//自动递增的主键
    private String code;
    /**
     * @Temporal 主要用来指明 java.util.Date 或 java.util.Calendar 类型的属性具体
     * 与数据库（date、time、timestamp）3 个类型中的哪一个进行映射
```

```
     */
    @Temporal(value = TemporalType.DATE)
    private Calendar birthday;
    private String address;
    /**
     * optional = false 设置 person 属性值不能为 null，也就是身份证必须有对应的主人。
     * mappedBy = "idCard"与 Person 类中的 idCard 属性一致
     *拥有 mappedBy 注解的实体类为关系的被维护端
     */
    @OneToOne(
            optional = true,
            fetch = FetchType.LAZY,
            targetEntity = Person.class,
            mappedBy = "idCard",
            cascade = CascadeType.ALL
            )
    private Person person;//对应的人
    //省略 get 方法和 set 方法
}
```

2）创建数据访问层

创建名为 com.ch.ch8_2.repository 的包，并在该包中创建名为 IdCardRepository 和 PersonRepository 的接口。

IdCardRepository 的核心代码如下：

```
public interface IdCardRepository extends JpaRepository<IdCard, Integer>{
    /**
     * 根据人员 ID 查询身份信息（关联查询，根据 person 属性的 id）
     * 相当于 JPQL 语句：select ic from IdCard ic where ic.person.id = ?1
     */
    public IdCard findByPerson_id(Integer id);
    /**
     * 根据地址和身份证号查询身份信息
     * 相当于 JPQL 语句：select ic from IdCard ic where ic.address = ?1 and ic.code =?2
     */
    public List<IdCard> findByAddressAndCode(String address, String code);
}
```

按照 Spring Data JPA 的规则，查询两个有关联关系的对象，可以通过方法名中的“_”来标识。例如根据人员 ID 查询身份信息 findByPerson_id。JPQL（Java Persistence Query Language）是一种和 SQL 非常类似的中间性和对象化查询语言，它最终被编译成针对不同底层数据库的 SQL 查询，从而屏蔽不同数据库的差异。JPQL 语句可以是 select 语句、update 语句或 delete 语句，它们都通过 Query 接口封装执行。JPQL 的具体学习内容不是本书涉及的内容，需要学习的读者请参考相关内容学习。

PersonRepository 的核心代码如下：

```
public interface PersonRepository extends JpaRepository<Person, Integer>{
    /**
     * 根据身份证 ID 查询人员信息（关联查询，根据 idCard 属性的 id）
     * 相当于 JPQL 语句：select p from Person p where p.idCard.id = ?1
```

```
     */
    public Person findByIdCard_id(Integer id);
    /**
     * 根据人名和性别查询人员信息
     * 相当于 JPQL 语句：select p from Person p where p.pname = ?1 and p.psex = ?2
     */
    public List<Person> findByPnameAndPsex(String pname, String psex);
}
```

3）创建业务层

创建名为 com.ch.ch8_2.service 的包，并在该包中创建名为 PersonAndIdCardService 的接口和接口实现类 PersonAndIdCardServiceImpl。PersonAndIdCardService 的代码略。

PersonAndIdCardServiceImpl 的核心代码如下：

```
@Service
public class PersonAndIdCardServiceImpl implements PersonAndIdCardService{
    @Autowired
    private IdCardRepository idCardRepository;
    @Autowired
    private PersonRepository personRepository;
    @Override
    public void saveAll() {
        //保存身份证
        IdCard ic1 = new IdCard();
        ic1.setCode("123456789");
        ic1.setAddress("北京");
        Calendar c1 = Calendar.getInstance();
        c1.set(2019, 8, 13);
        ic1.setBirthday(c1);
        IdCard ic2 = new IdCard();
        ic2.setCode("000123456789");
        ic2.setAddress("上海");
        Calendar c2 = Calendar.getInstance();
        c2.set(2019, 8, 14);
        ic2.setBirthday(c2);
        IdCard ic3 = new IdCard();
        ic3.setCode("1111123456789");
        ic3.setAddress("广州");
        Calendar c3 = Calendar.getInstance();
        c3.set(2019, 8, 15);
        ic3.setBirthday(c3);
        List<IdCard> idCards = new ArrayList<IdCard>();
        idCards.add(ic1);
        idCards.add(ic2);
        idCards.add(ic3);
        idCardRepository.saveAll(idCards);
        //保存人员
        Person p1 = new Person();
        p1.setPname("陈恒 1");
        p1.setPsex("男");
        p1.setPage(88);
        p1.setIdCard(ic1);
```

```
        Person p2 = new Person();
        p2.setPname("陈恒 2");
        p2.setPsex("女");
        p2.setPage(99);
        p2.setIdCard(ic2);
        Person p3 = new Person();
        p3.setPname("陈恒 3");
        p3.setPsex("女");
        p3.setPage(18);
        p3.setIdCard(ic3);
        List<Person> persons = new ArrayList<Person>();
        persons.add(p1);
        persons.add(p2);
        persons.add(p3);
        personRepository.saveAll(persons);
    }
    @Override
    public List<Person> findAllPerson() {
        return personRepository.findAll();
    }
    @Override
    public List<IdCard> findAllIdCard() {
        return idCardRepository.findAll();
    }
    /**
     * 根据人员 ID 查询身份信息（关联查询）
     */
    @Override
    public IdCard findByPerson_id(Integer id) {
        return idCardRepository.findByPerson_id(id);
    }
    @Override
    public List<IdCard> findByAddressAndCode(String address, String code) {
        return idCardRepository.findByAddressAndCode(address, code);
    }
    /**
     * 根据身份证 ID 查询人员信息（关联查询）
     */
    @Override
    public Person findByIdCard_id(Integer id) {
        return personRepository.findByIdCard_id(id);
    }
    @Override
    public List<Person> findByPnameAndPsex(String pname, String psex) {
        return personRepository.findByPnameAndPsex(pname, psex);
    }
    @Override
    public IdCard getOneIdCard(Integer id) {
        return idCardRepository.getOne(id);
    }
    @Override
    public Person getOnePerson(Integer id) {
```

```
        return personRepository.getOne(id);
    }
}
```

4）创建控制器类

创建名为 com.ch.ch8_2.controller 的包，并在该包中创建名为 TestOneToOneController 的控制器类。

TestOneToOneController 的核心代码如下：

```
@RestController
public class TestOneToOneController {
    @Autowired
    private PersonAndIdCardService personAndIdCardService;
    @RequestMapping("/save")
    public String save() {
        personAndIdCardService.saveAll();
        return "人员和身份保存成功！";
    }
    @RequestMapping("/findAllPerson")
    public List<Person> findAllPerson() {
        return  personAndIdCardService.findAllPerson();
    }
    @RequestMapping("/findAllIdCard")
    public List<IdCard>  findAllIdCard() {
        return personAndIdCardService.findAllIdCard();
    }
    /**
     * 根据人员 ID 查询身份信息（关联查询）
     */
    @RequestMapping("/findByPerson_id")
    public IdCard findByPerson_id(Integer id) {
        return personAndIdCardService.findByPerson_id(id);
    }
    @RequestMapping("/findByAddressAndCode")
    public List<IdCard> findByAddressAndCode(String address, String code){
        return personAndIdCardService.findByAddressAndCode(address, code);
    }
    /**
     * 根据身份证 ID 查询人员信息（关联查询）
     */
    @RequestMapping("/findByIdCard_id")
    public Person findByIdCard_id(Integer id) {
        return personAndIdCardService.findByIdCard_id(id);
    }
    @RequestMapping("/findByPnameAndPsex")
    public List<Person> findByPnameAndPsex(String pname, String psex) {
        return personAndIdCardService.findByPnameAndPsex(pname, psex);
    }
    @RequestMapping("/getOneIdCard")
    public IdCard getOneIdCard(Integer id) {
        return personAndIdCardService.getOneIdCard(id);
    }
```

```
    @RequestMapping("/getOnePerson")
    public Person getOnePerson(Integer id) {
        return personAndIdCardService.getOnePerson(id);
    }
}
```

5）运行

首先，运行 Ch82Application 主类。然后，访问 http://localhost:8080/ch8_2/save/。

http://localhost:8080/ch8_2/save/成功运行后，在 MySQL 的 springbootjpa 数据库中创建名为 idcard_table 和 person_table 的数据库表（实体类成功加载后就已创建好数据表），并分别插入 3 条记录。

通过“http://localhost:8080/ch8_2/findByIdCard_id?id=1”查询身份证 ID 为 1 的人员信息（关联查询），运行效果如图 8.7 所示。

```
{
  "id" : 1,
  "pname" : "陈恒1",
  "psex" : "男",
  "page" : 88
}
```

图 8.7　查询身份证 ID 为 1 的人员信息

通过“http://localhost:8080/ch8_2/findByPerson_id?id=1”查询人员 ID 为 1 的身份证信息（关联查询），运行效果如图 8.8 所示。

```
{
  "id" : 1,
  "code" : "123456789",
  "birthday" : "2019-09-12T16:00:00.000+00:00",
  "address" : "北京",
  "person" : {
    "id" : 1,
    "pname" : "陈恒1",
    "psex" : "男",
    "page" : 88
  }
}
```

图 8.8　查询人员 ID 为 1 的身份证信息

❷ @OneToMany 和@ManyToOne

在实际生活中，作者和文章是一对多的双向关系。那么在 Spring Data JPA 中，如何描述一对多的双向关系呢？

在 Spring Data JPA 中，使用@OneToMany 和@ManyToOne 表示一对多的双向关联。例如，一端（Author）使用@OneToMany，多端（Article）使用@ManyToOne。

在 JPA 规范中，一对多的双向关系由多端（如 Article）来维护。就是说多端为关系的维护端，负责关系的增删改查；一端则为关系的被维护端，不能维护关系。

一端（Author）使用@OneToMany 注解的 mappedBy="author"属性表明一端（Author）是

关系的被维护端。多端（Article）使用@ManyToOne 和@JoinColumn 注解属性 author，@ManyToOne 表明 Article 是多端，@JoinColumn 设置在 article 表的关联字段（外键）上。

【例 8-3】使用 Spring Data JPA 实现 Author 与 Article 的一对多关系映射。

在 ch8_2 应用中实现例 8-3，具体实现步骤如下。

1）添加 hibernate-validator 依赖

因为在持久化实体类中，使用 hibernate-validator 约束数据表，所以需要在 ch8_2 应用的 pom.xml 文件中添加 hibernate-validator 依赖。具体代码如下：

```
<dependency>
    <groupId>org.hibernate.validator</groupId>
    <artifactId>hibernate-validator</artifactId>
</dependency>
```

2）创建持久化实体类

在 com.ch.ch8_2.entity 包中，创建名为 Author 和 Article 的持久化实体类。

Author 的核心代码如下：

```
@Entity
@Table(name = "author_table")
@JsonIgnoreProperties(value = { "hibernateLazyInitializer"})
public class Author implements Serializable{
    private static final long serialVersionUID = 1L;
    @Id
    @GeneratedValue(strategy = GenerationType.IDENTITY)
    private int id;
    //作者名
    private String aname;
    //文章列表，作者与文章是一对多的关系
   @OneToMany(
        mappedBy = "author",
        cascade=CascadeType.ALL,
        targetEntity = Article.class,
        fetch=FetchType.LAZY
        )
   private List<Article> articleList;
    //省略 set 方法和 get 方法
}
```

Article 的代码如下：

```
@Entity
@Table(name = "article_table")
@JsonIgnoreProperties(value = { "hibernateLazyInitializer"})
public class Article  implements Serializable{
    private static final long serialVersionUID = 1L;
    @Id
    @GeneratedValue(strategy = GenerationType.IDENTITY)
    private int id;
    //标题
    @NotEmpty(message = "标题不能为空")
   @Size(min = 2, max = 50)
```

```
    @Column(nullable = false, length = 50)
    private String title;
    //文章内容
    @Lob  //大对象，映射为 MySQL 的 Long 文本类型
    @Basic(fetch = FetchType.LAZY)
    @NotEmpty(message = "内容不能为空")
    @Size(min = 2)
    @Column(nullable = false)
    private String content;
    //所属作者,文章与作者是多对一的关系
    @ManyToOne(cascade={CascadeType.MERGE,CascadeType.REFRESH},optional=false)
    //可选属性 optional=false，表示 author 不能为空。删除文章,不影响用户
    @JoinColumn(name="id_author_id")//设置在 article 表中的关联字段（外键）
    @JsonIgnore
    private Author author;
    //省略 set 方法和 get 方法
}
```

3）创建数据访问层

在 com.ch.ch8_2.repository 包中，创建名为 AuthorRepository 和 ArticleRepository 的接口。AuthorRepository 的核心代码如下：

```
public interface AuthorRepository extends JpaRepository<Author, Integer>{
    /**
     * 根据文章标题包含的内容，查询作者(关联查询)
     * 相当于 JPQL 语句：select a from Author a inner join a.articleList t where
     t.title like %?1%
     */
    public Author findByArticleList_titleContaining(String title);
}
```

ArticleRepository 的核心代码如下：

```
public interface ArticleRepository extends JpaRepository<Article, Integer>{
    /**
     * 根据作者 id 查询文章信息(关联查询，根据 author 属性的 id)
     * 相当于 JPQL 语句：select a from Article a where a.author.id = ?1
     */
    public List<Article> findByAuthor_id(Integer id);
    /**
     * 根据作者名查询文章信息(关联查询，根据 author 属性的 aname)
     * 相当于 JPQL 语句：select a from Article a where a.author.aname = ?1
     */
    public List<Article> findByAuthor_aname(String aname);
}
```

4）创建业务层

在 com.ch.ch8_2.service 包中，创建名为 AuthorAndArticleService 的接口和接口实现类 AuthorAndArticleServiceImpl。AuthorAndArticleService 接口代码略。

AuthorAndArticleServiceImpl 的核心代码如下：

```
@Service
public class AuthorAndArticleServiceImpl implements AuthorAndArticleService{
```

```
    @Autowired
    private AuthorRepository authorRepository;
    @Autowired
    private ArticleRepository articleRepository;
    @Override
    public void saveAll() {
        //保存作者(先保存一的一端)
        Author a1 = new Author();
        a1.setAname("陈恒 1");
        Author a2 = new Author();
        a2.setAname("陈恒 2");
        ArrayList<Author> allAuthor = new ArrayList<Author>();
        allAuthor.add(a1);
        allAuthor.add(a2);
        authorRepository.saveAll(allAuthor);
        //保存文章
        Article at1 = new Article();
        at1.setTitle("JPA 的一对多 111");
        at1.setContent("其实一对多映射关系很常见 111。");
        //设置关系
        at1.setAuthor(a1);
        Article at2 = new Article();
        at2.setTitle("JPA 的一对多 222");
        at2.setContent("其实一对多映射关系很常见 222。");
        //设置关系
        at2.setAuthor(a1);//文章 2 与文章 1 作者相同
        Article at3 = new Article();
        at3.setTitle("JPA 的一对多 333");
        at3.setContent("其实一对多映射关系很常见 333。");
        //设置关系
        at3.setAuthor(a2);
        Article at4 = new Article();
        at4.setTitle("JPA 的一对多 444");
        at4.setContent("其实一对多映射关系很常见 444。");
        //设置关系
        at4.setAuthor(a2);//文章 3 与文章 4 作者相同
        ArrayList<Article> allAt = new ArrayList<Article>();
        allAt.add(at1);
        allAt.add(at2);
        allAt.add(at3);
        allAt.add(at4);
        articleRepository.saveAll(allAt);
    }
    @Override
    public List<Article> findByAuthor_id(Integer id) {
        return articleRepository.findByAuthor_id(id);
    }
    @Override
    public List<Article> findByAuthor_aname(String aname) {
        return articleRepository.findByAuthor_aname(aname);
    }
    @Override
```

```
    public Author findByArticleList_titleContaining(String title) {
        return authorRepository.findByArticleList_titleContaining(title);
    }
}
```

5）创建控制器类

在 com.ch.ch8_2.controller 包中，创建名为 TestOneToManyController 的控制器类。TestOneToManyController 的核心代码如下：

```
@RestController
public class TestOneToManyController {
    @Autowired
    private AuthorAndArticleService authorAndArticleService;
    @RequestMapping("/saveOneToMany")
    public String save() {
        authorAndArticleService.saveAll();
        return "作者和文章保存成功！";
    }
    @RequestMapping("/findArticleByAuthor_id")
    public List<Article> findByAuthor_id(Integer id) {
        return authorAndArticleService.findByAuthor_id(id);
    }
    @RequestMapping("/findArticleByAuthor_aname")
    public List<Article> findByAuthor_aname(String aname){
        return authorAndArticleService.findByAuthor_aname(aname);
    }
    @RequestMapping("/findByArticleList_titleContaining")
    public Author findByArticleList_titleContaining(String title) {
        return authorAndArticleService.findByArticleList_titleContaining(title);
    }
}
```

6）运行

首先，运行 Ch82Application 主类。然后，访问 http://localhost:8080/ch8_2/saveOneToMany/。

http://localhost:8080/ch8_2/saveOneToMany/成功运行后，在 MySQL 的 springbootjpa 数据库中创建名为 author_table 和 article_table 的数据库表，并在 author_table 表中插入两条记录，同时在 article_table 表中插入四条记录。

通过“http://localhost:8080/ch8_2/findArticleByAuthor_id?id=2”查询作者 id 为 2 的文章列表（关联查询），运行效果如图 8.9 所示。

```
[ {
  "id" : 3,
  "title" : "JPA的一对多333",
  "content" : "其实一对多映射关系很常见333。"
}, {
  "id" : 4,
  "title" : "JPA的一对多444",
  "content" : "其实一对多映射关系很常见444。"
} ]
```

图 8.9　查询作者 ID 为 2 的文章列表

通过“http://localhost:8080/ch8_2/findArticleByAuthor_aname?aname=陈恒 1”查询作者名为“陈恒 1”的文章列表（关联查询），运行效果如图 8.10 所示。

```
[ {
  "id" : 1,
  "title" : "JPA的一对多111",
  "content" : "其实一对多映射关系很常见111。"
}, {
  "id" : 2,
  "title" : "JPA的一对多222",
  "content" : "其实一对多映射关系很常见222。"
} ]
```

图 8.10　查询作者名为“陈恒 1”的文章列表

通过“http://localhost:8080/ch8_2/findByArticleList_titleContaining?title=对多 1”查询文章标题包含“对多 1”的作者（关联查询），运行效果如图 8.11 所示。

```
{
  "id" : 1,
  "aname" : "陈恒1",
  "articleList" : [ {
    "id" : 1,
    "title" : "JPA的一对多111",
    "content" : "其实一对多映射关系很常见111。"
  }, {
    "id" : 2,
    "title" : "JPA的一对多222",
    "content" : "其实一对多映射关系很常见222。"
  } ]
}
```

图 8.11　查询文章标题包含“对多 1”的作者

❸ @ManyToMany

在实际生活中，用户和权限是多对多的关系。一个用户可以有多个权限，一个权限也可以被很多用户拥有。

在 Spring Data JPA 中使用@ManyToMany 注解多对多的映射关系，由一个关联表来维护。关联表的表名默认是：主表名+下画线+从表名（主表是指关系维护端对应的表，从表是指关系被维护端对应的表）。关联表只有两个外键字段，分别指向主表 ID 和从表 ID。字段的名称默认为：主表名+下画线+主表中的主键列名，从表名+下画线+从表中的主键列名。需要注意的是，多对多关系中一般不设置级联保存、级联删除、级联更新等操作。

【例 8-4】使用 Spring Data JPA 实现用户（User）与权限（Authority）的多对多关系映射。

在 ch8_2 应用中实现例 8-4，具体实现步骤如下。

1）创建持久化实体类

在 com.ch.ch8_2.entity 包中，创建名为 User 和 Authority 的持久化实体类。

User 的代码如下：

```
@Entity
@Table(name = "user")
```

```
    @JsonIgnoreProperties(value = { "hibernateLazyInitializer"})
    public class User implements Serializable{
        private static final long serialVersionUID = 1L;
        @Id
        @GeneratedValue(strategy = GenerationType.IDENTITY)
        private int id;
        private String username;
        private String password;
        @ManyToMany
        @JoinTable(name="user_authority",joinColumns=@JoinColumn(name="user_id"),
        inverseJoinColumns = @JoinColumn(name = "authority_id"))
        /**1. 关系维护端,负责多对多关系的绑定和解除
        2. @JoinTable 注解的 name 属性指定关联表的名字，joinColumns 指定外键的名字,关联到
关系维护端(User)
        3. inverseJoinColumns 指定外键的名字，需要关联的关系，称为被维护端(Authority)
        4. 其实可以不使用@JoinTable 注解，默认生成的关联表名称为主表表名+下画线+从表表名，
即表名为 user_authority
        关联到主表的外键名：主表名+下画线+主表中的主键列名，即 user_id。
        关联到从表的外键名：主表中用于关联的属性名+下画线+从表的主键列名，即 authority_id。
主表是关系维护端对应的表，从表是关系被维护端对应的表
        */
        private List<Authority> authorityList;
        //省略 get 方法和 set 方法
    }
```

Authority 的核心代码如下：

```
    @Entity
    @Table(name = "authority")
    @JsonIgnoreProperties(value = { "hibernateLazyInitializer"})
    public class Authority implements Serializable{
        private static final long serialVersionUID = 1L;
        @Id
        @GeneratedValue(strategy = GenerationType.IDENTITY)
        private int id;
        @Column(nullable = false)
        private String name;
        @ManyToMany(mappedBy = "authorityList")
        @JsonIgnore
        private List<User> userList;
        //省略 get 和 set 方法
    }
```

2）创建数据访问层

在 com.ch.ch8_2.repository 包中，创建名为 UserRepository 和 AuthorityRepository 的接口。UserRepository 的核心代码如下：

```
    public interface UserRepository extends JpaRepository<User, Integer>{
        /**
         * 根据权限 id 查询拥有该权限的用户（关联查询）
         * 相当于 JPQL 语句: select u from User u inner join u.authorityList a where a.id = ?1
         */
```

```
    public List<User> findByAuthorityList_id(int id);
    /**
     * 根据权限名查询拥有该权限的用户（关联查询）
     * 相当于 JPQL 语句：select u from User u inner join u.authorityList a where a.name = ?1
     */
    public List<User> findByAuthorityList_name(String name);
}
```

AuthorityRepository 的核心代码如下：

```
public interface AuthorityRepository extends JpaRepository<Authority, Integer>{
    /**
     * 根据用户 id 查询用户所拥有的权限（关联查询）
     * 相当于 JPQL 语句：select a from Authority a inner join a.userList u where u.id = ?1
     */
    public List<Authority> findByUserList_id(int id);
    /**
     * 根据用户名查询用户所拥有的权限（关联查询）
     * 相当于 JPQL 语句：select a from Authority a inner join a.userList u where u.username = ?1
     */
    public List<Authority> findByUserList_Username(String username);
}
```

3）创建业务层

在 com.ch.ch8_2.service 包中，创建名为 UserAndAuthorityService 的接口和接口实现类 UserAndAuthorityServiceImpl。UserAndAuthorityService 接口代码略。

UserAndAuthorityServiceImpl 的核心代码如下：

```
@Service
public class UserAndAuthorityServiceImpl implements UserAndAuthorityService{
    @Autowired
    private AuthorityRepository authorityRepository;
    @Autowired
    private UserRepository userRepository;
    @Override
    public void saveAll() {
        //添加权限 1
        Authority at1 = new Authority();
        at1.setName("增加");
        authorityRepository.save(at1);
        //添加权限 2
        Authority at2 = new Authority();
        at2.setName("修改");
        authorityRepository.save(at2);
        //添加权限 3
        Authority at3 = new Authority();
        at3.setName("删除");
        authorityRepository.save(at3);
        //添加权限 4
        Authority at4 = new Authority();
        at4.setName("查询");
        authorityRepository.save(at4);
```

```
        //添加用户 1
        User u1 = new User();
        u1.setUsername("陈恒 1");
        u1.setPassword("123");
        ArrayList<Authority> authorityList1 = new ArrayList<Authority>();
        authorityList1.add(at1);
        authorityList1.add(at2);
        authorityList1.add(at3);
        u1.setAuthorityList(authorityList1);
        userRepository.save(u1);
        //添加用户 2
        User u2 = new User();
        u2.setUsername("陈恒 2");
        u2.setPassword("234");
        ArrayList<Authority> authorityList2 = new ArrayList<Authority>();
        authorityList2.add(at2);
        authorityList2.add(at3);
        authorityList2.add(at4);
        u2.setAuthorityList(authorityList2);
        userRepository.save(u2);
    }
    @Override
    public List<User> findByAuthorityList_id(int id) {
        return userRepository.findByAuthorityList_id(id);
    }
    @Override
    public List<User> findByAuthorityList_name(String name) {
        return userRepository.findByAuthorityList_name(name);
    }
    @Override
    public List<Authority> findByUserList_id(int id) {
        return authorityRepository.findByUserList_id(id);
    }
    @Override
    public List<Authority> findByUserList_Username(String username) {
        return authorityRepository.findByUserList_Username(username);
    }
}
```

4）创建控制器类

在 com.ch.ch8_2.controller 包中，创建名为 TestManyToManyController 的控制器类。TestManyToManyController 的核心代码如下：

```
@RestController
public class TestManyToManyController {
    @Autowired
    private UserAndAuthorityService userAndAuthorityService;
    @RequestMapping("/saveManyToMany")
    public String save() {
        userAndAuthorityService.saveAll();
        return "权限和用户保存成功！";
    }
```

```
    @RequestMapping("/findByAuthorityList_id")
    public List<User> findByAuthorityList_id(int id) {
        return userAndAuthorityService.findByAuthorityList_id(id);
    }
    @RequestMapping("/findByAuthorityList_name")
    public List<User> findByAuthorityList_name(String name) {
        return userAndAuthorityService.findByAuthorityList_name(name);
    }
    @RequestMapping("/findByUserList_id")
    public List<Authority> findByUserList_id(int id) {
        return userAndAuthorityService.findByUserList_id(id);
    }
    @RequestMapping("/findByUserList_Username")
    public List<Authority> findByUserList_Username(String username) {
        return userAndAuthorityService.findByUserList_Username(username);
    }
}
```

5）运行

首先，运行 Ch82Application 主类。然后，访问 http://localhost:8080/ch8_2/saveManyToMany/。

http://localhost:8080/ch8_2/saveManyToMany/成功运行后，可以通过"http://localhost:8080/ch8_2/findByAuthorityList_id?id=1"查询拥有 id 为 1 的权限的用户列表（关联查询）；通过"http://localhost:8080/ch8_2/findByAuthorityList_name?name=修改"查询拥有"修改"权限的用户列表（关联查询）；通过"http://localhost:8080/ch8_2/findByUserList_id?id=2"查询 id 为 2 的用户的权限列表（关联查询）；通过"http://localhost:8080/ch8_2/findByUserList_Username?username=陈恒 2"查询用户名为"陈恒 2"的用户的权限列表（关联查询）。

在 8.1.2 节和本节中的查询方法必须严格按照 Spring Data JPA 的查询关键字命名规范进行查询方法命名。那么，如何摆脱查询关键字和关联查询命名规范约束呢？可以通过@Query、@NamedQuery 直接定义 JPQL 语句进行数据的访问操作。

视频讲解

8.1.4 @Query 和@Modifying 注解

❶ @Query 注解

使用@Query 注解可以将 JPQL 语句直接定义在数据访问接口方法上，并且接口方法名不受查询关键字和关联查询命名规范约束。示例代码如下：

```
public interface AuthorityRepository extends JpaRepository<Authority, Integer>{
    /**
     * 根据用户名查询用户所拥有的权限（关联查询）
     */
    @Query("select a from Authority a inner join a.userList u where u.username = ?1")
    public List<Authority> findByUserListUsername(String username);
}
```

使用@Query 注解定义 JPQL 语句，可以直接返回 List<Map<String, Object>>对象。示例代码如下：

```
/**
 * 根据作者 id 查询文章信息（标题和内容）
 */
```

```
@Query("select new Map(a.title as title, a.content as content) from Article
a where a.author.id = ?1 ")
public List<Map<String, Object>> findTitleAndContentByAuthorId(Integer id);
```

使用@Query 注解定义 JPQL 语句，之前的方法是使用参数位置（“?1”指代的是获取方法形参列表中第 1 个参数值，1 代表的是参数位置，以此类推）来获取参数值。除此之外，Spring Data JPA 还支持使用名称来获取参数值，使用格式为“:参数名称”。示例代码如下：

```
/**
* 根据作者名和作者 id 查询文章信息
*/
@Query("select a from Article a where a.author.aname = :aname1 and a.author.id = :id1 ")
public List<Article> findArticleByAuthorAnameAndId(@Param("aname1") String
aname, @Param("id1") Integer id);
```

❷ @Modifying 注解

可以使用@Modifying 和@Query 注解组合定义在数据访问接口方法上，进行更新查询操作。示例代码如下：

```
    /**
     * 根据作者 id 删除作者
     */
    @Transactional
    @Modifying
    @Query("delete from Author a where a.id = ?1")
    public int deleteAuthorByAuthorId(int id);
```

下面通过实例讲解@Query 和@Modifying 注解的使用方法。

【例 8-5】@Query 和@Modifying 注解的使用方法。

首先，为例 8-5 创建基于 Spring Data JPA 依赖的 Spring Boot Web 应用 ch8_3。ch8_3 应用的数据库、pom.xml 以及 application.properties 与 ch8_2 应用基本一样，不再赘述。

其他内容具体实现步骤如下。

1）创建持久化实体类

创建名为 com.ch.ch8_3.entity 的包，并在该包中创建名为 Article 和 Author 的持久化实体类。具体代码分别与 ch8_2 应用的 Article 和 Author 的代码一样，不再赘述。

2）创建数据访问层

创建名为 com.ch.ch8_3.repository 的包，并在该包中创建名为 ArticleRepository 和 AuthorRepository 的接口。

ArticleRepository 的核心代码如下：

```
public interface ArticleRepository extends JpaRepository<Article, Integer>{
    /**
     * 根据作者 id 查询文章信息（标题和内容）
     */
    @Query("select new Map(a.title as title, a.content as content) from Article
    a where a.author.id = ?1 ")
    public List<Map<String, Object>> findTitleAndContentByAuthorId(Integer id);
    /**
     * 根据作者名和作者 id 查询文章信息
```

```
    */
    @Query("select a from Article a where a.author.aname = :aname1 and a.author.id = :id1 ")
    public List<Article> findArticleByAuthorAnameAndId(@Param("aname1") String aname,
    @Param("id1") Integer id);
    /**
     * 根据作者 id 删除作者对应的文章
     */
    @Transactional
    @Modifying
    @Query("delete from Article a where a.author.id = :id1 ")
    public int deleteArticleByAuthorId(@Param("id1") Integer id);
}
```

AuthorRepository 的核心代码如下：

```
public interface AuthorRepository extends JpaRepository<Author, Integer>{
    /**
     * 根据文章标题包含的内容，查询作者（关联查询）
     */
    @Query("select a from Author a inner join a.articleList t where t.title like %?1%")
    public Author findAuthorByArticleListtitleContaining(String title);
}
```

3）创建业务层

创建名为 com.ch.ch8_3.service 的包，并在该包中创建名为 AuthorAndArticleService 的接口和接口实现类 AuthorAndArticleServiceImpl。AuthorAndArticleService 接口代码略。

AuthorAndArticleServiceImpl 的核心代码如下：

```
@Service
public class AuthorAndArticleServiceImpl implements AuthorAndArticleService{
    @Autowired
    private AuthorRepository authorRepository;
    @Autowired
    private ArticleRepository articleRepository;
    @Override
    public List<Map<String, Object>> findTitleAndContentByAuthorId(Integer id) {
        return articleRepository.findTitleAndContentByAuthorId(id);
    }
    @Override
    public List<Article> findArticleByAuthorAnameAndId(String aname, Integer id) {
        return articleRepository.findArticleByAuthorAnameAndId(aname, id);
    }
    @Override
    public Author findAuthorByArticleListtitleContaining(String title) {
        return authorRepository.findAuthorByArticleListtitleContaining(title);
    }
    @Override
    public int deleteArticleByAuthorId(int id) {
        return articleRepository.deleteArticleByAuthorId(id);
    }
}
```

4）创建控制器类

创建 com.ch.ch8_3.controller 的包，并在该包中创建名为 TestOneToManyController 的控制器类。

TestOneToManyController 的代码如下：

```
@RestController
public class TestOneToManyController {
    @Autowired
    private AuthorAndArticleService authorAndArticleService;
    @RequestMapping("/findTitleAndContentByAuthorId")
    public List<Map<String, Object>> findTitleAndContentByAuthorId(Integer id) {
        return authorAndArticleService.findTitleAndContentByAuthorId(id);
    }
    @RequestMapping("/findArticleByAuthorAnameAndId")
    public List<Article> findArticleByAuthorAnameAndId(String aname, Integer id) {
        return authorAndArticleService.findArticleByAuthorAnameAndId(aname, id);
    }
    @RequestMapping("/findAuthorByArticleListtitleContaining")
    public Author findAuthorByArticleListtitleContaining(String title) {
        return authorAndArticleService.findAuthorByArticleListtitleContaining(title);
    }
    @RequestMapping("/deleteArticleByAuthorId")
    public int deleteArticleByAuthorId(int id) {
        return authorAndArticleService.deleteArticleByAuthorId(id);
    }
}
```

5）运行

首先，运行 Ch83Application 主类。然后，通过“http://localhost:8080/ch8_3/findTitleAndContentByAuthorId?id=1”查询作者 id 为 1 的文章标题和内容；通过“http://localhost:8080/ch8_3/findArticleByAuthorAnameAndId?aname=陈恒 2&&id=2”查询作者名为“陈恒 2”且作者 id 为 2 的文件列表；通过“http://localhost:8080/ch8_3/findAuthorByArticleListtitleContaining?title=对多 1”查询文章标题包含“对多 1”的作者；通过“http:// localhost:8080/ch8_3/deleteArticleByAuthorId?id=1”删除 id 为 1 的作者的文章。

▶ 8.1.5　排序与分页查询

视频讲解

在实际应用开发中，排序与分页查询是必需的。幸运的是 Spring Data JPA 充分考虑了排序与分页查询的场景，为我们提供了 Sort 类、Page 接口以及 Pageable 接口。

例如，数据访问接口如下：

```
public interface AuthorRepository extends JpaRepository<Author, Integer>{
    List<Author> findByAnameContaining(String aname, Sort sort);
}
```

那么，在 Service 层可以这样使用排序：

```
public List<Author> findByAnameContaining(String aname, String sortColumn) {
    //按 sortColumn 降序排序
    return authorRepository.findByAnameContaining(aname, Sort.by(Direction.DESC,
```

```
        sortColumn));
}
```

可以使用 Pageable 接口的实现类 PageRequest 的 of 方法构造分页查询对象，示例代码如下：

```
Page<Author> pageData = authorRepository.findAll(PageRequest.of(page-1, size,
    Sort.by(Direction.DESC, "id")));
```

其中，Page 接口可以获得当前页面的记录、总页数、总记录数等信息，示例代码如下：

```
//获得当前页面的记录
List<Author> allAuthor = pageData.getContent();
model.addAttribute("allAuthor",allAuthor);
//获得总记录数
model.addAttribute("totalCount", pageData.getTotalElements());
//获得总页数
model.addAttribute("totalPage", pageData.getTotalPages());
```

下面通过实例讲解 Spring Data JPA 的排序与分页查询的使用方法。

【例 8-6】排序与分页查询的使用方法。

首先，为例 8-6 创建基于 Thymeleaf 和 Spring Data JPA 依赖的 Spring Boot Web 应用 ch8_4。ch8_4 应用的数据库、pom.xml、application.properties 以及静态资源等内容与 ch8_1 应用基本一样，不再赘述。

其他内容具体实现步骤如下。

❶ 创建持久化实体类

创建名为 com.ch.ch8_4.entity 的包，并在该包中创建名为 Article 和 Author 的持久化实体类。具体代码分别与 ch8_2 应用的 Article 和 Author 的代码一样，不再赘述。

❷ 创建数据访问层

创建名为 com.ch.ch8_4.repository 的包，并在该包中创建名为 AuthorRepository 的接口。

AuthorRepository 的核心代码如下：

```
public interface AuthorRepository extends JpaRepository<Author, Integer>{
    /**
     * 查询作者名含有 name 的作者列表，并排序
     */
    List<Author> findByAnameContaining(String aname, Sort sort);
}
```

❸ 创建业务层

创建名为 com.ch.ch8_4.service 的包，并在该包中创建名为 ArticleAndAuthorService 的接口和接口实现类 ArticleAndAuthorServiceImpl。

ArticleAndAuthorService 的核心代码如下：

```
public interface ArticleAndAuthorService {
    /**
     * name 代表作者名的一部分（模糊查询），sortColumn 代表排序列
     */
    List<Author> findByAnameContaining(String aname, String sortColumn);
```

```
    /**
     * 分页查询作者，page 代表第几页
     */
    public String findAllAuthorByPage(Integer page, Model model);
}
```

ArticleAndAuthorServiceImpl 的核心代码如下：

```
@Service
public class ArticleAndAuthorServiceImpl implements ArticleAndAuthorService{
    @Autowired
    private AuthorRepository authorRepository;
    @Override
    public List<Author> findByAnameContaining(String aname, String sortColumn) {
        //按 sortColumn 降序排序
        return authorRepository.findByAnameContaining(aname, Sort.by(Direction.DESC,
        sortColumn));
    }
    @Override
    public String findAllAuthorByPage(Integer page, Model model) {
        if(page == null) {//第一次访问 findAllAuthorByPage 方法时
            page = 1;
        }
        int size = 2;//每页显示 2 条
        //分页查询，of 方法的第一个参数代表第几页（比实际小 1），
        //第二个参数代表页面大小，第三个参数代表排序规则
        Page<Author> pageData =
         authorRepository.findAll(PageRequest.of(page-1, size, Sort.by(Direction.DESC,
         "id")));
        //获得当前页面数据并转换成 List<Author>，转发到视图页面显示
        List<Author> allAuthor = pageData.getContent();
        model.addAttribute("allAuthor",allAuthor);
        //共多少条记录
        model.addAttribute("totalCount", pageData.getTotalElements());
        //共多少页
        model.addAttribute("totalPage", pageData.getTotalPages());
        //当前页
        model.addAttribute("page", page);
        return "index";
    }
}
```

❹ 创建控制器类

创建 com.ch.ch8_4.controller 的包，并在该包中创建名为 TestSortAndPage 的控制器类。TestSortAndPage 的核心代码如下：

```
@Controller
public class TestSortAndPage {
    @Autowired
    private ArticleAndAuthorService articleAndAuthorService;
    @RequestMapping("/findByAnameContaining")
    @ResponseBody
    public List<Author> findByAnameContaining(String aname, String sortColumn){
```

```
        return articleAndAuthorService.findByAnameContaining(aname, sortColumn);
    }
    @RequestMapping("/findAllAuthorByPage")
    /**
     * @param page 第几页
     */
    public String findAllAuthorByPage(Integer page, Model model){
        return articleAndAuthorService.findAllAuthorByPage(page, model);
    }
}
```

❺ 创建 View 视图页面

在 src/main/resources/templates 目录下，创建视图页面 index.html。

index.html 的具体代码如下：

```
<!DOCTYPE html>
<html xmlns:th="http://www.thymeleaf.org">
<head>
<meta charset="UTF-8">
<title>显示分页查询结果</title>
<link rel="stylesheet" th:href="@{css/bootstrap.min.css}" />
<link rel="stylesheet" th:href="@{css/bootstrap-theme.min.css}" />
</head>
<body>
    <div class="panel panel-primary">
        <div class="panel-heading">
            <h3 class="panel-title">Spring Data JPA 分页查询</h3>
        </div>
    </div>
    <div class="container">
        <div class="panel panel-primary">
            <div class="panel-body">
                <div class="table table-responsive">
                    <table class="table table-bordered table-hover">
                        <tbody class="text-center">
                            <tr th:each="author:${allAuthor}">
                                <td><span th:text="${author.id}"></span></td>
                                <td><span th:text="${author.aname}"></span></td>
                            </tr>
                            <tr>
                            <td colspan="2" align="right">
                                <ul class="pagination">
                                <li><a>第<span th:text="${page}"></span>页</a></li>
                                <li><a>共<span th:text="${totalPage}"></span>页</a></li>
                                <li><a>共<span th:text="${totalCount}"></span>
                                条</a></li>
                                <li>
        <a th:href="@{findAllAuthorByPage(page=${page-1})}" th:if="${page != 1}">上一页</a>
                                </li>
                                <li>
<a th:href="@{findAllAuthorByPage(page=${page+1})}" th:if="${page != totalPage}">下一页</a>
                                    </li>
                                </ul>
```

```
                        </td>
                    </tr>
                    </tbody>
                </table>
            </div>
        </div>
    </div>
  </div>
</body>
</html>
```

❻ 运行

首先，运行 Ch84Application 主类。然后，通过“http://localhost:8080/ch8_4/findByAnameContaining?aname=陈&sortColumn=id”查询作者名含有“陈”的作者列表，并按照 id 降序排列。运行效果如图 8.12 所示。

```
[ {
  "id" : 2,
  "aname" : "陈恒2",
  "articleList" : [ {
    "id" : 3,
    "title" : "JPA的一对多333",
    "content" : "其实一对多映射关系很常见333。"
  }, {
    "id" : 4,
    "title" : "JPA的一对多444",
    "content" : "其实一对多映射关系很常见444。"
  } ]
}, {
  "id" : 1,
  "aname" : "陈恒1",
  "articleList" : [ ]
} ]
```

图 8.12　查询作者名含有“陈”的作者列表，并按照 id 降序排列

通过 http://localhost:8080/ch8_4/findAllAuthorByPage 分页查询作者，并按照 id 降序排列。运行效果如图 8.13 所示。

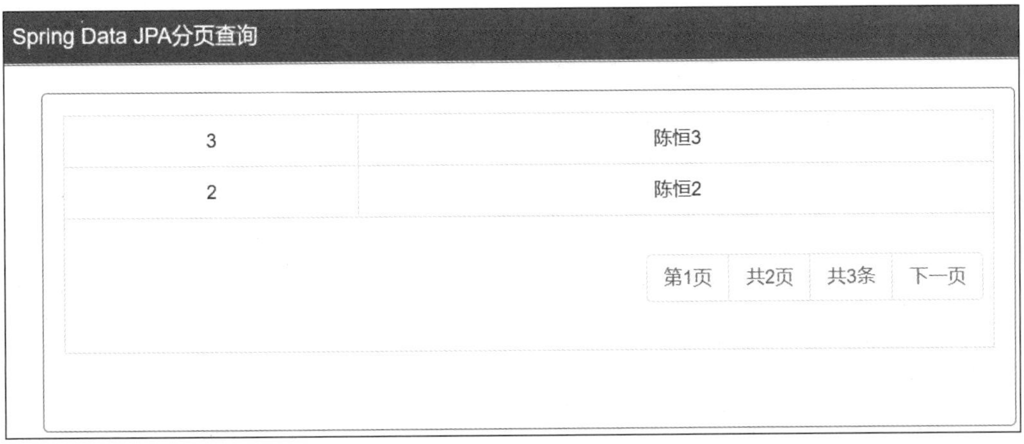

Spring Data JPA分页查询

3	陈恒3
2	陈恒2

第1页　共2页　共3条　下一页

图 8.13　分页查询作者

视频讲解

8.2 Spring Boot 整合 MyBatis

在第 3 章已经学习 SSM 框架整合开发的流程，那么 Spring Boot 如何整合 MyBatis 呢？下面通过实例讲解如何在 Spring Boot 应用中使用 MyBatis 框架操作数据库（基于 XML 的映射配置）。

【例 8-7】在 Spring Boot 应用中使用 MyBatis 框架操作数据库（基于 XML 的映射配置）。具体实现步骤如下。

❶ 创建 Spring Boot Web 应用

在创建 Spring Boot Web 应用 ch8_5 时，选择 MyBatis Framework 依赖，如图 8.14 所示。在该应用中，操作的数据库与 8.1.2 节一样，都是 springbootjpa。操作的数据表是 user 表。

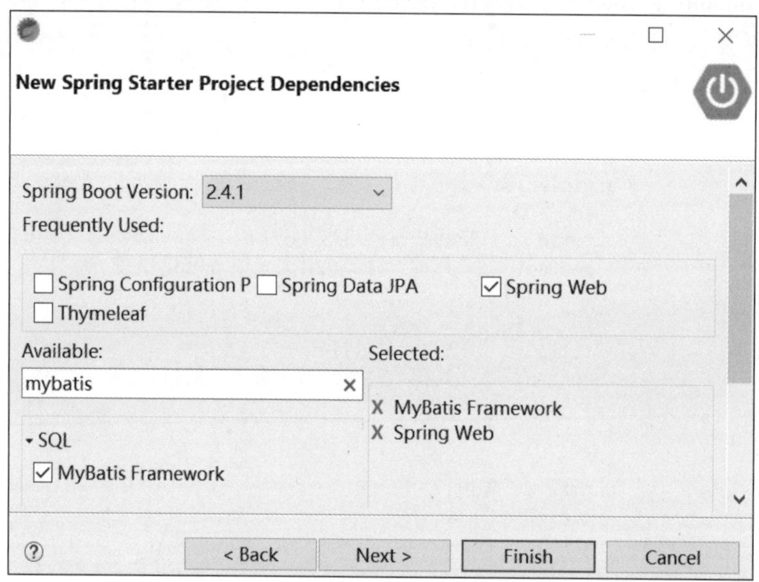

图 8.14　选中 MyBatis Framework

❷ 修改 pom.xml 文件

在 pom.xml 文件中添加 MySQL 连接器依赖。

❸ 设置 Web 应用 ch8_5 的上下文路径及数据源配置信息

在 ch8_5 的 application.properties 文件中配置如下内容：

```
server.servlet.context-path=/ch8_5
###
##数据源信息
###
#数据库地址
spring.datasource.url=jdbc:mysql://localhost:3306/springbootjpa?characterEncoding=utf8
#数据库用户名
spring.datasource.username=root
#数据库密码
spring.datasource.password=root
#数据库驱动
spring.datasource.driver-class-name=com.mysql.jdbc.Driver
```

```
#设置包别名（在 Mapper 映射文件中直接使用实体类名）
mybatis.type-aliases-package=com.ch.ch8_5.entity
#告诉系统到哪里去找 mapper.xml 文件（映射文件）
mybatis.mapperLocations=classpath:mappers/*.xml
#在控制台输出 SQL 语句日志
logging.level.com.ch.ch8_5.repository=debug
#让控制器输出的 JSON 字符串格式更美观
spring.jackson.serialization.indent-output=true
```

❹ 创建实体类

创建名为 com.ch.ch8_5.entity 的包，并在该包中创建 MyUser 实体类。具体代码如下：

```
package com.ch.ch8_5.entity;
public class MyUser {
    private Integer id;
    private String username;
    private String password;
    //省略 get 方法和 set 方法
}
```

❺ 创建数据访问接口

创建名为 com.ch.ch8_5.repository 的包，并在该包中创建 MyUserRepository 接口。MyUserRepository 的核心代码如下：

```
/*
 * @Repository 可有可无，但有时提示依赖注入找不到（不影响运行），
 * 加上后可以消去依赖注入的报错信息。
 * 这里不再需要@Mapper，是因为在启动类中使用@MapperScan 注解，
 * 将数据访问层的接口都注解为 Mapper 接口的实现类，
 * @Mapper 与@MapperScan 二者用其一即可
 */
@Repository
public interface MyUserRepository {
        public List<MyUser> findAll();
}
```

❻ 创建 Mapper 映射文件

在 src/main/resources 目录下，创建名为 mappers 的包，并在该包中创建 SQL 映射文件 MyUserMapper.xml。具体代码如下：

```
<?xml version="1.0" encoding="UTF-8" ?>
<!DOCTYPE mapper
PUBLIC "-//mybatis.org//DTD Mapper 3.0//EN"
"http://mybatis.org/dtd/mybatis-3-mapper.dtd">
<mapper namespace="com.ch.ch8_5.repository.MyUserRepository">
    <select id="findAll"  resultType="MyUser">
        select * from user
    </select>
</mapper>
```

❼ 创建业务层

创建名为 com.ch.ch8_5.service 的包，并在该包中创建 MyUserService 接口和 MyUserServiceImpl 实现类。MyUserService 的代码略。

MyUserServiceImpl 的核心代码如下：

```
@Service
public class MyUserServiceImpl implements MyUserService{
    @Autowired
    private MyUserRepository myUserRepository;
    @Override
    public List<MyUser> findAll() {
        return myUserRepository.findAll();
    }
}
```

❽ 创建控制器类 MyUserController

创建名为 com.ch.ch8_5.controller 的包，并在该包中创建控制器类 MyUserController。MyUserController 的代码如下：

```
@RestController
public class MyUserController {
    @Autowired
    private MyUserService myUserService;
    @RequestMapping("/findAll")
    public List<MyUser> findAll(){
        return myUserService.findAll();
    }
}
```

❾ 在应用程序的主类中扫描 Mapper 接口

在应用程序的 Ch85Application 主类中，使用@MapperScan 注解扫描 MyBatis 的 Mapper 接口。核心代码如下：

```
@SpringBootApplication
//配置扫描 MyBatis 接口的包路径
@MapperScan(basePackages={"com.ch.ch8_5.repository"})
public class Ch85Application {
    public static void main(String[] args) {
        SpringApplication.run(Ch66Application.class, args);
    }
}
```

❿ 运行

首先，运行 Ch85Application 主类。然后，访问 http://localhost:8080/ch8_5/findAll。运行效果如图 8.15 所示。

```
[ {
  "id" : 1,
  "username" : "陈恒1",
  "password" : "123"
}, {
  "id" : 2,
  "username" : "陈恒2",
  "password" : "234"
} ]
```

图 8.15　查询所有用户信息

8.3　REST

视频讲解

本节将介绍 REST 风格接口，并通过 Spring Boot 实现 RESTful。

8.3.1　REST 简介

REST 即表现层状态转化（Representational State Transfer，REST），是 Roy Thomas Fielding 博士在他 2000 年的博士论文中提出来的一种软件架构风格。它是一种针对网络应用的设计和开发方式，可以降低开发的复杂性，提高系统的可伸缩性。目前在三种主流的 Web 服务实现方案中，因为 REST 模式的 Web 服务与复杂的 SOAP 和 XML-RPC 对比来讲明显地更加简洁，越来越多的 Web 服务开始采用 REST 风格设计和实现。

REST 是一组架构约束条件和原则。这些约束有：

（1）使用客户/服务器模型。客户和服务器之间通过一个统一的接口来互相通信。

（2）层次化的系统。在一个 REST 系统中，客户端并不会固定地与一个服务器打交道。

（3）无状态。在一个 REST 系统中，服务端并不会保存有关客户的任何状态。也就是说，客户端自身负责用户状态的维持，并在每次发送请求时都需要提供足够的信息。

（4）可缓存。REST 系统需要能够恰当地缓存请求，以尽量减少服务端和客户端之间的信息传输，提高性能。

（5）统一的接口。一个 REST 系统需要使用一个统一的接口来完成子系统之间以及服务器与用户之间的交互。这使得 REST 系统中的各个子系统可以独自完成演化。

满足这些约束条件和原则的应用程序或设计就是 RESTful。需要注意的是，REST 是设计风格而不是标准。REST 通常基于 HTTP、URI、XML 以及 HTML 这些现有的广泛流行的协议和标准。

理解 RESTful 架构，应该先理解 Representational State Transfer 这个词组到底是什么意思，它的每一个词表达了什么含义。

- 资源（Resources）

"表现层状态转化"中的"表现层"其实指的是"资源"的"表现层"。

"资源"是网络上的一个实体，或者说是网络上的一个具体信息。"资源"可以是一段文本、一张图片、一段视频，总之就是一个具体的实体。可以使用一个 URI（Uniform Resource Identifier，统一资源定位符）指向资源，每种资源对应一个特定的 URI。当需要获取资源时，访问它的 URI 即可，因此 URI 是每个资源的地址或独一无二的标识符。REST 风格的 Web 服务是通过一个简洁清晰的 URI 来提供资源链接，客户端通过对 URI 发送 HTTP 请求获得这些资源，而获取和处理资源的过程让客户端应用的状态发生改变。

- 表现层（Representation）

"资源"是一种信息实体，可以有多种外在的表现形式。将"资源"呈现出来的形式称为它的"表现层"。例如，文本可以使用 TXT 格式表现，也可以使用 XML 格式、JSON 格式表现。

- 状态转化（State Transfer）

客户端访问一个网站，就代表了它和服务器的一个互动过程。在这个互动过程中，将涉及数据和状态的变化。我们知道 HTTP 协议是一个无状态的通信协议，这意味着所有状态都保存在服务器端。因此，如果客户端操作服务器，需要通过某种手段（如 HTTP 协议）让服务器端发生"状态变化"。而这种转化是建立在表现层之上的，所以就是"表现层状态转化"。

在流行的各种 Web 框架中，包括 Spring Boot，都支持 REST 开发。REST 并不是一种技术或者规范，而是一种架构风格，包括如何标识资源、如何标识操作接口及操作的版本、如何标识操作的结果等，主要内容如下。

❶ 使用“api”作为上下文

在 REST 架构中，建议使用“api”作为上下文。示例如下：

```
http://localhost:8080/api
```

❷ 增加一个版本标识

在 REST 架构中，可以通过 URL 标识版本信息。示例如下：

```
http://localhost:8080/api/v1.0
```

❸ 标识资源

在 REST 架构中，可以将资源名称放到 URL 中。示例如下：

```
http://localhost:8080/api/v1.0/user
```

❹ 确定 HTTP Method

HTTP 协议有 5 个常用的表示操作方式的动词：GET、POST、PUT、DELETE、PATCH。它们分别对应 5 种基本操作：GET 用来获取资源；POST 用来增加资源（也可以用于更新资源）；PUT 用来更新资源；DELETE 用来删除资源；PATCH 用来更新资源的部分属性。示例如下：

1）新增用户

```
POST http://localhost:8080/api/v1.0/user
```

2）查询 id 为 123 的用户

```
GET http://localhost:8080/api/v1.0/user/123
```

3）更新 id 为 123 的用户

```
PUT http://localhost:8080/api/v1.0/user/123
```

4）删除 id 为 123 的用户

```
DELETE http://localhost:8080/api/v1.0/user/123
```

❺ 确定 HTTP Status

服务器向用户返回的状态码和提示信息，常用的如下。

（1）200 OK - [GET]：服务器成功返回用户请求的数据。

（2）201 CREATED - [POST/PUT/PATCH]：用户新建或修改数据成功。

（3）202 Accepted - [*]：表示一个请求已经进入后台排队（异步任务）。

（4）204 NO CONTENT - [DELETE]：用户删除数据成功。

（5）400 INVALID REQUEST - [POST/PUT/PATCH]：用户发出的请求有错误，服务器没有进行新建或修改数据的操作。

（6）401 Unauthorized - [*]：表示用户没有权限（令牌、用户名、密码错误）。

（7）403 Forbidden - [*] 表示用户得到授权（与 401 错误相对），但是访问是被禁止的。

（8）404 NOT FOUND - [*]：用户发出的请求针对的是不存在的记录，服务器没有进行操作。

（9）406 Not Acceptable - [GET]：用户请求的格式不可得（例如用户请求 JSON 格式，但是只有 XML 格式）。

（10）410 Gone -[GET]：用户请求的资源被永久删除，且不会再得到。

（11）422 Unprocessable entity - [POST/PUT/PATCH]：当创建一个对象时，发生一个验证错误。

（12）500 INTERNAL SERVER ERROR - [*]：服务器发生错误，用户将无法判断发出的请求是否成功。

▶ 8.3.2　Spring Boot 整合 REST

在 Spring Boot 的 Web 应用中，自动支持 REST。也就是说，只要 spring-boot-starter-web 依赖在 pom.xml 中，就支持 REST。

【例 8-8】 一个 RESTful 应用示例。

假如在 ch8_2 应用的控制器类 TestOneToManyController 中有如下处理方法：

```
@RequestMapping("/findArticleByAuthor_id1/{id}")
public List<Article> findByAuthor_id1(@PathVariable("id") Integer id) {
    return authorAndArticleService.findByAuthor_id(id);
}
```

那么，可以使用如下所示的 REST 风格的 URL 访问上述处理方法：

```
http://localhost:8080/ch8_2/findArticleByAuthor_id1/2
```

在例 8-8 中使用了 URL 模板模式映射@RequestMapping("/findArticleByAuthor_id1/{id}")，其中，{XXX}为占位符，通过在处理方法中使用@PathVariable 获取{XXX}中的 XXX 变量值。@PathVariable 用于将请求 URL 中的模板变量映射到功能处理方法的参数上。如果{XXX}中的变量名 XXX 和形参名称一致，则@PathVariable 不用指定名称。

▶ 8.3.3　Spring Data REST

Spring Data JPA 基于 Spring Data 的 repository 之上，可以将 repository 自动输出为 REST 资源。目前，Spring Data REST 支持将 Spring Data JPA、Spring Data MongoDB、Spring Data Neo4j、Spring Data GemFire 以及 Spring Data Cassandra 的 repository 自动转换成 REST 服务。

Spring Boot 对 Spring Data REST 的自动配置位于 org.springframework.boot.autoconfigure.data.rest 包中。

通过 SpringBootRepositoryRestConfigurer 类的源码可以得出，Spring Boot 已经自动配置了 RepositoryRestConfiguration，所以在 Spring Boot 应用中使用 Spring Data REST 只需引入 spring-boot-starter-data-rest 的依赖即可。

下面通过实例讲解 Spring Data REST 的构建过程。

【例 8-9】 Spring Data REST 的构建过程。

具体实现步骤如下。

❶ 创建 Spring Boot 应用 ch8_6

创建 Spring Boot 应用 ch8_6，依赖为 Spring Data JPA 和 Rest Repositories。

❷ 修改 pom.xml 文件，添加 MySQL 依赖

修改应用 ch8_6 的 pom.xml 文件，添加 MySQL 依赖。

❸ 设置应用 ch8_6 的上下文路径及数据源配置信息

在 ch8_6 的 application.properties 文件中配置如下内容：

```
server.servlet.context-path=/api
#数据库地址
spring.datasource.url=jdbc:mysql://localhost:3306/springbootjpa?characterEncoding=utf8
#数据库用户名
spring.datasource.username=root
#数据库密码
spring.datasource.password=root
#数据库驱动
spring.datasource.driver-class-name=com.mysql.jdbc.Driver
#指定数据库类型
spring.jpa.database=MYSQL
#指定是否在日志中显示 SQL 语句
spring.jpa.show-sql=true
#指定自动创建、更新数据库表等配置，update 表示如果数据库中存在持久化类对应的表就不创建，
#不存在就创建
spring.jpa.hibernate.ddl-auto=update
#让控制器输出的 JSON 字符串格式更美观
spring.jackson.serialization.indent-output=true
```

❹ 创建持久化实体类 Student

创建名为 com.ch.ch8_6.entity 的包，并在该包中创建名为 Student 的持久化实体类。核心代码如下：

```
@Entity
@Table(name = "student_table")
public class Student implements Serializable{
    private static final long serialVersionUID = 1L;
    @Id
    @GeneratedValue(strategy = GenerationType.IDENTITY)
    private int id;//主键
    private String sno;
    private String sname;
    private String ssex;
    public Student() {
        super();
    }
    public Student(int id, String sno, String sname, String ssex) {
        super();
        this.id = id;
        this.sno = sno;
        this.sname = sname;
        this.ssex = ssex;
    }
    //省略 set 方法和 get 方法
}
```

❺ 创建数据访问层

创建名为 com.ch.ch8_6.repository 的包，并在该包中创建名为 StudentRepository 的接口，该接口继承 JpaRepository 接口。核心代码如下：

```
public interface StudentRepository extends JpaRepository<Student, Integer>{
    /**
     * 自定义接口查询方法，暴露为 REST 资源
     */
```

```
    @RestResource(path = "snameStartsWith", rel = "snameStartsWith")
    List<Student> findBySnameStartsWith(@Param("sname") String sname);
}
```

在上述数据访问接口中，定义了 findBySnameStartsWith，并使用@RestResource 注解将该方法暴露为 REST 资源，snameStartsWith 为请求路径。

至此，基于 Spring Data 的 REST 资源服务已经构建完毕，接下来就是使用 REST 客户端测试此服务。

8.3.4　REST 服务测试

在 Web 和移动端开发时，常常会调用服务器端的 RESTful 接口进行数据请求，为了调试，一般会先用工具进行测试，通过测试后才开始在开发中使用。本节将介绍如何使用 Google Chrome 的 Postman REST Client 进行 8.3.3 节的 RESTful 接口请求测试。

❶ 获得列表数据

在 RESTful 架构中，每个网址代表一种资源（resource），所以网址中不能有动词，只能有名词，而且所用的名词往往与实体名对应。一般来说，数据库中的表都是同种记录的“集合”（collection），所以 API 中的名词也应该使用复数，如 students。

运行 ch8_6 的主类 Ch86Application 后，手工在 student_table 中添加几条学生信息后，在 Postman REST Client 中，使用 GET 方式访问 http://localhost:8080/api/students 请求路径获得所有学生信息，结果如图 8.16 所示。

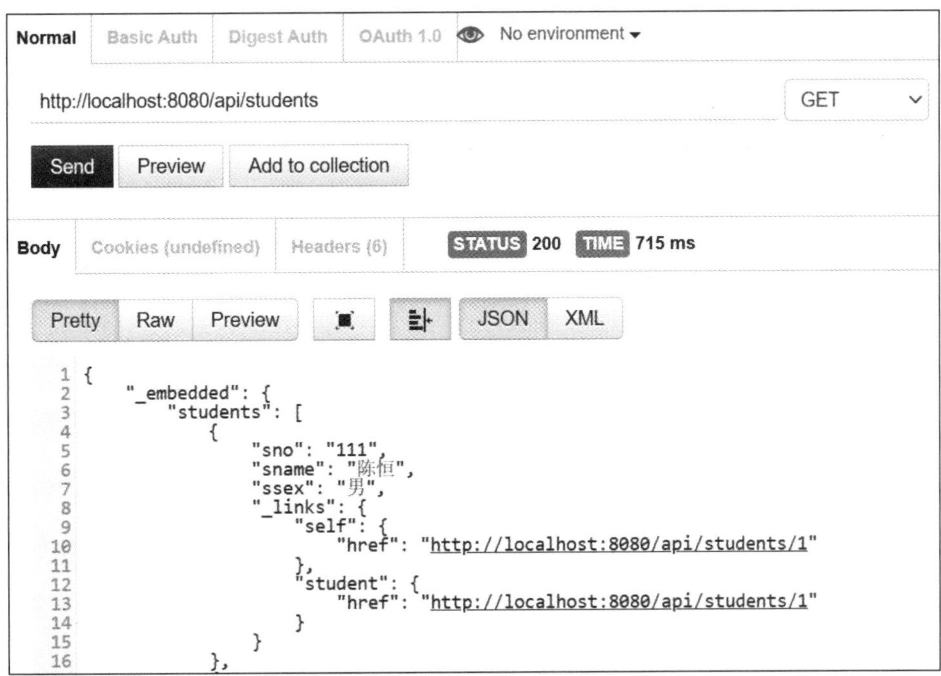

图 8.16　GET 所有学生信息

❷ 获得单一对象

在 Postman REST Client 中，使用 GET 方式访问“http://localhost:8080/api/students/1”请求可获得 id 为 1 的学生信息。

❸ 查询

在 Postman REST Client 中，search 调用自定义的接口查询方法。因此，可以使用 GET 访问“http://localhost:8080/api/students/search/snameStartsWith?sname=陈”请求路径调用 List<Student> findBySnameStartsWith(@Param("sname") String sname)接口方法，获得姓名前缀为“陈”的学生信息。

❹ 分页查询

在 Postman REST Client 中，使用 GET 方式访问“http://localhost:8080/api/students/?page=0&size=2”请求路径获得第一页的学生信息（page=0 即第一页，size=2 即每页数量为 2）。

❺ 排序

在 Postman REST Client 中，使用 GET 方式访问“http://localhost:8080/api/students/?sort=sno,desc”请求路径获得按照 sno 属性倒序的列表。

❻ 保存

在 Postman REST Client 中，发起 POST 方式请求 http://localhost:8080/api/students 实现新增功能，将要保存的数据放置在请求体中，数据类型为 JSON，如图 8.17 和图 8.18 所示。

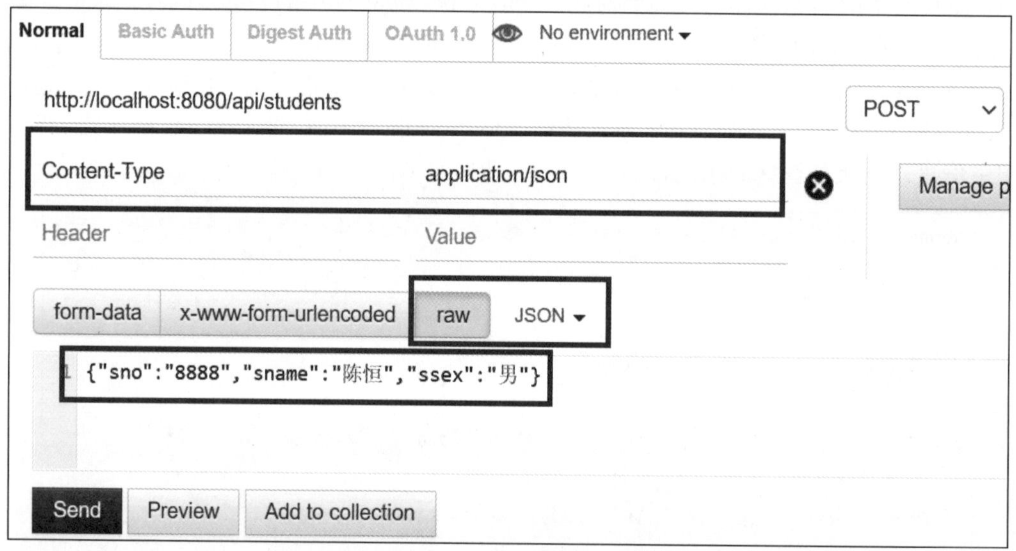

图 8.17　发起 POST 请求实现新增功能

```
Pretty  Raw  Preview  JSON  XML
{
    "sno": "8888",
    "sname": "陈恒",
    "ssex": "男",
    "_links": {
        "self": {
            "href": "http://localhost:8080/api/students/6"
        },
        "student": {
            "href": "http://localhost:8080/api/students/6"
        }
    }
}
```

图 8.18　保存成功

从图 8.18 可以看出，保存成功后，新数据的 id 为 6。

❼ 更新

假如需要更新新增的 id 为 5 的数据，可以在 Postman REST Client 中使用 PUT 方式访问 “http://localhost:8080/api/students/5”，修改提交的数据，如图 8.19 和图 8.20 所示。

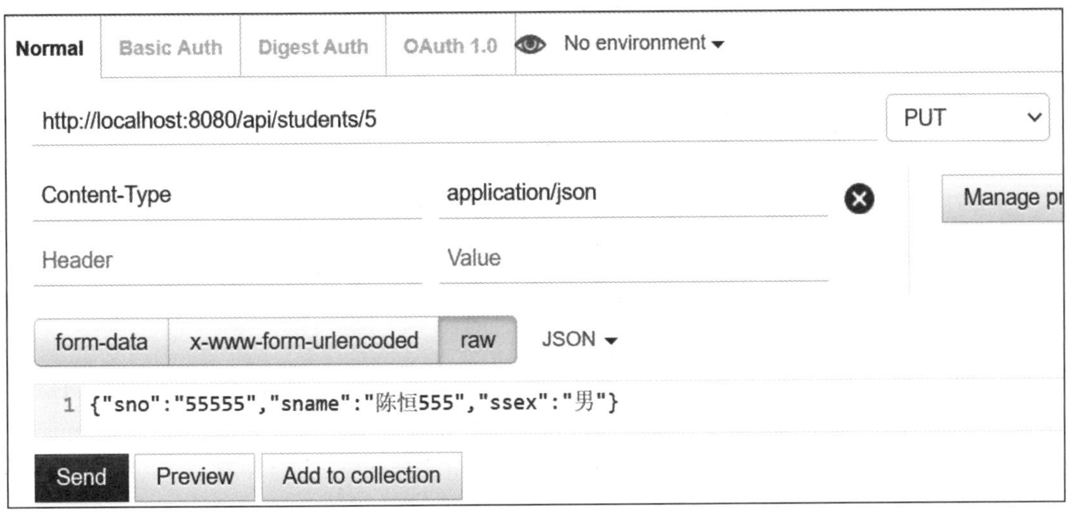

图 8.19　发起 PUT 请求实现更新功能

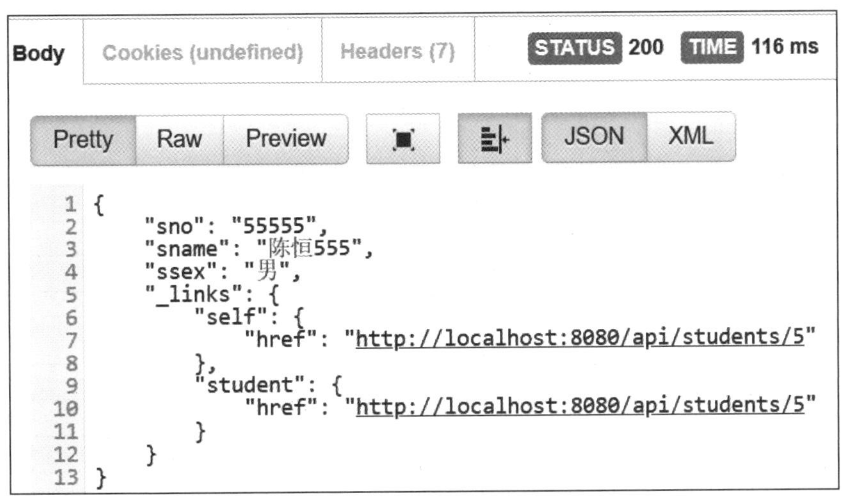

图 8.20　更新成功

❽ 删除

假如需要删除新增的 id 为 5 的数据，可以在 Postman REST Client 中使用 DELETE 方式访问 “http://localhost:8080/api/students/5”。

8.4　MongoDB

视频讲解

MongoDB 是一个基于分布式文件存储的 NoSQL 数据库，由 C++语言编写，旨在为 Web 应用提供可扩展的高性能数据存储解决方案。

MongoDB 是一个介于关系数据库和非关系数据库之间的产品，是非关系数据库中功能最丰富、最像关系数据库的。它支持的数据结构非常松散，类似于 JSON 的 BSON（Binary JSON，二进制 JSON）格式，因此可以存储比较复杂的数据类型。Mongo 最大的特点是它支持的查询语言非常强大，其语法有点类似于面向对象的查询语言，几乎可以实现类似关系数据库单表查询的绝大部分功能，而且还支持对数据建立索引。

本节不会介绍太多关于 MongoDB 数据库本身的知识，主要介绍 Spring Boot 对 MongoDB 的支持，以及基于 Spring Boot 和 MongoDB 的实例。

▶ 8.4.1 安装 MongoDB

可以从官方网站 https://www.mongodb.com/download-center/community 下载自己操作系统对应的 MongoDB 版本，作者编写本书时使用的 MongoDB 是 mongodb-win32-x86_64-2012plus-4.2.0-signed.msi。成功下载后，双击 mongodb-win32-x86_64-2012plus-4.2.0-signed.msi，按照默认安装即可。

可以使用 MongoDB 的图形界面管理工具 MongoDB Compass 可视化操作 MongoDB 数据库。可以使用 mongodb-win32-x86_64-2012plus-4.2.0-signed.msi 自带的 MongoDB Compass，也可以从官方网站 https://www.mongodb.com/download-center/compass 下载。

▶ 8.4.2 Spring Boot 整合 MongoDB

❶ Spring 对 MongoDB 的支持

Spring 对 MongoDB 的支持主要是通过 Spring Data MongoDB 实现的。Spring Data MongoDB 提供了如下功能。

1）对象/文档映射注解

Spring Data MongoDB 提供的对象/文档映射注解如表 8.2 所示。

表 8.2 Spring Data MongoDB 提供的对象/文档映射注解

注 解	含 义
@Document	映射领域对象与 MongoDB 的一个文档
@Id	映射当前属性是文档对象 ID
@DBRef	当前属性将参考其他文档
@Field	为文档的属性定义名称
@Version	将当前属性作为版本

2）MongoTemplate

与 JdbcTemplate 一样，Spring Data MongoDB 也提供了一个 MongoTemplate，并提供了数据访问的方法。

3）Repository

类似于 Spring Data JPA，Spring Data MongoDB 也提供了 Repository 的支持，使用方式和 Spring Data JPA 一样，示例如下：

```
public interface PersonRepository extends MongoRepository<Person, String>{
}
```

❷ Spring Boot 对 MongoDB 的支持

Spring Boot 对 MongoDB 的自动配置位于 org.springframework.boot.autoconfigure.mongo

包中，主要配置了数据库连接、MongoTemplate，可以在配置文件中使用以 spring.data.mongodb 为扩展名的属性来配置 MongoDB 的相关信息。Spring Boot 对 MongoDB 提供了一些默认属性，如默认端口号为 27017，默认服务器为 localhost，默认数据库为 test，默认无用户名和无密码访问方式，并默认开启了对 Repository 的支持。因此，在 Spring Boot 应用中，只需引入 spring-boot-starter-data-mongodb 依赖即可按照默认配置操作 MongoDB 数据库。

8.4.3　增、删、改、查

本节通过实例，讲解如何在 Spring Boot 应用中对 MongoDB 数据库进行增、删、改、查。

【例 8-10】在 Spring Boot 应用中，对 MongoDB 数据库进行增、删、改、查。

具体实现步骤如下。

❶ 创建基于 spring-boot-starter-data-mongodb 依赖的 Spring Boot Web 应用 ch8_7

创建基于 spring-boot-starter-data-mongodb 依赖的 Spring Boot Web 应用 ch8_7。

❷ 配置 application.properties 文件

在应用 ch8_7 中，使用 MongoDB 的默认数据库连接。所以，不需要在 application.properties 文件中配置数据库连接信息。application.properties 文件的其他内容配置如下：

```
server.servlet.context-path=/ch8_7
#让控制器输出的 JSON 字符串格式更美观
spring.jackson.serialization.indent-output=true
```

❸ 创建领域模型

创建名为 com.ch.ch8_7.domain 的包，并在该包中创建领域模型 Person（人）以及 Person 去过的 Location（地点）。在 Person 类中，使用@Document 注解对 Person 领域模型和 MongoDB 的文档进行映射。

Person 的代码如下：

```
package com.ch.ch8_7.domain;
import java.util.ArrayList;
import java.util.List;
import org.springframework.data.annotation.Id;
import org.springframework.data.mongodb.core.mapping.Document;
import org.springframework.data.mongodb.core.mapping.Field;
@Document
public class Person {
    @Id
    private String pid;
    private String pname;
    private Integer page;
    private String psex;
    @Field("plocs")
    private List<Location> locations = new ArrayList<Location>();
    public Person() {
        super();
    }
    public Person(String pname, Integer page, String psex) {
        super();
        this.pname = pname;
```

```
        this.page = page;
        this.psex = psex;
    }
    //省略 set 方法和 get 方法
}
```

Location 的代码如下：

```
package com.ch.ch8_7.domain;
public class Location {
    private String locName;
    private String year;
    public Location() {
        super();
    }
    public Location(String locName, String year) {
        super();
        this.locName = locName;
        this.year = year;
    }
}
```

❹ 创建数据访问接口

创建名为 com.ch.ch8_7.repository 的包，并在该包中创建数据访问接口 PersonRepository，该接口继承 MongoRepository 接口。PersonRepository 接口的代码如下：

```
package com.ch.ch8_7.repository;
import java.util.List;
import org.springframework.data.mongodb.repository.MongoRepository;
import org.springframework.data.mongodb.repository.Query;
import com.ch.ch8_7.domain.Person;
public interface PersonRepository extends MongoRepository<Person, String>{
    Person findByPname(String pname);//支持方法名查询，方法名命名规范参照表 8.1
    @Query("{'psex':?0}")//JSON 字符串
    List<Person> selectPersonsByPsex(String psex);
}
```

❺ 创建控制器层

由于本实例业务简单，直接在控制器层调用数据访问层。创建名为 com.ch.ch8_7.controller 的包，并在该包中创建控制器类 TestMongoDBController。

TestMongoDBController 的核心代码如下：

```
@RestController
public class TestMongoDBController {
    @Autowired
    private PersonRepository personRepository;
    @RequestMapping("/save")
    public List<Person> save() {
        List<Location> locations1 = new ArrayList<Location>();
        Location loc1 = new Location("北京","2019");
        Location loc2 = new Location("上海","2018");
        locations1.add(loc1);
```

```
        locations1.add(loc2);
        List<Location> locations2 = new ArrayList<Location>();
        Location loc3 = new Location("广州","2017");
        Location loc4 = new Location("深圳","2016");
        locations2.add(loc3);
        locations2.add(loc4);
        List<Person> persons = new ArrayList<Person>();
        Person p1 = new Person("陈恒1", 88, "男");
        p1.setLocations(locations1);
        Person p2 = new Person("陈恒2", 99, "女");
        p2.setLocations(locations2);
        persons.add(p1);
        persons.add(p2);
        return personRepository.saveAll(persons);
    }
    @RequestMapping("/findByPname")
    public Person findByPname(String pname) {
        return personRepository.findByPname(pname);
    }
    @RequestMapping("/selectPersonsByPsex")
    public List<Person> selectPersonsByPsex(String psex) {
        return personRepository.selectPersonsByPsex(psex);
    }
    @RequestMapping("/updatePerson")
    public Person updatePerson(String oldPname, String newPname) {
        Person p1 = personRepository.findByPname(oldPname);
        if(p1 != null)
            p1.setPname(newPname);
        return personRepository.save(p1);
    }
    @RequestMapping("/deletePerson")
    public void updatePerson(String pname) {
        Person p1 = personRepository.findByPname(pname);
        personRepository.delete(p1);
    }
}
```

❻ 运行

首先，运行 Ch87Application 主类。然后，访问“http://localhost:8080/ch8_7/save”测试保存数据。

保存成功后，使用 MongoDB 的图形界面管理工具 MongoDB Compass 打开查看已保存的数据，如图 8.21 所示。

通过“http://localhost:8080/ch8_7/selectPersonsByPsex?psex=女”查询性别为“女”的文档数据。

通过“http://localhost:8080/ch8_7/updatePerson?oldPname=陈恒 1&newPname=陈恒 111”将人名“陈恒 1”修改成人名“陈恒 111”。

通过“http://localhost:8080/ch8_7/deletePerson?pname=陈恒 111”将人名为“陈恒 111”的文档数据删除。

至此，通过 Spring Boot Web 应用对 MongoDB 数据库的操作演示完毕。

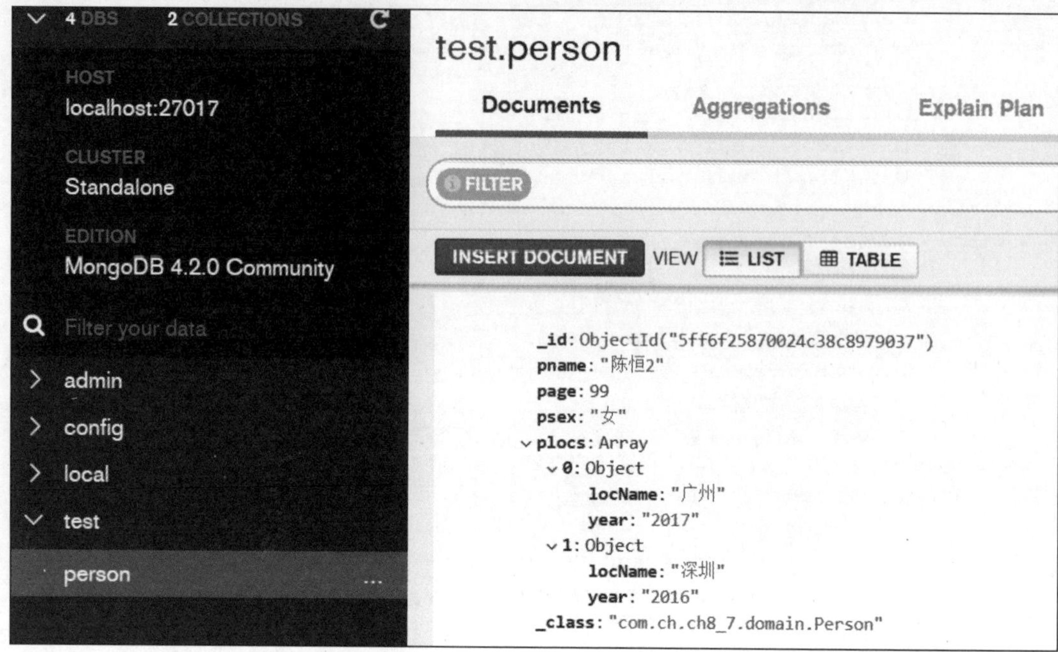

图 8.21　查看已保存的数据

视频讲解

8.5　Redis

Redis 是一个开源的使用 ANSI C 语言编写、支持网络、可基于内存亦可持久化的日志型、Key-Value 数据库，并提供多种语言的 API。它支持字符串、哈希表、列表、集合、有序集合、位图、地理空间信息等数据类型，同时也可以作为高速缓存和消息队列代理。但是，Redis 在内存中存储数据，因此，存放在 Redis 中的数据不应该大于内存容量，否则会导致操作系统性能降低。

本节主要介绍 Spring Boot 对 Redis 的支持，以及基于 Spring Boot 和 Redis 的实例，而不会介绍太多的关于 Redis 数据库本身的知识。

8.5.1　安装 Redis

❶ 下载 Redis

编写本书时，Redis 官方网站只提供 Linux 版本的下载。因此，只能通过 https://github.com/MSOpenTech/redis/tags 从 github 上下载 Redis，本书下载的版本是 Redis-x64-3.2.100.zip。在运行中输入“cmd”，然后把目录指向解压的 Redis 目录，如图 8.22 所示。

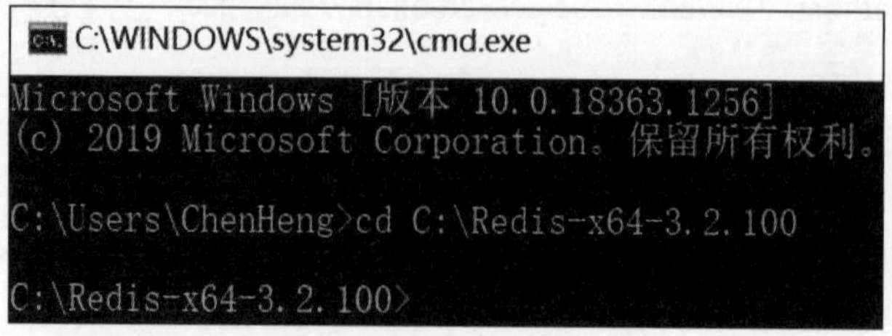

图 8.22　将目录指向解压的 Redis 目录

❷ 启动 Redis 服务

使用 redis-server redis.windows.conf 命令行启动 Redis 服务，出现如图 8.23 所示的显示内容，表示成功启动 Redis 服务。

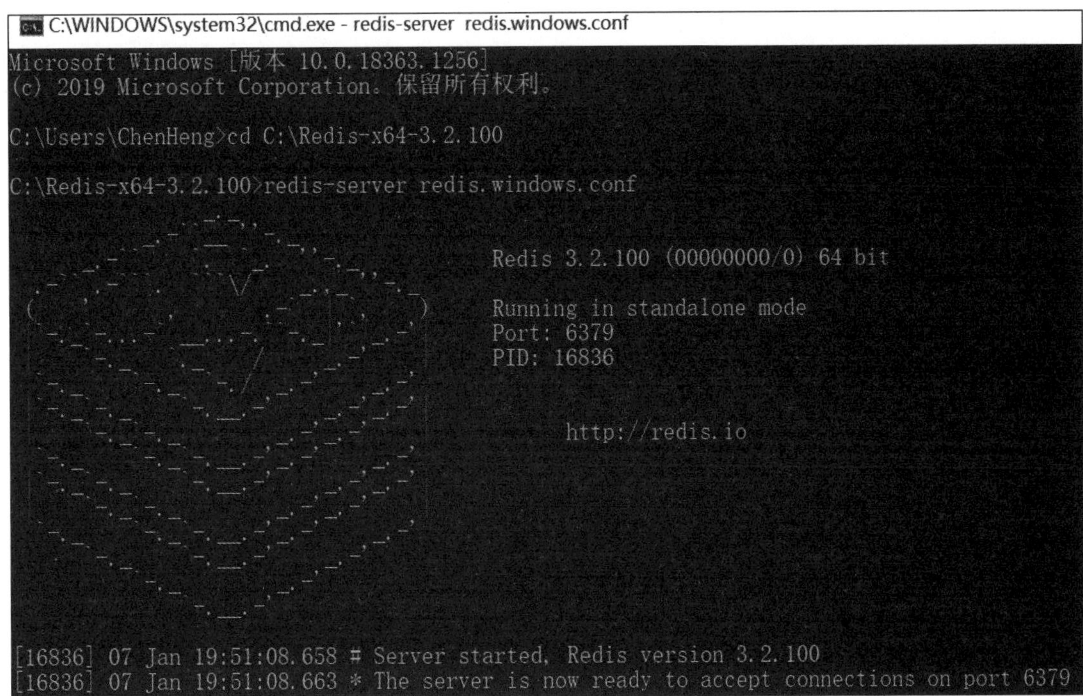

图 8.23　启动 Redis 服务

图 8.23 虽然启动了 Redis 服务，但如果关闭 cmd 窗口，Redis 服务就消失。所以需要把 Redis 设置成 Windows 下的服务。

关闭 cmd 重新打开 cmd，进入 Redis 解压目录。执行设置服务命令：redis-server --service-install redis.windows-service.conf --loglevel verbose，如图 8.24 所示。

```
C:\WINDOWS\system32\cmd.exe
Microsoft Windows [版本 10.0.18363.1256]
(c) 2019 Microsoft Corporation。保留所有权利。

C:\Users\ChenHeng>cd C:\Redis-x64-3.2.100

C:\Redis-x64-3.2.100>redis-server --service-install redis.windows-service.conf --loglevel verbose

C:\Redis-x64-3.2.100>
```

图 8.24　将 Redis 服务设置成 Windows 下的服务

图 8.24 中没有报错，表示成功设置成 Windows 下的服务。刷新计算机管理的服务，将看到 Redis 服务。

❸ 常用的 Redis 服务命令

卸载服务：redis-server --service-uninstall

开启服务：redis-server --service-start

停止服务：redis-server --service-stop

❹ 操作测试 Redis

如图 8.25 所示，启动 Redis 服务后，首先，进入 Redis 目录使用 redis-cli.exe -h 127.0.0.1 -p 6379 命令创建一个地址为 127.0.0.1、端口号为 6379 的 Redis 数据库服务；然后，使用 set key value 和 get key 命令保存和获得数据。

```
C:\Redis-x64-3.2.100>redis-cli.exe -h 127.0.0.1 -p 6379
127.0.0.1:6379> set uname chenheng
OK
127.0.0.1:6379> get uname
"chenheng"
127.0.0.1:6379>
```

图 8.25　测试 Redis

▶8.5.2　Spring Boot 整合 Redis

❶ Spring Data Redis

Spring 对 Redis 的支持是通过 Spring Data Redis 实现的。Spring Data Redis 提供了 RedisTemplate 和 StringRedisTemplate 两个模板进行数据操作，其中，StringRedisTemplate 只针对键值都是字符串类型的数据进行操作。

RedisTemplate 和 StringRedisTemplate 模板提供的主要数据访问方法如表 8.3 所示。

表 8.3　RedisTemplate 和 StringRedisTemplate 模板提供的主要数据访问方法

方　法	说　明
opsForValue()	操作只有简单属性的数据
opsForList()	操作含有 List 的数据
opsForSet()	操作含有 Set 的数据
opsForZSet()	操作含有 ZSet（有序的 Set）的数据
opsForHash()	操作含有 Hash 的数据

❷ Serializer

当数据存储到 Redis 时，键和值都是通过 Spring 提供的 Serializer 序列化到数据的。RedisTemplate 默认使用 JdkSerializationRedisSerializer 序列化，StringRedisTemplate 默认使用 StringRedisSerializer 序列化。

❸ Spring Boot 的支持

Spring Boot 对 Redis 的支持位于 org.springframework.boot.autoconfigure.data.redis 包下。

在 RedisAutoConfiguration 配置类中，默认配置了 RedisTemplate 和 StringRedisTemplate，我们可以直接使用 Redis 存储数据。

在 RedisProperties 类中，可以使用以 spring.redis 为扩展名的属性在 application.properties 中配置 Redis。主要属性默认配置如下：

```
spring.redis.database = 0          #数据库名 db0
spring.redis.host = localhost      #服务器地址
spring.redis.port = 6379           #连接端口号
spring.redis.max-idle = 8          #连接池的最大连接数
spring.redis.min-idle = 0          #连接池的最小连接数
spring.redis.max-active = 8        #在给定时间连接池可以分配的最大连接数
```

```
spring.redis.max-wait = -1          #当池被耗尽时，抛出异常之前连接分配应该阻塞的最大时
                                    #间量（以毫秒为单位）。使用负值表示无限期地阻止
```

从上述默认属性值可以看出，Spring Boot 默认配置了数据库名为 db0、服务器地址为 localhost、端口号为 6379 的 Redis。

因此，在 Spring Boot 应用中，只要引入 spring-boot-starter-data-redis 依赖就可以使用默认配置的 Redis 进行数据操作。

▶ 8.5.3　使用 StringRedisTemplate 和 RedisTemplate

本节通过实例讲解如何在 Spring Boot 应用中使用 StringRedisTemplate 和 RedisTemplate 模板操作 Redis 数据库。

【例 8-11】在 Spring Boot 应用中使用 StringRedisTemplate 和 RedisTemplate 模板操作 Redis 数据库。

具体实现步骤如下。

❶ 创建基于 spring-boot-starter-data-redis 依赖的 Spring Boot Web 应用 ch8_8

创建基于 spring-boot-starter-data-redis 依赖的 Spring Boot Web 应用 ch8_8。

❷ 配置 application.properties 文件

在 ch8_8 应用中，使用 Redis 的默认数据库连接。所以，不需要在 application.properties 文件中配置数据库连接信息。

❸ 创建实体类

创建名为 com.ch.ch8_8.entity 的包，并在该包中创建名为 Student 的实体类。该类必须实现序列化接口，这是因为使用 Jackson 做序列化需要一个空构造。

Student 的代码如下：

```
package com.ch.ch8_8.entity;
import java.io.Serializable;
public class Student implements Serializable{
    private static final long serialVersionUID = 1L;
    private String sno;
    private String sname;
    private Integer sage;
    public Student() {
        super();
    }
    public Student(String sno, String sname, Integer sage) {
        super();
        this.sno = sno;
        this.sname = sname;
        this.sage = sage;
    }
    //省略 get 方法和 set 方法
}
```

❹ 创建数据访问层

创建名为 com.ch.ch8_8.repository 的包，并在该包中创建名为 StudentRepository 的类，该类使用@Repository 注解标注为数据访问层。

StudentRepository 的核心代码如下：

```
@Repository
public class StudentRepository{
    @SuppressWarnings("unused")
    @Autowired
    private StringRedisTemplate stringRedisTemplate;
    @SuppressWarnings("unused")
    @Autowired
    private RedisTemplate<Object, Object> redisTemplate;
    /**
     * 使用@Resource 注解指定 stringRedisTemplate，可注入基于字符串的简单属性操作方法
     * ValueOperations<String, String> valueOpsStr = stringRedisTemplate.opsForValue();
     */
    @Resource(name="stringRedisTemplate")
    ValueOperations<String, String> valueOpsStr;
    /**
     * 使用@Resource 注解指定 redisTemplate，可注入基于对象的简单属性操作方法
     * ValueOperations<Object, Object> valueOpsObject = redisTemplate.opsForValue();
     */
    @Resource(name="redisTemplate")
    ValueOperations<Object, Object> valueOpsObject;
    /**
     * 保存字符串到 redis
     */
    public void saveString(String key, String value) {
        valueOpsStr.set(key, value);
    }
    /**
     * 保存对象到 redis
     */
    public void saveStudent(Student stu) {
        valueOpsObject.set(stu.getSno(), stu);
    }
    /**
     * 保存 List 数据到 redis
     */
    public void saveMultiStudents(Object key, List<Student> stus) {
        valueOpsObject.set(key, stus);
    }
    /**
     * 从 redis 中获得字符串数据
     */
    public String getString(String key) {
        return valueOpsStr.get(key);
    }
    /**
     * 从 redis 中获得对象数据
     */
    public Object getObject(Object key) {
        return valueOpsObject.get(key);
    }
}
```

❺ 创建控制器层

由于本实例业务简单，直接在控制器层调用数据访问层。创建名为 com.ch.ch8_8.controller 的包，并在该包中创建控制器类 TestRedisController。

TestRedisController 的核心代码如下：

```
@RestController
public class TestRedisController {
    @Autowired
    private StudentRepository studentRepository;
    @RequestMapping("/save")
    public void save() {
        studentRepository.saveString("uname", "陈恒");
        Student s1 = new Student("111","陈恒 1",77);
        studentRepository.saveStudent(s1);
        Student s2 = new Student("222","陈恒 2",88);
        Student s3 = new Student("333","陈恒 3",99);
        List<Student>  stus = new ArrayList<Student>();
        stus.add(s2);
        stus.add(s3);
        studentRepository.saveMultiStudents("mutilStus",stus);
    }
    @RequestMapping("/getUname")
    public String getUname(String key) {
        return studentRepository.getString(key);
    }
    @RequestMapping("/getStudent")
    public Student getStudent(String key) {
        return (Student)studentRepository.getObject(key);
    }
    @SuppressWarnings("unchecked")
    @RequestMapping("/getMultiStus")
    public List<Student> getMultiStus(String key) {
        return (List<Student>)studentRepository.getObject(key);
    }
}
```

❻ 运行测试

运行 Ch88Application 主类后，通过“http://localhost:8080/save”存储数据。

通过“http://localhost:8080/getUname?key=uname”查询 key 为 uname 的字符串值，如图 8.26 所示。

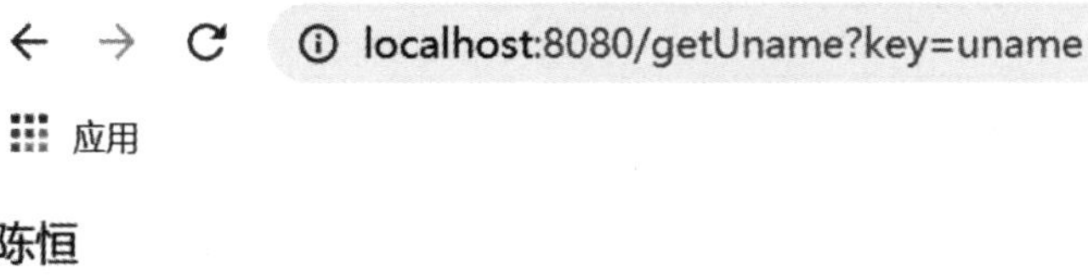

图 8.26 根据 key 查询字符串值

通过“http://localhost:8080/getStudent?key=111”查询 key 为 111 的 Student 对象值，如图 8.27 所示。

localhost:8080/getStudent?key=111

应用

{"sno":"111","sname":"陈恒1","sage":77}

图 8.27　根据 key 查询简单对象

通过“http://localhost:8080/getMultiStus?key=multiStus”查询 key 为 multiStus 的 List 集合，如图 8.28 所示。

应用

[{"sno":"222","sname":"陈恒2","sage":88},{"sno":"333","sname":"陈恒3","sage":99}]

图 8.28　根据 key 查询 List 集合

视频讲解

8.6　数据缓存 Cache

本节将介绍 Spring Boot 应用中 Cache 的一般概念，Spring Cache 对 Cache 进行了抽象，提供了 CacheManager 和 Cache 接口，并提供了@Cacheable、@CachePut、@CacheEvict、@Caching、@CacheConfig 等注解。Spring Boot 应用基于 Spring Cache，既提供了基于内存实现的缓存管理器用于单体应用系统，也集成了 Redis、EhCache 等缓存服务器用于大型系统或分布式系统。

8.6.1　Spring 缓存支持

Spring 框架定义了 org.springframework.cache.CacheManager 和 org.springframework.cache.Cache 接口统一不同的缓存技术。针对不同的缓存技术，需要实现不同的 CacheManager。例如，使用 EhCache 作为缓存技术时，需要注册实现 CacheManager 的 Bean，示例代码如下：

```
@Bean
public EhCacheCacheManager cacheManager(CacheManager ehCacheCacheManager) {
    return new EhCacheCacheManager(ehCacheCacheManager);
}
```

CacheManager 的常用实现如表 8.4 所示。

表 8.4　CacheManager 的常用实现

CacheManager	描　述
SimpleCacheManager	使用简单的 Collection 存储缓存，主要用于测试
NoOpCacheManager	仅用于测试，不会实际存储缓存
ConcurrentMapCacheManager	使用 ConcurrentMap 存储缓存，Spring 默认采用此技术存储缓存
EhCacheCacheManager	使用 EhCache 作为缓存技术
JCacheCacheManager	支持 JCache（JSR-107）标准实现作为缓存技术，如 Apache Commons JCS
RedisCacheManager	使用 Redis 作为缓存技术
HazelcastCacheManager	使用 Hazelcast 作为缓存技术

一旦配置好 Spring 缓存支持，就可以在 Spring 容器管理的 Bean 中使用缓存注解（基于 AOP 原理），一般情况下，都是在业务层（Service 类）使用这些注解。

❶ @Cacheable

@Cacheable 可以标记在一个方法上，也可以标记在一个类上。当标记在一个方法上时表示该方法是支持缓存的，当标记在一个类上时则表示该类所有的方法都是支持缓存的。对于一个支持缓存的方法，在方法执行前，Spring 先检查缓存中是否存在方法返回的数据，如果存在，则直接返回缓存数据；如果不存在，则调用方法并将方法返回值存入缓存。

@Cacheable 注解经常使用 value、key、condition 等属性。

value：缓存的名称，指定一个或多个缓存名称。如@Cacheable(value="mycache")或者@Cacheable(value={"cache1","cache2"})。该属性与 cacheNames 属性意义相同。

key：缓存的 key，可以为空，如果指定，需要按照 SpEL 表达式编写；如果不指定，则默认按照方法的所有参数进行组合。如@Cacheable(value="testcache",key="#student.id")。

condition：缓存的条件，可以为空，如果指定，需要按照 SpEL 编写，返回 true 或者 false，只有为 true 时才进行缓存。如@Cacheable(value="testcache",condition="#student.id>2")。该属性与 unless 相反，条件成立时，不进行缓存。

❷ @CacheEvict

@CacheEvict 是用来标注在需要清除缓存元素的方法或类上的。当标记在一个类上时，表示其中所有方法的执行都会触发缓存的清除操作。@CacheEvict 可以指定的属性有 value、key、condition、allEntries 和 beforeInvocation。其中，value、key 和 condition 的语义与@Cacheable 对应的属性类似。

allEntries：是否清空所有缓存内容，默认为 false，如果指定为 true，则方法调用后将立即清空所有缓存。如@CacheEvict(value="testcache", allEntries=true)。

beforeInvocation：是否在方法执行前就清空，默认为 false，如果指定为 true，则在方法还没有执行时就清空缓存。默认情况下，如果方法执行抛出异常，则不会清空缓存。

❸ @CachePut

@CachePut 也可以声明一个方法支持缓存功能，与@Cacheable 不同的是使用@CachePut 标注的方法在执行前不会检查缓存中是否存在之前执行过的结果，而是每次都会执行该方法，并将执行结果以键值对的形式存入指定的缓存中。

@CachePut 也可以标注在类和方法上。@CachePut 的属性与@Cacheable 的属性一样。

❹ @Caching

@Caching 注解可以在一个方法或者类上同时指定多个 Spring Cache 相关的注解。其拥有三个属性：cacheable、put 和 evict，分别用于指定@Cacheable、@CachePut 和@CacheEvict。示例如下：

```
@Caching(
cacheable = @Cacheable("cache1"),
evict = { @CacheEvict("cache2"),@CacheEvict(value = "cache3", allEntries = true) }
)
```

❺ @CacheConfig

所有的 Cache 注解都需要提供 Cache 名称，如果每个 Service 方法上都包含相同的 Cache 名称，可能写起来重复，此时，可以使用@CacheConfig 注解作用在类上，对当前缓存进行一些公共设置。

▶ 8.6.2 Spring Boot 缓存支持

在 Spring 中使用缓存技术的关键是配置缓存管理器 CacheManager，而 Spring Boot 为我们自动配置了多个 CacheManager 的实现。Spring Boot 的 CacheManager 的自动配置位于 org.springframework.boot.autoconfigure.cache 包中。

从 org.springframework.boot.autoconfigure.cache 包中可以看出，Spring Boot 自动配置了 EhCacheCacheConfiguration、GenericCacheConfiguration、HazelcastCacheConfiguration、HazelcastJCacheCustomizationConfiguration、InfinispanCacheConfiguration、JcacheCacheConfiguration、NoOpCacheConfiguration、RedisCacheConfiguration 和 SimpleCacheConfiguration。默认情况下，Spring Boot 使用的是 SimpleCacheConfiguration，即使用 ConcurrentMapCacheManager。Spring Boot 支持以 spring.cache 为扩展名的属性来进行缓存的相关配置。

在 Spring Boot 应用中，使用缓存技术只需在应用中引入相关缓存技术的依赖，并在配置类中使用@EnableCaching 注解开启缓存支持即可。

下面通过实例讲解如何在 Spring Boot 应用中使用默认的缓存技术 ConcurrentMapCacheManager。

【例 8-12】在 Spring Boot 应用中使用默认的缓存技术 ConcurrentMapCacheManager。

具体实现步骤如下。

❶ 创建基于 spring-boot-starter-cache 和 spring-boot-starter-data-jpa 依赖的 Spring Boot Web 应用 ch8_9

创建基于 spring-boot-starter-cache 和 spring-boot-starter-data-jpa 依赖的 Spring Boot Web 应用 ch8_9。

❷ 配置 application.properties 文件

在应用 ch8_9 中，使用 Spring Data JPA 访问 MySQL 数据库。所以，在 application.properties 文件中配置数据库连接信息，但因为使用默认的缓存技术 ConcurrentMapCacheManager，所以不需要缓存的相关配置。配置信息如下：

```
server.servlet.context-path=/ch8_9
#数据库地址
spring.datasource.url=jdbc:mysql://localhost:3306/springbootjpa?character
Encoding=utf8
#数据库用户名
spring.datasource.username=root
#数据库密码
spring.datasource.password=root
#数据库驱动
spring.datasource.driver-class-name=com.mysql.jdbc.Driver
#指定数据库类型
spring.jpa.database=MYSQL
#指定是否在日志中显示 SQL 语句
spring.jpa.show-sql=true
#指定自动创建、更新数据库表等配置，update 表示如果数据库中存在持久化类对应的表就不创建，
#不存在就创建
spring.jpa.hibernate.ddl-auto=update
```

```
#让控制器输出的 JSON 字符串格式更美观
spring.jackson.serialization.indent-output=true
```

❸ 修改 pom.xml 文件，添加 MySQL 连接依赖

因为在该应用中访问的数据库是 MySQL，所以需要将 MySQL 连接器依赖添加到 pom.xml 文件中。

❹ 创建持久化实体类

创建名为 com.ch.ch8_9.entity 的包，并在该包中创建持久化实体类 Student，该类与应用 ch8_6 中的 Student 类一样，不再赘述。

❺ 创建数据访问接口

创建名为 com.ch.ch8_9.repository 的包，并在该包中创建名为 StudentRepository 的数据访问接口。

StudentRepository 的核心代码如下：

```
public interface StudentRepository extends JpaRepository<Student, Integer>{}
```

❻ 创建业务层

创建名为 com.ch.ch8_9.service 的包，并在该包中创建 StudentService 接口和该接口的实现类 StudentServiceImpl。StudentService 的代码略。

StudentServiceImpl 的核心代码如下：

```
@Service
public class StudentServiceImpl implements StudentService{
    @Autowired
    private StudentRepository studentRepository;
    @Override
    @CachePut(value = "student", key="#student.id")
    public Student saveStudent(Student student) {
        Student s = studentRepository.save(student);
        System.out.println("为 key=" + student.getId() + "数据做了缓存");
        return s;
    }
    @Override
    @CacheEvict(value = "student", key="#student.id")
    public void deleteCache(Student student) {
        System.out.println("删除了 key=" + student.getId() + "的数据缓存");
    }
    @Override
    @Cacheable(value = "student")
    public Student selectOneStudent(Integer id) {
        Student s = studentRepository.getOne(id);
        System.out.println("为 key=" + id + "数据做了缓存");
        return s;
    }
}
```

在上述 Service 的实现类中，使用@CachePut 注解将新增的或更新的数据保存到缓存，其中缓存名为 student，数据的 key 是 student 的 id；使用@CacheEvict 注解从缓存 student 中删

除 key 为 student 的 id 的数据；使用@Cacheable 注解将 key 为 student 的 id 的数据缓存到名为 student 的缓存中。

❼ 创建控制器层

创建名为com.ch.ch8_9.controller的包，并在该包中创建名为TestCacheController的控制器类。TestCacheController 的核心代码如下：

```
@RestController
public class TestCacheController {
    @Autowired
    private StudentService studentService;
    @RequestMapping("/savePut")
    public Student save(Student student) {
        return studentService.saveStudent(student);
    }
    @RequestMapping("/selectAble")
    public Student select(Integer id) {
        return studentService.selectOneStudent(id);
    }
    @RequestMapping("/deleteEvict")
    public String deleteCache(Student student) {
        studentService.deleteCache(student);
        return "ok";
    }
}
```

❽ 开启缓存支持

在应用的启动类 Ch89Application 中，使用@EnableCaching 注解开启缓存支持，核心代码如下：

```
@EnableCaching
@SpringBootApplication
public class Ch86Application {
    …
}
```

❾ 运行测试

1）测试@Cacheable

启动应用程序的主类后，手工向数据表 student_table 添加 4 条记录。第一次访问 http://localhost:8080/ch8_9/selectAble?id=4，将调用方法查询数据库，并将查询到的数据存储到缓存 student 中。此时控制台输出结果如图 8.29 所示，页面显示数据如图 8.30 所示。

```
2021-01-03 11:12:49.720  INFO 12192 --- [nic
2021-01-03 11:12:49.720  INFO 12192 --- [nic
2021-01-03 11:12:49.722  INFO 12192 --- [nic
为key=4数据做了缓存
Hibernate: select student0_.id as id1_0_0_,
```

图 8.29　第一次访问查询控制台输出结果

```
{
  "id" : 4,
  "sno" : "444",
  "sname" : "陈恒4",
  "ssex" : "女"
}
```

图 8.30　第一次访问查询页面数据

再次访问 http://localhost:8080/ch8_9/selectAble?id=4，控制台没有输出“为 key=4 数据做了缓存”以及 Hibernate 的查询语句，这表明没有调用查询方法，页面数据直接从数据缓存中获得。

2）测试@CachePut

重启应用程序的主类，访问 http://localhost:8080/ch8_9/savePut?sname=陈恒 5&sno=555&ssex=男，此时控制台输出结果如图 8.31 所示，页面数据如图 8.32 所示。

```
2021-01-03 11:15:14.753  INFO 10144 --- [nio-
2021-01-03 11:15:14.753  INFO 10144 --- [nio-
2021-01-03 11:15:14.755  INFO 10144 --- [nio-
Hibernate: insert into student_table (sname,
为key=5数据做了缓存
```

图 8.31　测试@CachePut 控制台输出结果

```
{
  "id" : 5,
  "sno" : "555",
  "sname" : "陈恒5",
  "ssex" : "男"
}
```

图 8.32　测试@CachePut 页面数据

这时访问 http://localhost:8080/ch8_9/selectAble?id=5，控制台无输出，从缓存直接获得数据，页面数据如图 8.32 所示。

3）测试@CacheEvict

重启应用程序的主类，首先，访问 http://localhost:8080/ch8_9/selectAble?id=1，为 key 为 1 的数据做缓存，再次访问 http://localhost:8080/ch8_9/selectAble?id=1，确认数据已从缓存中获取。然后，访问 http://localhost:8080/ch8_9/deleteEvict?id=1，从缓存 student 中删除 key 为 1 的数据，此时控制台输出结果如图 8.33 所示。

```
2021-01-03 11:16:44.107  INFO 14556 --- [nic
为key=1数据做了缓存
Hibernate: select student0_.id as id1_0_0_,
删除了key=1的数据缓存
```

图 8.33　测试@CacheEvict 删除缓存数据

最后，再次访问 http://localhost:8080/ch8_9/selectAble?id=1，此时重新做了缓存，控制台输出结果如图 8.34 所示。

```
2021-01-03 11:16:44.107  INFO 14556 --- [nio
为key=1数据做了缓存
Hibernate: select student0_.id as id1_0_0_,
删除了key=1的数据缓存
为key=1数据做了缓存
Hibernate: select student0_.id as id1_0_0_,
```

图 8.34　测试@CacheEvict 重做缓存数据

8.6.3　使用 Redis Cache

在 Spring Boot 中使用 Redis Cache，只要添加 spring-boot-starter-data-redis 依赖即可。下面通过实例测试 Redis Cache。

【例 8-13】在例 8-11 的基础上，测试 Redis Cache。

具体实现步骤如下。

❶ 使用@Cacheable 注解修改控制器方法

将控制器中的方法 getUname 修改如下：

```
@RequestMapping("/getUname")
@Cacheable(value = "myuname")
public String getUname(String key) {
    System.out.println("测试缓存");
    return studentRepository.getString(key);
}
```

❷ 使用@EnableCaching 注解开启缓存支持

在应用程序的主类 Ch88Application 中使用@EnableCaching 注解开启缓存支持。

❸ 测试 Redis Cache

启动应用程序的主类后，多次访问“http://localhost:8080/getUname?key=uname”，但“测试缓存”字样在控制台仅打印一次，页面查询结果不变。这说明，只有第一次访问时调用了查询方法，后面多次访问都是从缓存直接获得数据。

8.7 本章小结

本章是本书的重点章节，重点讲解了 Spring Data JPA、Spring Boot 整合 MyBatis 以及数据缓存 Cache。通过本章的学习，掌握 Spring Boot 访问数据库的解决方案。

习 题 8

1. 在 Spring Boot 应用中，数据缓存技术解决了什么问题？
2. Spring 框架提供了哪些缓存注解？这些注解如何使用？
3. 在 Spring Data JPA 中，如何实现一对一、一对多、多对多关联查询？请举例说明。

第 9 章 电子商务平台的设计与实现（Spring Boot + MyBatis + Thymeleaf）

视频讲解

学习目的与要求

本章通过一个小型的电子商务平台的设计与实现，讲述如何使用 Spring Boot + Thymeleaf + MyBatis 开发一个 Web 应用，其中主要涉及的技术包括 Spring 与 Spring MVC 框架技术、MyBatis 持久层技术、Thymeleaf 表现层技术。通过本章的学习，掌握基于 Thymeleaf + MyBatis 的 Spring Boot Web 应用开发的流程、方法以及技术。

主要内容

- ❖ 系统设计
- ❖ 数据库设计
- ❖ 系统管理
- ❖ 组件设计
- ❖ 系统实现

本章系统使用 Spring Boot + Thymeleaf + MyBatis 实现各个模块，Web 服务器使用内嵌的 Servlet 容器，数据库采用 MySQL 5.5，集成开发环境为 STS 或 IDEA。

9.1 系统设计

电子商务平台分为两个子系统，一是后台管理子系统，一是电子商务子系统。下面分别说明这两个子系统的功能需求与模块划分。

▶ 9.1.1 系统功能需求

❶ 后台管理子系统

后台管理子系统要求管理员登录成功后，才能对商品进行管理，包括添加商品、查询商品、修改商品以及删除商品。除商品管理外，管理员还需要对商品类型、注册用户以及用户的订单等进行管理。

❷ 电子商务子系统

1）非注册用户

非注册用户或未登录用户具有浏览首页、查看商品详情以及搜索商品的功能。

2）用户

成功登录的用户除具有未登录用户具有的功能外，还具有购买商品、查看购物车、收藏商品、查看订单、查看收藏以及查看用户个人信息的功能。

9.1.2 系统模块划分

❶ 后台管理子系统

管理员登录成功后，进入后台管理主页面（selectGoods.html），可以对商品、商品类型、注册用户以及用户的订单进行管理。后台管理子系统的模块划分如图 9.1 所示。

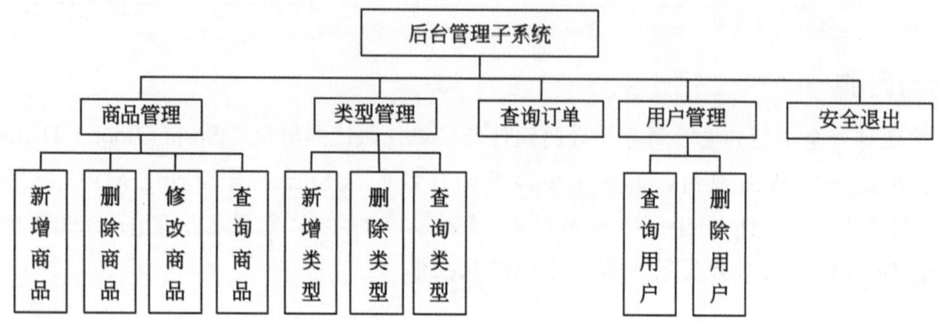

图 9.1 后台管理子系统

❷ 电子商务子系统

非注册用户只可以浏览商品、搜索商品，不能购买商品、收藏商品、查看购物车、查看用户中心、我的订单和我的收藏。成功登录的用户可以实现电子商务子系统的所有功能，包括购买商品、支付等功能。电子商务子系统的模块划分如图 9.2 所示。

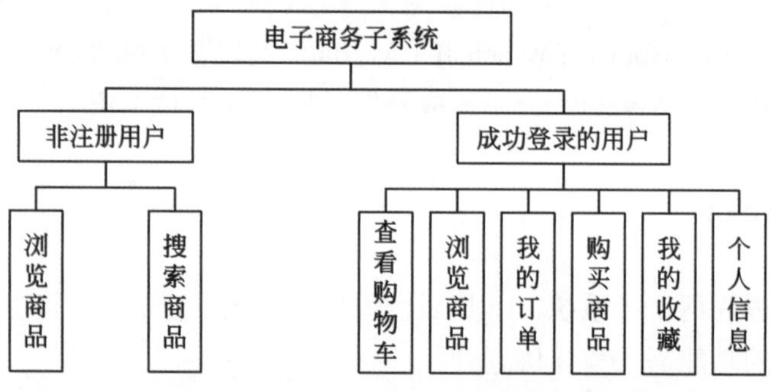

图 9.2 电子商务子系统

9.2 数据库设计

系统采用加载纯 Java 数据库驱动程序的方式连接 MySQL 5.5 数据库。在 MySQL 5.5 中创建数据库 ch9，并在 ch9 中创建 8 张与系统相关的数据表：ausertable、busertable、carttable、focustable、goodstable、goodstype、orderdetail 和 orderbasetable。

9.2.1 数据库概念结构设计

根据系统设计与分析，可以设计出如下数据结构。

❶ 管理员

管理员包括管理员 ID、用户名和密码。管理员的用户名和密码由数据库管理员预设，不需要注册。

❷ 用户

用户包括用户 ID、E-mail 和密码。注册用户的邮箱不能相同，用户 ID 唯一。

❸ 商品类型

商品类型包括类型 ID 和类型名称。

❹ 商品

商品包括商品编号、商品名称、原价、现价、库存、图片以及类型。其中，商品编号唯一，类型与“3. 商品类型”关联。

❺ 购物车

购物车包括购物车 ID、用户 ID、商品编号以及购买数量。其中，购物车 ID 唯一，用户 ID 与“2. 用户”关联，商品编号与“4. 商品”关联。

❻ 商品收藏

商品收藏包括 ID、用户 ID、商品编号以及收藏时间。其中，ID 唯一，用户 ID 与“2. 用户”关联，商品编号与“4. 商品”关联。

❼ 订单基础

订单基础包括订单编号、用户 ID、订单金额、订单状态以及下单时间。其中，订单编号唯一，用户 ID 与“2. 用户”关联。

❽ 订单详情

订单详情包括订单编号、商品编号以及购买数量。其中，订单编号与“7. 订单基础”关联，商品编号与“4. 商品”关联。

根据以上的数据结构，结合数据库设计的特点，可以画出如图 9.3 所示的数据库概念结构图。

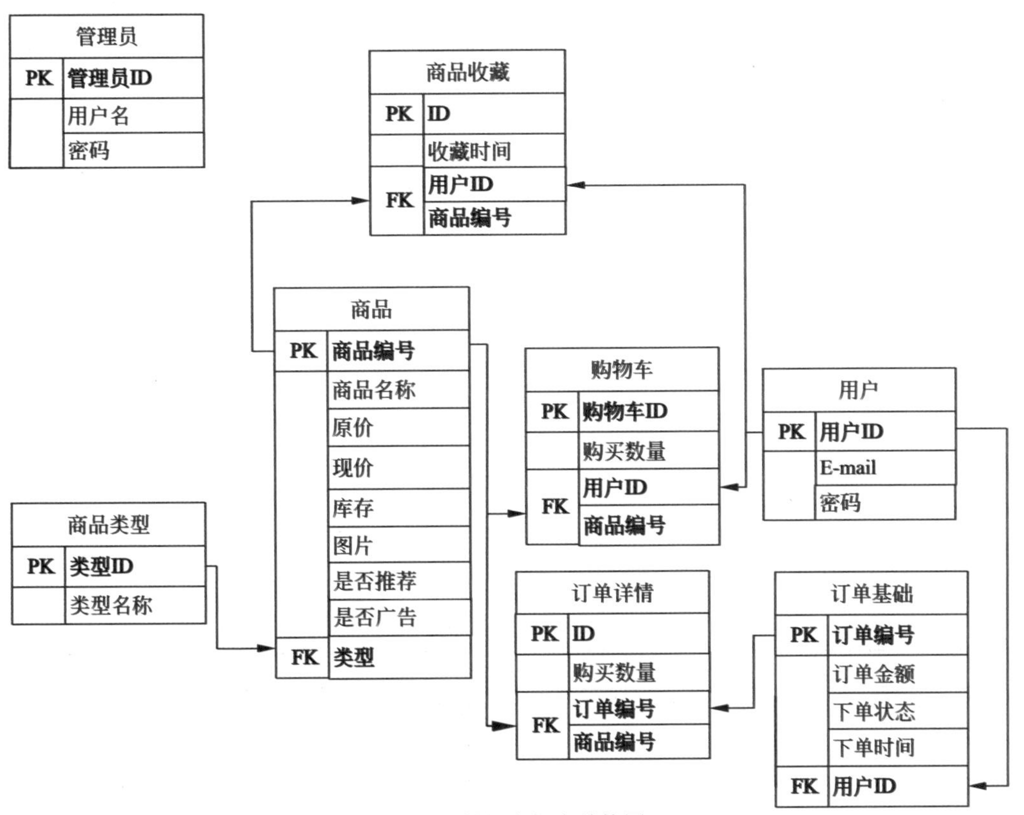

图 9.3　数据库概念结构图

▶ 9.2.2 数据逻辑结构设计

将数据库概念结构图转换为MySQL数据库所支持的实际数据模型，即数据库的逻辑结构。管理员信息表（ausertable）的设计如表9.1所示。

表9.1 管理员信息表

字段	含义	类型	长度	是否为空
id	管理员ID（PK 自增）	int	11	no
aname	用户名	varchar	50	no
apwd	密码	varchar	50	no

用户信息表（busertable）的设计如表9.2所示。

表9.2 用户信息表

字段	含义	类型	长度	是否为空
id	用户ID（PK 自增）	int	11	no
bemail	E-mail	varchar	50	no
bpwd	密码	varchar	50	no

商品类型表（goodstype）的设计如表9.3所示。

表9.3 商品类型表

字段	含义	类型	长度	是否为空
id	类型ID（PK 自增）	int	11	no
typename	类型名称	varchar	50	no

商品信息表（goodstable）的设计如表9.4所示。

表9.4 商品信息表

字段	含义	类型	长度	是否为空
id	商品编号（PK 自增）	int	11	no
gname	商品名称	varchar	50	no
goprice	原价	double		no
grprice	现价	double		no
gstore	库存	int	11	no
gpicture	图片	varchar	50	
isRecommend	是否推荐	tinyint	2	no
isAdvertisement	是否广告	tinyint	2	no
goodstype_id	类型（FK）	int	11	no

购物车表（carttable）的设计如表9.5所示。

表9.5 购物车表

字段	含义	类型	长度	是否为空
id	购物车ID（PK 自增）	int	11	no
busertable_id	用户ID（FK）	int	11	no
goodstable_id	商品编号（FK）	int	11	no
shoppingnum	购买数量	int	11	no

商品收藏表（focustable）的设计如表 9.6 所示。

表 9.6　商品收藏表

字段	含义	类型	长度	是否为空
id	ID（PK 自增）	int	11	no
goodstable_id	商品编号（FK）	int	11	no
busertable_id	用户 ID（FK）	int	11	no
focustime	收藏时间	datetime		no

订单基础表（orderbasetable）的设计如表 9.7 所示。

表 9.7　订单基础表

字段	含义	类型	长度	是否为空
id	订单编号（PK 自增）	int	11	no
busertable_id	用户 ID（FK）	int	11	no
amount	订单金额	double		no
status	订单状态	tinyint	4	no
orderdate	下单时间	datetime		no

订单详情表（orderdetail）的设计如表 9.8 所示。

表 9.8　订单详情表

字段	含义	类型	长度	是否为空
id	ID（PK 自增）	int	11	no
orderbasetable_id	订单编号（FK）	int	11	no
goodstable_id	商品编号（FK）	int	11	no
shoppingnum	购买数量	int	11	no

9.2.3　创建数据表

根据 9.2.2 节的逻辑结构，创建数据表。篇幅有限，创建数据表的代码请读者参考本书提供的源代码 ch9.sql。

9.3　系统管理

9.3.1　添加相关依赖

新建一个基于 Thymeleaf + MyBatis 的 Spring Boot Web 应用 eBusiness，在 eBusiness 应用中开发本系统。除了 STS 快速创建基于 Thymeleaf + MyBatis 的 Spring Boot Web 应用自带的 spring-boot-starter-thymeleaf、mybatis-spring-boot-starter 和 spring-boot-starter-web 依赖外，还需要向 eBusiness 应用的 pom.xml 文件中添加上传文件依赖 commons-fileupload、表单验证依赖 hibernate-validator 以及 MySQL 连接器依赖，具体见本书提供的源代码 eBusiness 的 pom.xml 文件。

9.3.2　HTML 页面及静态资源管理

系统由后台管理和电子商务两个子系统组成，为了方便管理，两个子系统的 HTML 页面分开存放。在 src/main/resources/templates/admin 目录下存放与后台管理子系统相关的 HTML

页面；在 src/main/resources/templates/user 目录下存放与电子商务子系统相关的 HTML 页面；在 src/main/resources/static 目录下存放与整个系统相关的 BootStrap 及 jQuery。由于篇幅受限，本章仅附上部分 HTML 和 Java 文件的核心代码，具体代码请读者参考本书提供的源代码 eBusiness。

❶ 后台管理子系统

管理员在浏览器的地址栏中输入 http://localhost:8080/eBusiness/admin/toLogin 访问登录页面，登录成功后，进入后台商品管理主页面（adminGoods.html）。adminGoods.html 的运行效果如图 9.4 所示。

类型管理▾ 商品管理▾ 查询订单 用户管理 安全退出

商品ID	商品名称	商品类型	操作
47	衣服66	家电	详情 修改 删除
46	衣服155	服装	详情 修改 删除
45	衣服144	服装	详情 修改 删除
44	衣服33	服装	详情 修改 删除
43	衣服22	服装	详情 修改 删除

第1页 共3页 下一页

添加商品

图 9.4　后台商品管理主页面

❷ 电子商务子系统

注册用户或游客在浏览器的地址栏中输入 http://localhost:8080/eBusiness，可以访问电子商务子系统的首页（index.html）。index.html 的运行效果如图 9.5 所示。

图 9.5　电子商务子系统的首页

▶ 9.3.3　应用的包结构

❶ com.ch.ebusiness 包

该包中包括应用的主程序类 EBusinessApplication、统一异常处理类 GlobalExceptionHandleController 以及自定义异常类 NoLoginException。

❷ com.ch.ebusiness.controller 包

系统的控制器类都在该包中，后台管理相关的控制器类在 admin 子包中，电子商务相关的控制器类在 before 子包中。

❸ com.ch.ebusiness.entity 包

实体类存放在该包中。

❹ com.ch.ebusiness.repository 包

该包中存放的 Java 接口程序实现数据库的持久化操作。每个接口方法与 SQL 映射文件中的 ID 相同。后台管理相关的数据库操作在 admin 子包中；电子商务相关的数据库操作在 before 子包中。

❺ com.ch.ebusiness.service 包

service 包中有两个子包：admin 和 before。admin 子包存放后台管理相关业务层的接口与实现类；before 子包存放电子商务相关业务层的接口与实现类。

❻ com.ch.ebusiness.util 包

该包中存放的是系统的工具类。

▶ 9.3.4　配置文件

在配置文件 application.properties 中，配置了数据源、实体类包别名、映射文件位置、SQL 日志、页面热部署及文件上传等信息，具体如下：

```
server.servlet.context-path=/eBusiness
###
##数据源信息配置
###
#数据库地址
spring.datasource.url=jdbc:mysql://localhost:3306/shop?characterEncoding=utf8
#数据库用户名
spring.datasource.username=root
#数据库密码
spring.datasource.password=root
#数据库驱动
spring.datasource.driver-class-name=com.mysql.jdbc.Driver
#设置包别名（在 Mapper 映射文件中直接使用实体类名）
mybatis.type-aliases-package=com.ch.ebusiness.entity
#告诉系统到哪里去找 mapper.xml 文件（映射文件）
mybatis.mapperLocations=classpath:mappers/*.xml
#在控制台输出 SQL 语句日志
logging.level.com.ch.ebusiness.repository=debug
#关闭 Thymeleaf 模板引擎缓存（使页面热部署），默认是开启的
spring.thymeleaf.cache=false
#上传文件时，默认单个上传文件大小是 1MB，max-file-size 设置单个上传文件大小
spring.servlet.multipart.max-file-size=50MB
```

```
#默认总上传文件大小是 10MB，max-request-size 设置总上传文件大小
spring.servlet.multipart.max-request-size=500MB
```

9.4 组件设计

本系统的组件包括管理员登录权限验证控制器、前台用户登录权限验证控制器、验证码、统一异常处理以及工具类。

9.4.1 管理员登录权限验证

从系统分析得知，管理员成功登录后，才能管理商品、商品类型、用户、订单等功能模块。因此，本系统需要对这些功能模块的操作进行管理员登录权限控制。在 com.ch.ebusiness.controller.admin 包中创建 AdminBaseController 控制器类，该类中有一个@ModelAttribute 注解的方法 isLogin。isLogin 方法的功能是判断管理员是否已成功登录。需要进行管理员登录权限控制的控制器类继承 AdminBaseController 类即可，因为带有@ModelAttribute 注解的方法首先被控制器执行。AdminBaseController 控制器类的核心代码如下：

```
@Controller
public class AdminBaseController {
    @ModelAttribute
    public void isLogin(HttpSession session) throws NoLoginException {
        if(session.getAttribute("auser") == null){
            throw new NoLoginException("没有登录");
        }
    }
}
```

9.4.2 前台用户登录权限验证

从系统分析得知，用户成功登录后，才能购买商品、收藏商品、查看购物车、查询我的订单以及修改个人信息。与管理员登录权限验证同理，在 com.ch.ebusiness.controller.before 包中创建 BeforeBaseController 控制器类，该类中有一个@ModelAttribute 注解的方法 isLogin。isLogin 方法的功能是判断前台用户是否已成功登录。需要进行前台用户登录权限控制的控制器类继承 BeforeBaseController 类即可。BeforeBaseController 控制器类的代码与 AdminBaseController 基本一样，为节省篇幅，不再赘述。

9.4.3 验证码

本系统验证码的使用步骤如下。

❶ 创建产生验证码的控制器类

在 com.ch.ebusiness.controller.before 包中，创建产生验证码的控制器类 ValidateCodeController，具体代码参见本书提供的源程序 eBusiness。

❷ 使用验证码

在需要验证码的 HTML 页面中，调用产生验证码的控制器显示验证码，示例代码片段如下：

```
<img th:src="@{/validateCode}" id="mycode">
```

▶ 9.4.4　统一异常处理

系统对未登录异常、数据库操作异常以及程序未知异常进行了统一异常处理。具体步骤如下。

❶ 创建未登录自定义异常

创建未登录自定义异常 NoLoginException，代码如下：

```
package com.ch.ebusiness;
public class NoLoginException extends Exception{
    private static final long serialVersionUID = 1L;
    public NoLoginException() {
        super();
    }
    public NoLoginException(String message) {
        super(message);
    }
}
```

❷ 创建统一异常处理类

使用注解 @ControllerAdvice 和 @ExceptionHandler 创建统一异常处理类 GlobalExceptionHandleController。使用注解@ControllerAdvice 的类是一个增强的 Controller 类，在增强的控制器类中使用@ExceptionHandler 注解的方法对所有控制器类进行统一异常处理。核心代码如下：

```
@ControllerAdvice
public class GlobalExceptionHandleController {
    @ExceptionHandler(value=Exception.class)
    public String exceptionHandler(Exception e, Model model) {
        String message = "";
        //数据库异常
        if (e instanceof SQLException) {
            message = "数据库异常";
        } else if (e instanceof NoLoginException) {
            message = "未登录异常";
        } else {//未知异常
            message =  "未知异常";
        }
        model.addAttribute("mymessage",message);
        return "myError";
    }
}
```

▶ 9.4.5　工具类

本系统使用的工具类有两个：MD5Util 和 MyUtil。MD5Util 工具用来对明文密码加密；MyUtil 工具包含文件重命名和获得用户信息两个功能。MD5Util 和 MyUtil 的代码参见本书提供的源程序 eBusiness。

9.5 后台管理子系统的实现

管理员登录成功后，可以对商品及商品类型、注册用户以及用户的订单进行管理。本节将详细讲解管理员的功能实现。

▶ 9.5.1 管理员登录

管理员输入用户名和密码后，系统将对管理员的用户名和密码进行验证。如果用户名和密码同时正确，则成功登录，进入后台商品管理主页面（adminGoods.html）；如果用户名或密码有误，则提示错误。实现步骤如下。

❶ 编写视图

login.html 页面是提供登录信息输入的页面，效果如图 9.6 所示。

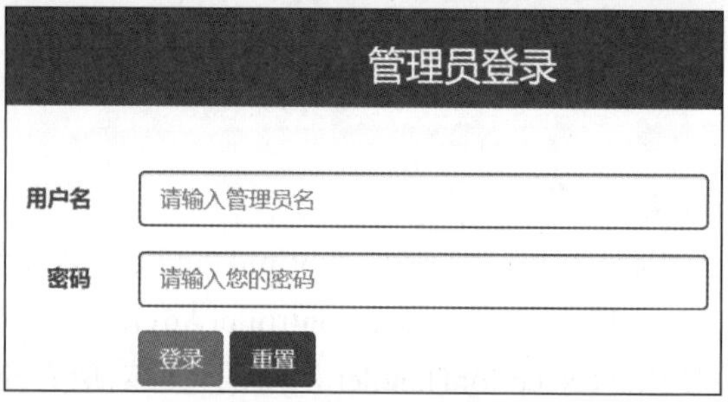

管理员登录

用户名 请输入管理员名

密码 请输入您的密码

登录 重置

图 9.6 管理员登录页面

在 src/main/resources/templates/admin 目录下创建 login.html。该页面的代码参见本书提供的源程序 eBusiness。

❷ 编写控制器层

视图 Action 的请求路径为“admin/login”，系统根据请求路径和@RequestMapping 注解找到对应控制器类 com.ch.ebusiness.controller.admin 的 login 方法处理登录。在控制器类的 login 方法中调用 com.ch.ebusiness.service.admin.AdminService 接口的 login 方法处理登录。登录成功后，首先将登录人信息存入 session，然后转发到查询商品请求方法；登录失败回到本页面。控制器层的相关代码如下：

```
@Controller
@RequestMapping("/admin")
public class AdminController {
    @Autowired
    private AdminService adminService;
    @RequestMapping("/toLogin")
    public String toLogin(@ModelAttribute("aUser") AUser aUser) {
        return "admin/login";
    }
    @RequestMapping("/login")
    public String login(@ModelAttribute("aUser") AUser aUser, HttpSession session,
    Model model) {
        return adminService.login(aUser, session, model);
    }
}
```

❸ 编写 Service 层

Service 层由接口 com.ch.ebusiness.service.admin.AdminService 和接口的实现类 com.ch.ebusiness.service.admin.AdminServiceImpl 组成。Service 层是功能模块实现的核心，Service 层调用数据访问层（Repository）进行数据库操作。管理员登录的业务处理方法 login 的代码如下：

```
public String login(AUser aUser, HttpSession session, Model model) {
    List<AUser> list = adminRepository.login(aUser);
    if(list.size() > 0) {//登录成功
        session.setAttribute("auser", aUser);
        return "forward:/goods/selectAllGoodsByPage?currentPage=1&act=select";
    }else {//登录失败
        model.addAttribute("errorMessage", "用户名或密码错误！");
        return "admin/login";
    }
}
```

❹ 编写 SQL 映射文件

数据访问层（Repository）仅由@Repository 注解的接口组成，接口方法与 SQL 映射文件中 SQL 语句的 id 相同。管理员登录的 SQL 映射文件为 src/main/resources/mappers 目录下的 AdminMapper.xml，实现的 SQL 语句如下：

```
<select id="login" parameterType="AUser" resultType="AUser">
    select * from ausertable where aname = #{aname} and apwd = #{apwd}
</select>
```

9.5.2 类型管理

管理员登录成功后，管理商品类型。类型管理分为添加类型和删除类型，如图 9.7 所示。

图 9.7 类型管理

❶ 添加类型

添加类型实现步骤如下。

1）编写视图

单击图 9.7 中的“添加类型”超链接（type/toAddType），打开如图 9.8 所示的添加类型页面。

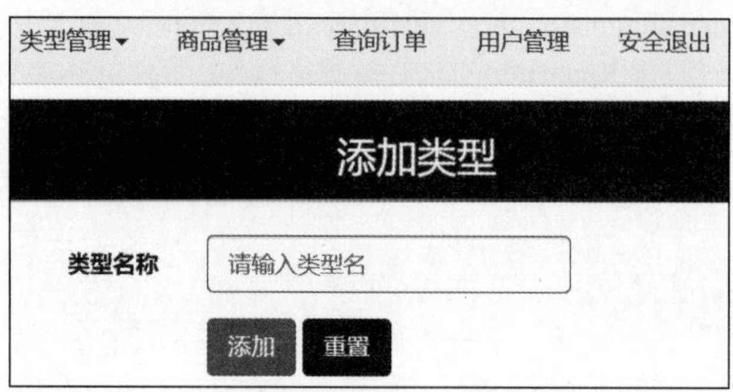

图 9.8　添加类型页面

在 src/main/resources/templates/admin 目录下，创建添加类型页面 addType.html。该页面的代码参见本书提供的源程序 eBusiness。

2）编写控制器层

此功能共有两个处理请求：“添加类型”超链接 type/toAddType 和视图 Action 的请求路径 type/addType。系统根据 @RequestMapping 注解找到对应控制器类 com.ch.ebusiness.controller.admin.TypeController 的 toAddType 和 addType 方法处理请求。在控制器类的处理方法中调用 com.ch.ebusiness.service.admin.TypeService 接口的 addType 方法处理业务。控制器层的相关代码如下：

```
@RequestMapping("/toAddType")
public String toAddType(@ModelAttribute("goodsType") GoodsType goodsType) {
    return "admin/addType";
}
@RequestMapping("/addType")
public String addType(@ModelAttribute("goodsType") GoodsType goodsType) {
    return typeService.addType(goodsType);
}
```

3）编写 Service 层

添加类型 type/addType 的业务处理方法 addType 的代码如下：

```
@Override
public String addType(GoodsType goodsType) {
    typeRepository.addType(goodsType);
    return "redirect:/type/selectAllTypeByPage?currentPage=1";
}
```

4）编写 SQL 映射文件

实现添加类型 type/addType 的 SQL 语句如下（位于 src/main/resources/mappers/TypeMapper.xml 文件中）：

```
<insert id="addType" parameterType="GoodsType">
    insert into goodstype (id, typename) values(null, #{typename})
</insert>
```

❷ 删除类型

删除类型的实现步骤如下。

1）编写视图

单击图 9.7 中的“类型管理”超链接（type/selectAllTypeByPage?currentPage=1），打开如图 9.7 所示的查询页面。

在 src/main/resources/templates/admin 目录下，创建“查询类型”页面 selectGoodsType.html。该页面的代码参见本书提供的源程序 eBusiness。

2）编写控制器层

此功能模块共有两个处理请求：“查询类型”超链接 type/selectAllTypeByPage?currentPage=1 与视图“删除”的请求路径 type/deleteType。系统根据@RequestMapping 注解找到对应控制器类 com.ch.ebusiness.controller.admin.TypeController 的 selectAllTypeByPage 和 delete 方法处理请求。在控制器类的处理方法中调用 com.ch.ebusiness.service.admin.TypeService 接口的 selectAllTypeByPage 和 delete 方法处理业务。控制器层的相关代码如下：

```
@RequestMapping("/selectAllTypeByPage")
public String selectAllTypeByPage(Model model, int currentPage) {
    return typeService.selectAllTypeByPage(model, currentPage);
}
@RequestMapping("/deleteType")
@ResponseBody//返回字符串数据而不是视图
public String delete(int id) {
    return typeService.delete(id);
}
```

3）编写 Service 层

超链接 type/selectAllTypeByPage?currentPage=1 的业务处理方法 selectAllTypeByPage 的代码如下：

```
public String selectAllTypeByPage(Model model, int currentPage) {
    //共多少个类型
    int totalCount = typeRepository.selectAll();
    //计算共多少页
    int pageSize = 2;
    int totalPage = (int)Math.ceil(totalCount*1.0/pageSize);
    List<GoodsType> typeByPage =
     typeRepository.selectAllTypeByPage((currentPage-1)*pageSize, pageSize);
    model.addAttribute("allTypes", typeByPage);
    model.addAttribute("totalPage", totalPage);
    model.addAttribute("currentPage", currentPage);
    return "admin/selectGoodsType";
}
```

删除 type/deleteType 的业务处理方法 delete 的代码如下：

```
public String delete(int id) {
    List<Goods> list = typeRepository.selectGoods(id);
    if(list.size() > 0) {
        //该类型下有商品不允许删除
        return "no";
    }else {
        typeRepository.deleteType(id);
        //删除后回到查询页面
        return "/type/selectAllTypeByPage?currentPage=1";
    }
}
```

4）编写 SQL 映射文件

实现超链接 type/selectAllTypeByPage?currentPage=1 的 SQL 语句如下：

```
<select id="selectAll"  resultType="integer">
    select count(*) from goodstype
</select>
<!-- 分页查询 -->
<select id="selectAllTypeByPage"  resultType="GoodsType">
    select * from goodstype  limit #{startIndex}, #{perPageSize}
</select>
```

实现删除 type/deleteType 的 SQL 语句如下：

```
<!-- 删除类型 -->
<delete id="deleteType" parameterType="integer">
    delete from goodstype where id=#{id}
</delete>
<!-- 查询该类型下是否有商品 -->
<select id="selectGoods" parameterType="integer" resultType="Goods">
    select * from goodstable where goodstype_id = #{goodstype_id}
</select>
```

▶ 9.5.3 添加商品

单击图 9.4 中的“添加商品”超链接，打开如图 9.9 所示的添加商品页面。添加商品的实现步骤如下。

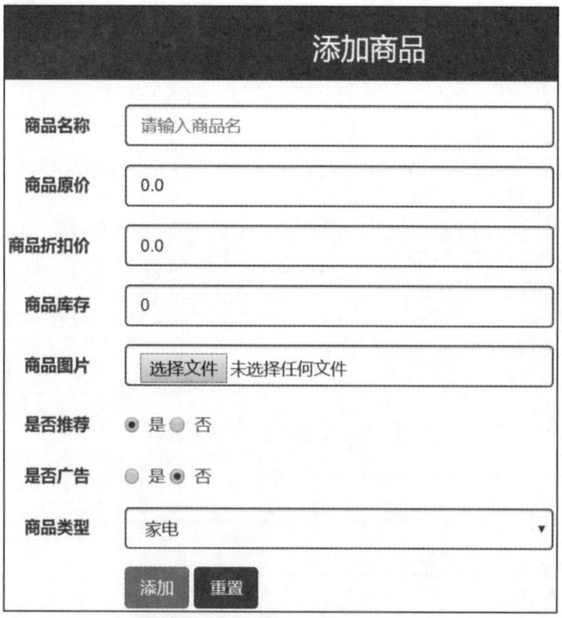

图 9.9　添加商品页面

❶ 编写视图

在 src/main/resources/templates/admin 目录下，创建添加商品页面 addGoods.html。该页面的代码参见本书提供的源程序 eBusiness。

❷ 编写控制器层

此功能模块共有两个处理请求："添加商品"超链接 goods/toAddGoods 与视图"添加"的请求路径 goods/addGoods?act=add。系统根据@RequestMapping 注解找到对应控制器类 com.ch.ebusiness.controller.admin.GoodsController 的 toAddGoods 和 addGoods 方法处理请求。在控制器类的处理方法中调用 com.ch.ebusiness.service.admin.GoodsService 接口的 toAddGoods 和 addGoods 方法处理业务。控制器层的相关代码如下：

```
    @RequestMapping("/toAddGoods")
    public String toAddGoods(@ModelAttribute("goods") Goods goods, Model model) {
        goods.setIsAdvertisement(0);
        goods.setIsRecommend(1);
        return goodsService.toAddGoods(goods, model);
    }
    @RequestMapping("/addGoods")
    public String addGoods(@ModelAttribute("goods") Goods goods, HttpServletRequest
request, String act) throws IllegalStateException, IOException {
        return goodsService.addGoods(goods, request, act);
    }
```

❸ 编写 Service 层

添加商品的 Service 层相关代码如下：

```
    @Override
    public String addGoods(Goods goods, HttpServletRequest  request, String act)
throws IllegalStateException, IOException {
        MultipartFile myfile = goods.getFileName();
        //如果选择了上传文件，将文件上传到指定的目录 images
        if(!myfile.isEmpty()) {
            //上传文件路径（生产环境）
            //String path = request.getServletContext().getRealPath("/images/");
            //获得上传文件原名
            //上传文件路径（开发环境）
            String path = "C:\\ workspace-spring-tool-suite-4-4.9.0.RELEASE\\
            eBusiness\\src\\main\\resources\\static\\images";
            //获得上传文件原名
            String fileName = myfile.getOriginalFilename();
            //对文件重命名
            String fileNewName = MyUtil.getNewFileName(fileName);
            File filePath = new File(path + File.separator + fileNewName);
            //如果文件目录不存在，创建目录
            if(!filePath.getParentFile().exists()) {
                filePath.getParentFile().mkdirs();
            }
            //将上传文件保存到一个目标文件中
            myfile.transferTo(filePath);
            //将重命名后的图片名存到 goods 对象中，添加时使用
            goods.setGpicture(fileNewName);
        }
        if("add".equals(act)) {
            int n = goodsRepository.addGoods(goods);
            if(n > 0)//成功
```

```
            return "redirect:/goods/selectAllGoodsByPage?currentPage=1&act=select";
        //失败
        return "admin/addGoods";
    }else {//修改
        int n = goodsRepository.updateGoods(goods);
        if(n > 0)//成功
            return "redirect:/goods/selectAllGoodsByPage?currentPage=1&act=
            updateSelect";
        //失败
        return "admin/UpdateAGoods";
    }
}
@Override
public String toAddGoods(Goods goods, Model model) {
    model.addAttribute("goodsType", goodsRepository.selectAllGoodsType());
    return "admin/addGoods";
}
```

❹ 编写 SQL 映射文件

添加商品的 SQL 语句如下：

```
<!-- 添加商品 -->
    <insert id="addGoods" parameterType="Goods">
        insert into goodstable (id,gname,goprice,grprice,gstore,gpicture,
        isRecommend,isAdvertisement,goodstype_id)
        values (null, #{gname}, #{goprice}, #{grprice}, #{gstore}, #{gpicture},
        #{isRecommend}, #{isAdvertisement}, #{goodstype_id})
    </insert>
    <!-- 查询商品类型 -->
    <select id="selectAllGoodsType"  resultType="GoodsType">
        select * from goodstype
    </select>
```

▶ 9.5.4 查询商品

管理员登录成功后，进入如图 9.4 所示的后台商品管理主页面，在主页面中单击“详情”超链接，显示如图 9.10 所示的商品详情页面。

图 9.10　商品详情页面

❶ 编写视图

在 src/main/resources/templates/admin 目录下，创建商品管理主页面 adminGoods.html，该页面显示查询商品、修改商品查询以及删除商品查询的结果。其代码如下：

```
<!DOCTYPE html>
<!DOCTYPE html>
<html xmlns:th="http://www.thymeleaf.org">
<head>
<base th:href="@{/}">
<meta charset="UTF-8">
<title>主页</title>
<link rel="stylesheet" href="css/bootstrap.min.css" />
<script src="js/jquery.min.js"></script>
<script type="text/javascript" th:inline="javascript">
    function deleteGoods(tid){
        $.ajax(
            {
                //请求路径，要注意的是 url 和 th:inline="javascript"
                url : [[@{/goods/delete}]],
                //请求类型
                type : "post",
                //data 表示发送的数据
                data : {
                    id : tid
                },
                //成功响应的结果
                success : function(obj){//obj 响应数据
                    if(obj == "no"){
                        alert("该商品有关联不允许删除！");
                    }else{
                        if(window.confirm("真的删除该商品吗？")){
                            //获取路径
                            varpathName=window.document.location.pathname;
                            //截取，得到项目名称
                    var projectName=pathName.substring(0,pathName.substr(1).
                    indexOf('/')+1);
                            window.location.href = projectName + obj;
                        }
                    }
                },
                error : function() {
                    alert("处理异常！");
                }
            }
        );
    }
    </script>
</head>
<body>
    <!-- 加载 header.html -->
    <div th:include="admin/header"></div>
```

```
    <br><br><br>
    <div class="container">
        <div class="panel panel-primary">
            <div class="panel-heading">
                <h3 class="panel-title">商品列表</h3>
            </div>
            <div class="panel-body">
                <div class="table table-responsive">
                    <table class="table table-bordered table-hover">
                        <tbody class="text-center">
                            <tr>
                                <th>商品 ID</th>
                                <th>商品名称</th>
                                <th>商品类型</th>
                                <th>操作</th>
                            </tr>
                            <tr th:each="gds:${allGoods}">
                                <td th:text="${gds.id}"></td>
                                <td th:text="${gds.gname}"></td>
                                <td th:text="${gds.typename}"></td>
                                <td>
            <a th:href="@{goods/detail(id=${gds.id},act=detail)}" target="_blank">详情</a>
            <a th:href="@{goods/detail(id=${gds.id},act=update)}" target="_blank">修改</a>
            <a th:href="'javascript:deleteGoods(' + ${gds.id} +')'" >删除</a>
                                </td>
                            </tr>
                            <tr>
                                <td colspan="4" align="right">
                                    <ul class="pagination">
                        <li><a>第<span th:text="${currentPage}" ></span>页</a></li>
                        <li><a>共<span th:text="${totalPage}" ></span>页</a></li>
                                        <li>
                                            <span th:if="${currentPage} != 1" >
        <a th:href="@{goods/selectAllGoodsByPage(currentPage=${currentPage - 1})}">上一页</a>
                                            </span>
                                    <span th:if="${currentPage} != ${totalPage}" >
        <a th:href="@{goods/selectAllGoodsByPage(currentPage=${currentPage + 1})}">下一页</a>
                                            </span>
                                        </li>
                                    </ul>
                                </td>
                            </tr>
    <tr><td colspan="4" align="center"><a th:href="@{goods/toAddGoods}">添加商品</a></td></tr>
                        </tbody>
                    </table>
                </div>
            </div>
        </div>
    </div>
</body>
</html>
```

在 src/main/resources/templates/admin 目录下，创建商品详情页面 detail.html。该页面的代码略。

❷ 编写控制器层

此功能模块共有两个处理请求：goods/selectAllGoodsByPage?currentPage=1 和 @{goods/detail(id=${gds.id},act=detail)}。系统根据@RequestMapping 注解找到对应控制器类 com.ch.ebusiness.controller.admin.GoodsController 的 selectAllGoodsByPage 和 detail 方法处理请求。在控制器类的处理方法中调用 com.ch.ebusiness.service.admin.GoodsService 接口的 selectAllGoodsByPage 和 detail 方法处理业务。控制器层的相关代码如下：

```
@RequestMapping("/selectAllGoodsByPage")
public String selectAllGoodsByPage(Model model, int currentPage) {
    return goodsService.selectAllGoodsByPage(model, currentPage);
}
@RequestMapping("/detail")
public String detail(Model model, Integer id, String act) {
    return goodsService.detail(model, id, act);
}
```

❸ 编写 Service 层

查询商品和查看详情的 Service 层相关代码如下：

```
@Override
public String selectAllGoodsByPage(Model model, int currentPage) {
    //共多少个商品
    int totalCount = goodsRepository.selectAllGoods();
    //计算共多少页
    int pageSize = 5;
    int totalPage = (int)Math.ceil(totalCount*1.0/pageSize);
    List<Goods> typeByPage = goodsRepository.selectAllGoodsByPage((currentPage-1)*
    pageSize, pageSize);
    model.addAttribute("allGoods", typeByPage);
    model.addAttribute("totalPage", totalPage);
    model.addAttribute("currentPage", currentPage);
    return "admin/adminGoods";
}
@Override
public String detail(Model model, Integer id, String act) {
    model.addAttribute("goods", goodsRepository.selectAGoods(id));
    if("detail".equals(act))
        return "admin/detail";
    else {
        model.addAttribute("goodsType", goodsRepository.selectAllGoodsType());
        return "admin/updateAGoods";
    }
}
```

❹ 编写 SQL 映射文件

查询商品和查看详情的 SQL 语句如下：

```
<select id="selectAllGoods"  resultType="integer">
    select count(*) from goodstable
```

```
</select>
<!-- 分页查询 -->
<select id="selectAllGoodsByPage"  resultType="Goods">
    select gt.*,gy.typename
     from goodstable gt,goodstype gy
    where gt.goodstype_id = gy.id
    order by id desc limit #{startIndex}, #{perPageSize}
</select>
<!-- 查询商品详情 -->
<select id="selectAGoods"  resultType="Goods">
    select gt.*, gy.typename
    from  goodstable gt,goodstype gy
    where  gt.goodstype_id = gy.id  and gt.id = #{id}
</select>
```

9.5.5 修改商品

单击图 9.4 中的“修改”超链接（goods/detail(id=${gds.id},act=update)），打开修改商品页面 updateAGoods.html，如图 9.11 所示。在图 9.11 中输入要修改的信息后，单击“修改”按钮，将商品信息提交给 goods/addGoods?act=update 处理。“修改商品”的实现步骤如下。

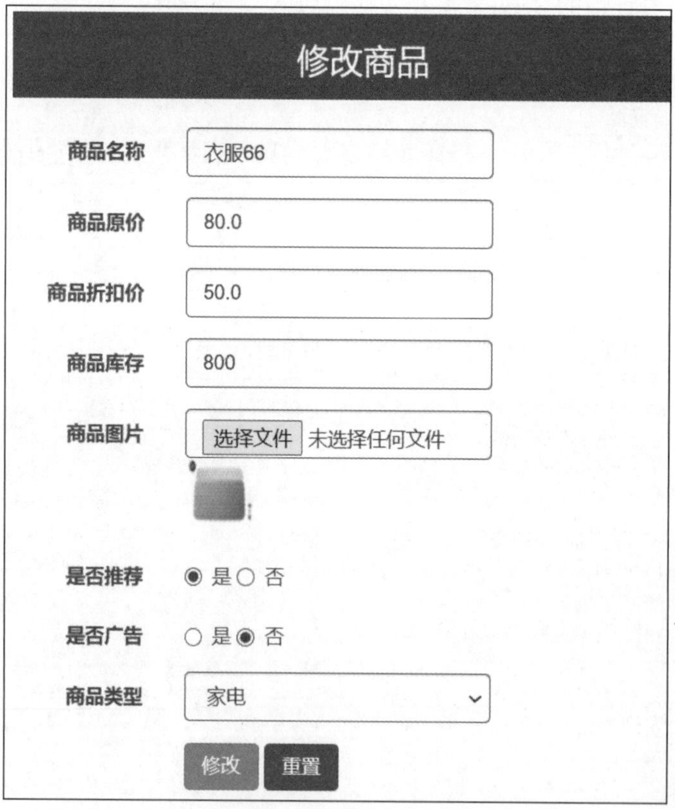

图 9.11　修改商品页面

❶ 编写视图

在 src/main/resources/templates/admin 目录下，创建修改商品页面 updateAGoods.html。updateAGoods.html 与添加商品页面内容基本一样，不再赘述。

❷ 编写控制器层

此功能模块共有两个处理请求：goods/detail(id=${gds.id},act=update)和 goods/addGoods?act=update。goods/detail(id=${gds.id},act=update)请求已在 9.5.4 节介绍，goods/addGoods?act=update 请求已在 9.5.3 节介绍。

❸ 编写 Service 层

同理，Service 层请参考 9.5.3 节和 9.5.4 节。

❹ 编写 SQL 映射文件

“修改商品”的 SQL 语句如下：

```
<!-- 修改一个商品 -->
<update id="updateGoods" parameterType="Goods">
    update goodstable set
        gname = #{gname},
        goprice = #{goprice},
        grprice = #{grprice},
        gstore = #{gstore},
        gpicture = #{gpicture},
        isRecommend = #{isRecommend},
        isAdvertisement = #{isAdvertisement},
        goodstype_id = #{goodstype_id}
     where id = #{id}
</update>
```

▶ 9.5.6　删除商品

单击图 9.4 中的“删除”超链接（'javascript:deleteGoods(' + ${gds.id} +')'），可实现单个商品的删除。成功删除（关联商品不允许删除）后，返回删除商品管理主页面。

❶ 编写控制器层

此功能模块的处理请求是 goods/delete。相关控制器层代码如下：

```
@RequestMapping("/delete")
@ResponseBody
public String delete(Integer id) {
    return goodsService.delete(id);
}
```

❷ 编写 Service 层

删除商品的相关业务处理代码如下：

```
@Override
public String delete(Integer id) {
    if(goodsRepository.selectCartGoods(id).size() > 0
            || goodsRepository.selectFocusGoods(id).size() > 0
            || goodsRepository.selectOrderGoods(id).size() > 0)
        return "no";
    else {
        goodsRepository.deleteAGoods(id);
        return "/goods/selectAllGoodsByPage?currentPage=1";
    }
}
```

❸ 编写 SQL 映射文件

“删除商品”功能模块的相关 SQL 语句如下：

```
<select id="selectFocusGoods" parameterType="integer" resultType="map">
    select * from focustable where goodstable_id = #{id}
</select>
<select id="selectCartGoods" parameterType="integer" resultType="map">
    select * from carttable where goodstable_id = #{id}
</select>
<select id="selectOrderGoods" parameterType="integer" resultType="map">
    select * from orderdetail where goodstable_id = #{id}
</select>
<delete id="deleteAGoods" parameterType="Integer">
    delete from goodstable where id=#{id}
</delete>
```

▶ 9.5.7 查询订单

在后台管理主页面中，单击“查询订单”超链接（selectOrder?currentPage=1），打开查询订单页面 allOrder.html，如图 9.12 所示。

类型管理▾ 商品管理▾ 查询订单 用户管理 安全退出

订单列表

订单ID	用户邮箱	订单金额	订单状态	下单日期
6	chenheng@126.com	2740.0	未支付	2021-01-04 05:11:23.0

第1页 共1页

图 9.12 查询订单页面

❶ 编写视图

在 src/main/resources/templates/admin 目录下，创建查询订单页面 allOrder.html。该页面的代码参见本书提供的源程序 eBusiness。

❷ 编写控制器层

此功能模块有一个处理请求：selectOrder?currentPage=1。系统根据@RequestMapping 注解找到对应控制器类 com.ch.ebusiness.controller.admin.UserAndOrderAndOutController 的 selectOrder 方法处理请求。在控制器类的处理方法中调用 com.ch.ebusiness.service.admin.UserAndOrderAndOutService 接口的 selectOrder 方法处理业务。相关控制器层代码如下：

```
@RequestMapping("/selectOrder")
public String selectOrder(Model model, int currentPage) {
    return userAndOrderAndOutService.selectOrder(model, currentPage);
}
```

❸ 编写 Service 层

“查询订单”功能模块的相关 Service 层代码如下：

```
@Override
public String selectOrder(Model model, int currentPage) {
    //共多少个订单
    int totalCount = userAndOrderAndOutRepository.selectAllOrder();
    //计算共多少页
    int pageSize = 5;
    int totalPage = (int)Math.ceil(totalCount*1.0/pageSize);
    List<Map<String, Object>> orderByPage = userAndOrderAndOutRepository.
    selectOrderByPage((currentPage-1)*pageSize, pageSize);
    model.addAttribute("allOrders", orderByPage);
    model.addAttribute("totalPage", totalPage);
    model.addAttribute("currentPage", currentPage);
    return "admin/allOrder";
}
```

❹ 编写 SQL 映射文件

“查询订单”功能模块的相关 SQL 语句如下：

```
<select id="selectAllOrder"  resultType="integer">
    select count(*) from orderbasetable
</select>
<!-- 分页查询 -->
<select id="selectOrderByPage"  resultType="map">
    select obt.*, bt.bemail from orderbasetable obt, busertable bt where obt.busertable_
    id = bt.id limit #{startIndex}, #{perPageSize}
</select>
```

▶ 9.5.8　用户管理

在后台管理主页面中，单击“用户管理”超链接（selectUser?currentPage=1），打开用户管理页面 allUser.html，如图 9.13 所示。

图 9.13　用户管理页面

单击图 9.13 中的“删除”超链接（'javascript:deleteUsers('+ ${u.id} +')'），可删除未关联的用户。

“用户管理”与 9.5.7 节“查询订单”的实现方式基本一样，不再赘述。

▶ 9.5.9 安全退出

在后台管理主页面中，单击“安全退出”超链接（loginOut），将返回后台登录页面。系统根据@RequestMapping 注解找到对应控制器类 com.ch.ebusiness.controller.admin.UserAndOrderAndOutController 的 loginOut 方法处理请求。在 loginOut 方法中执行 session.invalidate() 使 session 失效，并返回后台登录页面。具体代码如下：

```
@RequestMapping("/loginOut")
public String loginOut(@ModelAttribute("aUser") AUser aUser, HttpSession session) {
    session.invalidate();
    return "admin/login";
}
```

9.6 前台电子商务子系统的实现

游客具有浏览首页、查看商品详情和搜索商品等权限。成功登录的用户除具有游客所具有的权限外，还具有购买商品、查看购物车、收藏商品、查看我的订单以及查看用户信息的权限。本节将详细讲解前台电子商务子系统的实现。

▶ 9.6.1 导航栏及首页搜索

在前台每个 HTML 页面中都引入了一个名为 header.html 的页面，引入代码如下：

```
<div th:include="user/header"></div>
```

header.html 中的商品类型以及广告区域的商品信息都是从数据库中获取。header.html 页面的运行效果如图 9.14 所示。

图 9.14 导航栏

在导航栏的搜索框中输入信息，单击“搜索”按钮，将搜索信息提交给 search 请求处理，系统根据@RequestMapping 注解找到 com.ch.ebusiness.controller.before.IndexController 控制器类的 search 方法处理请求，并将搜索到的商品信息转发给 searchResult.html。searchResult.html 页面的运行效果如图 9.15 所示。

❶ 编写视图

该模块的视图涉及 src/main/resources/templates/user 目录下的两个 HTML 页面：header.html

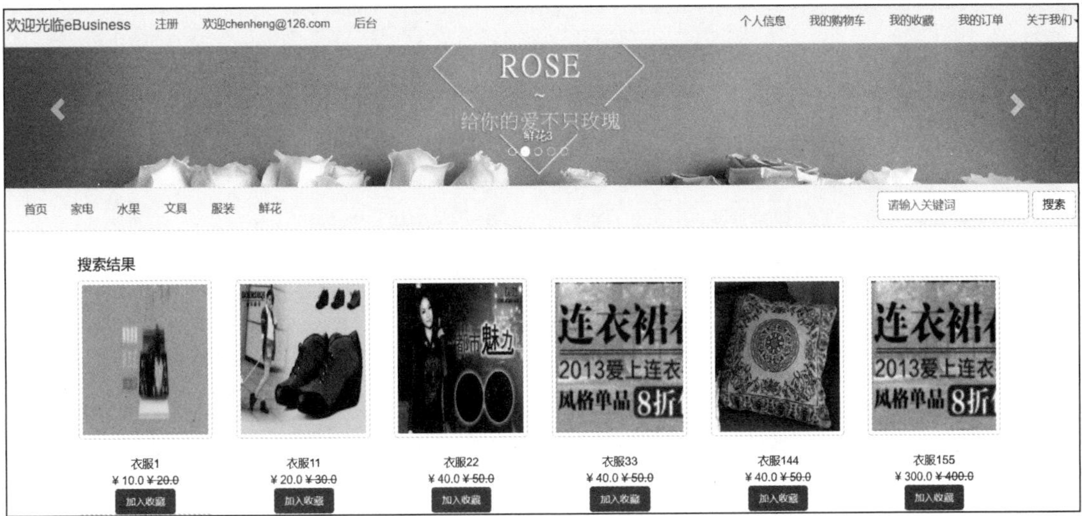

图 9.15　搜索结果

和 searchResult.html。header.html 和 searchResult.html 页面代码请参见本书提供的源程序 eBusiness。

❷ 编写控制器层

该功能模块的控制器层涉及 com.ch.ebusiness.controller.before.IndexController 控制器类的处理方法 search，具体代码如下：

```
@RequestMapping("/search")
public String search(Model model, String mykey) {
    return indexService.search(model, mykey);
}
```

❸ 编写 Service 层

该功能模块的 Service 层代码如下：

```
@Override
public String search(Model model, String mykey) {
    //广告区商品
    model.addAttribute("advertisementGoods", indexRepository.selectAdvertisementGoods());
    //导航栏商品类型
    model.addAttribute("goodsType", indexRepository.selectGoodsType());
    //商品搜索
    model.addAttribute("searchgoods", indexRepository.search(mykey));
    return "user/searchResult";
}
```

❹ 编写 SQL 映射文件

该功能模块涉及的 SQL 语句如下：

```
<!-- 查询广告商品 -->
<select id="selectAdvertisementGoods"  resultType="Goods">
    select
        gt.*, gy.typename
    from
        goodstable gt,goodstype gy
```

```
    where
        gt.goodstype_id = gy.id
        and gt.isAdvertisement = 1
    order by  gt.id desc limit 5
</select>
<!-- 查询商品类型 -->
<select id="selectGoodsType" resultType="GoodsType">
    select * from goodstype
</select>
<!-- 首页搜索 -->
<select id="search" resultType="Goods" parameterType="String">
    select gt.*, gy.typename from GOODSTABLE gt,GOODSTYPE gy where gt.goodstype_id = gy.id
    and gt.gname like concat('%',#{mykey},'%')
</select>
```

▶ 9.6.2 推荐商品及最新商品

推荐商品是根据商品表中的字段 isRecommend 值判断的。最新商品是以商品 ID 排序的，因为商品 ID 是用 MySQL 自动递增产生的。具体实现步骤如下。

❶ 编写视图

该模块的视图涉及 src/main/resources/templates/user 目录下的 index.html 页面，其核心代码如下：

```
<div class="container">
    <div>
        <h4>推荐商品</h4>
    </div>
    <div class="row">
        <div class="col-xs-6 col-md-2" th:each="rGoods:${recommendGoods}">
            <a th:href="'goodsDetail?id=' + ${rGoods.id}" class="thumbnail"><img
                alt="100%x180" th:src="'images/' + ${rGoods.gpicture}"
                style="height: 180px; width: 100%; display: block;">
            </a>
            <div class="caption" style="text-align: center;">
                <div>
                    <span th:text="${rGoods.gname}"></span>
                </div>
                <div>
                    <span style="color: red;">&yen;
                        <span th:text="${rGoods.grprice}"></span>
                    </span>
                    <span class="text-dark" style="text-decoration: line-through;"> &yen;
                        <span th:text="${rGoods.goprice}"></span>
                    </span>
                </div>
                <a th:href="'javascript:focus('+ ${rGoods.id} +')'" class="btn btn-primary"
                    style="font-size: 10px;">加入收藏</a>
            </div>
        </div>
    </div>
    <!-- ************************************************************ -->
```

```
    <div>
        <h4>最新商品</h4>
    </div>
    <div class="row">
        <div class="col-xs-6 col-md-2" th:each="lGoods:${lastedGoods}">
        <a th:href="'goodsDetail?id=' + ${lGoods.id}" class="thumbnail"> <img alt="100%x180"
                th:src="'images/' + ${lGoods.gpicture}"
                style="height: 180px; width: 100%; display: block;">
            </a>
            <div class="caption" style="text-align: center;">
                <div>
                    <span th:text="${lGoods.gname}"></span>
                </div>
                <div>
                    <span style="color: red;">&yen;
                        <span th:text="${lGoods.grprice}"></span>
                    </span>
                    <span class="text-dark" style="text-decoration: line-through;">&yen;
                        <span th:text="${lGoods.goprice}"></span>
                    </span>
                </div>
        <a th:href="'javascript:focus('+${lGoods.id}+')'" class="btn btn-primary"
style="font-size: 10px;">加入收藏</a>
            </div>
        </div>
    </div>
</div>
```

❷ 编写控制器层

该功能模块的控制器层涉及 com.ch.ebusiness.controller.before.IndexController 控制器类的处理方法 index，具体代码如下：

```
@RequestMapping("/")
public String index(Model model, Integer tid) {
    return indexService.index(model, tid);
}
```

❸ 编写 Service 层

该功能模块的 Service 层代码如下：

```
@Override
public String index(Model model, Integer tid) {
    if(tid == null)
        tid = 0;
    //广告区商品
    model.addAttribute("advertisementGoods", indexRepository.selectAdvertisementGoods());
    //导航栏商品类型
    model.addAttribute("goodsType", indexRepository.selectGoodsType());
    //推荐商品
    model.addAttribute("recommendGoods", indexRepository.selectRecommendGoods(tid));
    //最新商品
    model.addAttribute("lastedGoods", indexRepository.selectLastedGoods(tid));
    return "user/index";
}
```

❹ 编写 SQL 映射文件

该功能模块涉及的 SQL 语句如下：

```
<!-- 查询推荐商品 -->
<select id="selectRecommendGoods" resultType="Goods" parameterType="integer">
    select
        gt.*, gy.typename
    from
        goodstable gt,goodstype gy
    where
        gt.goodstype_id = gy.id
        and gt.isRecommend = 1
        <if test="tid != 0">
            and gy.id = #{tid}
        </if>
    order by  gt.id desc limit 6
</select>
<!-- 查询最新商品 -->
<select id="selectLastedGoods" resultType="Goods" parameterType="integer">
    select
        gt.*, gy.typename
    from
        goodstable gt,goodstype gy
    where
        gt.goodstype_id = gy.id
        <if test="tid != 0">
            and gy.id = #{tid}
        </if>
    order by  gt.id desc limit 6
</select>
```

▶ 9.6.3 用户注册

单击导航栏的“注册”超链接（user/toRegister），将打开注册页面 register.html，如图 9.16 所示。

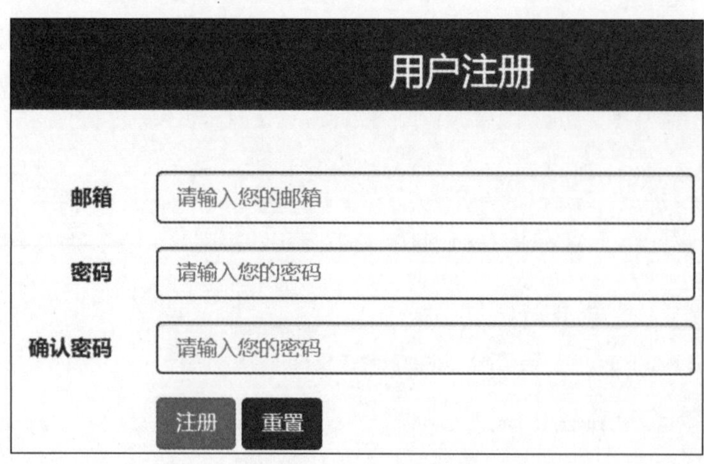

图 9.16　注册页面

输入用户信息，单击“注册”按钮，将用户信息提交给 user/register 处理请求，系统根据 @RequestMapping 注解找到 com.ch.ebusiness.controller.before.UserController 控制器类的 toRegister 和 register 方法处理请求。注册模块的实现步骤如下。

❶ 编写视图

该模块的视图涉及 src/main/resources/templates/user 目录下的 register.html，其代码与后台登录页面代码类似，不再赘述。

❷ 编写控制器层

该功能模块涉及 com.ch.ebusiness.controller.before.UserController 控制器类的 toRegister 和 register 方法。具体代码如下：

```
@RequestMapping("/toRegister")
public String toRegister(@ModelAttribute("bUser") BUser bUser) {
    return "user/register";
}
@RequestMapping("/register")
public String register(@ModelAttribute("bUser") @Validated BUser bUser,BindingResult rs) {
    if(rs.hasErrors()){//验证失败
       return "user/register";
    }
    return userService.register(bUser);
}
```

❸ 编写 Service 层

该功能模块的 Service 层代码如下：

```
@Override
public String isUse(BUser bUser) {
    if(userRepository.isUse(bUser).size() > 0) {
        return "no";
    }
    return "ok";
}
@Override
public String register(BUser bUser) {
    //对密码 MD5 加密
    bUser.setBpwd(MD5Util.MD5(bUser.getBpwd()));
    if(userRepository.register(bUser) > 0) {
        return "user/login";
    }
    return "user/register";
}
```

❹ 编写 SQL 映射文件

该功能模块涉及的 SQL 语句如下：

```
<select id="isUse" parameterType="BUser" resultType="BUser">
    select * from busertable where bemail = #{bemail}
</select>
<insert id="register" parameterType="BUser">
    insert into busertable (id, bemail, bpwd)  values(null, #{bemail}, #{bpwd})
</insert>
```

9.6.4 用户登录

用户注册成功后，跳转到登录页面 login.html，如图 9.17 所示。

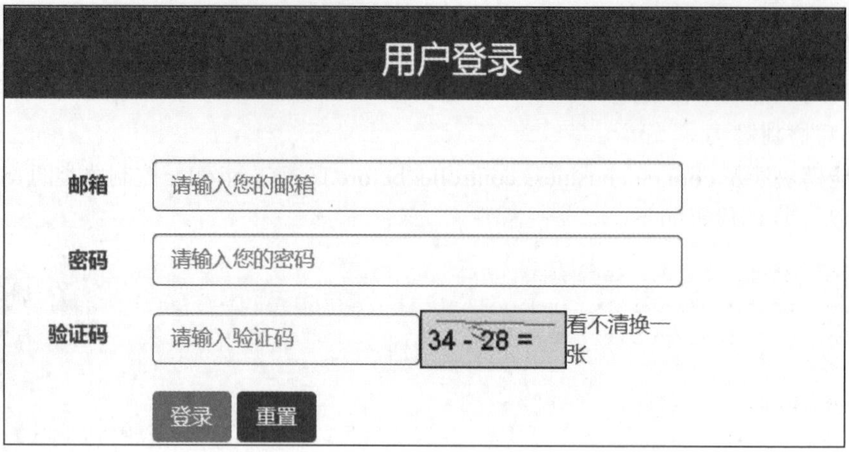

用户登录
邮箱 请输入您的邮箱
密码 请输入您的密码
验证码 请输入验证码 34 - 28 = 看不清换一张
登录 重置

图 9.17 登录页面

在图 9.17 中，输入信息后单击“登录”按钮，将用户输入的 E-mail、密码以及验证码提交给 user/login 请求处理。系统根据@RequestMapping 注解找到 com.ch.ebusiness.controller.before.UserController 控制器类的 login 方法处理请求。登录成功后，将用户的登录信息保存在 session 对象中，然后回到网站首页。具体实现步骤如下。

❶ 编写视图

该模块的视图涉及 src/main/resources/templates/user 目录下的 login.html。其代码与后台登录页面代码类似，不再赘述。

❷ 编写控制器层

该功能模块涉及 com.ch.ebusiness.controller.before.UserController 控制器类的 login 方法。具体代码如下：

```
@RequestMapping("/login")
public String login(@ModelAttribute("bUser") @Validated BUser bUser,
        BindingResult rs, HttpSession session, Model model) {
    if(rs.hasErrors()){//验证失败
        return "user/login";
    }
    return userService.login(bUser, session, model);
}
```

❸ 编写 Service 层

该功能模块的 Service 层代码如下：

```
@Override
public String login(BUser bUser, HttpSession session, Model model) {
    //对密码 MD5 加密
    bUser.setBpwd(MD5Util.MD5(bUser.getBpwd()));
    String rand = (String)session.getAttribute("rand");
    if(!rand.equalsIgnoreCase(bUser.getCode())) {
```

```
            model.addAttribute("errorMessage", "验证码错误！");
            return "user/login";
        }
        List<BUser> list = userRepository.login(bUser);
        if(list.size() > 0) {
            session.setAttribute("bUser", list.get(0));
            return "redirect:/";//到首页
        }
        model.addAttribute("errorMessage", "用户名或密码错误！");
        return "user/login";
    }
```

❹ 编写 SQL 映射文件

该功能模块的 SQL 语句如下：

```
<select id="login" parameterType="BUser" resultType="BUser">
    select * from busertable where bemail = #{bemail} and bpwd = #{bpwd}
</select>
```

9.6.5 商品详情

可以从推荐商品、最新商品、广告商品以及搜索商品结果等位置处，单击商品图片进入商品详情页面 goodsDetail.html，如图 9.18 所示。

图 9.18　商品详情页面

商品详情具体实现步骤如下。

❶ 编写视图

该模块的视图涉及 src/main/resources/templates/user 目录下的 goodsDetail.html，其核心代码如下：

```
<body>
    <!-- 加载 header.html -->
    <div th:include="user/header"></div>
    <div class="container">
        <div class="row">
```

```
            <div class="col-xs-6 col-md-3">
                <img
                    th:src="'images/' + ${goods.gpicture}"
                    style="height: 220px; width: 280px; display: block;">
            </div>
            <div class="col-xs-6 col-md-3">
                <p>商品名：<span th:text="${goods.gname}"></span></p>
                <p>
                    商品折扣价：<span style="color: red;">&yen;
                        <span th:text="${goods.grprice}"></span>
                    </span>
                </p>
                <p>
                    商品原价：
                    <span class="text-dark" style="text-decoration: line-through;"> &yen;
                        <span th:text="${goods.goprice}"></span>
                    </span>
                </p>
                <p>
                    商品类型：<span th:text="${goods.typename}"></span>
                </p>
                <p>
                    库存：<span id="gstore"  th:text="${goods.gstore}"></span>
                </p>
                <p>
                    <input type="text" size="12" class="form-control" placeholder=
                    "请输入购买量" id="buyNumber" name="buyNumber"/>
                    <input type="hidden" name="gid" id="gid" th:value="${goods.id}"/>
                </p>
                <p>
                    <a href="javascript:focus()" class="btn btn-primary"
                        style="font-size: 10px;">加入收藏</a>
                    <a href="javascript:putCart()" class="btn btn-success"
                        style="font-size: 10px;">加入购物车</a>
                </p>
            </div>
        </div>
    </div>
</body>
```

❷ 编写控制器层

该功能模块涉及 com.ch.ebusiness.controller.before.IndexController 控制器类的 goodsDetail 方法。具体代码如下：

```
@RequestMapping("/goodsDetail")
public String goodsDetail(Model model, Integer id) {
    return indexService.goodsDetail(model, id);
}
```

❸ 编写 Service 层

该功能模块的 Service 层代码如下：

```
@Override
public String goodsDetail(Model model, Integer id) {
    //广告区商品
    model.addAttribute("advertisementGoods", indexRepository.selectAdvertisementGoods());
    //导航栏商品类型
    model.addAttribute("goodsType", indexRepository.selectGoodsType());
    //商品详情
    model.addAttribute("goods", indexRepository.selectAGoods(id));
    return "user/goodsDetail";
}
```

❹ 编写 SQL 映射文件

该功能模块的 SQL 语句如下：

```
<!-- 查询商品详情 -->
<select id="selectAGoods"  resultType="Goods">
    select
        gt.*, gy.typename
    from
        goodstable gt,goodstype gy
    where
        gt.goodstype_id = gy.id
        and gt.id = #{id}
</select>
```

▶ 9.6.6　收藏商品

登录成功的用户可以在商品详情页面、首页以及搜索商品结果页面单击“加入收藏”按钮收藏该商品。此时，请求路径为 cart/focus（Ajax 实现）。系统根据@RequestMapping 注解找到 com.ch.ebusiness.controller.before.CartController 控制器类的 focus 方法处理请求。具体实现步骤如下。

❶ 编写控制器层

该功能模块涉及 com.ch.ebusiness.controller.before.CartController 控制器类的 focus 方法。具体代码如下：

```
@RequestMapping("/focus")
@ResponseBody
public String focus(@RequestBody Goods goods, Model model, HttpSession session) {
    return cartService.focus(model, session, goods.getId());
}
```

❷ 编写 Service 层

该功能模块的 Service 层代码如下：

```
@Override
public String focus(Model model, HttpSession session, Integer gid) {
    Integer uid = MyUtil.getUser(session).getId();
    List<Map<String,Object>>  list = cartRepository.isFocus(uid, gid);
    //判断是否已收藏
    if(list.size() > 0) {
        return "no";
```

```
    }else {
        cartRepository.focus(uid, gid);
        return "ok";
    }
}
```

❸ 编写 SQL 映射文件

该功能模块的 SQL 语句如下：

```
<!-- 处理加入收藏 -->
<select id="isFocus" resultType="map">
    select * from focustable where goodstable_id = #{gid} and busertable_id = #{uid}
</select>
<insert id="focus">
  insert into focustable (id, goodstable_id, busertable_id, focustime)
  values(null, #{gid}, #{uid}, now())
</insert>
```

9.6.7 购物车

单击商品详情页面中的“加入购物车”按钮或导航栏中的“我的购物车”超链接，打开“购物车列表”页面 cart.html，如图 9.19 所示。

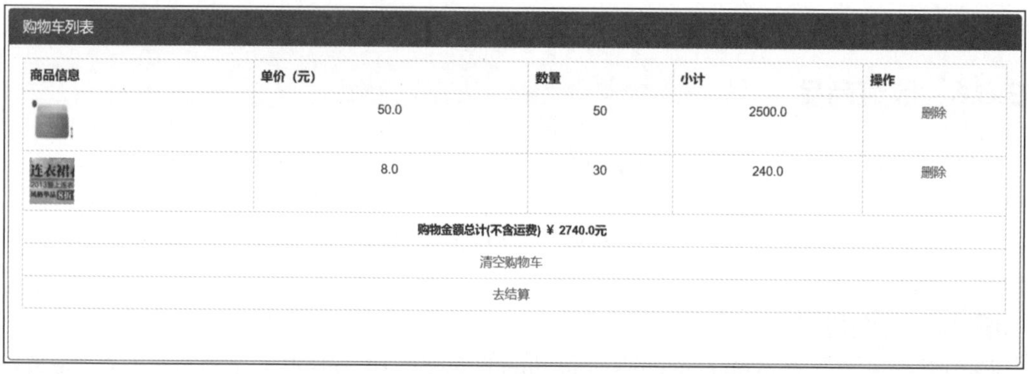

购物车列表

商品信息	单价（元）	数量	小计	操作
	50.0	50	2500.0	删除
	8.0	30	240.0	删除
购物金额总计(不含运费) ¥ 2740.0元				
清空购物车				
去结算				

图 9.19 “购物车列表”页面

与购物车有关的处理请求有 cart/putCart（加入购物车）、cart/clearCart（清空购物车）、cart/selectCart（查询购物车）和 cart/deleteCart（删除购物车）。系统根据@RequestMapping 注解分别找到 com.ch.ebusiness.controller.before.CartController 控制器类的 putCart、clearCart、selectCart、deleteCart 等方法处理请求。具体实现步骤如下。

❶ 编写视图

该模块的视图涉及 src/main/resources/templates/user 目录下的 cart.html，其代码如下：

```
<!DOCTYPE html>
<html xmlns:th="http://www.thymeleaf.org">
<head>
<base th:href="@{/}"><!-- 不用 base 就使用 th:src="@{/js/jquery.min.js} -->
<meta charset="UTF-8">
<title>购物车页面</title>
<link rel="stylesheet" href="css/bootstrap.min.css" />
```

```
<script src="js/jquery.min.js"></script>
<script type="text/javascript">
    function deleteCart(obj){
        if(window.confirm("确认删除吗？")){
            //获取路径
            var pathName=window.document.location.pathname;
            //截取，得到项目名称
            var projectName=pathName.substring(0,pathName.substr(1).indexOf('/')+1);
            window.location.href=projectName+"/cart/deleteCart?gid="+ obj;
        }
    }
    function clearCart(){
        if(window.confirm("确认清空吗？")){
            //获取路径
            var pathName=window.document.location.pathname;
            //截取，得到项目名称
            var projectName=pathName.substring(0,pathName.substr(1).indexOf('/')+1);
            window.location.href = projectName + "/cart/clearCart";
        }
    }
</script>
</head>
<body>
<div th:include="user/header"></div>
<div class="container">
    <div class="panel panel-primary">
        <div class="panel-heading">
            <h3 class="panel-title">购物车列表</h3>
        </div>
        <div class="panel-body">
            <div class="table table-responsive">
                <table class="table table-bordered table-hover">
                    <tbody class="text-center">
                        <tr>
                            <th>商品信息</th>
                            <th>单价（元）</th>
                            <th>数量</th>
                            <th>小计</th>
                            <th>操作</th>
                        </tr>
                        <tr th:each="cart:${cartlist}">
                            <td>
                                <a th:href="'goodsDetail?id=' + ${cart.id}">
                                <img th:src="'images/' + ${cart.gpicture}"
                                    style="height: 50px; width: 50px; display: block;">
                                </a>
                            </td>
                            <td th:text="${cart.grprice}"></td>
                            <td th:text="${cart.shoppingnum}"></td>
                            <td th:text="${cart.smallsum}"></td>
                            <td>
                                <a th:href="'javascript:deleteCart('+${cart.id}+')'">删除</a>
```

```
                            </td>
                        </tr>
                        <tr>
                            <td colspan="5">
            <font style="color: #a60401; font-size: 13px; font-weight: bold;
            letter-spacing: 0px;">
            购物金额总计(不含运费) ￥ <span th:text="${total}"></span>元
                            </font>
                            </td>
                        </tr>
                        <tr>
                            <td colspan="5">
                                <a href="javascript:clearCart()">清空购物车</a>
                            </td>
                        </tr>
                        <tr>
                            <td colspan="5">
                                <a href="cart/selectCart?act=toCount">去结算</a>
                            </td>
                        </tr>
                    </tbody>
                </table>
            </div>
        </div>
    </div>
</div>
</body>
</html>
```

❷ 编写控制器层

该功能模块涉及 com.controller.before.CartController 控制器类的 putCart、clearCart、selectCart、deleteCart 等方法。具体代码如下：

```
@RequestMapping("/putCart")
public String putCart(Goods goods, Model model, HttpSession session) {
    return cartService.putCart(goods, model, session);
}
@RequestMapping("/selectCart")
public String selectCart(Model model, HttpSession session, String act) {
    return cartService.selectCart(model, session, act);
}
@RequestMapping("/deleteCart")
public String deleteCart(HttpSession session, Integer gid) {
    return cartService.deleteCart(session, gid);
}
@RequestMapping("/clearCart")
public String clearCart(HttpSession session) {
    return cartService.clearCart(session);
}
```

❸ 编写 Service 层

该功能模块的 Service 层代码如下：

```
@Override
public String putCart(Goods goods, Model model, HttpSession session) {
    Integer uid = MyUtil.getUser(session).getId();
    //如果商品已在购物车，只更新购买数量
    if(cartRepository.isPutCart(uid, goods.getId()).size() > 0) {
        cartRepository.updateCart(uid, goods.getId(), goods.getBuyNumber());
    }else {//新增到购物车
        cartRepository.putCart(uid, goods.getId(), goods.getBuyNumber());
    }
    //跳转到查询购物车
    return "forward:/cart/selectCart";
}
@Override
public String selectCart(Model model, HttpSession session, String act) {
    List<Map<String, Object>> list=cartRepository.selectCart(MyUtil.getUser(session).getId());
    double sum = 0;
    for (Map<String, Object> map : list) {
        sum = sum + (Double)map.get("smallsum");
    }
    model.addAttribute("total", sum);
    model.addAttribute("cartlist", list);
    //广告区商品
    model.addAttribute("advertisementGoods", indexRepository.selectAdvertisementGoods());
    //导航栏商品类型
    model.addAttribute("goodsType", indexRepository.selectGoodsType());
    if("toCount".equals(act)) {//去结算页面
        return "user/count";
    }
    return "user/cart";
}
@Override
public String deleteCart(HttpSession session, Integer gid) {
    Integer uid = MyUtil.getUser(session).getId();
    cartRepository.deleteAgoods(uid, gid);
    return "forward:/cart/selectCart";
}
@Override
public String clearCart(HttpSession session) {
    cartRepository.clear(MyUtil.getUser(session).getId());
    return "forward:/cart/selectCart";
}
```

❹ 编写 SQL 映射文件

该功能模块的 SQL 语句如下：

```
<!-- 是否已添加购物车 -->
<select id="isPutCart" resultType="map">
    select * from carttable where goodstable_id=#{gid} and busertable_id=#{uid}
</select>
<!-- 添加购物车 -->
<insert id="putCart">
    insert into carttable (id, busertable_id, goodstable_id, shoppingnum)
```

```
    values(null, #{uid},#{gid},#{bnum})
</insert>
<!-- 更新购物车 -->
<update id="updateCart">
    update carttable set shoppingnum=shoppingnum+#{bnum} where busertable_id=
    #{uid} and goodstable_id=#{gid}
</update>
<!-- 查询购物车 -->
<select id="selectCart" parameterType="Integer" resultType="map">
    select gt.id, gt.gname, gt.gpicture, gt.grprice, ct.shoppingnum,
    ct.shoppingnum*gt.grprice smallsum
    from goodstable gt, carttable ct where gt.id=ct.goodstable_id and
    ct.busertable_id=#{uid}
</select>
<!-- 删除购物车 -->
<delete id="deleteAgoods">
    delete from carttable where busertable_id=#{uid} and goodstable_id=#{gid}
</delete>
<!-- 清空购物车 -->
<delete id="clear" parameterType="Integer">
    delete from carttable where busertable_id=#{uid}
</delete>
```

9.6.8 下单

在购物车页面单击“去结算”超链接，进入订单确认页面 count.html，如图 9.20 所示。

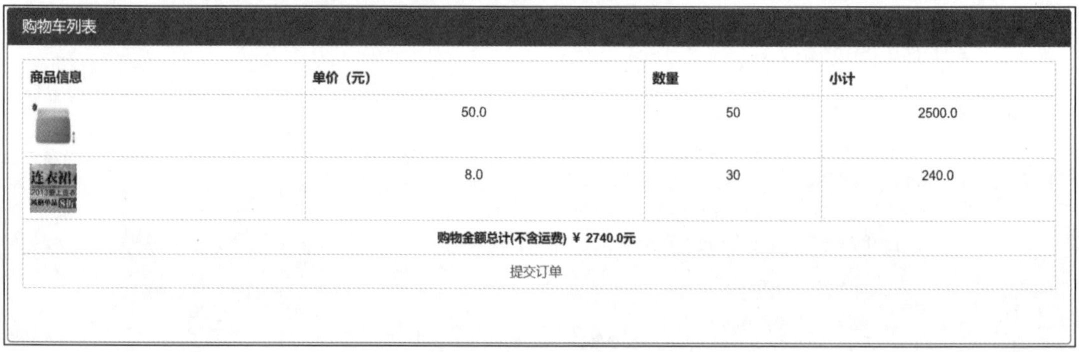

购物车列表

商品信息	单价（元）	数量	小计
	50.0	50	2500.0
	8.0	30	240.0
购物金额总计(不含运费) ¥ 2740.0元			
提交订单			

图 9.20　订单确认

在订单确认页面单击“提交订单”超链接，完成订单提交。订单完成时，提示页面效果如图 9.21 所示。

订单提交成功

您的订单编号为 6.

去支付

图 9.21　订单提交完成页面

单击图 9.21 中的“去支付”超链接完成订单支付。具体实现步骤如下。

❶ 编写视图

该模块的视图涉及 src/main/resources/templates/user 目录下的 count.html 和 pay.html。count.html 的代码与购物车页面代码基本一样，不再赘述。pay.html 的代码如下：

```
<!DOCTYPE html>
<html xmlns:th="http://www.thymeleaf.org">
<head>
<base th:href="@{/}"><!-- 不用 base 就使用 th:src="@{/js/jquery.min.js} -->
<meta charset="UTF-8">
<title>支付页面</title>
<link rel="stylesheet" href="css/bootstrap.min.css" />
<script src="js/jquery.min.js"></script>
<script type="text/javascript" th:inline="javascript">
    function pay(){
            $.ajax(
                {
                  //请求路径，要注意的是 url 和 th:inline="javascript"
                  url : [[@{/cart/pay}]],
                  //请求类型
                  type : "post",
                  contentType : "application/json",
                  //data 表示发送的数据
                  data : JSON.stringify({
                     id : $("#oid").text()
                  }),
                  //成功响应的结果
                  success : function(obj){//obj 响应数据
                     alert("支付成功");
                     //获取路径
                     var pathName=window.document.location.pathname;
                     //截取，得到项目名称
                  var projectName=pathName.substring(0,pathName.substr(1).indexOf('/')+1);
                     window.location.href = projectName;
                  },
                  error : function() {
                     alert("处理异常！");
                  }
                }
            );
    }
</script>
</head>
<body>
<div class="container">
    <div class="panel panel-primary">
        <div class="panel-heading">
            <h3 class="panel-title">订单提交成功</h3>
        </div>
        <div class="panel-body">
            <div>
    您的订单编号为<font color="red" size="5"> <span id="oid" th:text="${order.id}">
    </span></font>。<br><br>
```

```
                    <a href="javascript:pay()">去支付</a>
                </div>
            </div>
        </div>
    </div>
</body>
</html>
```

❷ 编写控制器层

该功能模块涉及 com.ch.ebusiness.controller.before.CartController 控制器类的 submitOrder 和 pay 方法。具体代码如下：

```
@RequestMapping("/submitOrder")
public String submitOrder(Order order, Model model, HttpSession session) {
    return cartService.submitOrder(order, model, session);
}
@RequestMapping("/pay")
@ResponseBody
public String pay(@RequestBody Order order) {
    return cartService.pay(order);
}
```

❸ 编写 Service 层

该功能模块的 Service 层代码如下：

```
@Override
@Transactional
public String submitOrder(Order order, Model model, HttpSession session) {
    order.setBusertable_id(MyUtil.getUser(session).getId());
    //生成订单
    cartRepository.addOrder(order);
    //生成订单详情
    cartRepository.addOrderDetail(order.getId(), MyUtil.getUser(session).getId());
    //减少商品库存
List<Map<String,Object>> listGoods =
cartRepository.selectGoodsShop(MyUtil.getUser(session).getId());
    for (Map<String, Object> map : listGoods) {
        cartRepository.updateStore(map);
    }
    //清空购物车
    cartRepository.clear(MyUtil.getUser(session).getId());
    model.addAttribute("order", order);
    return "user/pay";
}
@Override
public String pay(Order order) {
    cartRepository.pay(order.getId());
    return "ok";
}
```

❹ 编写 SQL 映射文件

该功能模块涉及的 SQL 语句如下：

```
    <!-- 添加一个订单，成功后将主键值回填给 id（实体类的属性）-->
    <insert id="addOrder" parameterType="Order" keyProperty="id" useGeneratedKeys="true">
        insert into orderbasetable (busertable_id, amount, status, orderdate) values
(#{busertable_id}, #{amount}, 0, now())
    </insert>
    <!-- 生成订单详情 -->
    <insert id="addOrderDetail">
        insert into orderdetail (orderbasetable_id, goodstable_id, shoppingnum) select
#{ordersn}, goodstable_id, shoppingnum from carttable where busertable_id = #{uid}
    </insert>
    <!-- 查询商品购买量，以便更新库存使用 -->
    <select id="selectGoodsShop"  parameterType="Integer" resultType="map">
        select shoppingnum gshoppingnum, goodstable_id gid from carttable where
busertable_id=#{uid}
    </select>
    <!-- 更新商品库存 -->
    <update id="updateStore" parameterType="map">
        update goodstable set gstore= gstore -#{gshoppingnum} where id=#{gid}
    </update>
    <!-- 支付订单 -->
    <update id="pay" parameterType="Integer">
        update orderbasetable set status=1 where id=#{ordersn}
    </update>
```

9.6.9　个人信息

成功登录的用户，在导航栏的上方单击“个人信息”超链接（cart/userInfo），进入用户修改密码页面 userInfo.html，如图 9.22 所示。

图 9.22　用户修改密码页面

具体实现步骤如下。

❶ 编写视图

该模块的视图涉及 src/main/resources/templates/user 目录下的 userInfo.html，其代码与登录页面类似，不再赘述。

❷ 编写控制器层

该功能模块涉及 com.ch.ebusiness.controller.before.CartController 控制器类的 userInfo 和 updateUpwd 方法。具体代码如下：

```
@RequestMapping("/userInfo")
public String userInfo() {
    return "user/userInfo";
}
@RequestMapping("/updateUpwd")
public String updateUpwd(HttpSession session, String bpwd) {
    return cartService.updateUpwd(session, bpwd);
}
```

❸ 编写 Service 层

该功能模块的 Service 层代码如下：

```
@Override
public String updateUpwd(HttpSession session, String bpwd) {
    Integer uid = MyUtil.getUser(session).getId();
    cartRepository.updateUpwd(uid, MD5Util.MD5(bpwd));
    return "forward:/user/toLogin";
}
```

❹ 编写 SQL 映射文件

该功能模块的 SQL 语句如下：

```
<!-- 修改密码 -->
<update id="updateUpwd">
    update busertable set bpwd=#{bpwd} where id=#{uid}
</update>
```

▶ 9.6.10 我的收藏

成功登录的用户，在导航栏的上方单击“我的收藏”超链接（cart/myFocus），进入用户收藏页面 myFocus.html，如图 9.23 所示。

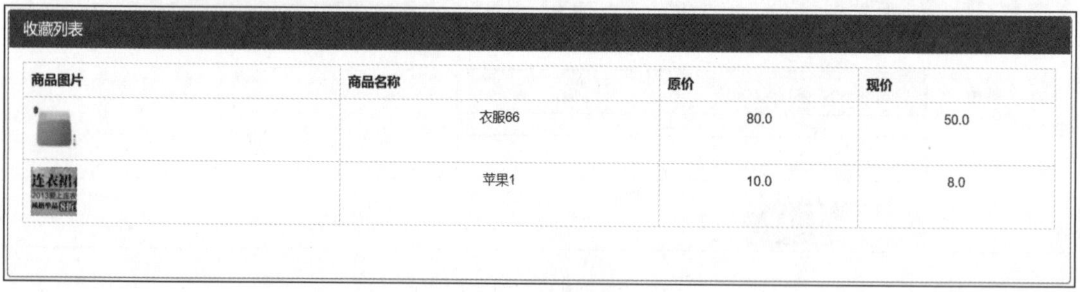

收藏列表

商品图片	商品名称	原价	现价
	衣服66	80.0	50.0
	苹果1	10.0	8.0

图 9.23 用户收藏页面

具体实现步骤如下。

❶ 编写视图

该模块的视图涉及 src/main/resources/templates/user 目录下的 myFocus.html，其代码参见本书提供的源程序 eBusiness。

❷ 编写控制器层

该功能模块涉及 com.ch.ebusiness.controller.before.CartController 控制器类的 myFocus 方法。具体代码如下：

```
@RequestMapping("/myFocus")
public String myFocus(Model model, HttpSession session) {
    return cartService.myFocus(model, session);
}
```

❸ 编写 Service 层

该功能模块的 Service 层代码如下：

```
@Override
public String myFocus(Model model, HttpSession session) {
    //广告区商品
    model.addAttribute("advertisementGoods", indexRepository.selectAdvertisementGoods());
    //导航栏商品类型
    model.addAttribute("goodsType", indexRepository.selectGoodsType());
    model.addAttribute("myFocus", cartRepository.myFocus(MyUtil.getUser(session).getId()));
    return "user/myFocus";
}
```

❹ 编写 SQL 映射文件

该功能模块的 SQL 语句如下：

```
<!-- 我的收藏 -->
<select id="myFocus" resultType="map"  parameterType="Integer">
    select gt.id, gt.gname, gt.goprice, gt.grprice, gt.gpicture from FOCUSTABLE ft, GOODSTABLE gt
     where ft.goodstable_id=gt.id and  ft.busertable_id = #{uid}
</select>
```

9.6.11　我的订单

成功登录的用户，在导航栏的上方单击“我的订单”超链接（cart/myOrder），进入用户订单页面 myOrder.html，如图 9.24 所示。

图 9.24　用户订单页面

单击图 9.24 中的“查看详情”超链接（'cart/orderDetail?id=' + ${order.id}），进入订单详情页面 orderDetail.html，如图 9.25 所示。

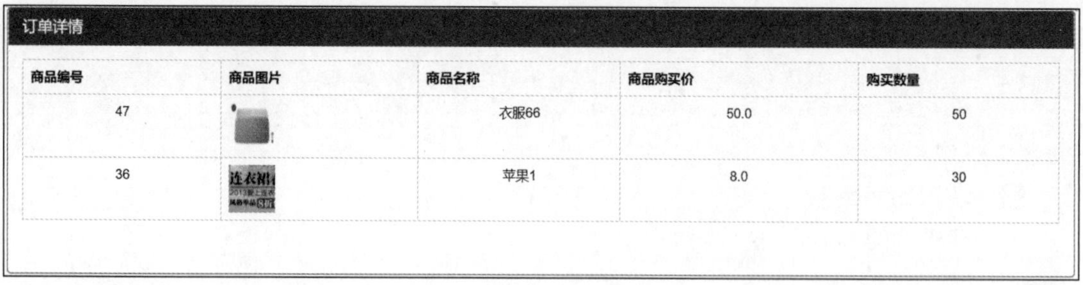

订单详情

商品编号	商品图片	商品名称	商品购买价	购买数量
47		衣服66	50.0	50
36		苹果1	8.0	30

图 9.25　订单详情页面

具体实现步骤如下。

❶ 编写视图

该模块的视图涉及 src/main/resources/templates/user 目录下的 myOrder.html 和 orderDetail.html。myOrder.html 和 orderDetail.html 的代码参见本书提供的源程序 eBusiness。

❷ 编写控制器层

该功能模块涉及 com.ch.ebusiness.controller.before.CartController 控制器类的 myOrder 和 orderDetail 方法。具体代码如下：

```
@RequestMapping("/myOder")
public String myOder(Model model, HttpSession session) {
    return cartService.myOder(model, session);
}
@RequestMapping("/orderDetail")
public String orderDetail(Model model, Integer id) {
    return cartService.orderDetail(model, id);
}
```

❸ 编写 Service 层

该功能模块的 Service 层代码如下：

```
@Override
public String myOder(Model model, HttpSession session) {
    //广告区商品
    model.addAttribute("advertisementGoods", indexRepository.selectAdvertisementGoods());
    //导航栏商品类型
    model.addAttribute("goodsType", indexRepository.selectGoodsType());
    model.addAttribute("myOrder", cartRepository.myOrder(MyUtil.getUser(session).getId()));
    return "user/myOrder";
}
@Override
public String orderDetail(Model model, Integer id) {
    model.addAttribute("orderDetail", cartRepository.orderDetail(id));
    return "user/orderDetail";
}
```

❹ 编写 SQL 映射文件

该功能模块的 SQL 语句如下：

```
<!-- 我的订单 -->
<select id="myOrder"  resultType="map" parameterType="Integer">
```

```
    select id, amount, busertable_id, status, orderdate  from ORDERBASETABLE
    where busertable_id = #{uid}
</select>
<!-- 订单详情 -->
<select id="orderDetail" resultType="map"  parameterType="Integer">
    select gt.id, gt.gname, gt.goprice, gt.grprice, gt.gpicture,
    odt.shoppingnum from  GOODSTABLE gt, ORDERDETAIL odt
    where  odt.orderbasetable_id=#{id} and gt.id=odt.goodstable_id
</select>
```

9.7　本章小结

本章讲述了电子商务平台通用功能的设计与实现。通过本章的学习，读者不仅可以掌握 Spring Boot 应用开发的流程、方法和技术，还可以熟悉电子商务平台的业务需求、设计以及实现。

习 题 9

1. 在本章的电子商务平台中，是如何控制管理员登录权限的？
2. 在本章的电子商务平台中，有几对关联数据表？

第10章 Spring Boot 的安全控制

学习目的与要求

本章首先重点讲解 Spring Security 安全控制机制，然后介绍 Spring Boot Security 操作实例。通过本章的学习，掌握如何使用 Spring Security 安全控制机制解决企业应用程序的安全问题。

主要内容

- ❖ Spring Security 快速入门
- ❖ Spring Boot Security 操作实例

在 Web 应用开发中，安全毋庸置疑是十分重要的，选择 Spring Security 来保护 Web 应用是一个非常好的选择。Spring Security 是 Spring 框架的一个安全模块，可以非常方便地与 Spring Boot 应用无缝集成。

视频讲解

10.1 Spring Security 快速入门

10.1.1 什么是 Spring Security

Spring Security 是一个专门针对 Spring 应用系统的安全框架，充分利用了 Spring 框架的依赖注入和 AOP 功能，为 Spring 应用系统提供安全访问控制解决方案。

在 Spring Security 安全框架中，有两个重要概念，即授权（Authorization）和认证（Authentication）。授权是确定用户在当前应用系统下所拥有的功能权限；认证是确认用户访问当前系统的身份。

10.1.2 Spring Security 的适配器

Spring Security 为 Web 应用提供了一个适配器类 WebSecurityConfigurerAdapter，该类实现了 WebSecurityConfigurer<WebSecurity>接口，并提供了两个 configure 方法用于认证和授权操作。

开发者创建自己的 Spring Security 适配器类是非常简单的，只需要定义一个继承 WebSecurityConfigurerAdapter 的类，并在该类中使用@Configuration 注解，就可以通过重写两个 configure 方法来配置所需要的安全配置。自定义适配器类的示例代码如下：

```
@Configuration
public class MySecurityConfigurerAdapter extends WebSecurityConfigurerAdapter{
    /**
     * 用户认证
     */
    @Override
```

```
    protected void configure(AuthenticationManagerBuilder auth) throws Exception {
    }
    /**
     * 请求授权
     */
    @Override
    protected void configure(HttpSecurity http) throws Exception {
    }
}
```

▶ 10.1.3　Spring Security 的用户认证

在 Spring Security 的适配器类中，通过重写 configure (AuthenticationManagerBuilder auth) 方法完成用户认证。

❶ 内存中的用户认证

使用 AuthenticationManagerBuilder 的 inMemoryAuthentication()方法可以添加在内存中的用户，并给用户指定角色权限。示例代码如下：

```
@Override
protected void configure(AuthenticationManagerBuilder auth) throws Exception {
    auth.inMemoryAuthentication().withUser("chenheng").password("123456").roles
      ("ADMIN","DBA");
    auth.inMemoryAuthentication().withUser("zhangsan").password("123456").roles
      ("USER");
}
```

上述示例代码中添加了两个用户，一个用户的用户名为 chenheng，密码为 123456，用户权限为 ROLE_ADMIN 和 ROLE_DBA；另一个用户的用户名为 zhangsan，密码为 123456，用户权限为 ROLE_USER。ROLE_是 Spring Security 保存用户权限时默认加上的。

❷ 通用的用户认证

在实际应用中，可以查询数据库获取用户和权限，这时需要自定义实现 org.springframework.security.core.userdetails.UserDetailsService 接口的类，并重写 public UserDetails loadUserByUsername(String username)方法查询对应的用户和权限。示例代码如下：

```
@Service
public class MyUserSecurityService implements UserDetailsService{
    @Autowired
    private MyUserRepository myUserRepository;
    /**
     * 通过重写 loadUserByUsername 方法查询对应的用户
     * UserDetails 是 Spring Security 的一个核心接口
     * UserDetails 定义了可以获取用户名、密码、权限等与认证信息相关的方法
     */
    @Override
    public UserDetails loadUserByUsername(String username) throws UsernameNotFoundException {
        //根据用户名（页面接收的用户名）查询当前用户
        MyUser myUser = myUserRepository.findByUsername(username);
        if(myUser == null) {
            throw new UsernameNotFoundException("用户名不存在");
        }
```

```
        //GrantedAuthority 代表赋予当前用户的权限（认证权限）
        List<GrantedAuthority> authorities = new ArrayList<GrantedAuthority>();
        //获得当前用户权限集合
        List<Authority> roles = myUser.getAuthorityList();
        //将当前用户的权限保存为用户的认证权限
        for (Authority authority : roles) {
            GrantedAuthority sg = new SimpleGrantedAuthority(authority.getName());
            authorities.add(sg);
        }
        //org.springframework.security.core.userdetails.User 是 Spring Security
        //内部的实现，专门用于保存用户名、密码、权限等与认证相关的信息
        User su = new User(myUser.getUsername(), myUser.getPassword(), authorities);
        return su;
    }
}
```

除此之外，还需要注册 MyUserSecurityService 完成用户认证，示例代码如下：

```
@Configuration
public class MySecurityConfigurerAdapter extends WebSecurityConfigurerAdapter{
    //依赖注入通用的用户服务类
    @Autowired
    private MyUserSecurityService myUserSecurityService;
    //依赖注入用户认证接口
    @Autowired
    private AuthenticationProvider authenticationProvider;
    /**
     * DaoAuthenticationProvider 是 AuthenticationProvider 的实现
     */
    @Bean
    public AuthenticationProvider authenticationProvider() {
        DaoAuthenticationProvider provide = new DaoAuthenticationProvider();
        //设置自定义认证方式，用户登录认证
        provide.setUserDetailsService(myUserSecurityService);
        return provide;
    }
    /**
     * 用户认证
     */
    @Override
    protected void configure(AuthenticationManagerBuilder auth) throws Exception {
        //设置认证方式
        auth.authenticationProvider(authenticationProvider);
    }
}
```

10.1.4 Spring Security 的请求授权

在 Spring Security 的适配器类中，通过重写 configure(HttpSecurity http)方法完成用户授权。在 configure(HttpSecurity http)方法中，使用 HttpSecurity 的 authorizeRequests()方法的子节点给指定用户授权访问 URL 模式。可以通过 antMatchers 方法使用 Ant 风格匹配 URL 路径。

匹配请求路径后，可以针对当前用户对请求进行安全处理。Spring Security 提供了许多安全处理方法，具体如表 10.1 所示。

表 10.1　Spring Security 的安全处理方法

方　法	用　　途
anyRequest()	匹配所有请求路径
access(String attribute)	Spring EL 表达式结果为 true 时可以访问
anonymous()	匿名可以访问
authenticated()	用户登录后可访问
denyAll()	用户不能访问
fullyAuthenticated()	用户完全认证可以访问（非 remember-me 下自动登录）
hasAnyAuthority(String...)	参数表示权限，用户权限与其中任一权限相同就可以访问
hasAnyRole(String...)	参数表示角色，用户角色与其中任一角色相同就可以访问
hasAuthority(String authority)	参数表示权限，用户权限与参数相同才可以访问
hasIpAddress(String ipaddressExpression)	参数表示 IP 地址，用户 IP 和参数匹配才可以访问
hasRole(String role)	参数表示角色，用户角色与参数相同才可以访问
permitAll()	任何用户都可以访问
rememberMe()	允许通过 remember-me 登录的用户访问

Spring Security 的请求授权示例代码如下：

```
@Override
protected void configure(HttpSecurity http) throws Exception {
    http.authorizeRequests()
    //首页、登录、注册页面、登录注册功能以及静态资源过滤掉，即可任意访问
    .antMatchers("/toLogin", "/toRegister", "/", "/login", "/register", "/css/**",
                "/fonts/**", "/js/**").permitAll()
    //这里默认追加 ROLE_，/user/**是控制器的请求匹配路径
    .antMatchers("/user/**").hasRole("USER")
    .antMatchers("/admin/**").hasAnyRole("ADMIN", "DBA")
    //其他所有请求登录后才能访问
    .anyRequest().authenticated()
    .and()
    //将输入的用户名与密码和授权的进行比较
    .formLogin()
    .loginPage("/login").successHandler(myAuthenticationSuccessHandler)
        .usernameParameter("username").passwordParameter("password")
        //登录失败
        .failureUrl("/login?error")
    .and()
    //注销行为可任意访问
    .logout().permitAll()
    .and()
    //指定异常处理页面
    .exceptionHandling().accessDeniedPage("/deniedAccess");
}
```

上述示例代码解释如下。

http.authorizeRequests()：开始进行请求权限设置。

antMatchers("/toLogin", "/toRegister", "/", "/login", "/register", "/css/**", "/fonts/**", "/js/**").permitAll()："/toLogin""/toRegister""/""/login""/register"等请求以及静态资源过滤掉，则任意用户可访问。

antMatchers("/user/**").hasRole("USER")：请求匹配"/user/**"，拥有 ROLE_USER 角色的用户可访问。

antMatchers("/admin/**").hasAnyRole("ADMIN", "DBA")：请求匹配"/admin/**"，拥有 ROLE_ADMIN 或 ROLE_DBA 角色的用户可访问。

anyRequest().authenticated()：其余所有请求都需要认证（用户登录后）才可访问。

formLogin()：开始设置登录操作。

loginPage("/login").successHandler(myAuthenticationSuccessHandler)：设置登录的访问地址以及登录成功处理操作。

usernameParameter("username").passwordParameter("password")：登录时接收参数 username 的值作为用户名，接收参数 password 的值作为密码。

failureUrl("/login?error")：指定登录失败后转向的页面和传递的参数。

logout().permitAll()：注销操作，所有用户均可访问。

exceptionHandling().accessDeniedPage("/deniedAccess")：指定异常处理页面。

10.1.5 Spring Security 的核心类

Spring Security 的核心类包括 Authentication、SecurityContextHolder、UserDetails 和 UserDetailsService、GrantedAuthority、DaoAuthenticationProvider、PasswordEncoder。

❶ Authentication

Authentication 用来封装用户认证信息的接口。在用户登录认证之前，Spring Security 将相关信息封装为一个 Authentication 具体实现类的对象，在登录认证成功后将生成一个信息更全面、包含用户权限等信息的 Authentication 对象，然后将该对象保存在 SecurityContextHolder 所持有的 SecurityContext 中，方便后续程序调用，如当前用户名、访问权限等。

❷ SecurityContextHolder

SecurityContextHolder 顾名思义是用来持有 SecurityContext 的类。SecurityContext 中包含当前认证用户的详细信息。Spring Security 使用一个 Authentication 对象描述当前用户的相关信息。例如，最常见的是获得当前登录用户的用户名和权限，示例代码如下：

```
/**
 * 获得当前用户名称
 */
private String getUname() {
    return SecurityContextHolder.getContext().getAuthentication().getName();
}
/**
 * 获得当前用户权限
 */
private String getAuthorities() {
    Authentication authentication = SecurityContextHolder.getContext().getAuthentication();
    List<String> roles = new ArrayList<String>();
    for (GrantedAuthority ga : authentication.getAuthorities()) {
        roles.add(ga.getAuthority());
```

```
        }
        return roles.toString();
    }
```

❸ UserDetails

UserDetails 是 Spring Security 的一个核心接口。该接口定义了一些可以获取用户名、密码、权限等与认证相关的信息的方法。通常需要在应用中获取当前用户的其他信息，如 E-mail、电话等。这时只包含认证相关的 UserDetails 对象就不能满足我们的需要了。我们可以实现自己的 UserDetails，在该实现类中定义一些获取用户其他信息的方法，这样就可以直接从当前 SecurityContext 的 Authentication 的 principal 中获取用户的其他信息。

Authentication.getPrincipal()的返回类型是 Object，但通常返回的其实是一个 UserDetails 的实例，通过强制类型转换可以将 Object 转换为 UserDetails 类型。

❹ UserDetailsService

UserDetails 是通过 UserDetailsService 的 loadUserByUsername(String username)方法加载的。UserDetailsService 也是一个接口，也需要实现自己的 UserDetailsService 来加载自定义的 UserDetails 信息。

登录认证时，Spring Security 将通过 UserDetailsService 的 loadUserByUsername(String username)方法获取对应的 UserDetails 进行认证，认证通过后将该 UserDetails 赋给认证通过的 Authentication 的 principal，然后将该 Authentication 保存在 SecurityContext 中。在应用中，如果需要使用用户信息，可以通过 SecurityContextHolder 获取存放在 SecurityContext 中的 Authentication 的 principal，即 UserDetails 实例。

❺ GrantedAuthority

Authentication 的 getAuthorities()方法可以返回当前 Authentication 对象拥有的权限（一个 GrantedAuthority 类型的数组），即当前用户拥有的权限。GrantedAuthority 是一个接口，通常是通过 UserDetailsService 进行加载，然后赋给 UserDetails。

❻ DaoAuthenticationProvider

在 Spring Security 安全框架中，默认使用 DaoAuthenticationProvider 实现 AuthenticationProvider 接口进行用户认证的处理。DaoAuthenticationProvider 进行认证时，需要一个 UserDetailsService 来获取用户信息 UserDetails。当然我们可以实现自己的 AuthenticationProvider，进而改变认证方式。

❼ PasswordEncoder

在 Spring Security 安全框架中，通过 PasswordEncoder 接口完成对密码的加密。Spring Security 对 PasswordEncoder 有多种实现，包括 MD5 加密、SHA-256 加密等，开发者只需直接使用即可。在 Spring Boot 应用中，使用 BCryptPasswordEncoder 加密是较好的选择。BCryptPasswordEncoder 使用 BCrypt 的强散列哈希加密实现，并可以由客户端指定加密强度，强度越高安全性越高。

▶ 10.1.6 Spring Security 的验证机制

Spring Security 的验证机制是由许多 Filter 实现的，Filter 将在 Spring MVC 前拦截请求，主要包括注销 Filter（LogoutFilter）、用户名密码验证 Filter（UsernamePasswordAuthenticationFilter）等内容。Filter 再交由其他组件完成细分的功能，最常用的 UsernamePasswordAuthenticationFilter 会持有一个 AuthenticationManager 引用，AuthenticationManager 是一个验证管理器，专门负

责验证。AuthenticationManager 持有一个 AuthenticationProvider 集合，AuthenticationProvider 是做验证工作的组件，验证成功或失败之后调用对应的 Handler（处理）。

10.2 Spring Boot 的支持

在 Spring Boot 应用中，只需引入 spring-boot-starter-security 依赖即可使用 Spring Security 安全框架，这是因为 Spring Boot 对 Spring Security 提供了自动配置功能。从 org.springframework.boot.autoconfigure.security.SecurityProperties 类中可以看到，使用以 spring.security 为前缀的属性配置了 Spring Security 的相关默认配置。因此，在实际应用开发中只需自定义一个类继承 WebSecurityConfigurerAdapter，无须使用@EnableWebSecurity 注解，即可自己扩展 Spring Security 的相关配置。

视频讲解

10.3 实际开发中的 Spring Security 操作实例

下面通过一个简单实例演示在 Spring Boot 应用中如何使用基于 Spring Data JPA 的 Spring Security 安全框架。

【例 10-1】在 Spring Boot 应用中，使用基于 Spring Data JPA 的 Spring Security 安全框架。

具体实现步骤如下。

❶ 创建 Spring Boot Web 应用 ch10_1

创建基于 Spring Data JPA、Thymeleaf 及 Spring Security 依赖的 Web 应用 ch10_1。

❷ 修改 pom.xml 文件，添加 MySQL 依赖

在 pom.xml 文件中添加如下依赖：

```
<dependency>
    <groupId>mysql</groupId>
    <artifactId>mysql-connector-java</artifactId>
    <version>5.1.45</version>
</dependency>
```

❸ 设置 Web 应用 ch10_1 的上下文路径及数据源配置信息

在应用 ch10_1 的 application.properties 文件中配置如下内容：

```
server.servlet.context-path=/ch10_1
#数据库地址
spring.datasource.url=jdbc:mysql://localhost:3306/springbootjpa?characterEncoding=utf8
#数据库用户名
spring.datasource.username=root
#数据库密码
spring.datasource.password=root
#数据库驱动
spring.datasource.driver-class-name=com.mysql.jdbc.Driver
#指定数据库类型
spring.jpa.database=MYSQL
#指定是否在日志中显示 SQL 语句
spring.jpa.show-sql=true
spring.jpa.hibernate.ddl-auto=update
```

```
#让控制器输出的 JSON 字符串格式更美观
spring.jackson.serialization.indent-output=true
spring.thymeleaf.cache=false
logging.level.org.springframework.security=trace
```

❹ 整理脚本样式静态文件

ch10_1 应用引入的 BootStrap、jQuery 等静态文件放置在 src/main/resources/static 目录下。

❺ 创建用户和权限持久化实体类

创建名为 com.ch.ch10_1.entity 的包，并在该包中创建持久化实体类 MyUser 和 Authority。MyUser 类用来保存用户数据，用户名唯一。Authority 用来保存权限信息。用户和权限是多对多的关系。

MyUser 的核心代码如下：

```
@Entity
@Table(name = "user")
@JsonIgnoreProperties(value = { "hibernateLazyInitializer"})
public class MyUser implements Serializable{
      private static final long serialVersionUID = 1L;
      @Id
    @GeneratedValue(strategy = GenerationType.IDENTITY)
    private int id;
    private String username;
    private String password;
    //这里不能是懒加载 lazy，否则在 MyUserSecurityService 的 loadUserByUsername
    //方法中无法获得权限
    @ManyToMany(cascade = {CascadeType.REFRESH}, fetch = FetchType.EAGER)
    @JoinTable(name = "user_authority",joinColumns = @JoinColumn(name = "user_id"),
    inverseJoinColumns = @JoinColumn(name = "authority_id"))
    private List<Authority> authorityList;
    //repassword 不映射到数据表
    @Transient
    private String repassword;
    //省略 set 和 get 方法
}
```

需要注意的是，在实际开发中 MyUser 还可以实现 org.springframework.security.core.userdetails.UserDetails 接口，实现该接口后即可成为 Spring Security 所使用的用户。本例为了区分 Spring Data JPA 的 pojo 和 Spring Security 的用户对象，并没有实现 UserDetails 接口，而是在实现 UserDetailsService 接口的类中进行绑定。

Authority 的代码如下：

```
@Entity
@Table(name = "authority")
@JsonIgnoreProperties(value = { "hibernateLazyInitializer"})
public class Authority implements Serializable{
    private static final long serialVersionUID = 1L;
    @Id
    @GeneratedValue(strategy = GenerationType.IDENTITY)
    private int id;
    @Column(nullable = false)
    private String name;
    @ManyToMany(mappedBy = "authorityList")
```

```
    @JsonIgnore
    private List<MyUser> userList;
    //省略 set 和 get 方法
}
```

❻ 创建数据访问层接口

创建名为 com.ch.ch10_1.repository 的包，并在该包中创建名为 MyUserRepository 和 AuthorityRepository 的接口，这两个接口继承了 JpaRepository 接口。

MyUserRepository 接口的核心代码如下：

```
public interface MyUserRepository extends JpaRepository<MyUser, Integer>{
    //根据用户名查询用户，方法命名符合 Spring Data JPA 规范
    MyUser findByUsername(String username);
}
```

AuthorityRepository 接口的具体代码如下：

```
public interface AuthorityRepository extends JpaRepository<Authority, Integer>{}
```

❼ 创建业务层

创建名为 com.ch.ch10_1.service 的包，并在该包中创建 UserService 接口和 UserServiceImpl 实现类。UserService 接口代码略。

UserServiceImpl 的核心代码如下：

```
@Service
public class UserServiceImpl implements UserService{
    @Autowired
    private MyUserRepository myUserRepository;
    @Autowired
    private AuthorityRepository authorityRepository;
    /**
     * 实现注册
     */
    @Override
    public String register(MyUser userDomain) {
        String username = userDomain.getUsername();
        List<Authority> authorityList = new ArrayList<Authority>();
        //管理员权限
        if("admin".equals(username)) {
            Authority a1 = new Authority();
            Authority a2 = new Authority();
            a1.setName("ROLE_ADMIN");
            a2.setName("ROLE_DBA");
            authorityList.add(a1);
            authorityList.add(a2);
        }else {//用户权限
            Authority a1 = new Authority();
            a1.setName("ROLE_USER");
            authorityList.add(a1);
        }
        //注册权限
        authorityRepository.saveAll(authorityList);
        userDomain.setAuthorityList(authorityList);
```

```
        //加密密码
        String secret = new BCryptPasswordEncoder().encode(userDomain.getPassword());
        userDomain.setPassword(secret);
        //注册用户
        MyUser mu = myUserRepository.save(userDomain);
        if(mu != null)//注册成功
            return "/login";
        return "/register";//注册失败
    }
    /**
     * 用户登录成功
     */
    @Override
    public String loginSuccess(Model model) {
        model.addAttribute("user", getUname());
        model.addAttribute("role", getAuthorities());
        return "/user/loginSuccess";
    }
    /**
     * 管理员登录成功
     */
    @Override
    public String main(Model model) {
        model.addAttribute("user", getUname());
        model.addAttribute("role", getAuthorities());
        return "/admin/main";
    }
    /**
     * 注销用户
     */
    @Override
    public String logout(HttpServletRequest request, HttpServletResponse response) {
        //获得用户认证信息
        Authentication authentication = SecurityContextHolder.getContext().getAuthentication();
        if(authentication != null) {
            //注销
            new SecurityContextLogoutHandler().logout(request, response, authentication);
        }
        return "redirect:/login?logout";
    }
    /**
     * 没有权限拒绝访问
     */
    @Override
    public String deniedAccess(Model model) {
        model.addAttribute("user", getUname());
        model.addAttribute("role", getAuthorities());
        return "deniedAccess";
    }
    /**
     * 获得当前用户名称
     */
```

```
    private String getUname() {
        return SecurityContextHolder.getContext().getAuthentication().getName();
    }
    /**
     * 获得当前用户权限
     */
    private String getAuthorities() {
        Authentication authentication = SecurityContextHolder.getContext().getAuthentication();
        List<String> roles = new ArrayList<String>();
        for (GrantedAuthority ga : authentication.getAuthorities()) {
            roles.add(ga.getAuthority());
        }
        return roles.toString();
    }
}
```

❽ 创建控制器类

创建名为 com.ch.ch10_1.controller 的包，并在该包中创建控制器类 TestSecurityController，核心代码如下：

```
@Controller
public class TestSecurityController {
    @Autowired
    private UserService userService;
    @RequestMapping("/")
    public String index() {
        return "/index";
    }
    @RequestMapping("/toLogin")
    public String toLogin() {
        return "/login";
    }
    @RequestMapping("/toRegister")
    public String toRegister(@ModelAttribute("userDomain") MyUser userDomain) {
        return "/register";
    }
    @RequestMapping("/register")
    public String register(@ModelAttribute("userDomain") MyUser userDomain) {
        return userService.register(userDomain);
    }
    @RequestMapping("/login")
    public String login() {
        //这里什么都不做，由 Spring Security 负责登录验证
        return "/login";
    }
    @RequestMapping("/user/loginSuccess")
    public String loginSuccess(Model model) {
        return userService.loginSuccess(model);
    }
    @RequestMapping("/admin/main")
    public String main(Model model) {
        return userService.main(model);
```

```
    }
    @RequestMapping("/logout")
    public String logout(HttpServletRequest request, HttpServletResponse response) {
        return userService.logout(request, response);
    }
    @RequestMapping("/deniedAccess")
    public String deniedAccess(Model model) {
        return userService.deniedAccess(model);
    }
}
```

❾ 创建应用的安全控制相关实现

创建名为 com.ch.ch10_1.security 的包，并在该包中创建 MyUserSecurityService、MyAuthenticationSuccessHandler 和 MySecurityConfigurerAdapter 类。

MyUserSecurityService 类实现了 UserDetailsService 接口，通过重写 loadUserByUsername (String username)方法查询对应的用户，并将用户名、密码、权限等与认证相关的信息封装在 UserDetails 对象中。

MyUserSecurityService 的核心代码如下：

```
/**
 * 获得对应的 UserDetails，保存与认证相关的信息
 */
@Service
public class MyUserSecurityService implements UserDetailsService{
    @Autowired
    private MyUserRepository myUserRepository;
    /**
     * 通过重写 loadUserByUsername 方法查询对应的用户
     * UserDetails 是 Spring Security 的一个核心接口
     * UserDetails 定义了可以获取用户名、密码、权限等与认证信息相关的方法
     */
    @Override
    public UserDetails loadUserByUsername(String username) throws UsernameNotFoundException {
        //根据用户名（页面接收的用户名）查询当前用户
        MyUser myUser = myUserRepository.findByUsername(username);
        if(myUser == null) {
            throw new UsernameNotFoundException("用户名不存在");
        }
        //GrantedAuthority 代表赋予当前用户的权限（认证权限）
        List<GrantedAuthority> authorities = new ArrayList<GrantedAuthority>();
        //获得当前用户权限集合
        List<Authority> roles = myUser.getAuthorityList();

        //将当前用户的权限保存为用户的认证权限
        for (Authority authority : roles) {
            GrantedAuthority sg = new SimpleGrantedAuthority(authority.getName());
            authorities.add(sg);
        }
        //org.springframework.security.core.userdetails.User 是 Spring Security
        //的内部实现，专门用于保存用户名、密码、权限等与认证相关的信息
        User su = new User(myUser.getUsername(), myUser.getPassword(), authorities);
```

```
        return su;
    }
}
```

MyAuthenticationSuccessHandler 类继承了 SimpleUrlAuthenticationSuccessHandler 类，并重写了 handle(HttpServletRequest request, HttpServletResponse response, Authentication authentication)方法，根据当前认证用户的角色指定对应的 URL。

MyAuthenticationSuccessHandler 的核心代码如下：

```
@Component
/**
 * 用户授权、认证成功处理类
 */
public class MyAuthenticationSuccessHandler extends SimpleUrlAuthenticationSuccessHandler{
    //Spring Security 的重定向策略
    private RedirectStrategy redirectStrategy = new DefaultRedirectStrategy();
    /**
     * 重写 handle 方法，通过 RedirectStrategy 重定向到指定的 URL
     */
    @Override
    protected void handle(HttpServletRequest request, HttpServletResponse
response, Authentication authentication) throws IOException, ServletException {
        //根据当前认证用户的角色返回适当的 URL
        String tagetURL = getTargetURL(authentication);
        //重定向到指定的 URL
        redirectStrategy.sendRedirect(request, response, tagetURL);
    }
    /**
     * 从 Authentication 对象中提取当前登录用户的角色，并根据其角色返回适当的 URL
     */
    protected String getTargetURL(Authentication authentication) {
        String url = "";
        //获得当前登录用户的权限（角色）集合
        Collection<? extends GrantedAuthority> authorities = authentication.getAuthorities();
        List<String> roles = new ArrayList<String>();
        //将权限（角色）名称添加到 List 集合
        for (GrantedAuthority au : authorities) {
            roles.add(au.getAuthority());
        }
        //判断不同角色的用户跳转到不同的 URL
        //这里的 URL 是控制器的请求匹配路径
        if(roles.contains("ROLE_USER")) {
            url = "/user/loginSuccess";
        }else if(roles.contains("ROLE_ADMIN")) {
            url = "/admin/main";
        }else {
            url = "/deniedAccess";
        }
        return url;
    }
}
```

MySecurityConfigurerAdapter 类继承了 WebSecurityConfigurerAdapter 类，并通过重写 configure

(AuthenticationManagerBuilder auth)方法实现用户认证，重写 configure(HttpSecurity http)方法实现用户授权操作。

MySecurityConfigurerAdapter 的核心代码如下：

```
/**
 * 认证和授权处理类
 */
@Configuration
public class MySecurityConfigurerAdapter extends WebSecurityConfigurerAdapter{
    //依赖注入通用的用户服务类
    @Autowired
    private MyUserSecurityService myUserSecurityService;
    //依赖注入加密接口
    @Autowired
    private PasswordEncoder passwordEncoder;
    //依赖注入用户认证接口
    @Autowired
    private AuthenticationProvider authenticationProvider;
    //依赖注入认证处理成功类，验证用户成功后处理不同用户跳转到不同的页面
    @Autowired
    private MyAuthenticationSuccessHandler myAuthenticationSuccessHandler;
    /**
     * BCryptPasswordEncoder 是 PasswordEncoder 的接口实现
     * 实现加密功能
     */
    @Bean
    public PasswordEncoder passwordEncoder() {
        return new BCryptPasswordEncoder();
    }
    /**
     * DaoAuthenticationProvider 是 AuthenticationProvider 的实现
     */
    @Bean
    public AuthenticationProvider authenticationProvider() {
        DaoAuthenticationProvider provide = new DaoAuthenticationProvider();
        //不隐藏用户未找到异常
        provide.setHideUserNotFoundExceptions(false);
        //设置自定义认证方式，用户登录认证
        provide.setUserDetailsService(myUserSecurityService);
        //设置密码加密程序认证
        provide.setPasswordEncoder(passwordEncoder);
        return provide;
    }
    /**
     * 用户认证
     */
    @Override
    protected void configure(AuthenticationManagerBuilder auth) throws Exception {
        System.out.println("configure(AuthenticationManagerBuilder auth) ");
        //设置认证方式
        auth.authenticationProvider(authenticationProvider);
    }
    /**
     * 请求授权
```

```
     * 用户授权操作
     */
    @Override
    protected void configure(HttpSecurity http) throws Exception {
        System.out.println("configure(HttpSecurity http)");
        http.authorizeRequests()
        //首页、登录、注册页面、登录注册功能以及静态资源过滤掉，即可任意访问
        .antMatchers("/toLogin", "/toRegister", "/", "/login", "/register",
"/css/**", "/fonts/**", "/js/**").permitAll()
        //这里默认追加 ROLE_，/user/**是控制器的请求匹配路径
        .antMatchers("/user/**").hasRole("USER")
        .antMatchers("/admin/**").hasAnyRole("ADMIN", "DBA")
        //其他所有请求登录后才能访问
        .anyRequest().authenticated()
        .and()
        //将输入的用户名与密码和授权的进行比较
        .formLogin()
            .loginPage("/login").successHandler(myAuthenticationSuccessHandler)
            .usernameParameter("username").passwordParameter("password")
            //登录失败
            .failureUrl("/login?error")
        .and()
        //注销行为可任意访问
        .logout().permitAll()
        .and()
        //指定异常处理页面
        .exceptionHandling().accessDeniedPage("/deniedAccess");
    }
}
```

⑩ 创建用于测试的视图页面

在 src/main/resources/templates 目录下创建应用首页（index.html）、注册（register.html）、登录（login.html）以及拒绝访问页面（deniedAccess.html）；在 src/main/resources/templates/admin 目录下创建管理员用户认证成功后访问的页面（main.html）；在 src/main/resources/templates/user 目录下创建普通用户认证成功后访问的页面（loginSuccess.html）。这些视图页面，请读者参见本书提供的源程序 ch10_1。

⑪ 测试应用

测试前建议删除数据库 springbootjpa 中的 authority、user 和 user_authority 数据表。

运行 Ch101Application 的主类启动项目。Spring Boot 应用启动后，观察控制台，发现 MySecurityConfigurerAdapter 的两个 configure 方法都已经被执行，说明自定义的用户认证和用户授权工作已经生效。控制台的输出结果如图 10.1 所示。

```
Ch101Application [Java Application]
2021-01-04 08:00:57.897  INFO 16564 --- [
configure(AuthenticationManagerBuilder auth)
configure(HttpSecurity http)
```

图 10.1　启动 ch10_1 应用时控制台的输出结果

在浏览器地址栏中输入“http://localhost:8080/ch10_1”和“/toLogin”“/toRegister”“/”“/login”“/register”等其中任何一个请求都将正常访问，其他请求都将被重定向到 http://localhost:8080/ch10_1/login 登录页面，因为没有登录，用户没有访问权限。

可以通过“http://localhost:8080/ch10_1/”访问首页面，如图 10.2 所示；然后，单击“去注册”超链接打开注册页面。成功注册用户后，打开登录页面进行用户登录。

图 10.2 首页面

如果在注册页面中输入的用户名不是 admin，那么就是注册了一个普通用户，其权限为 ROLE_USER；如果在注册页面中输入的用户名是 admin，那么就是注册了一个管理员用户，其权限为 ROLE_ADMIN 和 ROLE_DBA。

在登录页面中任意输入用户名和密码，单击“登录”按钮，提示用户名或密码错误，如图 10.3 所示。

图 10.3 用户名或密码错误

在登录页面中输入管理员用户名和密码，成功登录后打开管理员主页面，如图 10.4 所示。

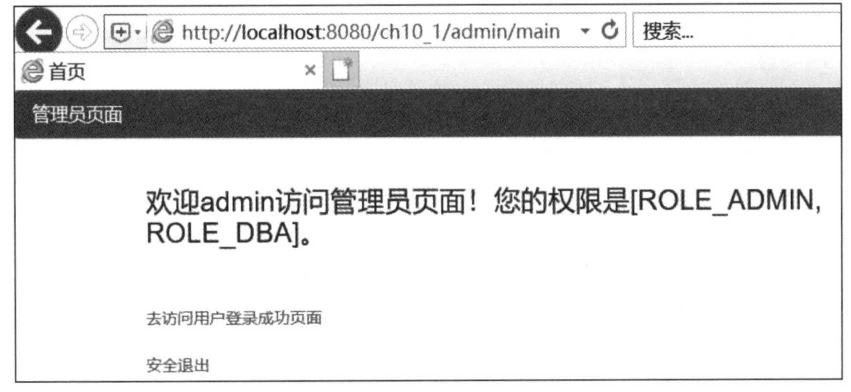

图 10.4 管理员成功登录页面

单击图 10.4 中的“去访问用户登录成功页面”超链接，显示拒绝访问页面，如图 10.5 所示。

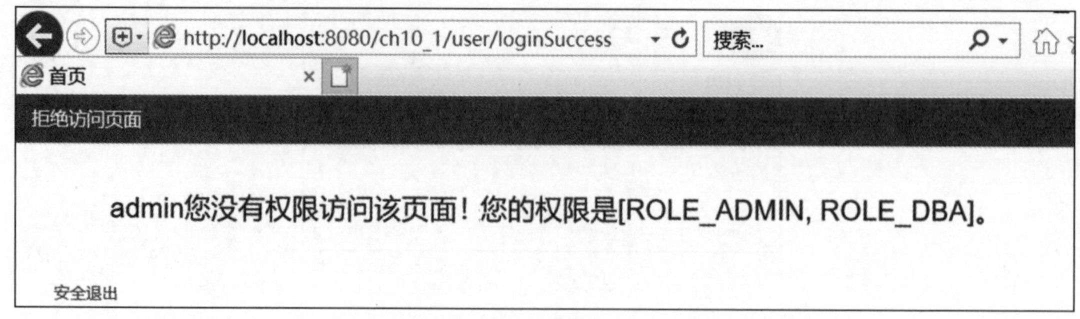

图 10.5　管理员被拒绝访问页面

在登录页面中输入普通用户名和密码，成功登录后打开用户成功登录页面，如图 10.6 所示。

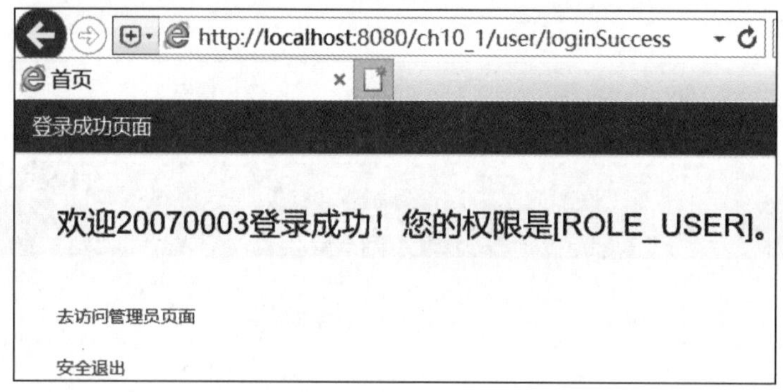

图 10.6　用户成功登录页面

单击图 10.6 中的“去访问管理员页面”超链接，显示拒绝访问页面，如图 10.7 所示。

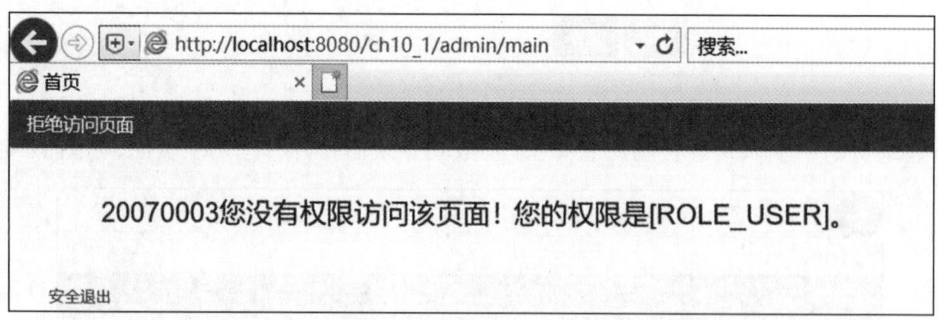

图 10.7　用户被拒绝访问页面

10.4　本章小结

本章首先介绍了 Spring Security 快速入门，然后详细介绍了实际开发中的 Spring Security 操作实例。通过本章的学习，读者应该了解 Spring Security 安全机制的基本原理，掌握在实际应用开发中使用 Spring Security 安全机制提供系统安全解决方案。

习 题 10

1．开发者如何自定义 Spring Security 的适配器？

2．Spring Security 的用户认证和请求授权是如何实现的？请举例说明。

第11章 Spring Boot的异步消息 

学习目的与要求

本章主要讲解了企业级消息代理 JMS 和 AMQP。通过本章的学习，理解异步消息通信原理，掌握异步消息通信技术。

主要内容

❖ JMS

❖ AMQP

当跨越多个微服务进行通信时，异步消息就显得至关重要了。例如在电子商务系统中，订单服务在下单时需要和库存服务进行通信，完成库存的扣减操作，这时就需要基于异步消息和最终一致性的通信方式来进行这样的操作，并且能够在发生故障时正常工作。

11.1 消息模型

异步消息的主要目的是解决跨系统的通信。所谓异步消息，即消息发送者无须等待消息接收者的处理及返回，甚至无须关心消息是否发送与接收成功。在异步消息中有两个极其重要的概念，即消息代理和目的地。当消息发送者发送消息后，消息将由消息代理管理，消息代理保证消息传递到目的地。

异步消息的目的地主要有两种形式，即队列和主题。队列用于点对点式的消息通信，即端到端通信（单接收者）；主题用于发布/订阅式的消息通信，即广播通信（多接收者）。

▶ 11.1.1 点对点式

在点对点式的消息通信中，消息代理获得发送者发送的消息后，将消息存入一个队列中，当有消息接收者接收消息时，将从队列中取出消息传递给接收者，这时从队列中清除该消息。

在点对点式的消息通信中，确保的是每一条消息只有唯一的发送者和接收者，但并不能说明只有一个接收者可以从队列中接收消息。这是因为队列中有多个消息，点对点式的消息通信只保证每一条消息只有唯一的发送者和接收者。

▶ 11.1.2 发布/订阅式

多接收者是消息通信中一种更加灵活的方式，而点对点式的消息通信只保证每一条消息只有唯一的接收者。这时可以使用发布/订阅式的消息通信解决多接收者的问题。和点对点式不同，发布/订阅式是消息发送者将消息发送到主题，而多个消息接收者监听这个主题。此时的消息发送者叫作发布者，接收者叫作订阅者。

11.2 企业级消息代理

视频讲解

异步消息传递技术常用的有 JMS 和 AMQP。JMS 是面向基于 Java 的企业应用的异步消息代理。AMQP 是面向所有应用的异步消息代理。

11.2.1 JMS

JMS（Java Messaging Service）即 Java 消息服务，是 Java 平台上有关面向消息中间件的技术规范，它便于消息系统中的 Java 应用程序进行消息交换，并且通过提供标准的产生、发送、接收消息的接口简化企业应用的开发。

❶ JMS 元素

JMS 由以下元素组成。

1）JMS 消息代理实现

连接面向消息中间件的，JMS 消息代理接口的一个实现。JMS 的消息代理实现可以是 Java 平台的 JMS 实现，也可以是非 Java 平台的面向消息中间件的适配器。开源的 JMS 实现有 Apache ActiveMQ、JBoss 社区所研发的 HornetQ、The OpenJMS Group 的 OpenJMS 等实现。

2）JMS 客户

生产或消费基于消息的 Java 应用程序或对象。

3）JMS 生产者

创建并发送消息的 JMS 客户。

4）JMS 消费者

接收消息的 JMS 客户。

5）JMS 消息

JMS 消息包括可以在 JMS 客户之间传递的数据对象。JMS 定义了五种不同的消息正文格式，以及调用的消息类型，允许你发送并接收一些不同形式的数据，提供现有消息格式的一些级别的兼容性。常见的消息格式有 StreamMessage（指 Java 原始值的数据流消息）、MapMessage（映射消息）、TextMessage（文本消息）、ObjectMessage（一个序列化的 Java 对象消息）、BytesMessage（字节消息）。

6）JMS 队列

一个容纳那些被发送的等待阅读的消息区域。与队列名字所暗示的意思不同，消息的接收顺序并不一定要与消息的发送顺序相同。一旦一个消息被阅读，该消息将被从队列中移走。

7）JMS 主题

一种支持发送消息给多个订阅者的机制。

❷ JMS 的应用接口

JMS 的应用接口包括以下接口类型。

1）ConnectionFactory 接口（连接工厂）

用户用来创建到 JMS 消息代理实现的连接的被管对象。JMS 客户通过可移植的接口访问连接，这样当下层的实现改变时，代码不需要进行修改。管理员在 JNDI 名字空间中配置连接工厂，这样，JMS 客户才能够查找到它们。根据目的地的不同，用户将使用队列连接工厂，或者主题连接工厂。

2）Connection 接口（连接）

连接代表了应用程序和消息服务器之间的通信链路。在获得了连接工厂后，就可以创建一个与 JMS 消息代理实现（提供者）的连接。根据不同的连接类型，连接允许用户创建会话，以发送和接收队列和主题到目的地。

3）Destination 接口（目的地）

目的地是一个包装了消息目的地标识符的被管对象，消息目的地是指消息发布和接收的地点，或者是队列，或者是主题。JMS 管理员创建这些对象，然后用户通过 JNDI 发现它们。和连接工厂一样，管理员可以创建两种类型的目的地，分别是点对点模型的队列和发布者/订阅者模型的主题。

4）Session 接口（会话）

会话表示一个单线程的上下文，用于发送和接收消息。由于会话是单线程的，所以消息是连续的，就是说消息是按照发送的顺序一个一个接收的。会话的好处是它支持事务。如果用户选择了事务支持，会话上下文将保存一组消息，直到事务被提交才发送这些消息。在提交事务之前，用户可以使用回滚操作取消这些消息。一个会话允许用户创建消息，生产者发送消息，消费者接收消息。

5）MessageConsumer 接口（消息消费者）

消息消费者是由会话创建的对象，用于接收发送到目的地的消息。消费者可以同步地（阻塞模式），或（非阻塞模式）接收队列和主题类型的消息。

6）MessageProducer 接口（消息生产者）

消息生产者是由会话创建的对象，用于发送消息到目的地。用户可以创建某个目的地的发送者，也可以创建一个通用的发送者，在发送消息时指定目的地。

7）Message 接口（消息）

消息是在消费者和生产者之间传送的对象，消息被从一个应用程序传送到另一个应用程序。一个消息有三个主要部分。

消息头（必需）：包含用于识别和为消息寻找路由的操作设置。

一组消息属性（可选）：包含额外的属性，支持其他消息代理实现和用户的兼容。可以创建定制的字段和过滤器（消息选择器）。

一个消息体（可选）：允许用户创建五种类型的消息（文本消息、映射消息、字节消息、流消息和对象消息）。

JMS 各接口角色间的关系如图 11.1 所示。

11.2.2 AMQP

AMQP（Advanced Message Queuing Protocol）即高级消息队列协议，是一个提供统一消息服务的应用层标准高级消息队列协议，是应用层协议的一个开放标准，为面向消息的中间件设计。基于此协议的客户端与消息中间件可传递消息，并不受客户端/中间件的不同产品、不同开发语言等条件的限制。AMQP 的技术术语如下。

AMQP 模型（AMQP Model）：一个由关键实体和语义表示的逻辑框架，遵从 AMQP 规范的服务器必须提供这些实体和语义。为了实现本规范中定义的语义，客户端可以发送命令来控制 AMQP 服务器。

连接（Connection）：一个网络连接，例如 TCP/IP 套接字连接。

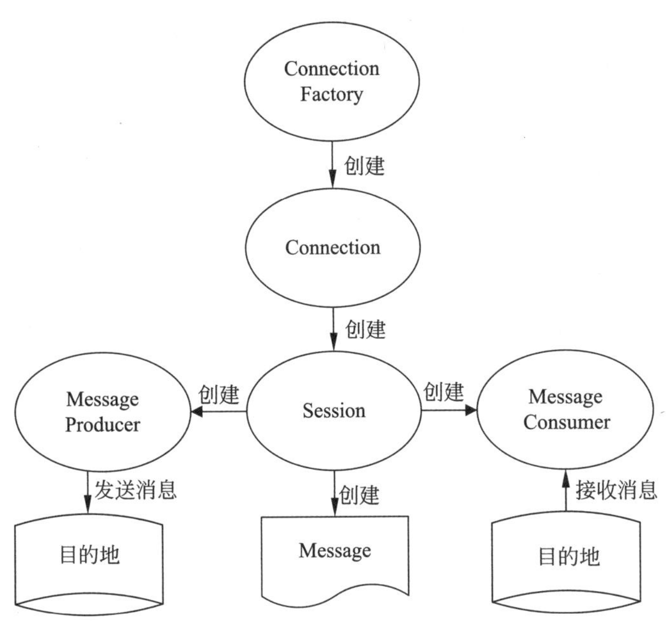

图 11.1　JMS 各角色间的关系

会话（Session）：端点之间的命名对话。在一个会话上下文中，保证“恰好传递一次”。

信道（Channel）：多路复用连接中的一条独立的双向数据流通道。信道为会话提供物理传输介质。

客户端（Client）：AMQP 连接或者会话的发起者。AMQP 是非对称的，客户端生产和消费消息，服务器存储和路由这些消息。

服务器（Server）：接受客户端连接，实现 AMQP 消息队列和路由功能的进程。服务器也称为“消息代理”。

端点（Peer）：AMQP 对话的任意一方。一个 AMQP 连接包括两个端点（一个是客户端，一个是服务器）。

搭档（Partner）：当描述两个端点之间的交互过程时，使用术语“搭档”来表示“另一个”端点的简记法。例如定义端点 A 和端点 B，当它们进行通信时，端点 B 是端点 A 的搭档，端点 A 是端点 B 的搭档。

片段集（Assembly）：段的有序集合，形成一个逻辑工作单元。

段（Segment）：帧的有序集合，形成片段集中一个完整子单元。

帧（Frame）：AMQP 传输的一个原子单元。一个帧是一个段中的任意分片。

控制（Control）：单向指令，AMQP 规范假设这些指令的传输是不可靠的。

命令（Command）：需要确认的指令，AMQP 规范规定这些指令的传输是可靠的。

异常（Exception）：在执行一个或者多个命令时可能发生的错误状态。

类（Class）：一批用来描述某种特定功能的 AMQP 命令或者控制。

消息头（Header）：描述消息数据属性的一种特殊段。

消息体（Body）：包含应用程序数据的一种特殊段。消息体对于服务器来说完全透明，也就是说服务器不能查看或者修改消息体。

消息内容（Content）：包含在消息体段中的消息数据。

交换器（Exchange）：服务器中的实体，用来接收生产者发送的消息并将这些消息路由给服务器中的队列。

交换器类型（Exchange Type）：基于不同路由语义的交换器类。

消息队列（Message Queue）：一个命名实体，用来保存消息直到发送给消费者。

绑定器（Binding）：消息队列和交换器之间的关联。

绑定器关键字（Binding Key）：绑定的名称。一些交换器类型可能使用这个名称作为定义绑定器路由行为的模式。

路由关键字（Routing Key）：一个消息头，交换器可以用这个消息头决定如何路由某条消息。

持久存储（Durable）：一种服务器资源，当服务器重启时，保存的消息数据不会丢失。

临时存储（Transient）：一种服务器资源，当服务器重启时，保存的消息数据会丢失。

持久化（Persistent）：服务器将消息保存在可靠磁盘存储中，当服务器重启时，消息不会丢失。

非持久化（Non-Persistent）：服务器将消息保存在内存中，当服务器重启时，消息可能丢失。

消费者（Consumer）：一个从消息队列中请求消息的客户端应用程序。

生产者（Producer）：一个向交换器发布消息的客户端应用程序。

虚拟主机（Virtual Host）：一批交换器、消息队列和相关对象。虚拟主机是共享相同的身份认证和加密环境的独立服务器域。客户端应用程序在登录到服务器之后，可以选择一个虚拟主机。

11.3 Spring Boot 的支持

11.3.1 JMS 的自动配置

Spring Boot 对 JMS 的自动配置位于 org.springframework.boot.autoconfigure.jms 包下，支持 JMS 的实现有 ActiveMQ 和 Artemis。

以 ActiveMQ 为例，Spring Boot 定义了 ActiveMQConnectionFactory 的 Bean 作为连接，并通过以 spring.activemq 为前缀的属性配置 ActiveMQ 的连接属性，主要包含以下属性。

```
spring.activemq.broker-url= tcp://localhost:61616  #消息代理路径
spring.activemq.user=
spring.activemq.password=
spring.activemq.in-memory=true
```

另外，Spring Boot 在 JmsAutoConfiguration 自动配置类中配置了 JmsTemplate，并且在 JmsAnnotationDrivenConfiguration 配置类中开启了注解式消息监听的支持，即自动开启 @EnableJms。

11.3.2 AMQP 的自动配置

Spring Boot 对 AMQP 的自动配置位于 org.springframework.boot.autoconfigure.amqp 包下，RabbitMQ 是 AMQP 的主要实现。

在 RabbitAutoConfiguration 自动配置类中配置了连接的 RabbitConnectionFactoryBean 和 RabbitTemplate，并且在 RabbitAnnotationDrivenConfiguration 配置类中开启了@EnableRabbit。从

RabbitProperties 类中可以看出，RabbitMQ 的配置可通过以 spring.rabbitmq 为前缀的属性进行配置，主要包含以下属性。

```
spring.rabbitmq.host=localhost  #RabbitMQ 服务器地址，默认为 localhost
spring.rabbitmq.port=5672       #RabbitMQ 端口，默认为 5672
spring.rabbitmq.username=guest  #默认用户名
spring.rabbitmq.password =guest #默认密码
```

11.4 异步消息通信实例

本节通过两个实例，讲解异步消息通信的实现过程。

11.4.1 JMS 实例

本节使用 JMS 的一种实现 ActiveMQ 讲解 JMS 实例。因此，需要事先安装 ActiveMQ（注意需要安装 JDK）。读者可访问 http://activemq.apache.org/下载符合自己要求的 ActiveMQ。编写本书时，作者下载了 Windows 版本的 apache-activemq-5.16.0-bin.zip。该版本的 ActiveMQ 解压缩即可完成安装。

解压缩后，双击 apache-activemq-5.16.0\bin\win64 下的 wrapper.exe 或 activemq.bat 启动 ActiveMQ，如图 11.2 所示。然后，通过 http://localhost:8161 运行 ActiveMQ 的管理界面，管理员账号和密码默认为 admin、admin，如图 11.3 所示。

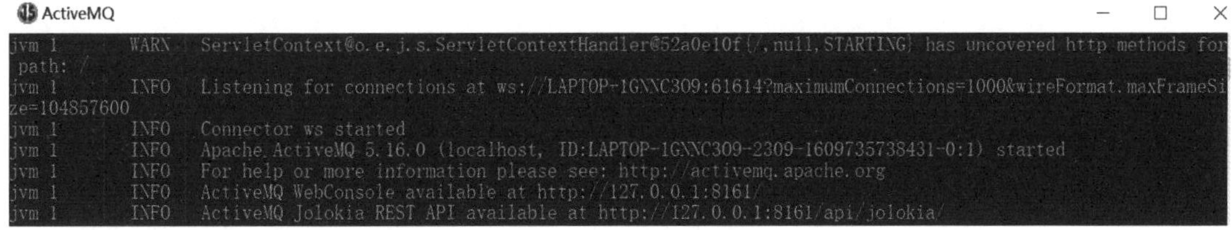

图 11.2 启动 ActiveMQ

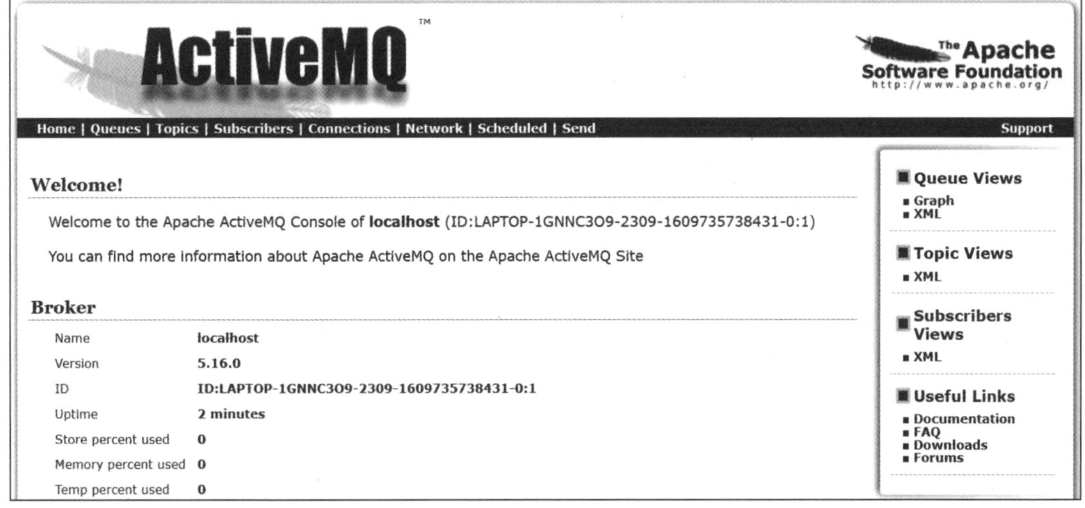

图 11.3 ActiveMQ 的管理界面

启动 ActiveMQ 服务后，下面通过实例讲解如何使用 JMS 的实现 ActiveMQ 进行两个应用系统间的点对点式通信。

【例 11-1】使用 JMS 的实现 ActiveMQ 进行两个应用系统间的点对点式通信。

具体实现步骤如下。

❶ 创建基于 Apache ActiveMQ5 依赖的 Spring Boot 应用 ch11_1Sender（消息发送者）

创建基于 Apache ActiveMQ5 依赖的 Spring Boot 应用 ch11_1Sender，该应用作为消息发送者。

❷ 配置 ActiveMQ 的消息代理地址

在应用 ch11_1Sender 的配置文件 application.properties 中，配置 ActiveMQ 的消息代理地址，具体如下：

```
spring.activemq.broker-url= tcp://localhost:61616
```

❸ 定义消息

在 com.ch.ch11_1Sender 包下，创建消息定义类 MyMessage，该类需要实现 MessageCreator 接口，并重写接口方法 createMessage 进行消息定义。核心代码如下：

```
public class MyMessage implements MessageCreator{
    @Override
    public Message createMessage(Session session) throws JMSException {
        MapMessage mapm = session.createMapMessage();
        ArrayList<String> arrayList = new ArrayList<String>();
        arrayList.add("陈恒 1");
        arrayList.add("陈恒 2");
        mapm.setObject("mesg1", arrayList);//只能存 Java 的基本对象
        mapm.setString("mesg2", "测试消息 2");
        return mapm;
    }
}
```

❹ 发送消息

在应用 ch11_1Sender 的主类 Ch111SenderApplication 中，实现 Spring Boot 的 CommandLineRunner 接口，并重写 run 方法，应用程序启动后执行 run 方法的代码。在该 run 方法中，使用 JmsTemplate 的 send 方法向目的地 mydestination 发送 MyMessage 的消息，也相当于在消息代理上定义了一个目的地叫 mydestination。核心代码如下：

```
@SpringBootApplication
public class Ch111SenderApplication implements CommandLineRunner{
    @Autowired
    private JmsTemplate jmsTemplate;
    public static void main(String[] args) {
        SpringApplication.run(Ch111SenderApplication.class, args);
    }
    /**
     * 这里为了方便操作使用 run 方法发送消息,
     * 当然完全可以使用控制器通过 Web 访问
     */
    @Override
```

```
    public void run(String... args) throws Exception {
        //new MyMessage()回调接口方法 createMessage 产生消息
        jmsTemplate.send("mydestination", new MyMessage());
    }
}
```

❺ 创建消息接收者

按照步骤 1 创建 Spring Boot 应用 ch11_1Receive，该应用作为消息接收者；并按照步骤 2 配置 ch11_1Receive 的 ActiveMQ 的消息代理地址。

❻ 定义消息监听器接收消息

在应用 ch11_1Receive 的 com.ch.ch11_1Receive 包中，创建消息监听器类 ReceiverMsg。在该类中使用@JmsListener 注解不停地监听目的地 mydestination 是否有消息发送过来，如果有就获取消息。核心代码如下：

```
@Component
public class ReceiverMsg {
    @JmsListener(destination="mydestination")
    public void receiverMessage(MapMessage mapm) throws JMSException {
        @SuppressWarnings("unchecked")
        ArrayList<String> arrayList = (ArrayList<String>)mapm.getObject("mesg1");
        System.out.println(arrayList);
        System.out.println(mapm.getString("mesg2"));
    }
}
```

❼ 运行测试

先启动消息接收者 ch11_1Receive 应用，启动 ch11_1Receive 应用后，单击图 11.3 中的 Queues，可看到如图 11.4 所示的界面。

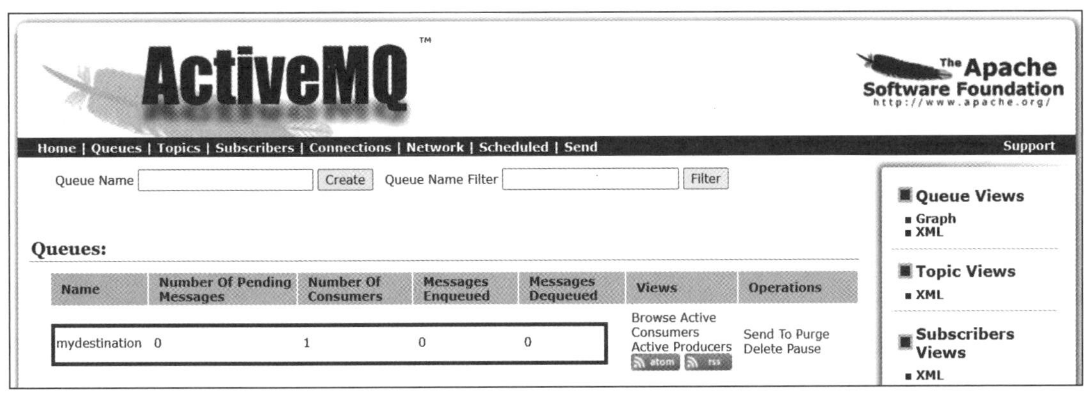

图 11.4 ActiveMQ 的 Queues

从图 11.4 可以看出，目的地 mydestination 有一个消费者正在等待接收消息。此时，启动消息发送者 ch11_1Sender 应用后，可在接收者 ch11_1Receive 应用的控制台上看到有消息打印，如图 11.5 所示。再去刷新图 11.4，可看到如图 11.6 所示的界面。

```
Ch111ReceiveApplication [Java
2021-01-04 13:02:02.039
2021-01-04 13:02:02.042
2021-01-04 13:02:03.041
[陈恒1, 陈恒2]
测试消息2
```

图 11.5 消息接收者接收的消息

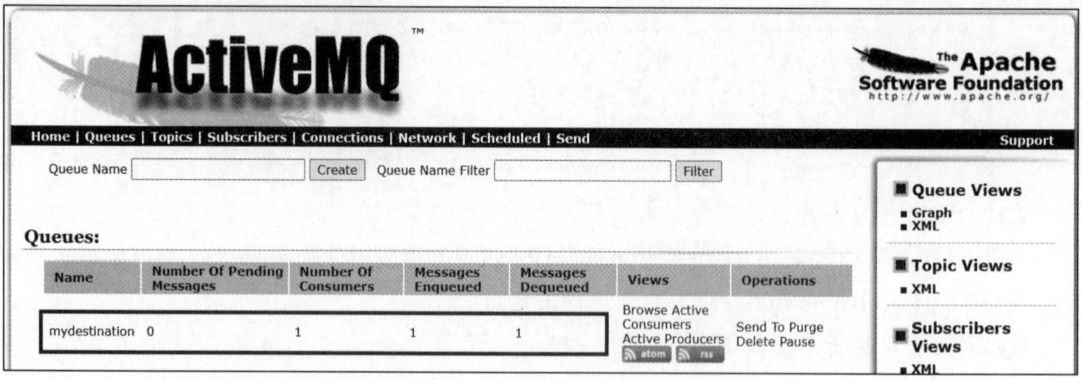

图 11.6　ActiveMQ 的 Queues

从图 11.6 可以看出，目的地 mydestination 有一个消息入列（表示发送成功）和一个消息出列（表示接收成功）。

▶ 11.4.2　AMQP 实例

本节使用 AMQP 的主要实现 RabbitMQ 讲解 AMQP 实例，因此需要事先安装 RabbitMQ。又因为 RabbitMQ 是基于 Erlang 语言开发的，所以安装 RabbitMQ 之前，先下载、安装 Erlang。Erlang 语言的下载地址为 https://www.erlang.org/downloads；RabbitMQ 的下载地址为 https://www.rabbitmq.com/download.html。本书采用的 Erlang 语言的安装包是 otp_win64_23.1.exe，RabbitMQ 的安装包是 rabbitmq-server-3.8.9.exe。

运行 Erlang 语言安装包 otp_win64_23.1.exe，一直单击 Next 按钮即可完成 Erlang 安装。安装 Erlang 后需要配置环境变量 ERLANG_HOME 以及在 path 中新增%ERLANG_HOME%\bin，如图 11.7 和图 11.8 所示。

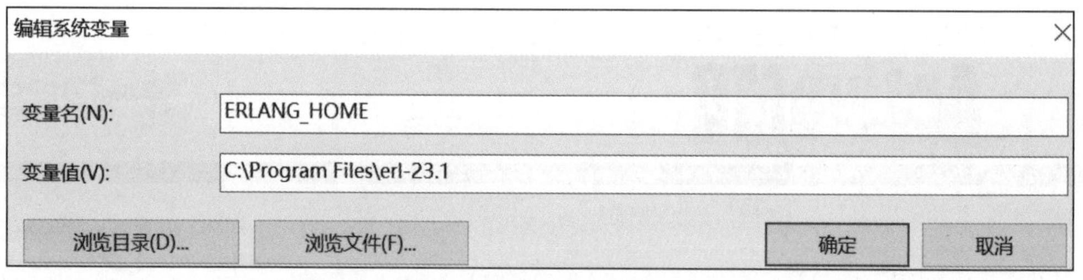

图 11.7　ERLANG_HOME

图 11.8　在 path 中新增 %ERLANG_HOME%\bin

运行 RabbitMQ 安装包 rabbitmq-server-3.8.9.exe，一直单击 Next 按钮即可完成 RabbitMQ 安装。安装 RabbitMQ 后需要配置环境变量 RABBITMQ_SERVER=C:\Program Files\RabbitMQ Server\rabbitmq_server-3.8.9 以及在 path 中新增%RABBITMQ_SERVER%\sbin，操作界面与图 11.7 和图 11.8 类似。

在 cmd 命令行窗口中，进入 RabbitMQ 的 sbin 目录，运行 rabbitmq-plugins.bat enable rabbitmq_management 命令，打开 RabbitMQ 的管理组件，如图 11.9 所示。

```
Microsoft Windows [版本 10.0.18363.1256]
(c) 2019 Microsoft Corporation。保留所有权利。

C:\Users\ChenHeng>cd C:\Program Files\RabbitMQ Server\rabbit

C:\Program Files\RabbitMQ Server\rabbitmq_server-3.8.9\sbin>
Enabling plugins on node rabbit@LAPTOP-1GNNC309:
rabbitmq_management
The following plugins have been configured:
  rabbitmq_management
  rabbitmq_management_agent
  rabbitmq_web_dispatch
Applying plugin configuration to rabbit@LAPTOP-1GNNC309...
Plugin configuration unchanged.

C:\Program Files\RabbitMQ Server\rabbitmq_server-3.8.9\sbin>
```

图 11.9　打开 RabbitMQ 的管理组件

以管理员方式打开 cmd 命令，运行 net start RabbitMQ 命令，提示 RabbitMQ 服务已经启动，如图 11.10 所示。

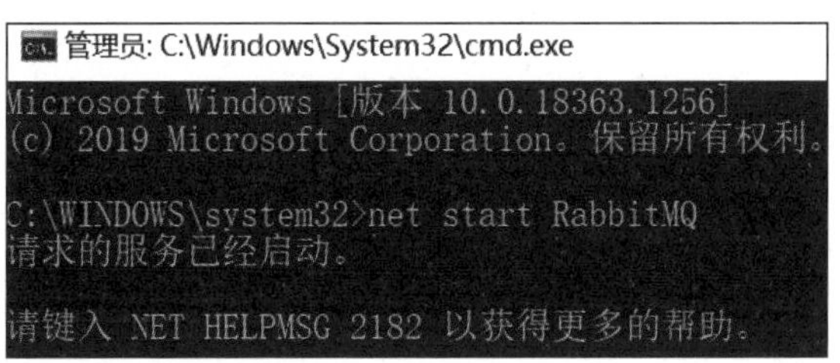

图 11.10　启动 RabbitMQ 服务

在浏览器地址栏中，输入"http://localhost:15672"，账号和密码默认为 guest/guest，进入 RabbitMQ 的管理界面，如图 11.11 所示。

图 11.11　RabbitMQ 的管理界面

至此，完成了 RabbitMQ 服务器的搭建。

在例 11-1 中，不管是消息发送者（生产者）还是消息接收者（消费者），都必须知道一个指定的目的地（队列）才能发送、获取消息。如果同一个消息要求每个消费者都处理的话，就需要发布/订阅式的消息分发模式。

下面通过实例讲解如何使用 RabbitMQ 实现发布/订阅式异步消息通信。在本例中，创建一个发布者应用、两个订阅者应用。该实例中的三个应用都是使用 Spring Boot 默认配置的 RabbitMQ，主机为 localhost，端口号为 5672，所以无须在配置文件中配置 RabbitMQ 的连接信息。另外，三个应用需要使用 Weather 实体类封装消息，并且使用 JSON 数据格式发布和订阅消息。

【例 11-2】使用 RabbitMQ 实现发布/订阅式异步消息通信。

具体实现步骤如下。

❶ 创建发布者应用 ch11_2Sender

创建发布者应用 ch11_2Sender，包括以下步骤。

（1）创建基于 RabbitMQ 依赖的 Spring Boot 应用 ch11_2Sender。

（2）在 ch11_2Sender 应用中创建名为 com.ch.ch11_2Sender.entity 的包，并在该包中创建 Weather 实体类，具体代码如下：

```
package com.ch.ch11_2Sender.entity;
import java.io.Serializable;
public class Weather implements Serializable{
    private static final long serialVersionUID = -8221467966772683998L;
    private String id;
    private String city;
    private String weatherDetail;
    //省略 set 和 get 方法
    @Override
    public String toString() {
        return "Weather [id=" + id + ", city=" + city + ", weatherDetail=" + weatherDetail + "]";
    }
}
```

（3）在应用 ch11_2Sender 的主类 Ch112SenderApplication 中，实现 Spring Boot 的 CommandLineRunner 接口，并重写 run 方法，应用程序启动后执行 run 方法的代码。在该 run 方法中，使用 RabbitTemplate 的 convertAndSend 方法将特定的路由 weather.message 和 Weather 消息对象发送到指定的交换机 weather-exchange。在发布消息前，需要使用 ObjectMapper 将 Weather 对象转换为 byte[]类型的 JSON 数据。核心代码如下：

```
@SpringBootApplication
public class Ch112SenderApplication implements CommandLineRunner{
    @Autowired
   private ObjectMapper objectMapper;
    @Autowired
    RabbitTemplate rabbitTemplate;
    public static void main(String[] args) {
        SpringApplication.run(Ch82SenderApplication.class, args);
    }
    /**
     * 定义发布者
     */
    @Override
    public void run(String... args) throws Exception {
```

```
        //定义消息对象
        Weather weather = new Weather();
        weather.setId("010");
        weather.setCity("北京");
        weather.setWeatherDetail("今天晴到多云，南风 5-6 级，温度 19-26℃");
        //指定 JSON 转换器,Jackson2JsonMessageConverter 默认将消息转换为 byte[]类型的消息
        rabbitTemplate.setMessageConverter(new Jackson2JsonMessageConverter());
        //objectMapper 将 weather 对象转换为 JSON 字节数组
        Message msg=MessageBuilder.withBody(objectMapper.writeValueAsBytes(weather))
                .setDeliveryMode(MessageDeliveryMode.NON_PERSISTENT)
            .build();
        // 消息唯一 ID
        CorrelationData correlationData = new CorrelationData(weather.getId());
        //使用已封装好的 convertAndSend(String exchange , String routingKey ,
        //Object message, CorrelationData correlationData)
        //将特定的路由 Key 和消息发送到指定的交换机
        rabbitTemplate.convertAndSend(
                "weather-exchange", //分发消息的交换机名称
                "weather.message", //用来匹配消息的路由 Key
                msg, //消息体
                correlationData);
    }
}
```

❷ 创建订阅者应用 ch11_2Receiver-1

创建订阅者应用 ch11_2Receiver-1，包括以下步骤。

（1）创建基于 RabbitMQ 依赖的 Spring Boot 应用 ch11_2Receiver-1。

（2）将 ch11_2Sender 中的 Weather 实体类复制到 com.ch.ch11_2Receiver1 包中。

（3）在 com.ch.ch11_2Receiver1 包中创建订阅者类 Receiver1，在该类中使用@RabbitListener 和@RabbitHandler 注解监听发布者并接收消息。核心代码如下：

```
/**
 * 定义订阅者 Receiver1
 */
@Component
public class Receiver1 {
    @Autowired
    private ObjectMapper objectMapper;
    @RabbitListener(
            bindings =
            @QueueBinding(
                //队列名 weather-queue1 保证和别的订阅者不一样，可以随机起名
                value = @Queue(value = "weather-queue1",durable = "true"),
                //weather-exchange 与发布者的交换机名相同
                exchange = @Exchange(value = "weather-exchange",durable = "true",type = "topic"),
                //weather.message 与发布者的消息的路由 Key 相同
                key = "weather.message"
            )
    )
    @RabbitHandler
    public void receiveWeather(@Payload byte[] weatherMessage) throws Exception{
        System.out.println("----------订阅者 Receiver1 接收到消息----------");
```

```
            //将 JSON 字节数组转换为 Weather 对象
            Weather w=objectMapper.readValue(weatherMessage, Weather.class);
            System.out.println("Receiver1 收到的消息内容: "+w);
        }
    }
```

❸ 创建订阅者应用 ch11_2Receiver-2

与创建订阅者应用 ch11_2Receiver-1 的步骤一样，这里不再赘述。但需要注意的是两个订阅者的队列名不同。

❹ 测试运行

首先，运行发布者应用 ch11_2Sender 的主类 Ch112SenderApplication。

然后，运行订阅者应用 ch11_2Receiver-1 的主类 Ch112Receiver1Application，此时接收到的消息如图 11.12 所示。

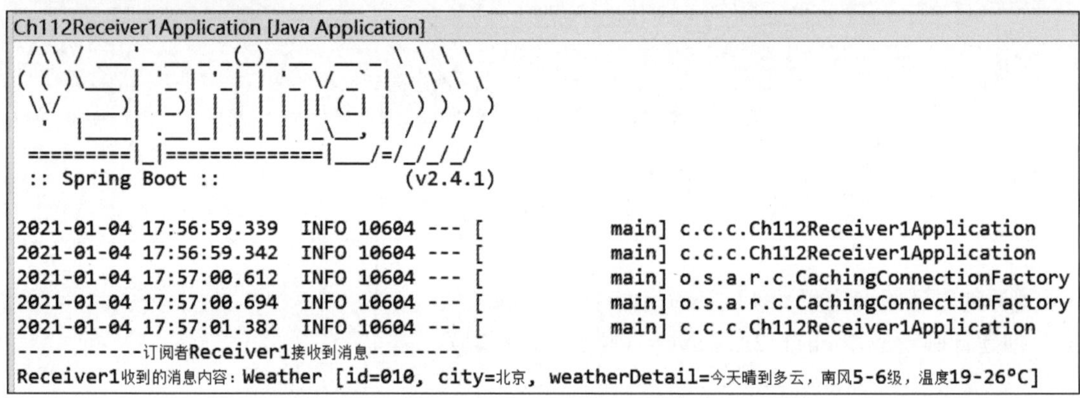

图 11.12　订阅者 ch11_2Receiver-1 接收到的消息

最后，运行订阅者应用 ch11_2Receiver-2 的主类 Ch112Receiver2Application，此时接收到的消息如图 11.13 所示。

```
Ch112Receiver2Application [Java Application]
 :: Spring Boot ::                (v2.4.1)

2021-01-04 18:00:29.277  INFO 16172 --- [           main] c.c.c.Ch112Receiver2Application
2021-01-04 18:00:29.281  INFO 16172 --- [           main] c.c.c.Ch112Receiver2Application
2021-01-04 18:00:30.810  INFO 16172 --- [           main] o.s.a.r.c.CachingConnectionFactory
2021-01-04 18:00:30.906  INFO 16172 --- [           main] o.s.a.r.c.CachingConnectionFactory
2021-01-04 18:00:31.015  INFO 16172 --- [           main] c.c.c.Ch112Receiver2Application
-----------订阅者Receiver2接收到消息--------
Receiver2收到的消息内容：Weather [id=010, city=北京, weatherDetail=今天晴到多云，南风5-6级，温度19-26°C]
```

图 11.13　订阅者 ch11_2Receiver-2 接收到的消息

从例 11-2 可以看出，一个发布者发布的消息可以被多个订阅者订阅，这就是所谓的发布/订阅式异步消息通信。

11.5　本章小结

本章主要介绍了多个应用系统间的异步消息。通过本章的学习，读者应该了解 Spring Boot 对 JMS 和 AMQP 的支持，掌握如何在实际应用开发中使用 JMS 或 AMQP 提供异步通信解决方案。

习 题 11

1. 在多个应用系统间的异步消息中，有哪些消息模型？
2. JMS 和 AMQP 有什么区别？

第 12 章 Spring Boot 的热部署与单元测试

学习目的与要求

本章主要讲解 Spring Boot 开发的热部署以及单元测试。通过本章的学习，掌握开发的热部署，理解单元测试的原理。

主要内容

- ❖ 开发的热部署
- ❖ Spring Boot 的单元测试

在实际应用开发过程中，业务变化、代码错误等发生时，难免修改程序。为了正确运行出修改的结果，往往需要重启应用，否则将不能看到修改后的效果。这一启动过程是非常浪费时间的，导致开发效率低。因此，有必要学习 Spring Boot 开发的热部署，自动实现应用的重启和部署，大大提高开发调试效率。

视频讲解

12.1 开发的热部署

开发热部署的目的是使应用自动重启和部署，提高开发效率。本节将讲解如何实现 Spring Boot 开发的热部署，包括前端模板引擎和后端程序的热部署。

12.1.1 模板引擎的热部署

在 Spring Boot 应用中，使用模板引擎的页面默认是开启缓存的，如果修改了页面内容，则刷新页面得不到修改后的页面效果。因此，可以在配置文件 application.properties 中关闭模板引擎的缓存。示例如下。

关闭 Thymeleaf 缓存的配置：

```
spring.thymeleaf.cache=false
```

关闭 FreeMarker 缓存的配置：

```
spring.freemarker.cache=false
```

关闭 Groovy 缓存的配置：

```
spring.groovy.template.cache=false
```

12.1.2 使用 spring-boot-devtools 进行热部署

在 Spring Boot 应用的 pom.xml 文件中添加 spring-boot-devtools 依赖即可实现页面和代码的热部署。

spring-boot-devtools 是一个为开发者服务的模块，最重要的功能是自动实现将修改的应用代码更新到最新的应用上。其工作原理是使用两个 ClassLoader，一个 ClassLoader 加载那些不会改变的类(如第三方JAR包);一个ClassLoader加载更新的类,称为Restart ClassLoader。于是在修改代码时，原来的 Restart ClassLoader 被丢弃，重新创建一个 Restart ClassLoader 加载更新的类，由于只加载部分修改的类，所以实现了较快的重启。

下面通过实例讲解如何使用 spring-boot-devtools 进行热部署。

【例 12-1】使用 spring-boot-devtools 进行热部署。

具体实现步骤如下。

❶ 创建基于 spring-boot-devtools 依赖的 Spring Boot Web 应用

创建基于 spring-boot-devtools 依赖的 Spring Boot Web 应用 ch12_1。

❷ 创建控制器类

在 com.ch.ch12_1 包中，创建控制器类 TestDevToolsController，核心代码如下：

```
@RestController
public class TestDevToolsController {
    @RequestMapping("/testDevTools")
    public String testDevTools() {
        return "test DevTools 111";
    }
}
```

❸ 测试运行

首先，运行 Ch121Application 主类，启动应用 ch12_1。然后，通过 http://localhost:8080/testDevTools 请求 TestDevToolsController 类中的 testDevTools 方法，运行效果如图 12.1 所示。

现在，将 testDevTools 方法中的 return 语句修改如下：

```
return "test DevTools 222";
```

无须重启应用 ch12_1，直接刷新 http://localhost:8080/testDevTools，运行效果如图 12.2 所示。

图 12.1　请求 testDevTools 方法

图 12.2　刷新页面效果

从例 12-1 看出，spring-boot-devtools 实现了代码修改后的热部署，同样，也可实现新增类、修改配置文件等热部署。

12.2　Spring Boot 的单元测试

视频讲解

测试是系统开发中非常重要的工作，单元测试在帮助开发人员编写高品质的程序、提升代码质量方面发挥了极大的作用。本节将介绍 Spring Boot 的单元测试。

Spring Boot 为测试提供了一个名为 spring-boot-starter-test 的 Starter。使用 STS 创建 Spring Boot 应用时，将自动添加 spring-boot-starter-test 依赖。这样在测试时，就没有必要再添加额外的 JAR 包。spring-boot-starter-test 主要提供了以下测试库。

（1）Junit：标准的单元测试 Java 应用程序。

（2）Spring Test&Spring Boot Test：针对 Spring Boot 应用程序的单元测试。

（3）Mockito：Java mocking 框架，用于模拟任何 Spring 管理的 Bean，例如在单元测试中模拟一个第三方系统 Service 接口返回的数据，而不去真正调用第三方系统。

（4）AssertJ：一个流畅的 assertion 库，同时也提供了更多的期望值与测试返回值的比较方式。

（5）JSONassert：对 JSON 对象或 JSON 字符串断言的库。

（6）JsonPath：提供类似 Xpath（一种在 XML 文档中查找信息的语言）那样的符号来获取 JSON 数据片段。

12.2.1 Spring Boot 单元测试程序模板

在 Spring Boot 应用中，使用一系列注解增强单元测试以支持 Spring Boot 测试。通常，Spring Boot 单元测试程序类似如下模板：

```
@SpringBootTest
public class GoodsServiceTest {
    //注入要测试的 service
    @Autowired
    private GoodsService goodsService;
    @Test
    public void testGoodsService() {
        //调用 GoodsService 的方法进行测试
    }
}
```

@SpringBootTest 用于 Spring Boot 应用测试，它默认根据包名逐级往上找，一直找到 Spring Boot 主程序（包含@SpringBootApplication 注解的类），并在单元测试时启动该主程序来创建 Spring 上下文环境。

12.2.2 测试 Service

单元测试 Service 代码与通过 Controller 调用 Service 代码相比，需要特别考虑该 Service 是否依赖其他还未开发完毕的 Service（第三方接口）。如果依赖其他还未开发完毕的 Service，需要使用 Mockito 来模拟未完成的 Service。

假设在 UserService 中依赖 CreditService（第三方接口）的 getCredit 方法获得用户积分，UserService 的定义如下：

```
@Service
public class UserServiceImpl implements UserService{
    @Autowired
    private CreditService creditService;
    @Autowired
    UserRepository userRepository;
    @Override
    public int getCredit(Integer uid){
        User user = userRepository.getOne(uid);
        if(user != null)
```

```
            return creditService.getCredit(uid);
        else
            return -1;
    }
}
```

那么，如何测试 UserService 呢？问题是单元测试不能实际调用 CreditService（因为 CreditService 是第三方系统），因此，在单元测试类需要使用 Mockito 的注解@MockBean 自动注入 Spring 管理的 Service，用来提供模拟实现，在 Spring 上下文中，CreditService 实现已经被模拟实现代替了。UserService 测试类的代码模板如下：

```
import org.mockito.BDDMockito;
import org.springframework.boot.test.mock.mockito.MockBean;
@SpringBootTest
@Transactional
public class UserServiceTest {
    //注入要测试的 service
    @Autowired
    private UserService userService;
    @MockBean
    private CreditService creditService;
    @Test
    public void testUserService() {
        int uid = 1;
        int expectedCredit = 50;
        /*given 是 BDDMockito 的一个静态方法，用来模拟一个 Service 方法调用返回，
anyInt()表示可以传入任何参数，willReturn 方法说明这个调用将返回 50*/
        BDDMockito.given(creditService.getCredit(anyInt())).willReturn(expectedCredit);
        int credit = userService.getCredit(uid);
        /*assert 定义测试的条件，expectedCredit 与 credit 相等时，assertEquals 方
        法保持沉默，不相等时抛出异常*/
        assertEquals(expectedCredit, credit);
    }
}
```

▶ 12.2.3　测试 Controller

在 Spring Boot 应用中，可以单独测试 Controller 代码，用来验证与 Controller 相关的 URL 路径映射、文件上传、参数绑定、参数校验等特性。可以通过@WebMvcTest 注解来完成 Controller 单元测试，当然也可以通过@SpringBootTest 测试 Controller。通过@WebMvcTest 注解测试 Controller 的代码模板如下：

```
import org.mockito.BDDMockito;
import org.springframework.boot.test.mock.mockito.MockBean;
//被测试的 Controller
@WebMvcTest(UserController.class)
public class UserControllerTest{
    //MockMvc 是 Spring 提供的专用于测试 Controller 类的功能
    @Autowired
    private MockMvc mvc;
```

```
    /*用@MockBean 模拟实现 UserService，这是因为在测试 Controller 时，Spring 容器并
      不会初始化@Service 注解的 Service 类*/
    @MockBean
    private UserService userService;
    @Test
    public void testMvc(){
        int uid = 1;
        int expectedCredit = 50;
        /*given 是 BDDMockito 的一个静态方法，用来模拟一个 Service 方法调用返回。这里
        模拟 userService */
        BDDMockito.given(userService.getCredit(uid)).willReturn(50);
        /*perform 完成一次 Controller 的调用，Controller 测试是一种模拟测试，实际上
        并未发起一次真正的HTTP请求;get方法模拟了一次Get请求,请求地址为/getCredit/{id},
        这里的{id}被其后的参数 uid 代替，因此请求路径是/getCredit/1；andExpect 表示
        请求期望的返回结果。*/
        mvc.perform(get("/getCredit/{id}", uid))
            .andExpect(content().string(String.valueOf(expectedCredit)));
    }
}
```

需要注意的是，在使用@WebMvcTest 注解测试 Controller 时，带有@Service 以及别的注解组件类不会自动被扫描注册为 Spring 容器管理的 Bean，而@SpringBootTest 注解告诉 Spring Boot 去寻找一个主配置类（一个带@SpringBootApplication 的类），并使用它来启动 Spring 应用程序上下文，注入所有 Bean。另外，还需要注意的是，MockMvc 用来在 Servlet 容器内对 Controller 进行单元测试，并未真正发起 HTTP 请求调用 Controller。

@WebMvcTest 用于从服务器端对 Controller 层进行统一测试；如果需要从客户端与应用程序交互，应该使用@SpringBootTest 做集成测试。

12.2.4 模拟 Controller 请求

MockMvc 的核心方法是：

```
public ResultActions perform(RequestBuilder requestBuilder)
```

RequestBuilder 类可以通过调用 MockMvcRequestBuilders 的 get、post、multipart 等方法模拟 Controller 请求。常用示例如下。

模拟一个 get 请求：

```
mvc.perform(get("/getCredit/{id}", uid));
```

模拟一个 post 请求：

```
mvc.perform(post("/getCredit/{id}", uid));
```

模拟文件上传：

```
mvc.perform(multipart("/upload").file("file", "文件内容".getBytes("UTF-8")));
```

模拟请求参数：

```
//模拟提交 errorMessage 参数
mvc.perform(get("/getCredit/{id}/{uname}", uid, uname).param("errorMessage",
"用户名或密码错误"));
```

```
//模拟提交 check
mvc.perform(get("/getCredit/{id}/{uname}", uid, uname).param("job", "收银员",
"IT" ));
```

12.2.5　比较 Controller 请求返回的结果

我们知道，MockMvc 的 perform 方法返回 ResultActions 实例，这个实例代表了请求 Controller 返回的结果。它提供了一系列 andExpect 方法对请求 Controller 返回的结果进行比较。示例如下：

```
mvc.perform(get("/getOneUser/10"))
     .andExpect(status().isOk())  //期望请求成功，即状态码为 200
     //期望返回内容是 application/json
     .andExpect(content().contentType(MediaType.APPLICATION_JSON))
     //使用 JsonPath 比较返回的 JSON 内容
     .andExpect(jsonPath("$.name").value("chenheng")); //检查返回内容
```

除了上述对请求 Controller 返回的结果进行比较，还有如下的常见结果比较。

❶ 比较返回的视图

```
mvc.perform(get("/getOneUser/10"))
     .andExpect(view().name("/userDetail"));
```

❷ 比较模型

```
mvc.perform(post("/addOneUser"))
     .andExpect(status().isOk())
     .andExpect(model().size(1))
     .andExpect(model().attributeExists("oneUser"))
     .andExpect(model().attribute("oneUser", "chenheng"))
```

❸ 比较转发或重定向

```
mvc.perform(post("/addOneUser"))
     .andExpect(forwardedUrl("/user/selectAll")); //或者 redirectedUrl("/user/selectAll")
```

❹ 比较返回的内容

```
andExpect(content().string("测试很好玩")); //比较返回的字符串
andExpect(content().xml(xmlContent)); //返回内容是 XML，并且与 xmlContent（变量）一样
andExpect(content().json(jsonContent));//返回内容是 JSON，并且与 jsonContent（变量）一样
```

12.2.6　测试实例

本节将演示一个简单的测试实例，分别使用@WebMvcTest 和@SpringBootTest 两种方式测试某一个控制器方法是否满足测试用例。

【例 12-2】使用@WebMvcTest 和@SpringBootTest 两种方式测试某一个控制器方法。

具体实现步骤如下。

❶ 创建基于 Spring Data JPA 依赖的 Web 应用 ch12_2

创建基于 Spring Data JPA 依赖的 Web 应用 ch12_2。

❷ 修改 pom.xml 文件，引入 MySQL 依赖

修改 pom.xml 文件，引入 MySQL 依赖，具体代码如下：

```
<dependency>
    <groupId>mysql</groupId>
    <artifactId>mysql-connector-java</artifactId>
    <version>5.1.45</version>
</dependency>
```

❸ 配置数据库连接等基本属性

修改配置文件 application.properties 的内容，配置数据库连接等基本属性，具体内容如下：

```
server.servlet.context-path=/ch12_2
#数据库地址
spring.datasource.url=jdbc:mysql://localhost:3306/springbootjpa?characterEncoding=utf8
#数据库用户名
spring.datasource.username=root
#数据库密码
spring.datasource.password=root
#数据库驱动
spring.datasource.driver-class-name=com.mysql.jdbc.Driver
####
#JPA 持久化配置
####
#指定数据库类型
spring.jpa.database=MYSQL
#指定是否在日志中显示 SQL 语句
spring.jpa.show-sql=true
#指定自动创建、更新数据库表等配置，update 表示如果数据库中存在持久化类对应的表就不创建，
#不存在就创建对应的表
spring.jpa.hibernate.ddl-auto=update
#让控制器输出的 JSON 字符串格式更美观
spring.jackson.serialization.indent-output=true
```

❹ 创建持久化实体类

创建名为 com.ch.ch12_2.entity 的包，并在该包中创建名为 Student 的持久化实体类，核心代码如下：

```
@Entity
@Table(name = "student_table")
/**解决 No serializer found for class org.hibernate.proxy.pojo.bytebuddy.ByteBuddyInterceptor 异常*/
@JsonIgnoreProperties(value = {"hibernateLazyInitializer"})
public class Student implements Serializable{
    private static final long serialVersionUID = 1L;
    @Id
    @GeneratedValue(strategy = GenerationType.IDENTITY)
    private int id;//主键
    private String sno;
    private String sname;
    private String ssex;
    public Student() {
        super();
```

```
    }
    public Student(int id, String sno, String sname, String ssex) {
        super();
        this.id = id;
        this.sno = sno;
        this.sname = sname;
        this.ssex = ssex;
    }
    //省略 get 方法和 set 方法
}
```

❺ 创建数据访问层

创建名为 com.ch.ch12_2.repository 的包，并在该包中创建数据访问接口 StudentRepository，具体代码如下：

```
public interface StudentRepository extends JpaRepository<Student, Integer>{}
```

❻ 创建控制器层

创建名为 com.ch.ch12_2.controller 的包，并在该包中创建控制器类 StudentController，核心代码如下：

```
@RestController
@RequestMapping("/student")
public class StudentController {
    @Autowired
    private StudentRepository studentRepository;
    /**
     * 保存学生信息
     */
    @PostMapping("/save")
    public String save(@RequestBody Student student) {
        studentRepository.save(student);
        return "success";
    }
    /**
     * 根据 id 查询学生信息
     */
    @GetMapping("/getOne/{id}")
    public Student getOne(@PathVariable("id") int id){
        return studentRepository.getOne(id);
    }
}
```

❼ 创建测试用例

分别使用@WebMvcTest 和@SpringBootTest 两种方式测试控制器类 StudentController 中的请求处理方法。

1）创建基于@WebMvcTest 的测试用例

使用@WebMvcTest 注解测试 Controller 时，带有@Service 以及别的注解组件类不会自动被扫描注册为 Spring 容器管理的 Bean。因此，Controller 所依赖的对象必须使用@MockBean 来模拟实现。

在src/test/java目录下的com.ch.ch12_2包中，创建基于@WebMvcTest的测试用例类WebMvcTestStudentController，代码如下：

```
package com.ch.ch12_2;
import static org.springframework.test.web.servlet.request.MockMvcRequestBuilders.get;
import static org.springframework.test.web.servlet.request.MockMvcRequestBuilders.post;
import static org.springframework.test.web.servlet.result.MockMvcResultHandlers.print;
import static org.springframework.test.web.servlet.result.MockMvcResultMatchers.jsonPath;
import static org.springframework.test.web.servlet.result.MockMvcResultMatchers.status;
import org.junit.jupiter.api.Test;
import org.mockito.BDDMockito;
import org.springframework.beans.factory.annotation.Autowired;
import org.springframework.boot.test.autoconfigure.web.servlet.WebMvcTest;
import org.springframework.boot.test.mock.mockito.MockBean;
import org.springframework.http.MediaType;
import org.springframework.test.web.servlet.MockMvc;
import com.ch.ch12_2.controller.StudentController;
import com.ch.ch12_2.entity.Student;
import com.ch.ch12_2.repository.StudentRepository;
import com.fasterxml.jackson.databind.ObjectMapper;
/*仅仅扫描这个StudentController类，即注入StudentController到Spring容器*/
@WebMvcTest(StudentController.class)
public class WebMvcTestStudentController {
    //MockMvc是Spring提供的专用于测试Controller类的功能
    @Autowired
    private MockMvc mvc;
    //因为在StudentController类中依赖StudentRepository,所以需要模拟实现StudentRepository
    @MockBean
    private StudentRepository studentRepository;
    @SuppressWarnings("deprecation")
    @Test
    public void saveTest() throws Exception {
        Student stu = new Student(1,"5555","陈恒","男");
        ObjectMapper mapper = new ObjectMapper();//把对象转换为JSON字符串
        mvc.perform(post("/student/save")
                .contentType(MediaType.APPLICATION_JSON_UTF8)//发送JSON数据格式
                .accept(MediaType.APPLICATION_JSON_UTF8)//接收JSON数据格式
                .content(mapper.writeValueAsString(stu))//传递JSON字符串参数
                )
        .andExpect(status().isOk())//状态响应码为200，如果不是抛出异常，测试不通过
        .andDo(print());//输出结果
    }
    @SuppressWarnings("deprecation")
    @Test
    public void getStudent() throws Exception {
        Student stu = new Student(1,"5555","陈恒","男");
        //模拟StudentRepository，getOne(1)将返回stu对象
        BDDMockito.given(studentRepository.getOne(1)).willReturn(stu);
        mvc.perform(get("/student/getOne/{id}", 1)
                .contentType(MediaType.APPLICATION_JSON_UTF8)
                .accept(MediaType.APPLICATION_JSON_UTF8)
                )
```

```
                    .andExpect(status().isOk())//状态响应码为200，如果不是抛出异常，测试不通过
                    .andExpect(jsonPath("$.sname").value("陈恒"))
                    .andDo(print());//输出结果
        }
    }
```

2）创建基于@SpringBootTest 的测试用例

@SpringBootTest 注解告诉 Spring Boot 去寻找一个主配置类（一个带@SpringBootApplication 的类），并使用它启动 Spring 应用程序的上下文，同时注入所有 Bean。

在 src/test/java 目录下的 com.ch.ch12_2 包中，创建基于@SpringBootTest 的测试用例类 SpringBootTestStudentController，代码如下：

```
package com.ch.ch12_2;
import static org.springframework.test.web.servlet.request.MockMvcRequestBuilders.get;
import static org.springframework.test.web.servlet.request.MockMvcRequestBuilders.post;
import static org.springframework.test.web.servlet.result.MockMvcResultHandlers.print;
import static org.springframework.test.web.servlet.result.MockMvcResultMatchers.jsonPath;
import static org.springframework.test.web.servlet.result.MockMvcResultMatchers.status;
import org.junit.jupiter.api.BeforeEach;
import org.junit.jupiter.api.Test;
import org.springframework.beans.factory.annotation.Autowired;
import org.springframework.boot.test.context.SpringBootTest;
import org.springframework.http.MediaType;
import org.springframework.test.web.servlet.MockMvc;
import org.springframework.test.web.servlet.setup.MockMvcBuilders;
import org.springframework.transaction.annotation.Transactional;
import org.springframework.web.context.WebApplicationContext;
import com.ch.ch12_2.entity.Student;
import com.fasterxml.jackson.databind.ObjectMapper;
@SpringBootTest(classes = Ch122Application.class)//应用的主程序
public class SpringBootTestStudentController {
    //注入 Spring 容器
    @Autowired
    private WebApplicationContext wac;
    //MockMvc 模拟实现对 Controller 的请求
    private MockMvc mvc;
    //在测试前，初始化 MockMvc 对象
    @BeforeEach
    public void initMockMvc() {
        mvc = MockMvcBuilders.webAppContextSetup(wac).build();
    }
    @SuppressWarnings("deprecation")
    @Test
    @Transactional
    public void saveTest() throws Exception {
        Student stu = new Student(1, "5555","陈恒","男");
        ObjectMapper mapper = new ObjectMapper();//把对象转换为 JSON 字符串
        mvc.perform(post("/student/save")
                .contentType(MediaType.APPLICATION_JSON_UTF8)//发送 JSON 数据格式
                .accept(MediaType.APPLICATION_JSON_UTF8)//接收 JSON 数据格式
                .content(mapper.writeValueAsString(stu))//传递 JSON 字符串参数
                )
        .andExpect(status().isOk())//状态响应码为 200，如果不是抛出异常，测试不通过
        .andDo(print());//输出结果
    }
```

```
    @SuppressWarnings("deprecation")
    @Test
    public void getStudent() throws Exception {
        mvc.perform(get("/student/getOne/{id}", 1)
                .contentType(MediaType.APPLICATION_JSON_UTF8)
                .accept(MediaType.APPLICATION_JSON_UTF8)
                )
                .andExpect(status().isOk())//状态响应码为200，如果不是抛出异常，测试不通过
                .andExpect(jsonPath("$.sname").value("陈恒"))
                .andDo(print());//输出结果
    }
}
```

❽ 运行

打开 WebMvcTestStudentController 测试类，右击鼠标，选择 Run As、Junit Test 命令，执行结果如图 12.3 所示。

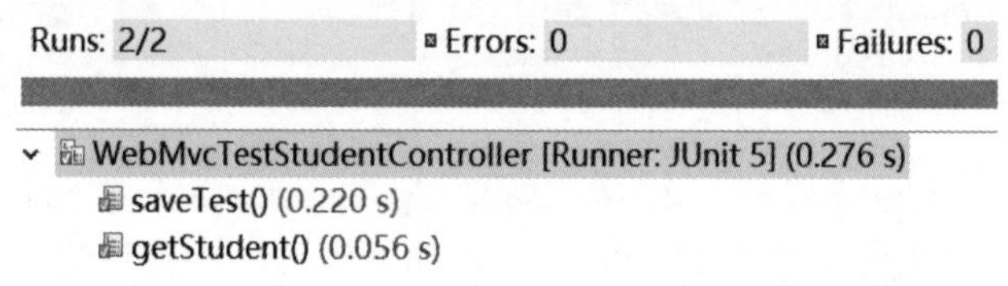

图 12.3　WebMvcTestStudentController 测试类的运行结果

打开 SpringBootTestStudentController 测试类，右击鼠标，选择 Run As、Junit Test 命令，执行结果如图 12.4 所示。

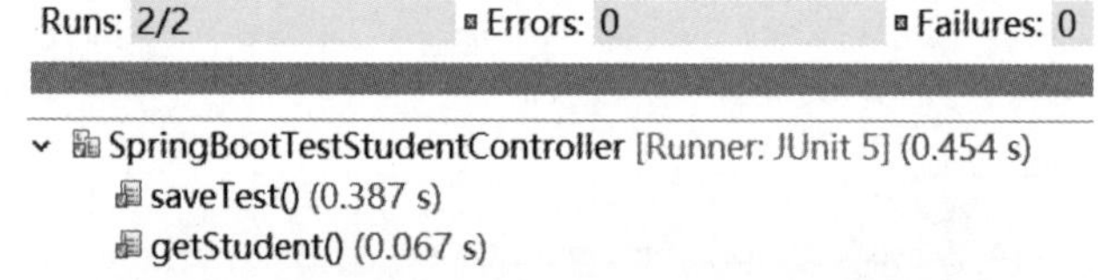

图 12.4　SpringBootTestStudentController 测试类的运行结果

从图 12.3 和图 12.4 可以看出两个测试用例达到测试预期效果。

12.3　本章小结

本章主要介绍了 Spring Boot 应用开发的热部署以及 Spring Boot 的单元测试。通过本章的学习，读者应该了解热部署的目的，掌握测试用例类的编写及原理。

习题 12

1. @SpringBootTest 和@WebMvcTest 的区别是什么？
2. 什么是热部署？在 Spring Boot 中如何进行热部署？

第 13 章 Spring Boot 应用的监控

学习目的与要求

本章主要讲解如何使用 Spring Boot 的 Actuator 功能完成 Spring Boot 应用的监控和管理。通过本章的学习，掌握如何通过 HTTP 进行 Spring Boot 的应用监控和管理。

主要内容

- ❖ 端点的分类与测试
- ❖ 自定义端点
- ❖ 自定义 HealthIndicator

Spring Boot 提供了 Actuator 功能，实现运行时的应用监控和管理功能。可以通过 HTTP、JMX（Java Management Extensions，Java 管理扩展）以及 SSH（远程脚本）进行 Spring Boot 的应用监控和管理。本章将学习如何通过 HTTP 进行 Spring Boot 的应用监控和管理。

在 Spring Boot 应用中，既然通过 HTTP 使用 Actuator 的监控和管理功能，那么在 pom.xml 文件中，除了引入 spring-boot-starter-web 之外，还需要引入 spring-boot-starter-actuator，具体代码如下：

```
<dependency>
    <groupId>org.springframework.boot</groupId>
    <artifactId>spring-boot-starter-actuator</artifactId>
</dependency>
```

13.1 端点的分类与测试

视频讲解

Spring Boot 提供了许多监控和管理功能的端点。根据端点的作用，可以将 Spring Boot 提供的原生端点分为三大类：应用配置端点、度量指标端点和操作控制端点。

13.1.1 端点的开启与暴露

在讲解端点的具体分类以及功能前，先通过实例查看 Spring Boot 默认暴露的端点。

【例 13-1】查看 Spring Boot 默认暴露的端点。

具体实现步骤如下。

❶ 创建基于 Spring Boot Actuator 依赖的 Web 应用 ch13_1

创建基于 Spring Boot Actuator 依赖的 Web 应用 ch13_1。

❷ 配置 JSON 输出格式

在 Web 应用 ch13_1 的配置文件 application.properties 中，配置 JSON 字符串的输出格式，具体如下：

```
#输出的 JSON 字符串格式更美观
spring.jackson.serialization.indent-output=true
```

❸ 启动主程序查看默认暴露的端点

启动 Web 应用 ch13_1 的主程序 Ch131Application 后，通过访问 http://localhost:8080/actuator 查看默认暴露的端点，运行效果如图 13.1 所示。

```
{
  "_links" : {
    "self" : {
      "href" : "http://localhost:8080/actuator",
      "templated" : false
    },
    "health" : {
      "href" : "http://localhost:8080/actuator/health",
      "templated" : false
    },
    "health-path" : {
      "href" : "http://localhost:8080/actuator/health/{*path}",
      "templated" : true
    },
    "info" : {
      "href" : "http://localhost:8080/actuator/info",
      "templated" : false
    }
  }
}
```

图 13.1 查看默认暴露的端点

从图 13.1 可以看出 Spring Boot 默认暴露了 health 和 info 两个端点。如果想暴露 Spring Boot 提供的所有端点，需要在配置文件 application.properties 中配置“management.endpoints.web.exposure.include=*”，配置后重启应用主程序，重新访问 http://localhost:8080/actuator 即可查看所有暴露的端点。

默认情况下，除了 shutdown 端点是关闭的，其他端点都是开启的。配置一个端点的开启，使用 management.endpoint..enabled 属性，如启用 shutdown 端点：

```
management.endpoint.shutdown.enabled=true
```

在配置文件中可使用 management.endpoints.web.exposure.include 属性列出暴露的端点，示例如下：

```
management.endpoints.web.exposure.include=info,health,env,beans
```

“*”可用来表示所有的端点，例如，除了 env 和 beans 端点外，通过 HTTP 暴露所有端点，示例如下：

```
management.endpoints.web.exposure.include=*
management.endpoints.web.exposure.exclude=env,beans
```

13.1.2 应用配置端点的测试

Spring Boot 采用了包扫描和自动化配置的机制来加载原本集中于 XML 文件中的各项配置内容，虽然这让代码变得非常简洁，但是整个应用的实例创建和依赖关系等信息都被离散到了各个配置类的注解上，使得分析整个应用中资源和实例的各种关系变得非常困难。而通过应用配置端点就可以帮助我们轻松地获取一系列关于 Spring 应用配置内容的详细报告，例如自动化配置的报告、Bean 创建的报告、环境属性的报告等。

❶ conditions

该端点在 1.x 版本中名为 autoconfig，该端点用来获取应用的自动化配置报告，其中包括所有自动化配置的候选项。同时还列出了每个候选项自动化配置的各个先决条件是否满足。所以，该端点可以帮助我们方便地找到一些自动化配置为什么没有生效的具体原因。该报告内容将自动化配置内容分为三部分：positiveMatches 中返回的是条件匹配成功的自动化配置；negativeMatches 中返回的是条件匹配不成功的自动化配置；unconditionalClasses 无条件配置类。启动并暴露该端点后，可通过 http://localhost:8080/actuator/conditions 测试访问。

❷ beans

该端点用来获取应用上下文中创建的所有 Bean，启动并暴露该端点后，可通过 http://localhost:8080/actuator/beans 测试访问。

每个 Bean 中都包含以下几个信息：外层的 key 是 Bean 的名称；aliases 是 Bean 的别名；scope 是 Bean 的作用域；type 是 Bean 的 Java 类型；resource 是 class 文件的具体路径；dependencies 是依赖的 Bean 名称。

❸ configprops

该端点用来获取应用中配置的属性信息报告，prefix 属性代表了属性的配置前缀，properties 代表了各个属性的名称和值，例如可以设置 spring.http.encoding.charset="UTF-8"。启动并暴露该端点后，可通过 http://localhost:8080/actuator/configprops 测试访问。

❹ env

该端点与 configprops 端点不同，它用来获取应用所有可用的环境属性报告，包括环境变量、JVM 属性、应用的配置、命令行中的参数等内容。启动并暴露该端点后，可通过 http://localhost:8080/actuator/env 测试访问。

❺ mappings

该端点用来返回所有 Spring MVC 的控制器映射关系报告。启动并暴露该端点后，可通过 http://localhost:8080/actuator/mappings 测试访问。

❻ info

该端点用来返回一些应用自定义的信息。默认情况下，该端点只会返回一个空的 json 内容。我们可以在 application.properties 配置文件中通过 info 前缀来设置一些属性。例如：

```
info.app.name=spring-boot-hello
info.app.version=v1.0.0
```

启动并暴露该端点后，可通过 http://localhost:8080/actuator/info 测试访问。

13.1.3 度量指标端点的测试

通过度量指标端点可获取应用程序运行过程中用于监控的度量指标，例如内存信息、线程信息、HTTP 请求统计等。

❶ metrics

该端点用来返回当前应用的各类重要度量指标，例如内存信息、线程信息、垃圾回收信息等。启动并暴露该端点后，可通过 http://localhost:8080/actuator/metrics 测试访问。

metrics 端点可以提供应用运行状态的完整度量指标报告，这项功能非常的实用，但是对于监控系统中的各项监控功能，它们的监控内容、数据收集频率都有所不同，如果我们每次都通过全量获取报告的方式来收集，略显粗暴。所以，还可以通过/metrics/{name}接口来更细粒度的获取度量信息，比如可以通过访问/metrics/jvm.memory.used 来获取当前 JVM 使用的内存数量。

❷ health

该端点用来获取应用的各类健康指标信息。在 spring-boot-starter-actuator 模块中自带实现了一些常用资源的健康指标检测器。这些检测器都是通过 HealthIndicator 接口实现，并且根据依赖关系的引入实现自动化装配，比如用于检测磁盘的 DiskSpaceHealthIndicator、检测 DataSource 连接是否可用的 DataSourceHealthIndicator 等。

有时我们可能还会用到一些 Spring Boot 的 Starter POMs 中还没有封装的产品来进行开发，例如：当使用 RocketMQ 作为消息代理时，由于没有自动化配置的检测器，所以需要自己实现一个用来采集健康信息的检测器。

启动并暴露该端点后，可通过 http://localhost:8080/actuator/health 测试访问。

结果中的“UP”表示健康，“DOWN”表示异常。

可以在配置文件中，配置属性 management.endpoint.health.show-details=always，将详细健康信息显示给所有用户。再次启动应用后，刷新 http://localhost:8080/actuator/health，显示健康指标详细信息。

❸ threaddump

该端点用来暴露程序运行中的线程信息。它使用 java.lang.management.ThreadMXBean 的 dumpAllThreads 方法来返回所有含有同步信息的活动线程详情。启动并暴露该端点后，可通过 http://localhost:8080/actuator/threaddump 测试访问。

❹ scheduledtasks

该端点统计应用程序中调度的任务。启动并暴露该端点后，可通过 http://localhost:8080/actuator/scheduledtasks 测试访问。

▶ 13.1.4 操作控制端点的测试

操作控制类端点拥有更强大的控制能力，如果使用它们，需要通过属性来配置开启。在原生端点中，只提供了一个用来关闭应用的端点：shutdown。我们可以通过如下配置开启它：management.endpoint.shutdown.enabled=true，在配置了上述属性之后，只需要访问该应用的 shutdown 端点就能实现关闭该应用的远程操作。由于开放关闭应用的操作本身是一件非常危险的事，所以真正在线上使用的时候，需要对其加入一定的保护机制，例如定制 Actuator 的端点路径、整合 Spring Security 进行安全校验等。

shutdown 端点不支持 get 提交，不可以直接在浏览器上访问，所以我们这里可以使用 Google Chrome 的 Postman REST Client 测试。用 post 方式访问 http://localhost:8080/actuator/shutdown，测试效果如图 13.2 所示。

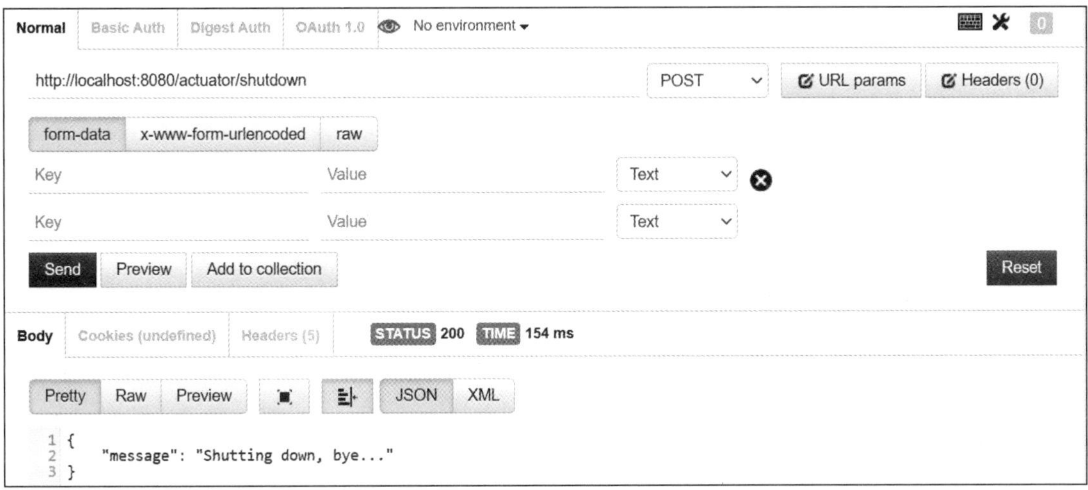

图 13.2　测试 shutdown 端点

13.2　自定义端点

视频讲解

当 Spring Boot 提供的端点不能满足我们的需求时，就需要自定义一个端点对应用进行监控。

Spring Boot 提供了注解@Endpoint 定义一个端点类，并在端点类的方法上使用@ReadOperation 注解来显示监控信息（对应 Get 请求），使用@WriteOperation 动态更新监控信息（对应 Post 请求）。

下面通过一个实例讲解如何自定义一个端点。该实例演示显示数据源的相关信息。

【例 13-2】自定义端点。

具体实现步骤如下。

❶ 创建基于 Spring Data JPA 和 Spring Boot Actuator 依赖的 Web 应用 ch13_2

创建基于 Spring Data JPA 和 Spring Boot Actuator 依赖的 Web 应用 ch13_2。

❷ 修改 pom.xml 文件添加 MySQL 连接器依赖

修改 pom.xml 文件添加 MySQL 连接器依赖。

❸ 配置数据源

因为该实例是监控数据源信息，所以需要在配置文件 application.properties 中配置数据源，具体内容如下：

```
#数据库地址
spring.datasource.url=jdbc:mysql://localhost:3306/springbootjpa?characterEncoding=utf8
#数据库用户名
spring.datasource.username=root
#数据库密码
spring.datasource.password=root
#数据库驱动
spring.datasource.driver-class-name=com.mysql.jdbc.Driver
#输出的 JSON 字符串格式更美观
spring.jackson.serialization.indent-output=true
```

❹ 自定义端点

创建名为com.ch.ch13_2.endPoint的包，并在该包中使用注解@Endpoint自定义端点类DataSourceEndpoint。在该端点类中，使用@ReadOperation注解显示数据源信息，使用@WriteOperation动态更新数据源信息。核心代码如下：

```
//注册为端点，id不能使用驼峰法（dataSource），需要以-分隔
@Endpoint(id = "data-source")
@Component
public class DataSourceEndpoint {
    //HikariDataSource提供多个监控信息
    HikariDataSource ds;
    public DataSourceEndpoint(HikariDataSource ds) {
        this.ds = ds;
    }
    @ReadOperation
    public Map<String, Object> info() {
        Map<String, Object> map = new HashMap<String, Object>();
        //连接池配置
        HikariConfigMXBean configBean = ds.getHikariConfigMXBean();
        map.put("max", configBean.getMaximumPoolSize());
        //连接池运行状态
        HikariPoolMXBean mxBean = ds.getHikariPoolMXBean();
        map.put("active", mxBean.getActiveConnections());
        map.put("idle", mxBean.getIdleConnections());
        //连接池无连接时，等待获取连接的线程个数
        map.put("wait", mxBean.getThreadsAwaitingConnection());
        return map;
    }
    @WriteOperation
    public void setMax(int max) {
        ds.getHikariConfigMXBean().setMaximumPoolSize(max);
    }
}
```

❺ 暴露端点

在配置文件application.properties中暴露端点，内容如下：

```
#暴露所有端点，包括data-source，也可以只暴露data-source端点
management.endpoints.web.exposure.include=*
```

❻ 测试端点

首先，启动应用程序主类Ch132Application，然后通过http://localhost:8080/actuator/data-source测试端点data-source，运行效果如图13.3所示。

使用Google Chrome的Postman REST Client发送POST请求http://localhost:8080/actuator/data-source?max=20，POST请求执行后，http://localhost:8080/actuator/data-source测试端点data-source，运行效果如图13.4所示。

从图13.4可以看出，通过POST请求调用端点的setMax方法修改了数据源信息。

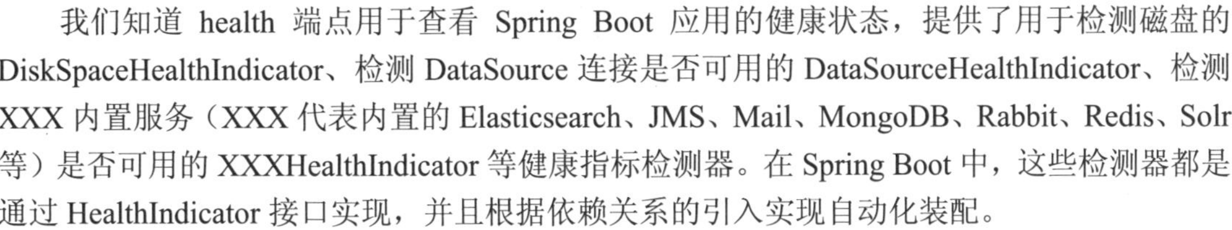

```
{
  "wait" : 0,
  "max" : 10,
  "idle" : 10,
  "active" : 0
}
```

图 13.3　测试端点 data-source

```
{
  "wait" : 0,
  "max" : 20,
  "idle" : 10,
  "active" : 0
}
```

图 13.4　刷新后的效果

13.3　自定义 HealthIndicator

视频讲解

我们知道 health 端点用于查看 Spring Boot 应用的健康状态，提供了用于检测磁盘的 DiskSpaceHealthIndicator、检测 DataSource 连接是否可用的 DataSourceHealthIndicator、检测 XXX 内置服务（XXX 代表内置的 Elasticsearch、JMS、Mail、MongoDB、Rabbit、Redis、Solr 等）是否可用的 XXXHealthIndicator 等健康指标检测器。在 Spring Boot 中，这些检测器都是通过 HealthIndicator 接口实现，并且根据依赖关系的引入实现自动化装配。

当 Spring Boot 自带的 HealthIndicator 接口实现类不能满足我们的需求时，就需要自定义 HealthIndicator 接口实现类。自定义 HealthIndicator 接口实现类很简单，只需要实现 HealthIndicator 接口，并重写接口方法 health，返回一个 Health 对象。

下面通过一个实例讲解如何自定义一个 HealthIndicator 接口实现类。

【例 13-3】自定义 HealthIndicator。

具体实现步骤如下。

❶ 创建 HealthIndicator 接口实现类 MyHealthIndicator

在应用 ch13_3 中，创建名为 com.ch.ch13_3.health 的包，并在该包中创建一个 HealthIndicator 接口实现类 MyHealthIndicator。在该类中重写接口方法 health，并使用@Component 注解将该类声明为组件对象。核心代码如下：

```
@Component
public class MyHealthIndicator implements HealthIndicator{
    @Override
    public Health health() {
        int errorCode = check();
        if(errorCode != 0) {
            //down 方法表示异常，withDetail 方法添加任意多的异常信息
            return Health.down().withDetail("message", "error:"+errorCode).build();
        }
        //up 方法表示健康
        return Health.up().build();
    }
    /**
     * 模拟返回一个错误状态
     */
    private int check() {
        return 1;
    }
}
```

❷ 将健康详细信息显示给所有用户

在配置文件 application.properties 中，配置将详细健康信息显示给所有用户。配置内容如下：

```
#将详细健康信息显示给所有用户
management.endpoint.health.show-details=always
```

❸ 测试运行

health 的对象名默认为类名去掉 HealthIndicator 扩展名，并且首字母小写，因此该例的 health 对象名为 my。

启动应用程序主类 Ch132Application，并通过 http://localhost:8080/actuator/health/my 测试运行，效果如图 13.5 所示。

```
{
  "status" : "DOWN",
  "details" : {
    "message" : "error:1"
  }
}
```

图 13.5　测试自定义 HealthIndicator

13.4　本章小结

本章主要介绍了 Spring Boot 的 Actuator 功能，并通过 HTTP 进行 Spring Boot 的应用监控和管理功能的测试。当 Spring Boot 自带的端点和 HealthIndicator 实现不能满足我们的需要时，可以自定义端点和 HealthIndicator 实现。因此，本章还介绍了如何自定义端点和 HealthIndicator 实现。

习 题 13

1．默认情况下，Spring Boot 暴露了哪几个端点？又如何暴露所有端点？
2．如何自定义端点？有哪几个步骤？
3．如何自定义 HealthIndicator 实现？有哪几个步骤？

第 14 章 Vue 3 基础 

学习目的与要求

本章主要讲解 Vue 3 的基础知识，包括 Vue 3 的安装、插值与表达式、计算属性、指令、生命周期、组件以及自定义指令等内容。通过本章的学习，掌握 Vue 3 的基础知识，为学习 Vue3 进阶知识做准备。

主要内容

- ❖ Vue 3 的安装
- ❖ Vue 3 的生命周期
- ❖ 插值与表达式
- ❖ 计算属性
- ❖ 指令
- ❖ 在 Vue 中动态使用样式
- ❖ 组件
- ❖ 自定义指令

Vue（读音/vju/，类似于 view）是一套构建用户界面的渐进式框架。与重量级框架不同的是，Vue 采用自底向上增量开发的设计。Vue 的核心库只关注视图层，它不仅易于上手，还便于与第三方库或既有项目整合。

Vue.js 是构建 Web 界面的 JavaScript 库，提供数据驱动的组件，有简单灵活的 API，使 MVVM（Model-View-ViewModel）更简单。MVVM 模式是由 MVC 衍生而来，当 View 变化时，将自动更新到 ViewModel（视图模型），反之亦然。View 和 ViewModel 之间通过双向绑定（data-binding）建立联系，如图 14.1 所示。

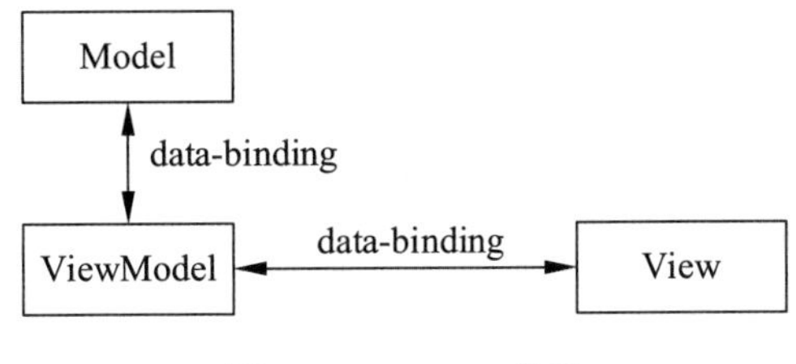

图 14.1 MVVM 关系

Vue.js 可以轻松构建 SPA（Single Web Application）应用程序，通过指令扩展 HTML，通过表达式将数据绑定到 HTML，最大程度解放 DOM 操作。

14.1 安装 Vue 3

将 Vue.js 添加到项目中有 4 种主要方式。

14.1.1 本地独立版本方法

可通过地址 https://unpkg.com/vue@next 将最新版本的 Vue.js 库（vue.global.js）下载到本地，编写本书时，最新版本是 3.0.5。然后，在页面上引入 Vue.js 库，示例代码如下：

```
<script src="js/vue.global.js"></script>
```

14.1.2 CDN 方法

读者在学习或开发时，在页面上可通过 CDN（Content Delivery Network，内容分发网络）引入最新版本的 Vue.js 库，示例代码如下：

```
<script src="https://unpkg.com/vue@next"></script>
```

对于生产环境，建议使用固定版本，以免因版本不同带来兼容性问题，示例代码如下：

```
<script src="https://unpkg.com/vue@3.0.5/dist/vue.global.js"></script>
```

14.1.3 NPM 方法

在用 Vue.js 构建大型应用时推荐使用 NPM 安装最新稳定版的 Vue.js，因为 NPM 能很好地和 webpack 模块打包器配合使用。示例如下：

```
$ npm install vue@next
```

14.1.4 命令行工具（CLI）方法

Vue.js 提供一个官方命令行工具（Vue CLI），为单页面应用快速搭建繁杂的脚手架。对于初学者不建议使用 NPM 和 Vue CLI 方法安装 Vue.js。NPM 和 Vue CLI 方法的安装过程将在本书后续内容中介绍。

视频讲解

14.2 使用 Visual Studio Code 开发第一个 Vue 程序

在前端开发工具的选择上，极少数大神使用记事本，而大多数程序员使用 JetBrains WebStorm 和 Visual Studio Code（VSCode）。JetBrains WebStorm 是收费的，本书推荐使用 VSCode。

14.2.1 安装 Visual Studio Code 及其插件

可通过 https://code.visualstudio.com 地址下载 VSCode，本书使用的安装文件是 VSCodeUserSetup-x64-1.52.1.exe。VSCode 中许多插件需要安装，例如安装 Vue 的插件 Vetur。打开 VSCode，单击左侧最下面一个图标，按照图 14.2 所示的步骤安装即可。

插件的安装方式与图 14.2 类似，不再赘述。在 VSCode 中，有关 Vue 的部分插件具体描述如下。

1）Vetur

此插件能够在.vue 文件中实现语法错误检查、语法高亮以及代码自动补全。

2）ESLint

此插件能够检测代码语法问题与格式问题，对项目代码风格统一至关重要。

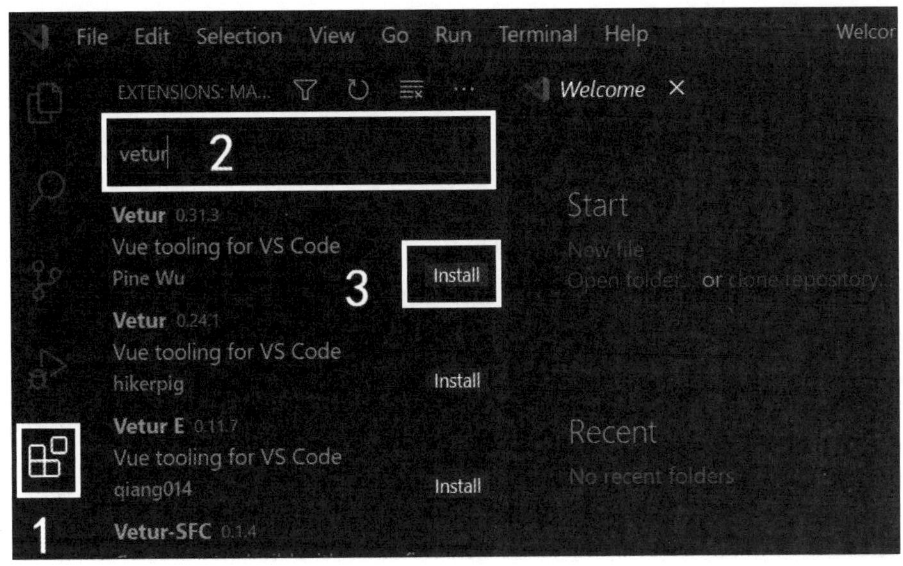

图 14.2 VSCode 的插件安装

3）EditorConfig for Visual Studio Code

EditorConfig 是一种被各种编辑器广泛支持的配置，使用此配置有助于项目在整个团队中保持一致的代码风格。

4）Path Intellisense

此插件能够在编辑器中输入路径时实现自动补全。

5）View In Browser

在 VSCode 中，使用浏览器预览运行静态文件。

6）Live Server

此插件很有用，安装后可以打开一个简单的服务器，而且还会自动更新。安装后，右击文件，出现一个名为 Open with Live Server 的选项，自动打开浏览器，默认端口号是 5500。

7）GitLens

此插件可查看 git 文件提交的历史。

8）Document This

此插件生成注释文档。

9）HTML CSS Support

在编写样式表时，自动补全功能，缩减编写时间。

10）JavaScript Snippet Pack

针对 JS 的插件，包含 JS 的常用语法关键字。

11）HTML Snippets

此插件包含 HTML 标签。

12）One Monokai Theme

此插件能够让编者选择自己喜欢的颜色主题编写代码。

13）vscode-icons

此插件能够让编者选择自己喜欢的图标主题。

14.2.2 创建第一个 Vue 应用

每个 Vue 应用都是通过用 createApp 函数创建一个新实例开始，具体语法如下：

```
const app = Vue.createApp({ /* 选项 */ })
```

传递给 createApp 的选项用于配置根组件（渲染的起点）。Vue 应用创建后，调用 mount 方法将 Vue 应用挂载到一个 DOM 元素（HTML 元素或 CSS 选择器）中，例如，如果把一个 Vue 应用挂载到<div id="app"></div>上，应传递#app。示例代码如下：

```
const HelloVueApp = {}//配置根组件
const vueApp = Vue.createApp(HelloVueApp)//创建 Vue 实例
const vm = vueApp.mount('#app')//将 Vue 实例挂载到#app
```

下面使用 VSCode 开发第一个 Vue 程序。

【例 14-1】使用 VSCode 新建一个名为 hellovue.html 的页面，在此页面中使用“<script src="js/vue.global.js"></script>”语句引入 Vue.js。hellovue.html 的具体代码如下：

```
<div id="hello-vue" class="demo">
    {{ message }}
</div>
<script src="js/vue.global.js"></script>
<script>
    const HelloVueApp = {
        data() {//Vue 实例的数据对象，ES6 语法，等价于 data: function () {}
            return {
                message: 'Hello Vue!!'
            }
        }
    }
    //每个 Vue 应用都是通过用 createApp 函数创建一个新的应用实例开始
    //mount 函数把一个 Vue 应用实例挂载到<div id="hello-vue"></div>
    Vue.createApp(HelloVueApp).mount('#hello-vue')
</script>
<style>
.demo {
    font-family: sans-serif;
}
</style>
```

从上述代码中可以看出，hellovue.html 文件内容由 HTML、JavaScript、CSS 三部分组成，所以读者在学习 Vue.js 之前，应该了解 HTML、JavaScript 以及 CSS 内容。

14.2.3 声明式渲染

Vue.js 的核心是采用简洁的模板将数据渲染到 DOM 中，例如在 14.2.2 节的 hellovue.html 文件中，通过模板<div id="hello-vue" class="demo">{{ message }}</div>声明将属性变量 message 的值“Hello Vue!!”渲染到页面显示。

Vue.js 框架在声明式渲染时，做的主要工作是将数据和 DOM 建立关联，一切皆响应。例如例 14-2 的 counter 属性每秒递增。

【例 14-2】使用 VSCode 新建一个名为 ch14_2.html 的页面，在该页面中使用时钟函数 setInterval 演示响应式程序。ch14_2.html 的具体代码如下：

```
<div id="counter" class="demo">
    <!--通过模板获取变量 counter 的值-->
    {{ counter }}
</div>
<script src="js/vue.global.js"></script>
<script>
    const CounterApp = {
        data() {
            return {
                counter: 0
            }
        },
        //mounted 是一个钩子函数，挂载到实例上后（初始化页面后）调用该函数，一般是第一个
        //业务逻辑在这里开始
        mounted() {
            setInterval(() => {
                this.counter++
            }, 1000)
        }
    }
    Vue.createApp(CounterApp).mount('#counter')
</script>
<style>
    .demo {
        font-family: sans-serif;
    }
</style>
```

▶ 14.2.4 Vue 生命周期

每个 Vue 实例在被创建时都要经过一系列的初始化过程，例如数据监听、编译模板、将实例挂载到 DOM 并在数据变化时更新 DOM 等，同时在这个过程中也会调用一些生命周期钩子的函数，在适当的时机执行我们的业务逻辑。

例如，created 钩子函数可用来在一个 Vue 实例被创建后执行代码（Vue 实例创建后被立即调用即 HTML 加载完成前）：

```
Vue.createApp({
    data() {
        return {
            message: '测试钩子函数'
        }
    },
    created() {
        //this 指向调用它的 Vue 实例
        console.log('message 是： ' + this.message) // "message 是：测试钩子函数"
    }
})
```

Vue 的生命周期共分 8 个阶段（如图 14.3 所示），即对应 8 个与 created 类似的钩子函数。

beforeCreate（创建前）：在 Vue 实例初始化后，数据观测和事件配置前调用，此时 el 和 data 并未初始化，因此无法访问 methods、data、computed 等方法和数据。

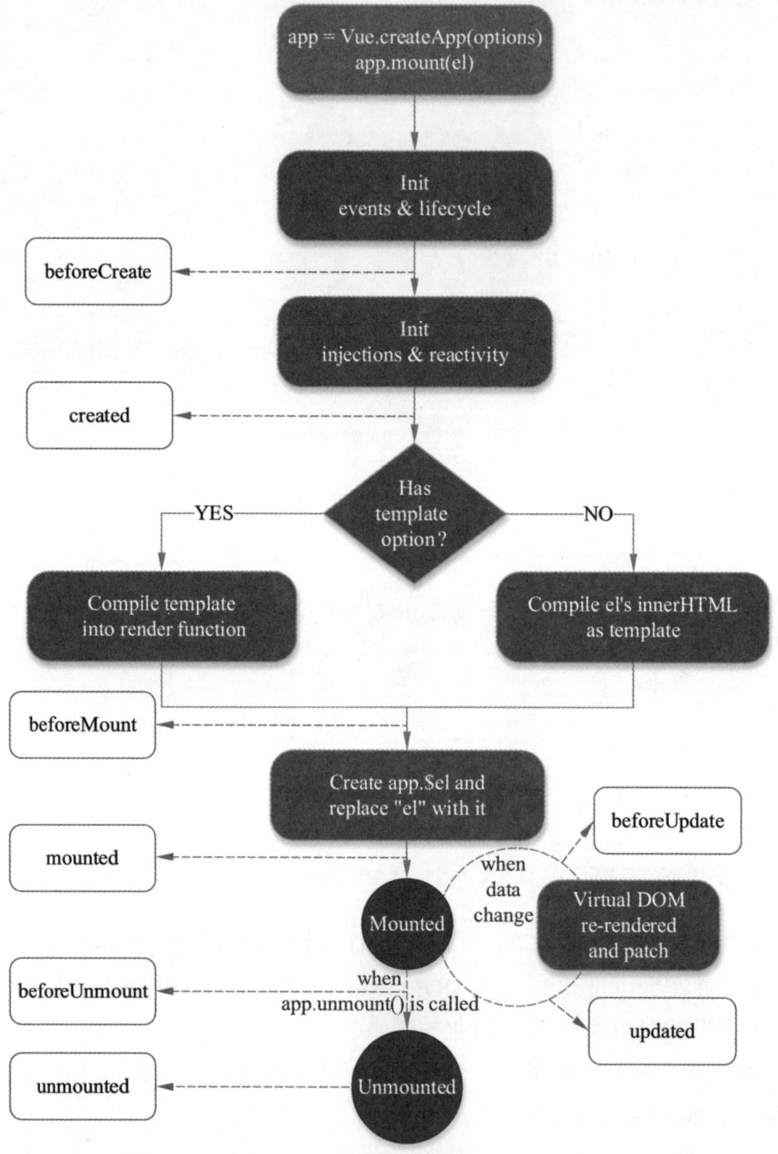

图 14.3　Vue 生命周期

created（创建后）：Vue 实例创建后被立即调用即 HTML 加载完成前。此时，Vue 实例已完成数据观测、属性和方法的运算、watch/event 事件回调、data 数据的初始化。然而，挂载阶段还没有开始，el 属性目前不可见。这是一个常用的生命周期钩子函数，可以调用 methods 中的方法、改变 data 中的数据、获取 computed 中的计算属性等，通常在此钩子函数中对实例进行预处理。

beforeMount(载入前)：挂载开始前被调用，Vue 实例已完成编译模板、把 data 里面的数据和模板生成 HTML、el 和 data 初始化，注意此时还没有挂载 HTML 到页面上。

mounted（载入后）：页面加载后调用该函数，这是一个常用的生命周期钩子函数，一般是第一个业务逻辑在此钩子开始，mounted 只会执行一次。

beforeUpdate（更新前）：在数据更新前被调用，发生在虚拟 DOM 重新渲染和打补丁之前，可以在该钩子中进一步更改状态，不会触发附加的重渲染过程。

updated（更新后）：在由数据更改导致虚拟 DOM 重新渲染和打补丁时调用，调用时，DOM 已经更新，所以可以执行依赖于 DOM 的操作，应该避免在此期间更改状态，这可能会导致更新无限循环。

beforeUnmount（销毁前）：Vue 实例销毁前调用（离开页面前调用），这是一个常用的生命周期钩子函数，一般在此时做一些重置的操作，例如清除定时器和监听的 DOM 事件。

unmounted（销毁后）：在实例销毁后调用，调用后，事件监听器被移出，所有子实例也被销毁。

图 14.3 展示了 Vue 实例的生命周期，现在不需要弄明白所有阶段的钩子函数，随着学习和使用的深入，慢慢理解它们。

14.3 插值与表达式

Vue 的插值表达式“{{ }}”的作用是读取 Vue 中的 data 数据，显示在视图中，数据更新，视图也随之更新。“{{ }}”里只能放表达式（有返回值），不能放语句，例如，{{ var a = 1 }}与{{ if (ok) { return message } }}都是无效的。

14.3.1 文本插值

数据绑定最常见的形式是使用“Mustache（小胡子）”语法（双花括号）的文本插值，它将绑定的数据实时显示出来。例如，例 14-2 中的{{ counter }}，无论何时，绑定的 Vue 实例的 counter 属性值发生改变，插值处的内容都将更新。

可通过使用 v-once 指令，执行一次性插值，即当数据改变时，插值处的内容不会更新。示例代码如下：

```
<span v-once>{{ counter }}</span>
```

14.3.2 原始 HTML 插值

“{{ }}”将数据解释为普通文本，而非 HTML 代码。当需要输出真正的 HTML 代码时，可使用 v-html 指令。动态渲染任意的 HTML 是非常危险的，因为很容易导致 XSS 攻击。最好只对可信内容使用 HTML 插值，绝不可将用户提供的 HTML 作为插值。v-html 指令示例代码如下。

假如 Vue 实例的 data 为：

```
data() {
    return {
        rawHtml: '<hr>'
    }
}
```

则“<p>无法显示 HTML 元素内容: {{ rawHtml }}</p>”显示的结果是<hr>；“<p>可正常显示 HTML 元素内容: <span v-html="rawHtml"></span></p>”显示的结果是一条水平线。

14.3.3 JavaScript 表达式

在前面的学习中，仅用表达式绑定简单的属性值。但实际上，对于所有的数据绑定，Vue.js 都提供了完全的 JavaScript 表达式支持。示例代码如下：

```
{{ number + 1 }}
{{ isLogin? 'True' : 'False' }}
{{ message.split('').reverse().join('')}}
```

视频讲解

14.4 计算属性和监听器

14.4.1 计算属性

使用模板内的表达式计算并显示数据非常便利，但是在模板中放入太多的逻辑会让模板难以维护。

【例 14-3】在模板中嵌套复杂的逻辑。例如，有一个嵌套数组对象：

```
Vue.createApp({
    data() {
        return {
            author: {
                name: '陈恒',
                books: [
                    'Java Web 开发从入门到实战',
                    'Java EE 框架整合开发入门到实战——Spring+Spring MVC+MyBatis',
                    'Spring Boot 从入门到实战',
                    'Spring MVC 开发技术指南（微课版）'
                ]
            }
        }
    }
}).mount('#computedProperty')
```

现在判断陈恒老师有没有写过书，模板代码如下：

```
<div id="computedProperty">
    <p>听说陈恒写书了？</p>
    <span>{{ author.books.length > 0 ? 'Yes' : 'No' }}</span>
</div>
```

此时，表达式不再是简单的声明，计算结果取决于三目运算的结果。在表达式中多次包含计算，问题将变得更糟。所以，对于包含复杂逻辑的表达式，应该使用计算属性。

在计算属性中可以完成各种复杂的逻辑，包括运算、函数调用等，只要最终返回一个结果即可。现在给例 14-3 声明一个计算属性 isPublished，代码如下：

```
Vue.createApp({
    data() {
        return {
            author: {
                name: '陈恒',
                books: [
                    'Java Web 开发从入门到实战',
                    'Java EE 框架整合开发入门到实战——Spring+Spring MVC+MyBatis',
                    'Spring Boot 从入门到实战',
                    'Spring MVC 开发技术指南（微课版）'
                ]
            }
        }
    },
    computed: {
        //计算属性默认调用 get 方法
        isPublished() {
            return this.author.books.length > 0 ? 'Yes' : 'No'
        }
    }
}).mount('#computedProperty')
```

可以像普通属性一样将数据绑定到模板中的计算属性，此时，模板代码修改如下：

```
<div id="computedProperty" class="demo">
    <p>听说陈恒写书了？</p>
    <span>{{ isPublished }}</span>
</div>
```

计算属性默认调用 getter 方法读取数据，不过在需要时也可以提供一个 setter 方法修改数据。例如，计算属性 aname 的定义如下：

```
computed: {
    aname: {
        get(){
            return this.author.name
        },
        set(sname){
            this.author.name = sname
        }
    }
}
```

计算属性和 data 达到的效果是一样的，但是计算属性是基于依赖进行缓存的，多次访问计算属性会立即返回之前的计算结果，而不必再次执行函数，只有计算属性数据发生改变才再次执行函数；而 data 是每次调用都会执行。

▶ 14.4.2 监听器

虽然计算属性在大多数情况下更合适，但有时也需要一个侦听器来响应数据的变化。Vue 通过 watch 选项提供监听数据属性的方法（方法名与属性名相同）来响应数据的变化。示例代码如下：

```
<div id="watch-example">
    <p>
        请问一个问题，包含英文字符?:
        <input v-model="question" />
    </p>
    <p>{{ answer }}</p>
</div>
<script src="js/vue.global.js"></script>
<script src="https://cdn.jsdelivr.net/npm/axios@0.12.0/dist/axios.min.js"></script>
<script>
    const watchExampleVM = Vue.createApp({
        data() {
            return {
                question: '',
                answer: '这是一个好问题。'
            }
        },
        watch: {//watch 选项提供监听数据属性的方法
            //question 方法名与数据属性名 question 一致
           question(newQuestion, oldQuestion) {//newQuestion 是改变后的值,
           oldQuestion 是没改变的值
                if (newQuestion.indexOf('?') > -1) {
                    //包含英文字符?时，执行 getAnswer()方法
                    this.getAnswer()
                }
            }
        },
        methods: {
            getAnswer() {
                this.answer = '让我想一想'//设置中间状态，即答案返回前
                axios
                    .get('https://yesno.wtf/api')//使用 axios 实现 AJAX 异步请求
                    .then(response => {
                        this.answer = response.data.answer
                    })
                    .catch(error => {
                        this.answer = '错误，不能访问 API. ' + error
                    })
            }
        }
    }).mount('#watch-example')
</script>
```

在上述示例中，使用 watch 选项执行异步操作，限制执行该操作的频率，并在得到最终结果前设置中间状态。这些都是计算属性无法做到的。

14.5 指令

视频讲解

Vue.js 指令都带有扩展名“v-”，它绑定一个表达式。Vue.js 内置许多指令，例如前面学习的 v-html、v-once。本节将介绍 Vue.js 中常用的内置指令。

14.5.1 v-bind 与 v-on 指令

❶ v-bind 指令

在 HTML 元素的属性中不能使用表达式动态更新属性值。幸运的是，Vue.js 提供了 v-bind 指令绑定 HTML 元素的属性，并可动态更新属性值。

【例 14-4】v-bind 指令应用示例。示例代码如下：

```
div id="app">
    <a v-bind:href="myurl.baiduUrl">去百度</a>
    <img v-bind:src="myurl.imgUrl"/>
    <!-- v-bind:可缩写为“:” -->
    <a :href="myurl.baiduUrl">去百度</a>
    <img :src="myurl.imgUrl"/>
</div>
<script src="js/vue.global.js"></script>
<script>
    Vue.createApp({
        data() {
            return {
                myurl: {
                    baiduUrl: 'https://www.baidu.com/',
                    imgUrl:'/images/ok.png'
                }
            }
        }
    }).mount('#app')
</script>
```

上述示例代码中，使用 v-bind 指令动态绑定了链接的 href 属性和图片的 src 属性，当数据变化时，href 属性值和 src 属性值也发生变化，即重新渲染。

❷ v-on 指令

可以用 v-on 指令给 HTML 元素添加一个事件监听器，通过该指令调用在 Vue 实例中定义的方法。下面使用 v-on 指令实现字符串反转。

【例 14-5】使用 v-on 指令实现字符串反转。示例代码如下：

```
<div id="event-handling">
    <p>{{ message }}</p>
<button v-on:click="reverseMessage">反转 Message</button>
<!-- v-on:可缩写为“@” -->
<button @click="reverseMessage">反转 Message</button>
</div>
<script src="js/vue.global.js"></script>
<script>
    const EventHandling = {
        data() {
            return {
```

```
                message: 'Hello Vue.js!'
            }
        },
        methods: {//方法定义
            reverseMessage() {
                this.message = this.message.split('').reverse().join('')
            }
        }
    }
    Vue.createApp(EventHandling).mount('#event-handling')
</script>
```

Vue.js 用特殊变量$event 访问原生的 DOM 事件，例如下面的实例阻止打开链接。

【**例 14-6**】阻止打开链接。代码如下：

```
<div id="event-handling">
    <a href="https://www.baidu.com/" @click="warn('考试期间禁止百度! ', $event)">去百度</a>
</div>
<script src="js/vue.global.js"></script>
<script>
    Vue.createApp({
        methods: {
            warn(message, event) {
                //event 访问原生的 DOM 事件
                event.preventDefault()
                alert(message)
            }
        }
    }).mount('#event-handling')
</script>
```

在事件处理中调用 event.preventDefault()或 event.stopPropagation()是非常常见的需求。尽管可以在方法中轻松实现这类需求，但方法最好只有纯粹的数据逻辑，而不是去处理 DOM 事件细节。为解决该问题，Vue.js 为 v-on 提供了事件修饰符。修饰符由以点开头的指令扩展名表示。Vue.js 支持的修饰符有.stop、.prevent、.capture、.self、.once 以及.passive。修饰符的用法是在@绑定的事件后加小圆点（.），再跟修饰符，具体如下：

<!-- 阻止单击事件-->

<a @click.stop="doThis"></a>

<!-- 提交事件不再重载页面 -->

<form @submit.prevent="onSubmit"></form>

<!-- 修饰符可以串联 -->

<a @click.stop.prevent="doThat"></a>

<!-- 只有修饰符 -->

<form @submit.prevent></form>

<!-- 添加事件监听器时使用事件捕获模式，即内部元素触发的事件先在此处理，然后才交由内部元素进行处理 -->

<div @click.capture="doThis">...</div>

<!-- 当事件在该元素自身触发时触发回调，即事件不是从内部元素触发的-->

<div @click.self="doThat">...</div>

```
<!--只触发一次 -->
<a @click.once="doThis"></a>
<!-- 滚动事件的默认行为 (即滚动行为)将会立即触发，而不会等待“onScroll”完成 -->
<div @scroll.passive="onScroll">...</div>
```

14.5.2　条件渲染指令 v-if 和 v-show

❶ v-if 指令

与 JavaScript 的条件语句 if、else、else if 类似，Vue.js 的条件指令也可以根据表达式的值渲染或销毁元素/组件。

【例 14-7】使用条件渲染指令判断成绩等级。代码如下：

```
<div id="event-handling">
    <div v-if="score >= 90">优秀</div>
    <div v-else-if="score >= 80">良好</div>
    <div v-else-if="score >= 70">中等</div>
    <div v-else-if="score >= 60">及格</div>
    <div v-else>不及格</div>
</div>
<script src="js/vue.global.js"></script>
<script>
    Vue.createApp({
        data() {
            return {
                score: 87
            }
        }
    }).mount('#event-handling')
</script>
```

从上述示例代码可以看出，v-else 元素必须紧跟在 v-if 或者 v-else-if 元素后面；v-else-if 元素必须紧跟在 v-if 或者 v-else-if 元素后面。

条件渲染指令必须添加到一个元素上。但是如果想包含多个元素呢？此时可以使用<template>元素（模板占位符）帮助包裹元素，并在上面使用 v-if。最终的渲染结果将不包含<template>元素。示例代码如下：

```
<template v-if="ok">
    <h1>Title</h1>
    <p>Paragraph 1</p>
    <p>Paragraph 2</p>
</template>
```

❷ v-show 指令

v-show 指令的用法基本与 v-if 一样，也是根据条件展示元素，例如“<h1 v-show="yes">一级标题</h1>”。不同的是，v-if 每次都会重新删除或创建元素，而带有 v-show 的元素始终会被渲染并保留在 DOM 中，只是切换元素的 display:none 样式。所以，v-if 有更高的切换消耗，而 v-show 有更高的初始渲染消耗。因此，如果需要频繁切换，v-show 较好；如果在运行时条件不大可能改变，v-if 较好。另外，v-show 不支持<template>元素，也不支持 v-else 元素。

【例 14-8】演示 v-if 与 v-show 的区别。代码如下：

```
<div id="event-handling">
    <div v-if="flag">一直显示</div>
    <div v-show="flag">反复无常</div>
    <button @click="flag=!flag">隐藏/显示</button>
</div>
<script src="js/vue.global.js"></script>
<script>
    Vue.createApp({
        data() {
            return {
                flag: true
            }
        }
    }).mount('#event-handling')
</script>
```

使用谷歌浏览器第一次运行程序时（按下 F12 键），页面初始化效果如图 14.4 所示。

图 14.4　页面初始化效果

单击“隐藏/显示”按钮后，页面如图 14.5 所示。

```
Elements    Console    Sources    Network
<html>
  <head></head>
··· ▼<body> == $0
    ▼<div id="event-handling" data-v-app>
        <!--v-if-->
        <div style="display: none;">反复无常</div>
        <button>隐藏/显示</button>
      </div>
      <script src="js/vue.global.js"></script>
    ▶<script>…</script>
    </body>
</html>
```

图 14.5　单击“隐藏/显示”按钮后的效果

从图 14.5 可以看出，通过 v-if 控制的元素，如果隐藏，从 DOM 中移除；而通过 v-show 控制的元素并没有真正移除，只是给其添加了 CSS 样式 display:none。

14.5.3　列表渲染指令 v-for

可以使用 v-for 指令遍历一个数组或对象，它的表达式需结合 in 来使用，形式为 item in items，其中，items 是源数据，而 item 是被迭代集合中元素的别名。v-for 还支持一个可选的参数作为当前项的索引。v-for 指令的常用方式如下。

❶ 遍历普通数组

```
<ul>
      <li v-for="(item,index) in items">
          {{index}} - {{ item }}
      </li>
</ul>
```

❷ 遍历对象数组

```
<ul>
    <li v-for="user in users">
        {{ user.uname }}
    </li>
</ul>
```

❸ 遍历对象属性

```
<li v-for="(value, key, index) in myObject">
    {{ ++index }}. {{ key }}: {{ value }}
</li>
```

❹ 迭代数字

```
<li v-for="i in 100">
     {{ i }}
</li>
```

【例 14-9】演示 v-for 指令的常用方式。代码如下：

```
<div id="myfor">
    <ul>
        <li v-for="(book, index) in books">
            {{++index}}. {{book}}
        </li>
    </ul>
    <ul>
        <li v-for="(author, index) in authors">
            {{++index}}. {{author.name}} - {{author.sex}}
        </li>
    </ul>
    <p v-for="(value, key, index) in userinfo">键是：{{key}}，值是：{{value}}，
    索引是：{{index}}</p>
    <p v-for="i in 5">这是第{{i}}段。</p>
</div>
```

```
<script src="js/vue.global.js"></script>
<script>
    Vue.createApp({
        data() {
            return {
                //普通数组
                books: ['Java Web 开发从入门到实战',
                 'Java EE 框架整合开发入门到实战——Spring+Spring MVC+MyBatis',
                 'Spring Boot 从入门到实战'],
                 //对象数组
                 authors:[{name: '陈恒', sex: '男'}, {name: '陈恒11', sex: '女'},
                         {name: '陈恒22', sex: '男'}],
                 //对象
                 userinfo:{
                     uname: '陈恒3',
                     age: 88
                 }
            }
        }
    }).mount('#myfor')
</script>
```

例 14-9 运行效果如图 14.6 所示。

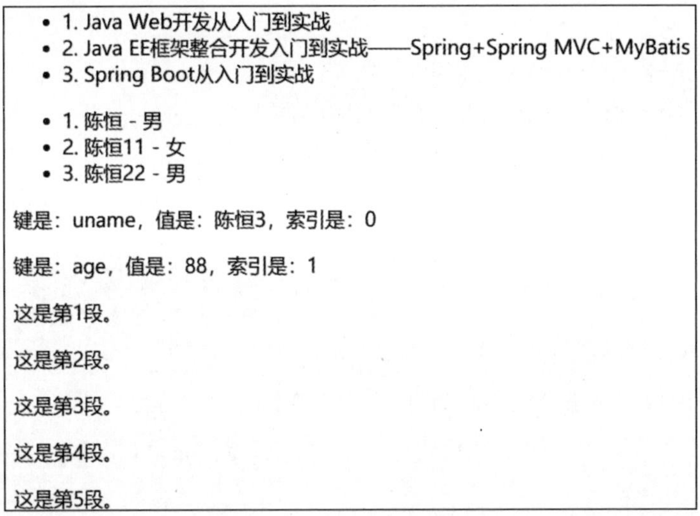

- 1. Java Web开发从入门到实战
- 2. Java EE框架整合开发入门到实战——Spring+Spring MVC+MyBatis
- 3. Spring Boot从入门到实战

- 1. 陈恒 - 男
- 2. 陈恒11 - 女
- 3. 陈恒22 - 男

键是：uname，值是：陈恒3，索引是：0

键是：age，值是：88，索引是：1

这是第1段。

这是第2段。

这是第3段。

这是第4段。

这是第5段。

图 14.6　例 14-9 运行效果

14.5.4　表单与 v-model

表单用于向服务器传输数据，较为常见的表单控件有单选、多选、下拉选择、输入框等，用表单控件可以完成数据的录入、校验、提交等。Vue.js 用 v-model 指令在表单<input>、<textarea>及<select>元素上创建双向数据绑定（Model 到 View 以及 View 到 Model）。使用 v-model 指令的表单元素将忽略该元素的 value、checked、selected 等属性初始值，而是将当前活动的 Vue 实例的数据作为数据来源。所以，使用 v-model 指令时，应通过 JavaScript 在 Vue 实例的 data 选项中声明初始值。

从 Model 到 View 的数据绑定，即 ViewModel 驱动将数据渲染到视图；从 View 到 Model 的数据绑定，即 View 中元素上的事件被触发后导致数据变更将通过 ViewModel 驱动修改数据层。下面通过一个实例演示 v-model 指令在表单元素上实现双向数据绑定。

【例 14-10】v-model 指令在表单元素上实现双向数据绑定。具体代码如下：

```
<div id="vmodel-databinding">
    用户名：<input v-model="uname" />
    <p>输入的用户名是：{{ uname }}</p>
    <textarea v-model="introduction"></textarea>
    <p>输入的个人简介是：{{ introduction }}</p>
    <p>
        备选歌手：
        <input type="checkbox" id="zhangsan" value="张三" v-model="singers" />
        <label for="zhangsan">张三</label>
        <input type="checkbox" id="lisi" value="李四" v-model="singers" />
        <label for="lisi">李四</label>
        <input type="checkbox" id="wangwu" value="王五" v-model="singers" />
        <label for="wangwu">王五</label>
        <input type="checkbox" id="chenheng" value="陈恒" v-model="singers" />
        <label for="chenheng">陈恒</label>
        <br />
        <span>你喜欢的歌手：{{ singers }}</span>
    </p>
    <p>
        性别：
        <input type="radio" id="male" value="男" v-model="sex" />
        <label for="male">男</label>
        <input type="radio" id="female" value="女" v-model="sex" />
        <label for="female">女</label>
        <br />
        <span>你的性别：{{ sex }}</span>
    </p>
    <p>
        <!--单选下拉-->
        备选国籍：
        <select v-model="single">
            <option v-for="option in options" :value="option.value">
                {{ option.text }}
            </option>
        </select><br>
        <span>你的国籍：{{ single }}</span>
    </p>
    <p>
        <!--多选下拉-->
        备选国家：
        <select v-model="moreselect" multiple>
            <option v-for="option in moreoptions" :value="option.value">
                {{ option.text }}
            </option>
        </select><br>
        <span>你去过的国家：{{ moreselect }}</span>
```

```
    </p>
</div>
<script src="js/vue.global.js"></script>
<script>
    Vue.createApp({
        data() {
            return {
                uname: '陈恒',
                introduction: '我是一个好少年',
                //多个复选框，绑定到同一个数组，歌手'陈恒'默认选择
                singers: ['陈恒'],
                sex: '女',//默认性别女
                //单选下拉列表绑定变量，国籍默认中国
                single: '中国',
                options: [
                    { text: '中国', value: '中国' },
                    { text: '日本', value: '日本' },
                    { text: '美国', value: '美国' }
                ],
                moreselect: ['中国'],//多选下拉列表绑定一个数组，默认去过中国
                moreoptions: [
                    { text: '中国', value: '中国' },
                    { text: '英国', value: '英国' },
                    { text: '日本', value: '日本' },
                    { text: '美国', value: '美国' }
                ],
            }
        }
    }).mount('#vmodel-databinding')
</script>
```

例 14-10 运行效果如图 14.7 所示。

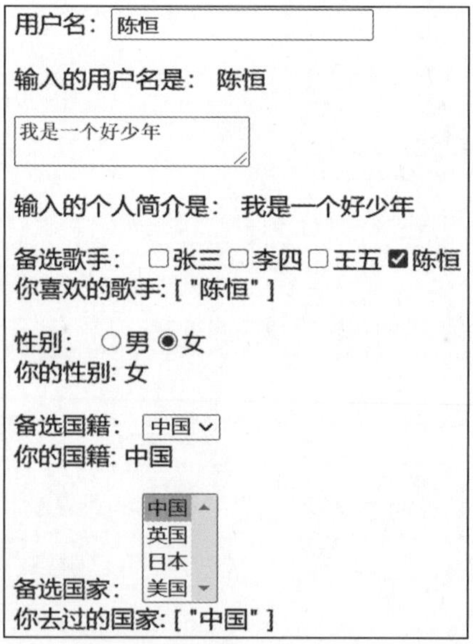

图 14.7　例 14-10 运行效果

默认情况下，v-model 在每次 input 事件触发后将输入框的值与数据同步。如果不想在每次 input 事件触发后同步，可以添加 lazy 修饰符，从而转为在 change 事件后进行同步。示例代码如下：

```
<!-- 在"change"时更新-->
<input v-model.lazy="msg"/>
```

如果需要将用户的输入值自动转为数值类型，可以给 v-model 添加 number 修饰符，示例代码如下：

```
<input v-model.number="age" type="number" />
```

如果需要将用户输入的首尾空格自动去除，可以给 v-model 添加 trim 修饰符，示例代码如下：

```
<input v-model.trim="msg" />
```

14.6　在 Vue 中动态使用样式

视频讲解

操作 HTML 元素的 class 和 style 属性动态改变其样式，是数据绑定的一个常见用法。因为 class 和 style 都是属性，所以可以用 v-bind 进行数据绑定。

14.6.1　绑定 class

❶ 对象语法

传给“:class”（v-bind:class 的简写）一个对象，可以动态地切换 class 属性值。示例代码如下：

```
<div :class="{ active: isActive }"></div>
```

可以在对象中传入更多字段动态切换多个 class。此外，“:class”指令也可以与普通的 class 属性同时存在。示例代码如下：

```
<div class="static" :class="{ active: isActive, 'text-danger': hasError }"></div>
```

❷ 数组语法

当需要多个 class 时，可以把一个数组与“:class”绑定，以应用一个 class 列表。示例代码如下：

```
<div :class="[activeClass, errorClass]"></div>
```

如果需要根据条件切换列表中的 class，可以使用三元表达式实现。示例代码如下：

```
<div :class="[isActive ? activeClass : '', errorClass]"></div>
```

❸ 在数组中嵌套对象

当有多个条件 class 时，在数组中使用三元表达式有些烦琐，所以在数组语法中也可以使用对象语法。示例代码如下：

```
<div :class="[{ active: isActive }, errorClass]"></div>
```

下面通过一个实例演示上述绑定 class 的方式。

【例 14-11】绑定 class 的几种方式。代码如下：

```
<div id="vbind-class">
    <div :class="mycolor">对象语法</div>
    <div class="static" :class="{ active: isActive, 'text-danger': hasError }">
    在对象中传入更多字段</div>
    <div :class="[activeClass, errorClass]">数组语法</div>
    <div :class="[isActive ? activeClass : '', errorClass]">在数组中使用三元表
    达式</div>
    <div :class="[{ active: isActive }, errorClass]">在数组中嵌套对象</div>
</div>
<script src="js/vue.global.js"></script>
<script>
    Vue.createApp({
        data() {
            return {
                mycolor: 'my',
                isActive: true,
                hasError: false,
                activeClass:'your',
                errorClass:'his'
            }
        }
    }).mount('#vbind-class')
</script>
<style>
    .my {
        background-color: red
    }
</style>
```

例 14-11 渲染结果（在谷歌浏览器中按下 F12 键）如图 14.8 所示。

```
▼<div id="vbind-class" data-v-app>
    <div class="my">对象语法</div>
    <div class="static active">在对象中传入更多字段</div>
    <div class="your his">数组语法</div>
    <div class="your his">在数组中使用三元表达式</div>
    <div class="active his">在数组中嵌套对象</div>
  </div>
```

图 14.8　例 14-11 渲染结果

▶ 14.6.2　绑定 style

使用 “:style” 可以给 HTML 元素绑定内联样式，方法与 “:class” 类似，也有对象语法和数组语法。“:style” 的对象语法十分直观——看起来像直接在元素上写 CSS，但其实是一个 JavaScript 对象。CSS 属性名可以用驼峰式或短横线分隔来命名。

下面通过一个实例演示绑定 style 的方式。

【例 14-12】绑定 style 的方式。代码如下：

```
<div id="vbind-style">
    <div :style="{ color: activeColor, fontSize: fontSize + 'px' }">绑定内联样式</div>
</div>
<script src="js/vue.global.js"></script>
<script>
    Vue.createApp({
        data() {
            return {
                activeColor: 'red',
                fontSize: 30
            }
        }
    }).mount('#vbind-style')
</script>
```

例 14-12 渲染结果如图 14.9 所示。

```
<div id="vbind-style" data-v-app>
  <div style="color: red; font-size: 30px;">绑定内联样式</div>
</div>
```

图 14.9　例 14-12 渲染结果

14.7　组件

组件（Component）是 Vue.js 最核心的功能，是可扩展的 HTML 元素（可看作自定义的 HTML 元素），是封装可重用的代码，同时也是 Vue 实例，可以接受与 Vue 相同的选项对象并提供相同的生命周期钩子。

组件系统是 Vue.js 中一个重要的概念，它提供了一种抽象，让我们可以使用独立可复用的小组件来构建大型应用，任意类型的应用界面都可以抽象为一个组件树。这种前端组件化方便 UI 组件的重用。

14.7.1　组件注册

视频讲解

为了能在 UI 模板中使用组件，必须先注册以便 Vue 识别。有两种组件的注册类型：全局注册和局部注册。

❶ 全局注册

组件可通过 component 方法实现全局注册。全局注册示例代码如下：

```
const app = Vue.createApp({})
app.component('component-a', {
    //选项
})
app.component('component-b', {
    //选项
})
app.component('component-c', {
```

```
        //选项
    })
    app.mount('#app')
```

app.component 的第一个参数 component-a 是组件的名称（自定义标签），组件名称推荐全部小写包含连字符（即有多个单词），避免与 HTML 元素相冲突。

注册后任何 Vue 实例都可以使用这些组件，示例代码如下：

```
<div id="app">
    <component-a></component-a>
    <component-b></component-b>
    <component-c></component-c>
</div>
```

下面通过一个实例演示全局组件用法。

【例 14-13】定义一个名为 button-counter 的全局组件，组件显示的内容为一个按钮。代码如下：

```
<template id="button-counter">
    <button @click="count++">You clicked me {{ count }} times.</button>
</template>
<div id="components-demo">
    <!--在模板中任意使用组件-->
    <!--每个组件都各自独立维护它的 count。因为每用一次组件，就会有一个它的新实例被创建。-->
    <button-counter></button-counter>
    <button-counter></button-counter>
    <button-counter></button-counter>
</div>
<script src="js/vue.global.js"></script>
<script>
    // 创建一个 Vue 应用
    const app = Vue.createApp({})
    // 定义一个名为 button-counter 的全局组件(注册)
    app.component('button-counter', {
        data() {
            return {
                count: 0
            }
        },
        //组件显示的内容
        template: '#button-counter'
    })
    app.mount('#components-demo')
</script>
```

上述组件定义的代码中，template 是定义组件显示的内容，必须被一个 HTML 元素包含（如<button>），否则无法渲染。除了 template 选项外，组件可以像 Vue 实例一样使用 data、computed、methods 选项，如上述 button-counter 组件。

❷ 局部注册

全局注册往往是不够理想的。例如，使用 webpack（后续讲解）构建系统，全局注册的组件即使不再使用，仍然被包含在最终的构建结果中，造成用户无意义的下载 JavaScript。在这些情况下，可以通过一个普通的 JavaScript 对象来定义组件：

```
const ComponentA = {
    /* ... */
}
const ComponentB = {
    /* ... */
}
const ComponentC = {
    /* ... */
}
```

然后，使用 Vue 实例的 components 选项局部注册组件：

```
const app = Vue.createApp({
    components: {
        'component-a': ComponentA,//component-a 为局部组件名称
        'component-b': ComponentB
    }
 })
```

局部注册的组件只在该组件作用域下有效。例如，希望 ComponentA 在 ComponentB 中可用，需要在 ComponentB 中使用 components 选项局部注册 ComponentA：

```
const ComponentA = {
/* ... */
}
const ComponentB = {
    components: {
        'component-a': ComponentA
    }
    //…
}
```

【例 14-14】局部组件用法演示。代码如下：

```
<div id="partcomponents-demo">
    <component-a></component-a>
</div>
<script src="js/vue.global.js"></script>
<script>
    const ComponentA = {
        template: '<span>这是私有组件！</span>'
    }
    // 创建一个 Vue 应用
    const app = Vue.createApp({
        components: {
            'component-a': ComponentA
        }
    })
    app.mount('#partcomponents-demo')
</script>
```

视频讲解

▶ 14.7.2 父组件向子组件传值

组件除了把模板内容复用外，更重要的是向组件传递数据。传递数据的过程就是由 props 实现的。在组件中，使用选项 props 来声明从父级组件接收的数据，props 的值可以是两种，一种是字符串数组，一种是对象。现在，先介绍数组类型的用法。

【例 14-15】构造两个数组 props，一个数组接收来自父级组件的数据 message（实现静态传递），一个数组接收来自父级组件的数据 id 和 title（实现动态传递），并将它们在组件模板中渲染。代码如下：

```
<!--父组件显示-->
<template id="parent">
    <h4>{{ message }}</h4>
     <!--使用 v-bind 将父组件 parent 的 data(posts)动态传递给 props，children 组件只
     能在 parent 中-->
    <children v-for="post in posts" :id="post.id" :title="post.title"></children>
    <!--将一个对象的所有属性都作为 prop 传入，与上面一句等价-->
    <children v-for="post in posts" v-bind="post" ></children>
</template>
<!--子组件显示-->
<template id="children">
    <h4>{{id}} : {{ title }}</h4>
</template>
<div id="message-post-demo">
    <!--静态传递字符串，父组件就是 Vue 当前的实例-->
    <parent message="来自父组件的消息"></parent>
</div>
<script src="js/vue.global.js"></script>
<script>
    const messageApp = Vue.createApp({})
    messageApp.component('parent', {
        data() {
            return {
                //posts 是对象数组
                posts: [
                    { id: 1, title: 'My journey with Vue' },
                    { id: 2, title: 'Blogging with Vue' },
                    { id: 3, title: 'Why Vue is so fun' }
                ]
            }
        },
        props: ['message'],//接收父组件 messageApp 传递的数据
        components: {//创建子组件 children
            'children':{
                props: ['id','title'],//接收父组件 parent 传递的数据
                template: '#children'
            }
        },
        template: '#parent'
    })
    messageApp.mount('#message-post-demo')
</script>
```

例 14-15 运行效果如图 14.10 所示。

127.0.0.1:5500/ch14_15.html

应用　常用

来自父组件的消息

1 : My journey with Vue

2 : Blogging with Vue

3 : Why Vue is so fun

1 : My journey with Vue

2 : Blogging with Vue

3 : Why Vue is so fun

图 14.10　例 14-15 运行效果

注意：如果不使用 v-bind 直接传递数字、布尔值、数组及对象，就是以字符串值传递。另外，使用 props 实现数据传递都是单向的，即父组件数据变化时，子组件中所有的 prop 将刷新为最新的值，但是反过来不行。

使用 props 实现数据传递的同时，还可以为 props 指定验证要求。一般当组件需要提供给别人使用时，最好进行数据验证。例如某个数据必须是数字类型，如果传入字符串，Vue 将在浏览器控制台中弹出警告。

为了定制 props 的验证方式，可以为 props 的值提供带有验证需求的对象，而不是字符串数组。下面通过一个验证实例讲解 props 验证。

【例 14-16】给组件的 props 提供带有验证需求的对象。具体代码如下：

```
<template id="validate">
    <div>
        <h4>{{ num }}</h4>
        <h4>{{ strnum }}</h4>
        <h4>{{ isrequired }}</h4>
        <h4>{{ numdefault }}</h4>
        <h4>{{ objectdefault }}</h4>
        <h4>{{ myfun }}</h4>
    </div>
</template>
<div id="validate-post-demo">
    <validate-post
        :num="200"
        :strnum="'sdf'"
        :isrequired="'abc'"
        :numdefault="300"
        :objectdefault="{a:'a'}"
        :myfun="'success'">
```

```
    </validate-post>
</div>
<script src="js/vue.global.js"></script>
<script>
    const messageApp = Vue.createApp({})
    messageApp.component('validate-post', {
        props: {
            //基础的类型检查 (null 和 undefined 会通过任何类型验证)
            num: Number,
            //多个可能的类型，字符串或数字
            strnum: [String, Number],
            //必填的字符串
            isrequired: {
                type: String,
                required: true
            },
            //带有默认值的数字
            numdefault: {
                type: Number,
                default: 100
            },
            //带有默认值的对象
            objectdefault: {
                type: Object,
                //对象或数组默认值必须从一个工厂函数获取
                default: function() {
                    return { message: 'hello' }
                }
            },
            //自定义验证函数
            myfun: {
                validator: function(value) {
                    //这个值必须匹配下列字符串中的一个
                    return ['success', 'warning', 'danger'].indexOf(value) !== -1
                }
            }
        },
        template: '#validate'
    })
    messageApp.mount('#validate-post-demo')
</script>
```

上述验证实例中，type 的类型可以是 String、Number、Boolean、Array、Object、Date、Function、Symbol 等数据类型。

视频讲解

▶ 14.7.3 子组件向父组件传值

可通过 props 从父组件向子组件传递数据，并且这种传递是单向的。当需要从子组件向父组件传递数据时，需要首先给子组件自定义事件并使用$emit(事件名, 要传递的数据)方法触发事件，然后父组件使用 v-on 或@监听子组件的事件。下面通过一个实例讲解自定义事件的使用方法。

【例 14-17】子组件触发两个事件，分别实现字体变大和变小。具体代码如下：

```
    <template id="blog">
        <!--0.1 是传递给父组件 blogApp 的数据，可以不填。然后当在父组件监听这个事件时，可以
通过$event 访问这个数据。如果事件处理函数是一个方法，那么这个数据将会作为第一个参数传入该方法
（如 onEnlargeText）-->
        <h4>{{id}} : {{ title }}</h4>
        <button @click="$emit('enlarge-text', 0.1)">变大</button>
        <button @click="$emit('ensmall-text', 0.1)">变小</button>
    </template>
    <div id="blog-post-demo">
        <div v-bind:style="{ fontSize: postFontSize + 'em' }">
         <!--将一个对象的所有属性作为 prop 传给子组件，@父组件监听事件并更新 postFontSize 值。-->
            <!--$event 接收子组件传递过来的数据 0.1-->
           <blog-post v-for="post in posts" v-bind="post" @ensmall-text="postFontSize -= $event"
           @enlarge-text="onEnlargeText"></blog-post>
        </div>
    </div>
    <script src="js/vue.global.js"></script>
    <script>
        const blogApp = Vue.createApp({
            data() {
                return {
                    //posts 是对象数组
                    posts: [
                        { id: 1, title: 'My journey with Vue' },
                        { id: 2, title: 'Blogging with Vue' },
                        { id: 3, title: 'Why Vue is so fun' }
                    ],
                    postFontSize: 1
                }
            },
            methods: {
                onEnlargeText(enlargeAmount) {
                    this.postFontSize += enlargeAmount
                }
            }
       })
        blogApp.component('blog-post', {//定义子组件
            props: ['id', 'title'],//接收父组件 blogApp 的两个参数 id 和 title
            template: '#blog'
        })
        blogApp.mount('#blog-post-demo')
    </script>
```

注意：上述代码中，事件名推荐使用短横线命名（例如 enlarge-text），这是因为 HTML 是不区分大小写的。如果事件名为 enlargeText，@enlargeText 将变成@enlargetext，事件 enlargeText 不可能被父组件监听到。

除了自定义事件实现子组件向父组件传值外，还可以在子组件上使用 v-model 向父组件传值，实现双向绑定。下面通过一个实例讲解在子组件上使用 v-model 向父组件传值。

【例 14-18】使用 v-model 实现子组件向父组件传值，并实现双向绑定。具体代码如下：

```
<template id="custom">
<!--为了让子组件正常工作，子组件内的 <input> 必须将其 value 属性绑定到一个名为modelValue 的 props 上，在其 input 事件被触发时，将新的值通过自定义的 update:modelValue 事件传递-->
    <input :value="modelValue" @input="$emit('update:modelValue', $event.target.value)" >
</template>
<div id="vmodel-post-demo">
    {{searchText}}<br>
    <custom-input v-model="searchText"></custom-input><br>
    <!--这两个子组件等价-->
<custom-input :model-value="searchText" @update:model-value="searchText = $event"></custom-input>
</div>
<script src="js/vue.global.js"></script>
<script>
    const blogApp = Vue.createApp({
        data() {
            return {
                searchText: '陈恒'
            }
        }
    })
    blogApp.component('custom-input', {
        props: ['modelValue'],
        template: '#custom'
    })
    blogApp.mount('#vmodel-post-demo')
</script>
```

视频讲解

▶ 14.7.4 提供/注入（组件链传值）

当需要将数据从父组件传递到子组件时，可以使用 props 实现。但有时有些子组件是深嵌套的，如果将 props 传递到整个组件链中，将很麻烦，更不可取。对于这种情况，可以使用 provide 和 inject 实现组件链传值。父组件可以作为其所有子组件的依赖项提供程序，而不管组件层次结构有多深，父组件有一个 provide 选项提供数据，子组件有一个 inject 选项使用这个数据。下面通过一个实例演示组件链传值的用法。

【例 14-19】创建 Vue 实例为祖先组件，并使用 provide 提供一个数据供其子孙组件 inject 使用。具体代码如下：

```
<template id="son">
    <div>{{ todos.length }}</div>
    <!--todo-son 是 todo-list 的私有组件-->
    <todo-son></todo-son>
</template>
<template id="grandson">
    <div>
        <!--使用注入的数据-->
        {{ todoLength }}
    </div>
</template>
```

```
<div id="vmodel-post-demo">
    <!--父组件 VUe 实例传递数据 todos 给子组件 todo-list-->
    <todo-list :todos="todos"></todo-list>
</div>
<script src="js/vue.global.js"></script>
<script>
    const app = Vue.createApp({
        data() {
            return {
                todos: ['Feed a cat', 'Buy tickets']
            }
        },
        provide() {//祖先组件 app 提供一个数据 todoLength
            return {
                todoLength: this.todos.length
            }
        }
    })
    app.component('todo-list', {
        props: ['todos'],
        components:{//在父组件 todo-list 中定义子组件 todo-son
            'todo-son': {
                inject: ['todoLength'],//孙组件注入数据 todoLength 供自己使用
                template: '#grandson'
            }
        },
        template: '#son'
    })
    app.mount('#vmodel-post-demo')
</script>
```

▶ 14.7.5　插槽

视频讲解

一个网页有时由多个模块组成，例如：

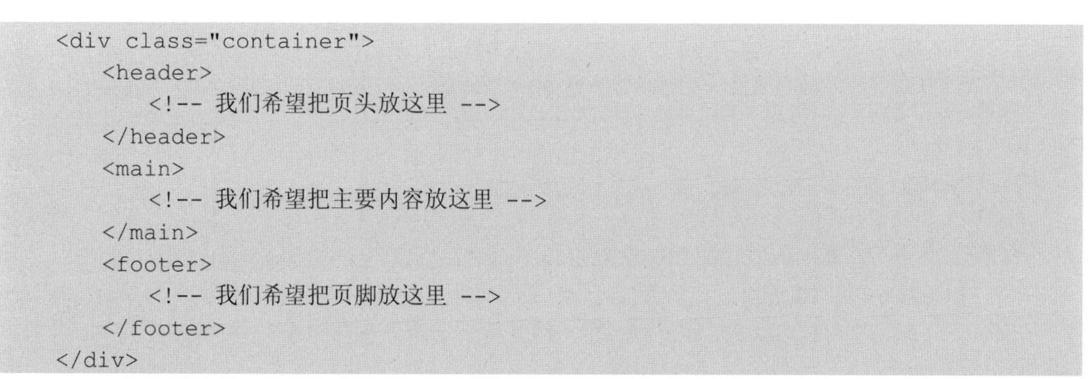

```
<div class="container">
    <header>
        <!-- 我们希望把页头放这里 -->
    </header>
    <main>
        <!-- 我们希望把主要内容放这里 -->
    </main>
    <footer>
        <!-- 我们希望把页脚放这里 -->
    </footer>
</div>
```

这时需要使用插槽混合父组件的内容与子组件的模板。那么插槽怎么使用呢？先来学习单插槽的使用。

❶ 单插槽 slot

在子组件模板中，可以使用插槽 slot 设置默认渲染内容。下面通过实例讲解单插槽的使用。

【例 14-20】使用插槽 slot 设置子组件的默认渲染内容。代码如下：

```
<template id="child">
    <slot>
        <p>插槽内容，默认内容！</p>
    </slot>
</template>
<div id="app">
    <child-com>
        <!--如果这里没有渲染内容，将渲染插槽中的默认内容-->
        <p>有我在 slot 就不显示！</p>
    </child-com>
</div>
<script src="js/vue.global.js"></script>
<script>
    const app = Vue.createApp({})
    app.component('child-com', {
        template: '#child'
    })
    app.mount('#app')
</script>
```

❷ 多个具名插槽

使用多个具名插槽可以实现混合渲染父组件的内容与子组件的模板。下面通过一个实例讲解具名插槽的使用方法。

【例 14-21】使用具名插槽实现一个页面由多个模块组成。代码如下：

```
<template id="child">
    <div>
        <div>
            <slot name="header">标题</slot>
        </div>
        <div>
            <slot>默认正文内容</slot>
        </div>
        <div>
            <slot name="footer">底部信息</slot>
        </div>
    </div>
</template>
<div id="app">
    <child-com>
        <!--显示插槽 header 的默认内容-->
        <h1 slot="header"></h1>
        <P>正文内容由我显示</P>
        <h1 slot="footer"></h1>
    </child-com>
</div>
<script src="js/vue.global.js"></script>
<script>
    const app = Vue.createApp({})
    app.component('child-com', {
```

```
        template: '#child'
    })
    app.mount('#app')
</script>
```

❸ 作用域插槽

有时让插槽能够访问组件中的数据是很有用的。作用域插槽更具代表性的用例是列表组件。下面通过实例演示作用域插槽的用法。

【例 14-22】使用作用域插槽实现列表组件渲染。具体代码如下：

```
<template id="blog">
    <ul>
        <li v-for="post in posts">
            <!--要使 post 可为父组件 blog-post（slot 为子组件）提供 slot 内容，
                可以添加一个<slot>元素并将 post 绑定为属性-->
            <slot :postgo="post"></slot>
        </li>
    </ul>
</template>
<div id="blog-post-demo">
   <blog-post>
       <!--绑定在<slot>元素上的属性 post 被称为插槽 props。
           可以使用带默认值的 v-slot 命令定义这个插槽 props 的名字-->
         <!--显示插槽内容-->
        <template v-slot:default="slotProps">
            {{ slotProps.postgo}}
        </template>
   </blog-post>
</div>
<script src="js/vue.global.js"></script>
<script>
    const blogApp = Vue.createApp({})
    blogApp.component('blog-post', {
        data() {
            return {
                posts: [
                    { id: 1, title: 'My journey with Vue' },
                    { id: 2, title: 'Blogging with Vue' },
                    { id: 3, title: 'Why Vue is so fun' }
                ]
            }
        },
        template: '#blog'
    })
    blogApp.mount('#blog-post-demo')
</script>
```

14.7.6　动态组件与异步组件

❶ 动态组件

在不同组件之间进行动态切换是常见的场景，例如在一个多标签的页面里进行内容的收纳和展现。Vue 可通过<component>元素动态挂载不同的组件，进行组件切换。示例代码如下：

```
<!-- is 属性选择挂载的组件，currentView 是已注册组件的名称或一个组件的选项对象 -->
<component :is="currentView"></component>
```

下面通过一个实例讲解动态组件的用法。

【例 14-23】通过<component>元素动态切换组件，在该实例中，有三个按钮代表标签，单击不同按钮展示不同组件的信息。具体代码如下：

```
<div id="app">
    <button @click="changeCom('add')">添加信息</button>
    <button @click="changeCom('update')">修改信息</button>
    <button @click="changeCom('delete')">删除信息</button>
    <component :is="currentCom"></component>
</div>
<script src="js/vue.global.js"></script>
<script>
    const blogApp = Vue.createApp({
        data() {
            return {
                currentCom: 'add'
            }
        },
        components:{//在组件选项中，定义三个局部组件供切换
            'add': {
                template: '<div>添加信息展示界面</div>'
            },
            'update': {
                template: '<div>修改信息展示界面</div>'
            },
            'delete': {
                template: '<div>删除信息展示界面</div>'
            }
        },
        methods:{
            changeCom(com) {
                this.currentCom = com
            }
        }
    })
    blogApp.mount('#app')
</script>
```

❷ 异步组件

在大型应用中，可能需要将应用分割成许多小的代码块，并且只在需要时才从服务器加载一个模块。这样可以避免一开始就把所有组件加载，浪费非必要的开销。Vue 有一个 defineAsyncComponent 方法将组件定义为一个工厂函数，动态地解析组件。Vue 只在组件需要渲染时触发工厂函数，并把结果缓存起来，以备再次渲染。下面通过一个实例讲解异步组件的用法。

【例 14-24】实现 5 秒钟后加载组件信息。具体代码如下：

```
<div id="app">
    <!--5 秒钟后才下载组件并展示-->
    <async-example></async-example>
</div>
<script src="js/vue.global.js"></script>
<script>
    const blogApp = Vue.createApp({})
    //定义异步组件
    const AsyncComp = Vue.defineAsyncComponent(() => //定义 defineAsyncComponent
                                                      //方法的返回值
        new Promise((resolve, reject) => {//返回 Promise 的工厂函数
            window.setTimeout(() => {//window.setTimeout 只是演示异步
                resolve({/*从服务器收到加载组件定义后，调用 Promise 的 resolve 方法异
                步下载组件，也可以调用 reject(reason)指示加载失败*/
                    template: '<div>5 秒钟后才展示我!</div>'
                })
            }, 5000)
        })
    )
    blogApp.component('async-example', AsyncComp)
    blogApp.mount('#app')
</script>
```

上述工厂函数返回 Promise 对象，当从服务器收到加载组件定义后，调用 Promise 的 resolve 方法异步下载组件，也可以调用 reject(reason)指示加载失败。这里，window.setTimeout 只是演示异步，具体的异步下载逻辑可由开发者自己决定。

此处只是简单演示一下异步组件用法，在进阶篇章中，将介绍打包编译工具 webpack 的用法，那时可以更优雅地实现异步组件（路由）。

14.7.7 使用 ref 获取 DOM 元素和组件引用

视频讲解

有时需要直接引用组件或 DOM 元素。为此，可以使用 ref 为子组件或 HTML 元素指定引用 ID。

引用 HTML 元素的示例代码如下：

```
<input ref="input" />
```

引用组件的示例代码如下：

```
<input ref="usernameInput" />
```

下面通过一个实例讲解 ref 的用法。

【例 14-25】ref 的用法。要求页面加载后，焦点聚焦在自定义组件上。具体代码如下：

```
<div id="app">
  <input type="text"/><br>
  <base-input></base-input><br>
  <input type="text"/>
</div>
<script src="js/vue.global.js"></script>
```

```
<script>
    const app = Vue.createApp({})
    app.component('base-input', {
        /*ref 引用子组件 usernameInput，也可以直接引用 HTML 元素（如 input）*/
        template: '<input ref="usernameInput" />',
        //直接引用 HTML 元素 input
        //template: '<input ref="input" />',
        methods: {
            focusInput() {
                this.$refs.usernameInput.focus()
                //this.$refs.input.focus()
            }
        },
        mounted() {//页面加载后调用该函数
            this.focusInput()
        },
        //如果前面直接引用 HTML 元素 input，去掉此处子组件的定义
        components: {
            'usernameInput':{
                template: ' <input type="text"/>'
            }
        }
    })
    app.mount('#app')
</script>
```

视频讲解

14.8 自定义指令

Vue 提供了功能丰富的内置指令，例如 v-model、v-show 等。这些内置指令可以满足大部分业务需求，但有时需要一些特殊功能，例如对普通 DOM 元素进行底层操作。幸运的是，Vue 允许自定义指令，实现特殊功能。

与组件类似，自定义指令的注册也分为全局注册和局部注册，例如注册一个名为 v-focus 的指令，用于输入元素（<input>和<textarea>）初始化时自动获得焦点。两种注册示例代码如下：

```
const app = Vue.createApp({})
// 注册一个全局自定义指令 v-focus
app.directive('focus', {
    //指令选项
})
// 注册一个局部自定义指令 v-focus
const app = Vue.createApp({
    directives: {
        focus: {
            //指令选项
        }
    }
}
```

一个自定义指令的选项是由以下几个钩子函数（均为可选）组成的。

❶ beforeMount

只调用一次，当指令第一次绑定到元素时调用，并进行初始化设置。

❷ mounted

在挂载绑定元素的父组件时调用。

❸ beforeUpdate

在元素本身更新前调用。

❹ updated

在元素本身更新后调用。

❺ beforeUnmount

在卸载绑定元素的父组件前调用。

❻ unmounted

只调用一次，当指令与元素解除绑定时调用。

可以根据业务需求，在不同的钩子函数中完成业务逻辑代码，例如下面的实例。

【例 14-26】自定义名为 v-focus 的指令，用于输入元素<input>（挂载绑定父组件调用 mounted 函数）初始化时自动获得焦点。具体代码如下：

```
<div id="app">
    <input v-focus type="text"/>
</div>
<script src="js/vue.global.js"></script>
<script>
const app = Vue.createApp({})
//自定义指令 focus，在模板中使用 v-focus
    app.directive('focus', {
        mounted(el) {
            el.focus()
        }
    })
    app.mount('#app')
</script>
```

上述代码中，el 为钩子函数的参数，除了 el 参数外，还有 binding、vnode 和 prevNnode 函数。它们的含义具体如下。

- el：指令所绑定的元素，可以用来直接操作 DOM。
- binding：一个对象，包含以下常用属性。
  - ➢ value：指令的绑定值，例如 v-my-directive="1 + 1" 中，绑定值为 2。
  - ➢ oldValue：指令绑定的前一个值，仅在 updated 钩子函数中可用。无论值是否改变都可用。
  - ➢ arg：传给指令的参数，可选。例如 v-my-directive:foo 中，参数为 foo。
  - ➢ modifiers：一个包含修饰符的对象。例如 v-my-directive.foo.bar 中，修饰符对象为 { foo: true, bar: true }。
- vnode：Vue 编译生成的虚拟节点，在 Vue 进阶中讲解。
- prevNnode：上一个虚拟节点，仅在 updated 钩子中可用。

下面通过一个实例讲解以上参数的用法。

【例 14-27】自定义一个名为 demo 的指令，并演示钩子函数参数的用法。具体代码如下：

```
<div id="app" >
    <!--msg 是传给指令的参数，a.b 是一个包含修饰符的对象-->
    <span v-demo:msg.a.b="message"></span>
</div>
<script src="js/vue.global.js"></script>
<script>
    const app = Vue.createApp({
        data() {
            return {
                message: 'hello!'
            }
        }
    })
    app.directive('demo', {
        beforeMount(el, binding, vnode) {
            //测试 binding 的属性
            //alert(Object.keys(binding))
            const keys =[]
            for (const i in vnode) {
                keys.push(i)
            }
            el.innerHTML =
            'value: '     + binding.value + '<br>' +
            'argument: '   + binding.arg+ '<br>' +
            'modifiers: ' + JSON.stringify(binding.modifiers) + '<br>' +
            'vnode keys: ' + keys.join(', ')
        }
    })
    app.mount('#app')
 </script>
```

例 14-27 运行结果如图 14.11 所示。

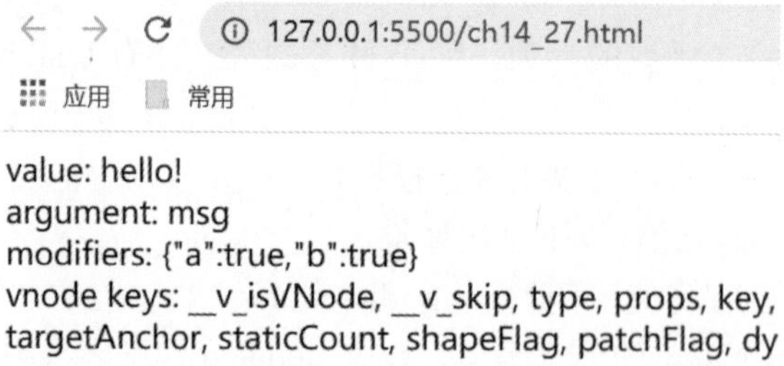

127.0.0.1:5500/ch14_27.html
应用 常用
value: hello!
argument: msg
modifiers: {"a":true,"b":true}
vnode keys: __v_isVNode, __v_skip, type, props, key, targetAnchor, staticCount, shapeFlag, patchFlag, dy

图 14.11 例 14-27 运行结果

14.9 本章小结

学习本章后，读者对 Vue 的插值与表达式、指令、组件系统等基础知识应有一个初步了解。组件是 Vue 最核心的功能，是非常抽象的概念，也是本章的重点。在以后的学习中，需要慢慢理解组件的内在机制。

习 题 14

1．为什么使用计算属性？请举例说明。

2．如何在表单元素上实现双向数据绑定？请举例说明。

3．如何注册组件？如何区分父子组件？你了解的组件传值有哪几种，它们是如何实现的？

4．简述 Vue 生命周期。

第 15 章　Vue 3 进阶

学习目的与要求

本章主要讲解 Vue 3 进阶知识，包括渲染函数、组合 API、webpack、Vue CLI、vue-router 以及 Vuex。通过本章的学习，将 Vue 3 进阶知识灵活应用于单页面综合程序中。

主要内容

- render 函数
- 组合 API
- webpack
- Vue CLI
- 路由 vue-router
- 状态管理与 Vuex

高效的开发离不开高效的项目构建工具以及实用的插件，本章除了学习 render 渲染函数和组合 API 外，还将学习 webpack、Vue CLI 以及 vue-router 插件和 Vuex 插件。

视频讲解

15.1　render 函数

render 函数与模板 template 一样，都是创建 HTML。本节学习 render 函数的具体用法。

15.1.1　什么是 render 函数

在多数情况下，Vue 推荐使用模板 template 创建 HTML。然而在一些应用场景中，需要使用 JavaScript 创建 HTML。这时可以用渲染函数，它比模板更方便。下面看一个应用场景，根据不同等级的锚点，显示不同的标题。

【例 15-1】根据不同等级的锚点，显示不同的标题。具体代码如下：

```
<div id="app">
    <anchored-heading :level="1" title="锚点1">Hello world111!</anchored-heading>
    <anchored-heading :level="2" title="锚点2">Hello world222!</anchored-heading>
    <anchored-heading :level="3" title="锚点3">Hello world333!</anchored-heading>
    <anchored-heading :level="4" title="锚点4">Hello world444!</anchored-heading>
    <anchored-heading :level="5" title="锚点5">Hello world555!</anchored-heading>
</div>
<template id="myanchored">
    <h1 v-if="level === 1">
        <a :href="'#' + title">
            <slot></slot>
        </a>
    </h1>
    <h2 v-else-if="level === 2">
```

```
        <a :href="'#' + title">
            <slot></slot>
        </a>
    </h2>
    <h3 v-else-if="level === 3">
        <a :href="'#' + title">
            <slot></slot>
        </a>
    </h3>
    <h4 v-else-if="level === 4">
        <a :href="'#' + title">
            <slot></slot>
        </a>
    </h4>
    <h5 v-else-if="level === 5">
        <a :href="'#' + title">
            <slot></slot>
        </a>
    </h5>
</template>
<script src="js/vue.global.js"></script>
<script>
    const app = Vue.createApp({})
    app.component('anchored-heading', {
        template: '#myanchored',
        props: {
            level: {
                type: Number,
                required: true
            },
            title: {
                type: String
            }
        }
    })
    app.mount('#app')
</script>
```

上述代码没有任何问题，但是缺点非常明显：代码冗长、重复率高。下面使用 render 函数改写，显得格外简练。

【例 15-2】使用 render 函数改写例 15-1。具体代码如下：

```
<div id="app">
    <anchored-heading :level="1" title="锚点 1">Hello world111!</anchored-heading>
    <anchored-heading :level="2" title="锚点 2">Hello world222!</anchored-heading>
    <anchored-heading :level="3" title="锚点 3">Hello world333!</anchored-heading>
    <anchored-heading :level="4" title="锚点 4">Hello world444!</anchored-heading>
    <anchored-heading :level="5" title="锚点 5">Hello world555!</anchored-heading>
</div>
<script src="js/vue.global.js"></script>
<script>
    const app = Vue.createApp({})
```

```
        app.component('anchored-heading', {
            render() {
                return Vue.h('h' + this.level,//tag参数
                    [//children参数
                        Vue.h(
                            'a',//tag参数
                            {//props参数
                                href: '#' + this.title
                            },
                            this.$slots.default()//children参数
                        )
                    ]
                )
            },
            props: {
                level: {
                    type: Number,
                    required: true
                },
                title: {
                    type: String
                }
            }
        })
        app.mount('#app')
    </script>
```

从上述代码可以看出，render 函数通过 Vue 的 h()函数创建虚拟 DOM，代码精简很多。第 14 章介绍的 slot，应用场景就是 render 函数。render 函数的神秘之处就是 Vue 的 h()函数，下面学习它的详细用法。

▶ 15.1.2 h()函数

h()函数是一个用于创建虚拟节点（VNode）的程序。也许可以更准确地将其命名为 createVNode()，为了便于使用简洁，称其为 h()。h()函数有 3 个参数。

❶ tag

tag 代表一个 HTML 标签（String）、一个组件（Object）、一个函数（Function）或者 null。使用 null 将渲染一个注释。这是一个必选参数。例如例 15-2 中的'a'。

❷ props

一个与 attribute、prop 和事件相对应的对象（Object），用于给创建的节点对象设置属性值，在模板中使用。这是一个可选参数。例如例 15-2 中的{href: '#' + this.title}。

❸ children

子 VNodes，使用 h()函数构建，或使用字符串获取“文本 VNode”（String|Array）或者有 slot 的对象（Object）。这是一个可选参数。例如例 15-2 中的代码：

```
[//children
    Vue.h(
        'a',// tag
        {// props
```

```
            href: '#' + this.title
        },
        this.$slots.default()//children
    )
]
```

对于大部分开发者来说，不会真正接触到 render 函数。因为开发时，基本是.vue 文件的开发模式，vue-loader 会编译模板到 render 函数。

15.2 组合 API

视频讲解

通过创建 Vue 组件，可以将接口的可重复部分及其功能提取到可重用的代码段中，从而使应用程序可维护且灵活。然而，当应用程序非常复杂（成百上千组件）时，再使用组件的选项（data、computed、methods、watch）组织逻辑，可能导致组件难以阅读和理解。如果能够将与同一个逻辑相关的代码配置在一起将有效解决逻辑复杂、可读性差等问题。这正是使用组合 API 的目的。

假设在一个大型应用程序中，有一个视图显示某个用户的仓库列表。除此之外，还希望应用搜索和筛选功能。处理此视图的组件代码逻辑如下所示：

```
// src/components/UserRepositories.vue
export default {
    components: { RepositoriesFilters, RepositoriesSortBy, RepositoriesList },
    props: {
        user: { type: String }
    },
    data() {
        return {
            repositories: [], // 1
            filters: { ... }, // 3
            searchQuery: ' ' // 2
        }
    },
    computed: {
        filteredRepositories() { ... }, // 3
        repositoriesMatchingSearchQuery() { ... }// 2
    },
    watch: {
        user: 'getUserRepositories' // 1
    },
    methods: {
        getUserRepositories() {
            // 使用 this.user 获取用户仓库
        }, // 1
        updateFilters() { ... }, // 3
    },
    mounted() {
        this.getUserRepositories() // 1
    }
}
```

上述组件有以下几个职责：假定的外部API获取该用户名的仓库，并在用户更改时刷新它；使用searchQuery字符串搜索存储库；使用filters对象筛选仓库。使用组件的选项（data、computed、methods、watch）组织逻辑，这种碎片化使理解和维护复杂组件变得困难。选项的分离掩盖了潜在的逻辑问题。此外，在处理单个逻辑关注点时，必须不断地"跳转"相关代码的选项块。下面使用组合API重新组织组件逻辑，具体代码如下：

```
// src/components/UserRepositories.vue
import { toRefs } from 'vue'
import useUserRepositories from '@/composables/useUserRepositories'
import useRepositoryNameSearch from '@/composables/useRepositoryNameSearch'
import useRepositoryFilters from '@/composables/useRepositoryFilters'
export default {
    components: { RepositoriesFilters, RepositoriesSortBy, RepositoriesList },
    props: {
        user: { type: String }
    },
    setup(props) {
        const { user } = toRefs(props)
        const { repositories, getUserRepositories } = useUserRepositories(user)
        const {
            searchQuery,
            repositoriesMatchingSearchQuery
        } = useRepositoryNameSearch(repositories)
        const {
            filters,
            updateFilters,
            filteredRepositories
        } = useRepositoryFilters(repositoriesMatchingSearchQuery)
        return {
            repositories: filteredRepositories,
            getUserRepositories,
            searchQuery,
            filters,
            updateFilters
        }
    }
}
```

目前，肯定是看不懂上述组件的代码逻辑，现在只是了解一下使用组合API的目的，等后面学习综合项目实战时再回头理解。

▶ 15.2.1 setup

Vue组件提供setup选项，供开发者使用组合API。setup选项在创建组件前执行，一旦props被解析，便充当组成API的入口点。由于在执行setup时尚未创建组件实例，因此在setup选项中没有this。这意味着，除了props之外，无法访问组件中声明的任何属性，包括本地状态、计算属性或方法。

setup选项是一个接受props和context的函数。此外，从setup返回的所有内容都将暴露给组件的其余部分（计算属性、方法、生命周期钩子、模板等）。

❶ setup 函数的参数

1）setup 函数中的第一个参数（props）

setup 函数中的 props 是响应式的，当传入新的属性时，它将被更新。但是，因为 props 是响应式的，不能响应式引用 props 的属性，因为它会消除属性的响应性。如果需要响应式引用 props 的属性，可通过使用 toRefs 来创建对 props 的属性的响应式引用。示例代码如下：

```
setup(props) {
    //使用 toRefs 创建对 props 的属性的响应式引用
    const { user } = toRefs(props)
    console.log(user.uname)
}
```

2）setup 函数中的第二个参数（context）

context 上下文是一个普通的 JavaScript 对象，它暴露组件的 3 个属性：attrs、slots 和 emit。示例代码如下：

```
setup(props, context) {
    // Attribute （非响应式对象）
    console.log(context.attrs)
    // 插槽 （非响应式对象）
    console.log(context.slots)
    // 触发事件 （方法）
    console.log(context.emit)
  }
```

context 是一个普通的 JavaScript 对象，也就是说，它不是响应式的，这意味着可以安全地对 context 使用 ES 解构。示例代码如下：

```
setup(props, { attrs, slots, emit }) {
    ...
}
```

attrs 和 slots 是有状态的对象，它们随组件本身的更新而更新。这意味着应该避免对它们进行解构，并始终以 attrs.x 或 slots.x 的方式引用属性。

❷ setup 函数的返回值

1）对象

如果 setup 返回一个对象，则可以在组件的模板中访问该对象的属性。下面通过一个实例讲解 setup 函数的使用方法。

【例 15-3】在该实例中，setup 函数返回一个对象。具体代码如下：

```
<template id="stesting">
    <!-- 模板中使用 readersNumber 对象会被自动开箱，所以不需要.value  -->
    <div>{{ readersNumber }} {{ book.title }}</div>
</template>
<div id="app">
    <setup-testing></setup-testing>
</div>
<script src="js/vue.global.js"></script>
<script>
    const app = Vue.createApp({})
```

```
    app.component('setup-testing', {
        setup() {
            //使用 ref 函数，对值创建一个响应式引用，并返回一个具有 value 属性的对象
            const readersNumber = Vue.ref(1000)
            //reactive()接收一个普通对象然后返回该对象的响应式代理，后面讲解
            const book = Vue.reactive({ title: '好书' })
            // 暴露给 template
            return {
                readersNumber,
                book
            }
        },
        template: '#stesting'
    })
    app.mount('#app')
</script>
```

2）渲染函数

setup 还可以返回一个渲染函数，该函数可以直接使用在同一作用域中声明的响应式状态。下面通过一个实例讲解 setup 返回渲染函数。

【例 15-4】实现例 15-3 的功能，要求 setup 返回渲染函数。具体代码如下：

```
<div id="app">
    <setup-testing></setup-testing>
</div>
<script src="js/vue.global.js"></script>
<script>
    const app = Vue.createApp({})
    app.component('setup-testing', {
        setup() {
            //使用 ref 函数，对值创建一个响应式引用，并返回一个具有 value 属性的对象
            const readersNumber = Vue.ref(1000)
            //reactive()接收一个普通对象，然后返回该对象的响应式代理，后面讲解
            const book = Vue.reactive({ title: '好书' })
            //返回渲染函数
            return () => Vue.h('div', [readersNumber.value, book.title])
        }
    })
    app.mount('#app')
</script>
```

❸ 在 setup 内部调用生命周期钩子函数

在 setup 内部，可通过在生命周期钩子函数前面加上 on 来访问组件的生命周期钩子函数。因为 setup 是围绕 beforeCreate 和 created 生命周期钩子函数运行的，所以不需要显式地定义它们。换句话说，在这些钩子函数中编写的任何代码都应该直接在 setup 函数中编写。这些 on 函数接受一个回调函数，当钩子函数被组件调用时将会被执行。示例代码如下：

```
setup() {
    // mounted
    onMounted(() => {
        console.log('Component is mounted!')
    })
}
```

▶ 15.2.2 响应性

非侵入性的响应性系统是 Vue 最独特的特性之一。本节将学习 Vue 响应性系统的底层细节。

响应性是一种允许以声明式的方式去适应变化的编程范例。例如，在某个 Excel 电子表格中，将数字 x 放在第一个单元格中，将数字 y 放在第二个单元格中，并要求自动计算 x + y 的值放在第三个单元格中。同时，如果更新数字 x 或 y，第三个单元格的值也会自动更新。Vue 如何追踪变化呢？在生成 Vue 实例时，使用带有 getter 和 setter 的处理程序遍历传入的 data，将其所有的 property 转换为 Proxy 对象。Proxy 对象对于用户来说是不可见的，但在内部，它使 Vue 能够在 property 的值被访问或修改的情况下进行依赖跟踪和变更通知。

❶ 声明响应式状态

要为 JavaScript 对象创建响应式状态，可以使用 reactive()方法。reactive()方法接收一个普通对象然后返回该对象的响应式代理。示例代码如下：

```
const book = Vue.reactive({ title: '好书' })
```

reactive()方法响应式转换是“深层的”，即影响对象内部所有嵌套的属性。基于 ES 的 Proxy 实现，返回的代理对象不等于原始对象。建议使用代理对象，避免依赖原始对象。例如，在例 15-3 中，使用代理对象 book。

❷ 使用 ref 创建独立的响应式值对象

ref 接受一个参数值并返回一个响应式且可改变的 ref 对象。ref 对象拥有一个指向内部值的单一属性.value。示例代码如下：

```
const readersNumber = Vue.ref(1000)
console.log(readersNumber.value) //1000
readersNumber.value++
console.log(readersNumber.value) // 1001
```

当 ref 作为渲染上下文的属性返回（即在 setup()返回的对象中）并在模板中使用时，它会自动开箱，无须在模板内额外书写.value。例如，在例 15-3 中，使用{{ readersNumber }}。

当嵌套在响应式对象中时，ref 才会自动开箱。从 Array 或者 Map 等原生集合类中访问 ref 时，不会自动开箱。示例代码如下：

```
const map = reactive(new Map([['foo', ref(0)]]))
// 这里需要.value
console.log(map.get('foo').value)
```

❸ 响应性计算

使用响应式计算 computed 方法有两种方式：传入一个 getter 函数，返回一个默认不可手动修改的 ref 对象；传入一个拥有 get 和 set 函数的对象，创建一个可手动修改的计算状态。

返回一个默认不可手动修改的 ref 对象的示例代码如下：

```
const count = ref(1)
const plusOne = computed(() => count.value + 1)
console.log(plusOne.value) // 2
plusOne.value++ // 错误!
```

返回一个可手动修改的 ref 对象的示例代码如下：

```
onst count = ref(1)
const plusOne = computed({
    get: () => count.value + 1,
    set: (val) => {
        count.value = val - 1
    },
})
plusOne.value = 1
console.log(count.value) // 0
```

❹ 响应性监听 watchEffect

可使用响应性监听 watchEffect 方法对响应性进行监听。该方法立即执行传入的一个函数，同时响应式追踪其依赖，并在其依赖变更时重新运行该函数。下面通过一个实例讲解响应性监听 watchEffect 方法的使用。

【例 15-5】响应性监听 watchEffect 方法的使用。具体代码如下：

```
<script src="js/vue.global.js"></script>
<script>
    const count = Vue.ref(0)
    Vue.watchEffect(() => console.log(count.value))
    // -> 打印出 0
    setTimeout(() => {
        count.value++
        // -> 打印出 1
    }, 100)
</script>
```

1）停止侦听

当 watchEffect 在组件的 setup()函数或生命周期钩子被调用时，侦听器会被链接到该组件的生命周期，并在组件卸载时自动停止。在一些情况下，也可以显式调用返回值以停止侦听。示例代码如下：

```
const stop = watchEffect(() => {
    /* ... */
})
//之后
stop()
```

2）清除副作用

有时副作用函数会执行一些异步的副作用，这些响应需要在其失效时清除（即完成前状态已改变）。所以侦听副作用传入的函数可以接收一个 onInvalidate 函数作为入参，用来注册清理失效时的回调。当副作用即将重新执行或侦听器被停止时，触发失效回调函数。示例代码如下：

```
watchEffect((onInvalidate) => {
    const token = performAsyncOperation(id.value)
    onInvalidate(() => {
        // id 改变时或停止侦听时
```

```
        // 取消之前的异步操作
        token.cancel()
    })
})
```

3）侦听器调试

onTrack 和 onTrigger 选项可用于调试一个侦听器的行为。当一个响应性对象属性或一个 ref 作为依赖被追踪时，将调用 onTrack；依赖项变更导致副作用被触发时，将调用 onTrigger。这两个回调都将接收一个包含有关所依赖项信息的调试器事件。onTrack 和 onTrigger 仅在开发模式下生效。示例代码如下：

```
watchEffect(
    () => {
        /* 副作用的内容 */
    },
    {
        onTrigger(e) {
            debugger 语句检查依赖关系
        },
    }
)
```

❺ 响应性侦听 watch

watch 需要侦听特定的数据源，并在回调函数中执行副作用。默认情况是懒执行的，也就是说仅在侦听的源变更时才执行回调。与 watchEffect 比较，watch 允许：懒执行副作用；更明确哪些状态的改变会触发侦听器重新运行副作用；访问侦听状态变化前后的值。

1）侦听单个数据源

侦听器的数据源可以是一个拥有返回值的 getter 函数，也可以是 ref。示例代码如下：

```
//侦听一个 getter
const state = reactive({ count: 0 })
watch(() => state.count,(count, prevCount) => {
        /* ... */
    }
)
// 直接侦听一个 ref
const count = ref(0)
watch(count, (count, prevCount) => {
    /* ... */
})
```

2）侦听多个数据源

也可以使用数组同时侦听多个源。示例代码如下：

```
watch([fooRef, barRef], ([foo, bar], [prevFoo, prevBar]) => {
    /* ... */
})
```

3）与 watchEffect 共享的行为

watch 和 watchEffect 在停止侦听、清除副作用、副作用刷新时机和侦听器调试等方面行为一致。

15.2.3 模板引用

在使用组合 API 时，响应式引用和模板引用的概念是统一的。为了获得对模板内元素或组件实例的引用，可以声明一个 ref 并从 setup()返回。下面通过一个实例讲解模板引用。

【例 15-6】在模板中使用 ref 引用响应式对象。具体代码如下：

```
<template id="st">
   <div ref="root">这是根元素</div>
</template>
<div id="app">
   <setup-testing></setup-testing>
</div>
<script src="js/vue.global.js"></script>
<script>
   const app = Vue.createApp({})
   app.component('setup-testing', {
      setup() {
         const root = Vue.ref(null)
         Vue.onMounted(() => {
            // DOM 元素将在初始渲染后分配给 ref
            console.log(root.value) // <div>这是根元素</div>
         })
         return {
            root
         }
      },
      template: "#st"
   })
   app.mount('#app')
</script>
```

在上述代码中，渲染上下文时暴露 root，并通过 ref="root"将其绑定到 div 作为其 ref。需要注意的是，模板引用在初始渲染后才能获得赋值。

Vue.js 的基本用法到本节就结束了，到目前为止，有关 Vue 的示例都是通过<script>引入 vue.global.js 来运行的，从 15.3 节开始，将陆续介绍前端工程化和 Vue 生态。

视频讲解

15.3 使用 webpack

本节简单介绍目前热门的 JavaScript 应用程序的静态模块打包工具 webpack。

15.3.1 webpack 介绍

webpack 根据模块的依赖关系进行静态分析，然后将这些模块按照指定的规则生成对应的静态资源。图 15.1 是来自 webpack 官方网站（https://webpack.js.org/）的模块化示意图。

图 15.1 的左边是业务中编写的各种类型文件，例如 typescript、jpg、less、css，还有后续将要学习的.vue 格式的文件。这些类型的文件通过特定的加载器（Loader）编译后，最终统一生成.js、.css、.jpg、.png 等静态资源文件。在 webpack 中，一张图片、一个 css 文件等都被称为模块，并彼此存在依赖关系。使用 webpack 的目的就是处理模块之间的依赖关系，并

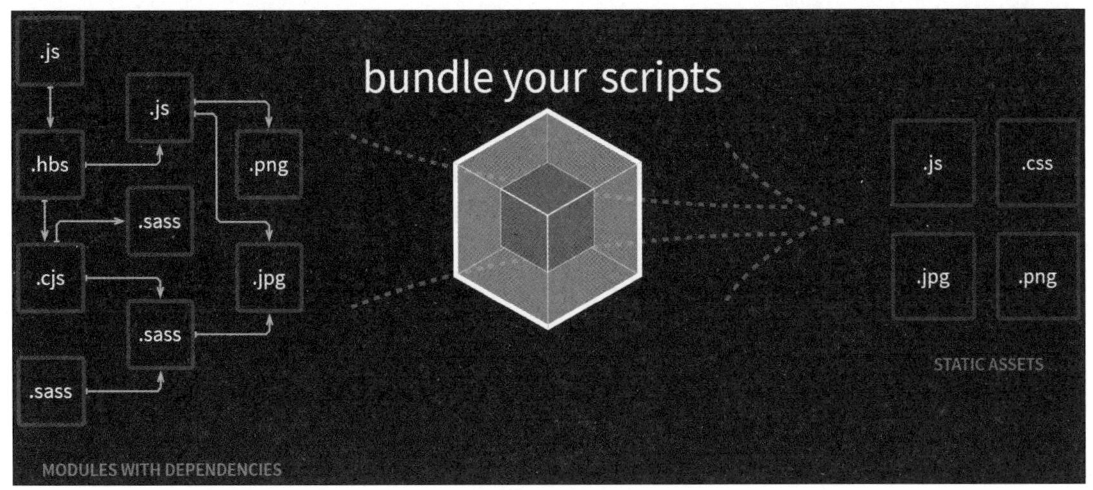

图 15.1　webpack 模块化示意图

将它们进行打包。

webpack 的主要适用场景是单页面应用（SPA），SPA 通常是由一个 HTML 文件和一堆按需加载的 JS 文件组成。接下来学习 webpack 的安装与使用。

▶ 15.3.2　安装 webpack 与 webpack-dev-server

本节使用 npm 安装 webpack，而 npm 是集成在 Node.js 中的，所以需要首先安装 Node.js。

❶ 安装 Node.js 和 npm

通过访问官网 https://nodejs.org/en/即可下载对应版本的 Node.js，本书下载的是“14.15.4 LTS”。

下载完成后运行安装包 node-v14.15.4-x64.msi，一路单击“下一步”按钮即可完成安装。然后在命令行窗口中输入“node -v”命令，检查是否安装成功，如图 15.2 所示。

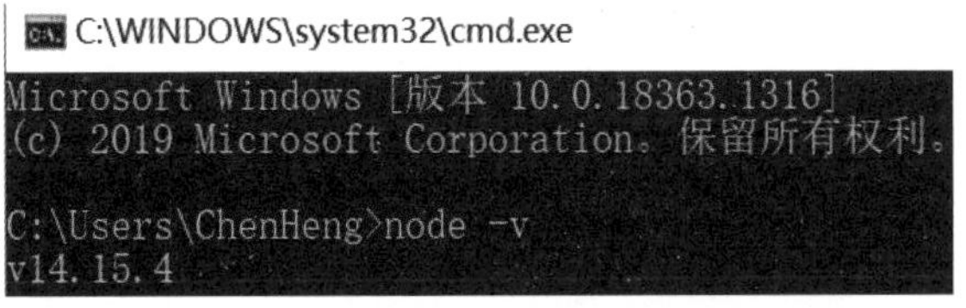

图 15.2　查看 Node.js 是否安装成功

图 15.2 中出现了版本号，说明 Node.js 已安装成功。同时，npm 包也已经安装成功。可以输入“npm -v”命令查看版本号；输入“npm -g install npm”命令将 npm 更新至最新版本，如图 15.3 所示。

❷ 安装 webpack

首先，创建一个目录，例如 C:\webpack-firstdemo，使用 VSCode 打开该目录，并进入 Terminal 控制台，如图 15.4 所示。

1）初始化配置

在图 15.4 中输入“npm init”命令初始化配置，该命令执行后，将有一系列选项，可以按 Enter 键快速确认，结束后将在 webpack-firstdemo 目录下生成一个 package.json 文件。

```
C:\WINDOWS\system32\cmd.exe
Microsoft Windows [版本 10.0.18363.1316]
(c) 2019 Microsoft Corporation。保留所有权利。

C:\Users\ChenHeng>npm -v
6.12.0

C:\Users\ChenHeng>npm -g install npm
npm WARN      npm does not support Node.js v14.15.4
npm WARN      You should probably upgrade to a newer version of node as we
npm WARN      can't make any promises that npm will work with this version.
npm WARN      Supported releases of Node.js are the latest release of 6, 8, 9, 10, 11, 12.
npm WARN      You can find the latest version at https://nodejs.org/
C:\Users\ChenHeng\AppData\Roaming\npm\npm -> C:\Users\ChenHeng\AppData\Roaming\npm\node_modul
C:\Users\ChenHeng\AppData\Roaming\npm\npx -> C:\Users\ChenHeng\AppData\Roaming\npm\node_modul
+ npm@6.14.11
added 26 packages from 7 contributors, removed 21 packages and updated 60 packages in 30.06s

C:\Users\ChenHeng>npm -v
6.14.11
```

图 15.3　查看与更新 npm

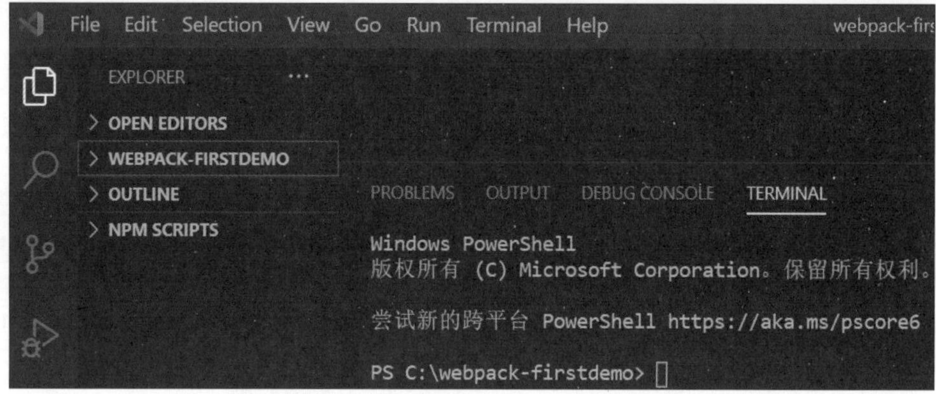

图 15.4　打开目录与 Terminal 控制台

2）安装 webpack

初始化配置后，在图 15.4 中输入“npm install webpack --save-dev”命令在本地局部（项目中）安装 webpack。--save-dev 将作为开发依赖来安装 webpack。安装成功后，在 package.json 文件中将增加一项配置：

```
"devDependencies": {
    "webpack": "^5.17.0",
}
```

❸ 安装 webpack-dev-server

webpack-dev-server 可以在开发环境中提供很多服务，例如启动一个服务器、热更新、接口代理等。在图 15.4 中输入“npm install webpack-dev-server --save-dev”命令在本地局部安装 webpack-dev-server。

❹ 安装 webpack-cli

在 webpack 4 以前，无须安装 webpack-cli 即可使用。在 webpack 4 以后，将 webpack 和 webpack-cli 分开处理，需要安装 webpack-cli。在图 15.4 中输入“npm install webpack-cli --save-dev”命令在本地局部安装 webpack-cli。

如果在 package.json 文件的 devDependencies 中包含 webpack、webpack-cli 和 webpack-dev-server，如图 15.5 所示，说明已成功安装。

▶ 15.3.3 webpack 配置文件

webpack 配置文件是一个名为 webpack.config.js 的.js 文件，架构的好坏都体现在该配置文件中。下面由浅入深，完成配置。

```
"devDependencies": {
  "webpack": "^5.17.0",
  "webpack-cli": "^4.4.0",
  "webpack-dev-server": "^3.11.2"
}
```

图 15.5 成功安装 webpack、webpack-cli 和 webpack-dev-server

【例 15-7】完成 webpack 的基本配置。具体步骤如下。

❶ 初始化配置

在 webpack-firstdemo 目录下创建一个名为 webpack.config.js 的.js 文件，并初始化配置内容：

```
const config = {
}
module.exports = config
```

❷ 添加快速启动 webpack-dev-server 脚本

在 package.json 的 scripts 里面添加一个快速启动 webpack-dev-server 服务的脚本：

```
"scripts": {
   "build":"webpack -p",
   "test": "echo \"Error: no test specified\" && exit 1",
   "dev": "webpack-dev-server --open Chrome.exe --config webpack.config.js"
}
```

在 Terminal 终端执行 npm run build 命令时，将执行 webpack -p 命令进行打包。

在 Terminal 终端执行 npm run dev 命令时，将执行 webpack-dev-server --open Chrome.exe --config webpack.config.js 命令。其中，--config 是指向 webpack-dev-server 读取配置文件的路径，这里指向上面步骤中创建的 webpack.config.js 文件；--open 将在执行命令时自动使用谷歌浏览器打开页面（如果 open 后面没有指定浏览器，则使用默认浏览器打开），默认地址是 127.0.0.1:8080，但 IP 和端口号可以修改，示例如下：

```
"dev": "webpack-dev-server --host 128.11.11.11 --port 9999 --open --config webpack.config.js"
```

❸ 配置入口和出口

在 webpack 配置中，最重要的也是必选的两项是入口（entry）和出口（output）。入口的作用是告诉 webpack 从哪里开始寻找依赖并编译；出口的作用是配置编译后的文件存储位置和文件名。

在 webpack-firstdemo 目录下创建一个名为 main.js 的空文件作为入口的文件，然后在 webpack.config.js 中进行入口和出口的配置：

```
const path = require('path')
const config = {
      entry: {
         main: './main'
      },
      output: {
```

```
            path: path.resolve(__dirname, 'dist'),
            publicPath: '/dist/',
            filename: 'bundle.js'
        },
        //开发模式:development，将保留开发时的一些必要信息
        //生产模式:production，尽力压缩
        //none:只打包
        mode: 'development'
    }
module.exports = config
```

上述配置中，entry 中的 main 是配置的入口，webpack 将从 main.js 文件开始工作。output 中的 path 选项用于存放打包后文件的输出目录，是必填项；publicPath 用于指定资源文件引用的目录，如果资源存放在 CDN 上，这里可以填 CDN 的网址；filename 用于指定输出文件的名称。所以这里配置的 output 意为打包后的文件将存储在 webpack-firstdemo/dist/bundle.js 文件中，在 html 中引入它即可。

❹ 创建 index.html 文件

在 webpack-firstdemo 目录下，新建一个名为 index.html 的文件，作为 SPA 的入口程序：

```
<div id="app">
   Hello Webpack!
</div>
<script type="text/javascript" src="/dist/bundle.js"></script>
```

❺ 在浏览器中打开 webpack 项目

在 Terminal 终端执行 npm run dev 命令，将会自动在浏览器中打开页面，如图 15.6 所示。

图 15.6 在浏览器中打开 webpack 项目

注意：执行 npm run dev 命令时，可能会报如下错误：

```
Cannot find module 'webpack-cli/bin/config-yargs'
```

出现上述错误的原因是 webpack-cli 的新版本与 webpack-dev-server 兼容性的问题，解决办法是：首先，执行 npm uninstall -g webpack-cli 命令，卸载 webpack-cli；然后，执行 npm install webpack-cli@3.3.12 --save-dev 命令，安装低版本的 webpack-cli；最后，执行 npm run dev 命令，将会自动在浏览器中打开页面。

15.3.4 加载器 Loaders 与插件 Plugins

❶ 加载器 Loaders

在 webpack 中，一切皆模块，例如.css、.js、.html、.jpg 等。对于不同的模块，需要使用不同的加载器（Loaders）来处理。Loaders 是 webpack 最强大的功能之一。webpack 通过使

用不同的 Loader，处理不同格式的文件，例如处理 CSS 样式文件，需要使用 style-loader 和 css-loader。下面通过 npm 安装它们：

```
npm install css-loader --save-dev
npm install style-loader --save-dev
```

Loaders 安装后，需要在 webpack.config.js 的 module 对象的 rules 属性中配置。rules 是一个数组配置规则，可以指定一系列 Loaders，每个 Loader 都必须包含 test 和 use 两个选项。配置规则告诉 webpack 符合 test 规定格式的文件，使用 use 后面的 Loader 处理。配置 style-loader 和 css-loader 的示例代码如下：

```
rules:[
     {
        test: /\.css$/,
        //webpack的loader的执行顺序是由右向左执行,先执行css-loader后执行style-loader
        use: [
           'style-loader',
           'css-loader'
          ]
     }
]
```

上述配置的意思是当 webpack 编译过程中遇到 require()或 import 语句导入扩展名为.css 的文件时，先将.css 文件通过 css-loader 转换，再通过 style-loader 转换，然后继续打包。配置 style-loader 和 css-loader 完成后，就可以在配置的入口文件 main.js 中使用 require()或 import 语句导入.css 文件了。

webpack 有许多功能强大的加载器，读者可通过官方网站 https://webpack.js.org/loaders/ 进行学习。

❷ 插件 Plugins

插件 Plugins 是实现 webpack 的自定义功能，可实现 Loaders 不能实现的复杂功能。使用 Plugins 丰富的自定义 API 以及生命周期事件，可以控制 webpack 打包流程的每个环节。现在用一个 mini-css-extract-plugin 插件将散落在 webpack-firstdemo 中的 css 提取出来，并生成一个 common.css 文件，最终在 index.html 中通过<link>的形式加载它。下面在例 15-7 的基础上实现该插件。

【例 15-8】实现 CSS 导出插件。具体实现过程如下。

1）创建.css 文件

在 C:\webpack-firstdemo 目录中创建 css 文件夹，并在该文件夹中创建 style.css，内容如下：

```
#app{
    font-size: 24px;
    color: #f50;
}
```

2）导入.css 文件

配置 style-loader 和 css-loader 的前提下，在配置的入口文件 main.js 中使用 import 语句导入.css 文件，具体如下：

```
import './css/style.css'
```

3）安装 mini-css-extract-plugin 插件

通过 npm install --save-dev mini-css-extract-plugin 命令安装 mini-css-extract-plugin 插件。

4）配置插件

在配置文件 webpack.config.js 中配置插件，完整的 webpack.config.js 配置如下：

```
const path = require('path')
//导入插件
const MiniCssExtractPlugin = require('mini-css-extract-plugin')
const config = {
    entry: {
        main:'./main.js'
    },
    output: {
        path: path.resolve(__dirname, 'dist'),
        publicPath: '/dist/',
        filename: 'bundle.js'
    },
    mode: 'development',
    plugins: [
        new MiniCssExtractPlugin({
            filename: 'common.css'//导出的文件名
        })
],
    // 添加的 module 里面的 rules
    module:{
        rules:[
            {
                test: /\.css$/i,
                use: [
                    {
                        loader: MiniCssExtractPlugin.loader,
                        options: {
                          publicPath: '/dist/',
                        },
                    },
                    'css-loader'
                ]
            }
        ]
    }
}
module.exports = config
```

5）使用 CSS

在 index.html 中，使用<link>元素引用 common.css，具体代码如下：

```
<head>
    <link rel="stylesheet" type="text/css" href="/dist/common.css">
</head>
<div id="app">
```

```
    Hello Webpack!
</div>
<script type="text/javascript" src="/dist/bundle.js"></script>
```

6）测试插件

在 Terminal 终端执行 npm run dev 命令，将会自动在浏览器中打开页面，如图 15.7 所示。从图 15.7 可以看出，导出 CSS 的插件已生效。

webpack 看似复杂，但它只不过是一个 js 配置文件，只要明白入口、出口、加载器和插件这 4 个概念，使用起来就不会那么困难。

图 15.7　使用 CSS 的结果

15.3.5　单文件组件与 vue-loader

Vue.js 是一个渐进式的 JavaScript 框架，在使用 webpack 构建 Vue 应用时，可以使用一种新的构建模式：.vue 单文件组件。

.vue 是 Vue.js 自定义的一种文件格式，一个.vue 文件就是一个单独的组件，在文件内封装了组件相关的代码：html、css、js。

.vue 文件由三部分组成：<template>、<style>、<script>，示例如下：

```
<template>
    html
</template>
<style>
    css
</style>
<script>
    js
</script>
```

可是，浏览器本身并不识别.vue 文件，所以必须对.vue 文件进行加载解析，此时需要 webpack 的 vue-loader 加载器。下面通过一个实例讲解如何使用 vue-loader 实现单文件组件。

【例 15-9】使用 vue-loader 实现单文件组件。

在例 15-8 的基础上完成例 15-9。具体实现过程如下。

❶ 安装开发依赖

使用 vue-loader 加载解析.vue 文件时，需要使用 vue-template-compiler 编译器将模板内容预编译为 JavaScript 渲染函数。安装 Vue、vue-loader 以及 vue-template-compiler 依赖时，要保证版本一致（编写本书时，还没有对应 Vue 3 的 vue-template-compiler，因此本节使用的是 Vue 2）。另外，还需要安装 babel-loader 加载器解析 ES 语法。进入 webpack-firstdemo 目录，按照以下命令安装依赖：

```
npm install -D vue vue-loader vue-template-compiler
npm install -D vue-style-loader
npm install -D babel-loader @babel/core @babel/preset-env
```

❷ 修改配置文件

在配置文件 webpack.config.js 中，添加 vue-loader 和 babel-loader 的配置。修改后的配置

文件内容如下：

```
const path = require('path')
const MiniCssExtractPlugin = require('mini-css-extract-plugin')
const VueLoaderPlugin = require('vue-loader/lib/plugin')
const config = {
    entry: {
        main:'./main.js'
    },
    output: {
        path: path.resolve(__dirname, 'dist'),
        publicPath: '/dist/',
        filename: 'bundle.js'
    },
    mode: 'development',
    module:{
        rules:[
            {
                test: /\.vue$/,
                loader: 'vue-loader'
            },
            {
                test: /\.css$/i,
                use: [
                    //vue 文件中的 css
                    'vue-style-loader',
                    {
                        loader: MiniCssExtractPlugin.loader,
                        options: {
                          publicPath: '/dist/',
                        },
                      },
                    'css-loader'
                ]
            },
            {
                test: /\.m?js$/,
                exclude: /node_modules/,
                use: {
                  loader: 'babel-loader',
                  options: {
                    presets: ['@babel/preset-env']
                  }
                }
              }
        ]
    },
    plugins: [
        // 请确保引入该插件！
        new VueLoaderPlugin(),
```

```
        new MiniCssExtractPlugin({
            filename: 'common.css'//导出的文件名
        })
    ]
}
module.exports = config
```

❸ 创建 app.vue 文件

在 webpack-firstdemo 目录下新建一个 app.vue 文件作为根实例组件，具体代码如下：

```
<template>
    <div>你好，{{vname}}</div>
</template>
<script>
    export default {
        data(){
            return {
                vname: 'Vue'
            }
        }
    }
</script>
<!--因为本项目使用 mini-css-extract-plugin 插件打包 css，加了 scoped 的这部分样式只
对当前组件（app.vue）有效-->
<style  scoped>
    div {
         color: red;
         font-size: 40pt;
    }
</style>
```

❹ 修改入口 main.js

.vue 是没有名称的组件，在父组件中使用时可以对它自定义。现在需要在 main.js 中使用它。修改后的 main.js 内容如下：

```
import './css/style.css'
//下面是 Vue 3 的写法
/*//导入 Vue 框架中的 createApp 方法，在 Vue 3 中不能全局导入 Vue
import {createApp} from 'vue'
//导入 app.vue 组件
import App from './app.vue'
//创建 Vue 根实例
createApp(App).mount("#app")
*/
//下面是 Vue 2 的写法
import Vue from 'vue'
//导入 app.vue 组件
import App from './app.vue'
//创建 Vue 根实例
const vm = new Vue({
    //指定 vm 实例要控制的页面区域
    el: '#app',
```

```
    //通过 render 函数，把指定的组件渲染到 el 区域中
    render: h => h(App)
    /**
     * 相当于 render: function(h){
     *     return h(app)
     * }
     */
})
```

❺ 运行测试

执行 npm run dev 命令，运行结果如图 15.8 所示。

图 15.8 例 15-9 的运行结果

下面在 webpack-firstdemo 目录中再新建一个.vue 文件 input.vue，具体代码如下：

```
<template>
  <div>
      <input v-model="uname">
      <p>输入的用户名是：{{ uname }}</p>
  </div>
</template>
<script>
export default {
  props: {
      uname: {
          type: String
      }
  }
}
</script>
```

在根实例 app.vue 组件中，导入 input.vue 组件，修改后的 app.vue 代码如下：

```
<template>
    <div>
        你好，{{vname}}
        <!--使用子组件 vInput 渲染-->
        <v-input></v-input>
    </div>
</template>
<script>
    import vInput from './input.vue'
    export default {
        data(){
            return {
                vname: 'Vue'
```

```
            }
        },
        components: {//vInput 作为根组件的子组件
            vInput
        }
    }
</script>
<style  scoped>
    div {
        color: red;
        font-size: 40pt;
    }
</style>
```

执行 npm run dev 命令，运行结果如图 15.9 所示。

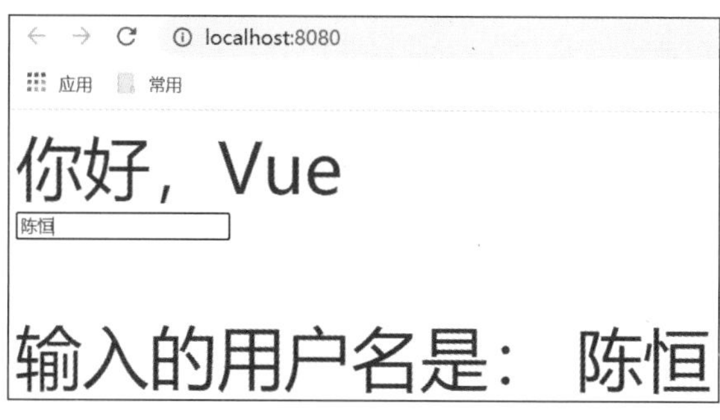

图 15.9　渲染两个组件内容

从图 15.9 可以看出，在 index.html 中渲染了多个组件内容，这就是一个简单的单页面应用，即仅有 index.html 页面。

webpack-firstdemo 目录是本节（15.3 节）的代码，读者可在目录下执行 npm install 命令自动安装所有的依赖，然后执行 npm run dev 命令启动服务。

15.4　路由 vue-router

视频讲解

vue-router 是 Vue.js 的官方路由管理器，使构建单页面应用变得更加容易。

▶ 15.4.1　什么是路由

学习 MVC 框架时，已经了解了控制器根据不同的 URL 请求对应不同的处理请求方法，这是后端路由完成的工作。那么什么是前端路由呢？

在 Web 前端单页面应用中，路由描述的是 URL 与 UI 之间的映射关系，这种映射是单向的，即 URL 变化引起 UI 更新（无须刷新页面）。

▶ 15.4.2　使用 Vue CLI 搭建 vue-router 项目

在 15.3 节学习搭建单页面应用程序时，安装了许多插件并编写了复杂的项目配置，大大降低了开发效率。为提高单页面应用程序的开发效率，从本节开始使用 Vue CLI（Vue 脚手架）搭建 Vue 项目。

Vue CLI 是一个基于 Vue.js 进行快速开发的完整系统，提供以下功能：

- 通过@vue/cli 实现交互式项目脚手架。
- 通过@vue/cli + @vue/cli-service-global 实现零配置原型开发。
- 一个运行时依赖@vue/cli-service，该依赖可升级，基于 webpack 构建，并带有合理的默认配置；可通过项目的配置文件进行配置；可通过插件进行扩展。
- 一个丰富的官方插件集合，集成了前端生态工具。
- 提供一套创建和管理 Vue.js 项目的用户界面。

Vue CLI 致力于将 Vue 生态工具基础标准化，确保各种构建工具平稳衔接，让开发者专注在撰写应用上，而不必纠结配置的问题。下面讲解如何安装 Vue CLI 与创建项目。具体步骤如下。

❶ 全局安装 Vue CLI

打开 cmd 命令行窗口，输入“npm install -g @vue/cli”命令全局安装 Vue 脚手架，输入“vue --version”命令查看版本（测试是否安装成功）。

❷ 打开用户界面

安装成功后，在命令行窗口继续输入“vue ui”命令打开一个浏览器窗口，并以图形化界面引导至项目创建的流程，如图 15.10 所示。

图 15.10 Vue CLI 图形化界面

❸ 创建项目

在图 15.10 中，单击“创建”进入创建新项目界面，如图 15.11 所示。

在图 15.11 中，输入并选择项目相关信息后，单击“下一步”按钮进入项目预设界面，选择“手动”，单击“下一步”按钮进入项目功能界面，在该界面中激活 Router 按钮，安装 vue-router 插件为本节后续内容做准备，如图 15.12 所示。

在图 15.12 中，单击“下一步”按钮进入项目配置界面，配置后单击“创建项目”按钮即可完成项目 router-demo 的创建（可能需要一定的创建时间），如图 15.13 所示。

❹ 使用 VSCode 打开项目

使用 VSCode 打开（使用菜单 File | Open Folder 选择项目目录）第 3 步创建的项目 router-demo。打开后，在 Terminal 终端输入“npm run serve”命令启动服务，如图 15.14 所示。

图 15.11 创建项目界面

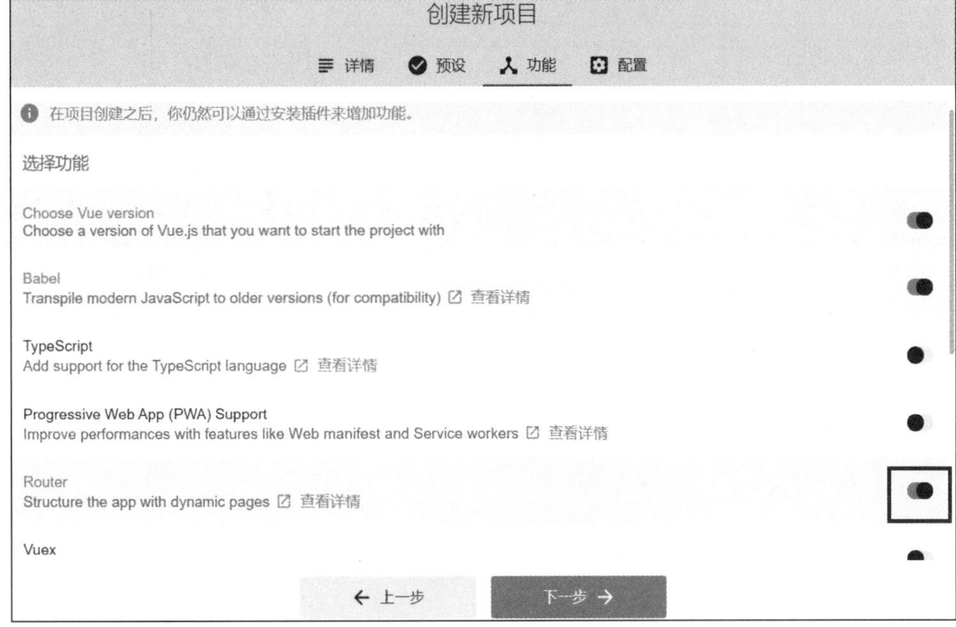

图 15.12 项目功能界面

创建新项目

详情　预设　功能　配置

Choose a version of Vue.js that you want to start the project with　3.x (Preview)

Use history mode for router? (Requires proper server setup for index fallback in production)
By using the HTML5 History API, the URLs don't need the '#' character anymore. 查看详情

Pick a linter / formatter config:
Checking code errors and enforcing an homogeneous code style is recommended.　ESLint with error prevention only

Pick additional lint features:
Lint on save
Lint and fix on commit□[31m (requires Git)□[39m

上一步　创建项目

图 15.13　项目配置界面

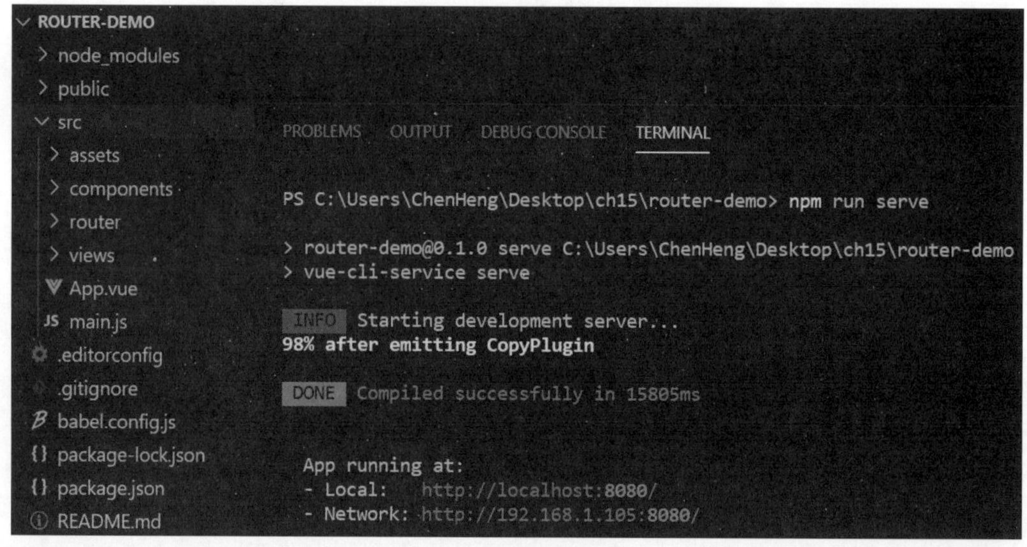

图 15.14　启动服务

❺ 运行项目

在浏览器地址栏中，访问 http://localhost:8080/即可运行项目，如图 15.15 所示。

通过 http://localhost:8080/访问时，打开的页面是 public 目录下的 index.html。index.html 是一个普通的 html 文件，让它与众不同的是“<div id="app"></div>”这句程序，下面有一行注释，构建的文件将会被自动注入，也就是说我们编写的其他内容都将在这个 div 中展示。另外，整个项目只有这一个 html 文件，所以这是一个单页面应用，当打开这个应用时，表面上可以看到很多页面，实际上它们都在这一个 div 中显示。

在 main.js 中创建了一个 Vue 对象。该 Vue 对象的挂载目标是“#app”（与 index.html 中

的 id="app"对应)；router 代表该对象包含 Vue Router，并使用项目中定义的路由（在 src/router 目录下的 index.js 文件中定义）。

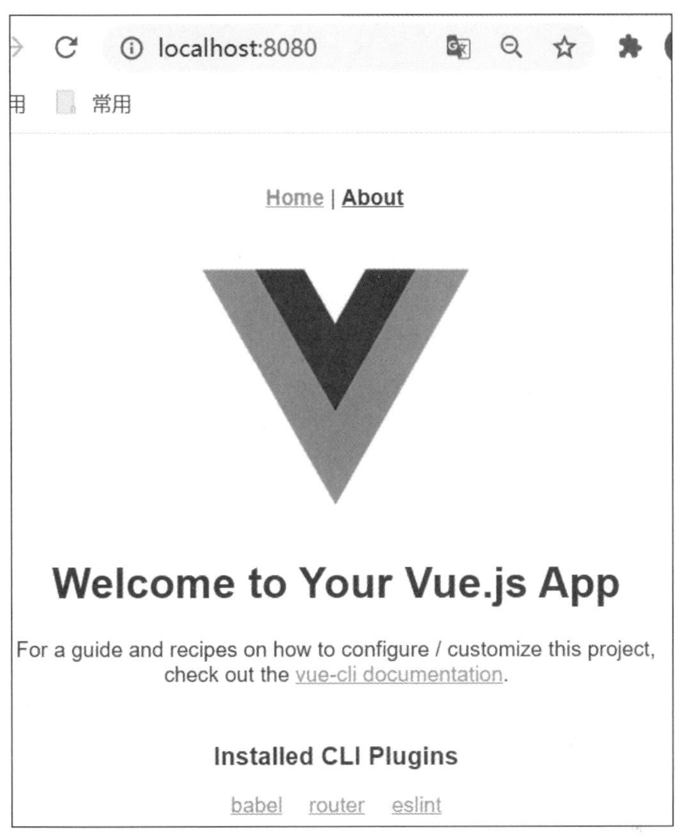

图 15.15　运行项目

综上所述，main.js 与 index.html 是项目启动的首加载页面资源与 js 资源，App.vue 则是 vue 页面资源的首加载项，称为根组件。vue 项目的具体执行过程如下：首先启动项目，找到 index.html 和 main.js，执行 main.js（入口程序），根据 import 加载 App.vue 根组件；然后将组件内容渲染到 index.html 中 id="app"的 DOM 元素上。

▶ 15.4.3　vue-router 基本用法

如图 15.14 所示，打开使用 Vue CLI 搭建好的 vue-router 项目 router-demo。在 router-demo/src 目录中找到 main.js，可以看到使用 use 方法加载 vue-router 插件：

```
import { createApp } from 'vue'
import App from './App.vue'
import router from './router'   //导入 router 目录中的 index.js，在该文件中进行路
                                //由的创建与配置
createApp(App).use(router).mount('#app')
```

下面通过一个实例讲解路由的配置。

【例 15-10】配置路由。具体实现步骤如下：

❶ 创建视图

在 src/views 目录下创建两个.vue 文件：MView1.vue 和 MView2.vue。

MView1.vue 的内容如下：

```
<template>
    <div>第一个页面</div>
</template>
```

Mview2.vue 的内容如下：

```
<template>
    <div>第二个页面</div>
</template>
```

❷ 配置路由

在 src/router 目录中找到路由配置文件 index.js。在该文件中制定路由匹配列表，即 URL 与 UI 的对应关系。每个路由映射一个组件。具体配置内容如下：

```
import { createRouter, createWebHistory } from 'vue-router'
import MView1 from '../views/MView1.vue'
import MView2 from '../views/MView2.vue'
const routes = [
    {
      path: '/',
      name: 'MView1',
      component: MView1
     },
     {
      path: '/MView2',
      name: 'MView2',
      component: MView2
     }
]
// 创建路由实例
const router = createRouter({
    // 开启 history 路由模式，通过“/”设置路径
    history: createWebHistory(process.env.BASE_URL),
    routes
})
export default router
```

❸ 修改根组件 App.vue

在 src 目录下找到 App.vue 文件，修改后的模板内容如下：

```
<template>
    <div id="nav">
        <router-link to="/">第一个页面</router-link> |
        <router-link to="/MView2">第二个页面</router-link>
    </div>
    <!--路由视图挂载所有路由组件-->
    <router-view/>
</template>
```

❹ 运行项目

在 Terminal 终端输入“npm run serve”命令启动服务，访问 http://localhost:8080/即可运行项目，如图 15.16 所示。

单击图 15.16 的超链接，进行路由切换，切换的是<router-view/>挂载的组件，其他内容不变。

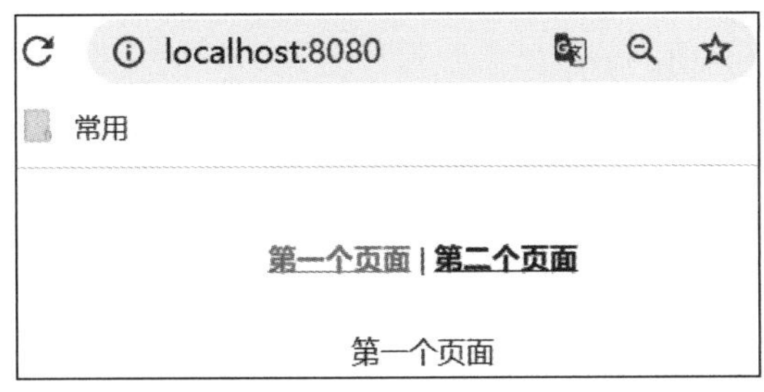

图 15.16　例 15-10 的运行结果

15.4.4　跳转与传参

❶ 跳转

vue-router 有两种跳转。第一种跳转是使用内置的<router-link>组件，默认渲染一个<a>标签，示例代码如下：

```
<div id="nav">
    <router-link to="/">第一个页面</router-link> |
    <router-link to="/MView2">第二个页面</router-link>
</div>
```

<router-link>组件与一般组件一样，to 是一个 prop，指定跳转的路径。使用<router-link>组件，在 HTML5 的 History 模式下将拦截点击，避免浏览器重新加载页面。<router-link>组件还有其他的常用属性。

（1）tag 属性：指定渲染的标签，例如<router-link to="/" tag="li">渲染的结果是<li>而不是<a>。

（2）replace 属性：使用 replace 不会留下 History 记录，所以导航后不能用后退键返回上一个页面，例如<router-link to="/" replace>。

vue-router 的第二种跳转需要在 JavaScript 中进行，类似于 window.location.href。这种方式需要使用 router 实例方法 push 或 replace。例如，在 MView1.vue 中，通过点击事件跳转，示例代码如下：

```
<template>
  <div>第一个页面</div>
  <button @click="goto">去第二个页面</button>
</template>
<script>
export default {
  methods: {
    goto () {
      // 也可以使用 replace 方法，与 replace 属性一样不会向 history 添加新记录
      this.$router.push('/MView2')
    }
  }
}
</script>
```

❷ 传参

路由传参一般有两种方式：query 和 params。不管哪种方式，都是通过修改 URL 来实现。

1）query 传参

query 传递参数的示例代码如下：

```
<router-link to="/?id=888&pwd=999">
```

通过$route.query 获取路由中的参数，示例代码如下：

```
<h4>id: {{$route.query.id}}</h4>
<h4>pwd: {{$route.query.pwd}}</h4>
```

2）params 传参

在路由规则中定义参数，修改路由规则的 path 属性，示例代码如下；

```
 {
path: '/:id/:pwd',
name: 'MView1',
component: MView1
}
<router-link to="/888/999">
```

通过$route.params 获取路由中的参数，示例代码如下：

```
<h4>id: {{$route.params.id}}</h4>
<h4>pwd: {{$route.params.pwd}}</h4>
```

▶ 15.4.5 路由钩子函数

在路由跳转时，可能需要一些权限判断或者其他操作，这时需要使用路由的钩子函数。路由钩子函数主要是给使用者在路由发生变化时进行一些特殊的处理而定义的函数。钩子函数分类具体如下。

❶ 全局前置钩子函数

可以使用 router.beforeEach 注册一个全局前置钩子函数（在跳转前执行），示例代码如下：

```
const router = new createRouter({ ... })
router.beforeEach((to, from, next) => {
        // ...
})
```

beforeEach 函数接收三个参数。

- to: Route：即将进入的目标路由对象。
- from: Route：当前导航正要离开的路由。
- next: Function：一定要调用该方法来解析 beforeEach 钩子。执行效果依赖 next 方法的调用参数。

next 参数的相关说明具体如下。

- next(): 执行管道中的下一个钩子。如果全部钩子执行完，则导航的状态就是 confirmed（确认的）。
- next(false)：中断当前的导航。如果浏览器的 URL 改变（可能是用户手动或者浏览器后退按钮），那么 URL 地址会重置到 from 路由对应的地址。

- next('/')或者 next({ path: '/' })：跳转到一个不同的地址。当前的导航被中断，然后进行一个新的导航。可以向 next 传递任意位置对象，且允许设置诸如 replace: true、name: 'home'之类的选项以及任何用在 router-link 的 to 属性或 router.push 中的选项。
- next(error)：如果传入 next 的参数是一个 Error 实例，则导航被终止且该错误被传递给 router.onError()注册过的回调。

确保 next()函数在任何给定的前置钩子中被严格调用一次。它可以出现多次，但是只能在所有的逻辑路径都不重叠的情况下，否则钩子永远都不会被解析或报错。例如，在用户未能验证身份时重定向到/login 的示例：

```
router.beforeEach((to, from, next) => {
    if (to.name !== 'Login' && !isAuthenticated)
        next({ name: 'Login' })
    else
        next()
})
```

❷ 全局后置钩子函数

也可以注册全局后置钩子函数，该钩子函数不接收 next 参数，也不会改变导航本身，在跳转之后判断。示例代码如下：

```
router.afterEach((to, from) => {
    // ...
})
```

❸ 某个路由的钩子函数

顾名思义，它是写在某个路由里的函数，本质上跟组件内的函数没有区别。示例代码如下：

```
const router = new VueRouter({
    routes: [
        {
            path: '/foo',
            component: Foo,
            beforeEnter: (to, from, next) => {
                // ...
            }
            beforeLeave: (to, from, next) => {
                // ...
            }
        }
    ]
})
```

❹ 组件内的钩子函数

可以在路由组件内直接定义路由导航钩子函数：beforeRouteEnter、beforeRouteUpdate、beforeRouteLeave。示例代码如下：

```
const Foo = {
    template: '...',
    beforeRouteEnter (to, from, next) {
```

```
        //在渲染该组件的对应路由被确认前调用
        //不能获取组件实例 'this'
        //因为当该钩子函数执行前，组件实例还没被创建
    },
    beforeRouteUpdate (to, from, next) {
        //在当前路由改变，但是该组件被复用时调用
        //举例来说，对于一个带有动态参数的路径/foo/:id，在/foo/1 和/foo/2 之间跳转的时候，
        //由于会渲染同样的 Foo 组件，因此组件实例会被复用，而这个钩子就会在此情况下被调用。
        //可以访问组件实例 'this'
    },
    beforeRouteLeave (to, from, next) {
        //导航离开该组件的对应路由时调用
        //可以访问组件实例 'this'
    }
}
```

router-demo 目录是本节（15.4 节）的代码，读者可在目录下执行 npm install 命令自动安装所有的依赖，然后执行 npm run serve 命令启动服务。

本节只是简单介绍钩子函数的分类与定义，具体应用在第 15.5.3 节中。

视频讲解

15.5 状态管理与 Vuex

Vuex 是一个专为 Vue.js 应用程序开发的状态管理模式。它采用集中式存储管理应用的所有组件的状态，并以相应的规则保证状态以一种可预测的方式发生变化。

▶ 15.5.1 状态管理与应用场景

状态管理管理的是全局状态，即全局变量。在较大型的项目中，有许多组件用到同一变量，例如，一个登录的状态，很多页面组件都需要这个信息。在这样的情景下，使用 Vuex 进行登录状态的统一管理就很方便。当然，虽然麻烦但也可以时刻在对应页面操作 cookie。所以，状态管理不是必需的，所有状态管理能做的，都能用其他方式实现，但是状态管理提供了统一管理的地方，操作方便，也更加明确。但一些状态只是父组件和子组件共享，不推荐使用状态管理实现，用$emit 和 props 即可简单实现。

▶ 15.5.2 Vuex 基本用法

本节参考 15.4.2 节使用 Vue CLI 搭建基于 Router 和 Vuex 功能（如图 15.12 所示）的项目 vuex-demo。Vuex 的用法与路由 vue-router 类似，在 main.js 中，通过 use()方法调用，示例代码如下：

```
import { createApp } from 'vue'
import App from './App.vue'
import router from './router'    //导入 router 目录中的 index.js，在该文件中进行路由
                                 //的创建与配置
import store from './store'      //导入 store 目录中的 index.js，在该文件中进行 Vuex
                                 //的创建与配置
createApp(App).use(store).use(router).mount('#app')
```

仓库 store 包含了应用的数据（状态）和操作过程。Vuex 中的数据都是响应式的，即任何组件使用同一 store 的数据时，只要 store 的数据变化，对应的组件立即更新。下面介绍 store 中的选项。

❶ state

需要状态跟踪（管理）的数据保存在 Vuex 选项的 state 字段内，例如要实现一个计算器，可以在 store 目录的 index.js 中定义一个数据 count，初始值为 0，示例代码如下：

```
import { createStore } from 'vuex'
export default createStore({
    state: { //state 用来存储数据变量
        count: 0
    }
})
```

那么，在任何组件内，都可以直接通过$store.state.count 读取数据，示例代码如下：

```
<template>
    <div>计数器值为 {{$store.state.count}}</div>
</template>
```

❷ mutations

在组件内，来自仓库 store 的数据只能读取，不能修改。改变 store 中数据的唯一办法是显示提交 mutations。mutations 是 Vuex 的一个选项，用来直接修改 store 中的数据。例如，给计数器添加两个 mutations，示例代码如下：

```
import { createStore } from 'vuex'
export default createStore({
    state: {//state 用来存储数据变量
        count: 0
    },
    mutations: {/*提交更新数据的方法，必须是同步的
        每个 mutation 都有一个更改状态的方法，并且接受 state 作为第一个参数，提交数据
          （可以是一个对象无限扩展数据）作为第二个参数*/
        add(state, data){
            state.count += data
        },
        sub(state, data){
            state.count -= data
        }
    }
})
```

在组件内，通过 this.$store.commit()方法执行 mutations，示例代码如下：

```
<template>
    <div>{{count}}</div>
    <button @click="myadd">+10</button>  
    <button @click="mysub">-5</button>
</template>
<script>
export default {
```

```
        computed: {
            count () {
                return this.$store.state.count
            }
        },
        methods: {
            myadd (){
                this.$store.commit('add', 10)//add 是 mutations 定义的方法
            },
            mysub (){
                this.$store.commit('sub', 5)
            }
        }
    }
    </script>
```

❸ getters

Vuex 允许在 store 中定义 getter（可以认为是 store 的计算属性）。注意，从 Vue 3.0 开始，getter 的结果不像计算属性那样被缓存。有时需要从 store 中的 state 派生出一些状态，例如对列表进行过滤并计数，示例代码如下：

```
import { createStore } from 'vuex'
export default createStore({
    state: {
        mylist: [100, 20, 30, 300, 200, 400]
    },
    getters: {
        //统计列表 mylist 中大于或等于 100 的项
        filterList (state) {
            return state.mylist.filter(item => item >= 100)
        }
    }
})
```

在组件中，可以通过$store.getters 访问 getters 中的方法，示例代码如下：

```
<template>
    <div>统计列表中大于 100 的数据：{{$store.getters.filterList}}</div>
</template>
```

❹ actions

actions 类似于 mutations，不同之处在于：actions 提交的是 mutations，而不是直接变更状态。actions 可以包含任意异步操作，示例代码如下：

```
import { createStore } from 'vuex'
export default createStore({
    state: {
        count: 0
    },
    mutations: {
        add(state, data){
            state.count  += data
```

```
        }
    },
    actions: {
        add (context) {
            //提交 mutations 中的 add 方法
            context.commit('add', 10)
        }
    }
})
```

在组件中，可以通过$store.dispatch()触发 actions 中的方法，示例代码如下：

```
<template>
    <button @click="$store.dispatch('add')"> actions+10 </button>
</template>
```

上述 actions 的用法感觉多此一举，就目前示例确实如此。但加上异步操作就不一样了，因为 mutations 是同步操作。假如现在要求 5 秒钟后提交 mutations，示例代码如下：

```
import { createStore } from 'vuex'
export default createStore({
    state: {
        count: 0
    },
    mutations: {
        add(state, data){
            state.count  += data
        }
    },
    actions: {
        addAsync ({ commit }) {
            setTimeout(() => {
                commit('add', 10)
            }, 5000)
        }
    }
})
```

在组件中，同样通过$store.dispatch()触发 actions 中的方法，示例代码如下：

```
<template>
    <button @click="$store.dispatch('addAsync')"> actions 异步+10 </button>
</template>
```

❺ modules

应用的所有状态都集中到 store 对象中的缺点是：当应用变得非常复杂时，store 可能变得相当臃肿。为了解决这个问题，Vuex 允许将 store 分割成模块（modules）。每个模块拥有自己的 state、mutation、action、getter，甚至是嵌套子模块——从上至下进行同样方式的分割。示例如下：

```
const moduleA = {
    state: () => ({ ... }),
    mutations: { ... },
```

```
    actions: { ... },
    getters: { ... }
}
const moduleB = {
    state: () => ({ ... }),
    mutations: { ... },
    actions: { ... }
}
const store = createStore({
    modules: {
        a: moduleA,
        b: moduleB
    }
})
store.state.a //moduleA 的 state
store.state.b //moduleB 的 state
```

以上 5 个选项比较常用的是 state 和 mutations，下面通过登录权限验证实例巩固 Vuex 的基本用法。

15.5.3 登录权限验证

登录权限验证实例要求如下：

- 在 App.vue 根组件中，通过<router-link>访问登录页面组件 Login.vue、主页面组件 Main.vue 以及 Home.vue 组件。
- 登录成功后，才能访问主页面组件 Main.vue 和 Home.vue。
- 在 main.js 中，使用路由 beforeEach((to,from,next)钩子函数实现登录权限验证。
- 演示 15.5.2 节的 Vuex 基本用法。

【例 15-11】登录权限验证实例。具体实现过程如下：

❶ 完善 App.vue

完善项目 vuex-demo 根组件 App.vue 的模板代码，具体如下：

```
<template>
    <div id="nav">
        <router-link to="/login">Login</router-link> |
        <router-link to="/main">Main</router-link> |
        <router-link to="/home">Home</router-link>
    </div>
    <router-view/>
</template>
```

❷ 配置路由

在 src/router 目录的 index.js 文件中配置路由，需要登录验证的路由使用 meta 数据标注。路由配置具体如下：

```
import { createRouter, createWebHistory } from 'vue-router'
import Login from '../views/Login.vue'
import Main from '../views/Main.vue'
import Home from '../views/Home.vue'
const routes = [
```

```
    {
        path: '/login',
        name: 'Login',
        component: Login
    },
    {
        path: '/home',
        name: 'Home',
        component: Home,
        meta:{auth:true}
    },
    {
        path: '/main',
        name: 'Main',
        component: Main,
        meta:{auth:true}//需要验证登录权限
    }
]
const router = createRouter({
    history: createWebHistory(process.env.BASE_URL),
    routes
})
export default router
```

❸ 配置状态管理

在 src/store 目录的 index.js 文件中配置 store，state 项有三个状态数据：count、mylist 和 isLogin。store 配置具体如下：

```
import { createStore } from 'vuex'
export default createStore({
    state: {//state 用来存储数据变量
        count: 0,
        mylist: [100, 20, 30, 300, 200, 400],
        //初始时 isLogin='0' 表示用户未登录
        isLogin: window.sessionStorage.getItem('user') == null ? '0' : window.
        sessionStorage.getItem('user')
    },
    mutations: {/*提交更新数据的方法，必须是同步的
    每个 mutation 都有一个更改状态的方法，并且接受 state 作为第一个参数，提交数据作为第
      二个参数（可以是一个对象无限扩展数据）。*/
        changeLogin(state, data) {
            state.isLogin = 1;
            window.sessionStorage.setItem('user', data)
        },
        add(state, data){
            state.count  += data
        },
        sub(state, data){
            state.count -= data
        }
    },
```

```
    getters: {
        //统计列表 mylist 中大于或等于 100 的项
        filterList (state) {
            return state.mylist.filter(item => item >= 100)
        }
    },
    actions: {
        add (context) {
            //提交 mutations 中的 add 方法
            context.commit('add', 10)
        },
        addAsync ({ commit }) {
            setTimeout(() => {
                commit('add', 10)
            }, 5000)
        }
    }
})
```

❹ 登录权限验证

在配置文件 main.js 中，使用路由钩子函数 beforeEach((to,from,next)实现登录权限验证。具体代码如下：

```
import { createApp } from 'vue'
import App from './App.vue'
import router from './router'    //导入 router 目录中的 index.js，在该文件中进行路
                                 //由的创建与配置
import store from './store'      //导入 store 目录中的 index.js，在该文件中进行
                                 //Vuex 的创建与配置
createApp(App).use(store).use(router).mount('#app')
router.beforeEach((to,from,next)=>{
    //如果路由器需要验证
    if(to.matched.some(m=>m.meta.auth)){
      //对路由进行验证
      if (store.state.isLogin == '0') {
        alert("您没有登录，无权访问！")
        /*未登录则跳转到登录界面，
        query:{ redirect: to.fullPath}表示把当前路由信息传递过去方便登录后跳转回来*/
        next({
          path: 'login',
          query: {redirect: to.fullPath}
        })
      } else { // 已经登录
        next()   // 正常跳转到设置好的页面
      }
    }else{
      next()
    }
  }
)
```

❺ 新建登录组件 Login.vue

在 views 目录中，新建登录组件 Login.vue。在该组件中使用$store.commit 触发 mutations，以便登录成功修改登录状态。Login.vue 的代码如下：

```
<template>
    <div>
    <h2>登录页面</h2>
    <form>
     用户名:<input type="text" v-model="uname" placeholder="请输入用户名"/><br><br>
     密码:  <input type="password" v-model="upwd" placeholder="请输入密码"/><br><br>
     <button type="button" @click="login"  :disabled="isDisable">登录</button>
     <button type="reset">重置</button>
     </form>
   </div>
</template>
<script>
export default {
   data () {
    return {
        isDisable:false,
        uname: '',
        upwd: ''
     }
   },
   methods: {
       login () {
           this.isDisable = true
            if (this.uname === 'zhangsan' && this.upwd == '123456') {
                alert('登录成功')
                //触发 changeLogin 为状态变量 isLogin 赋值 uname, commit 方法同步操作
                this.$store.commit('changeLogin', this.uname)
                let path = this.$route.query.redirect
                 /*未登录则跳转到登录界面,
                 query:{ redirect: to.fullPath}表示把当前路由信息传递过去方便登录后
                 跳转回来*/
                this.$router.replace({path: path === '/' || path === undefined ? '/main': path})
            }else {
                alert("用户名或密码错误! ")
                this.isDisable = false
            }
       }
   }
}
</script>
```

❻ 新建主页面组件

在 views 目录中，新建主页面组件 Main.vue。在该组件中使用$store 访问 state、getters、actions。Main.vue 的代码如下：

```
<template>
   <div>欢迎{{uname}}登录成功, 登录状态为{{$store.state.isLogin}}</div>
```

```
    <div>{{count}}</div>
    <button @click="myadd">+10</button>  
    <button @click="mysub">-5</button><br>
    <div>
       统计列表中大于 100 的数据：{{$store.getters.filterList}}
    </div><br>
    <button @click="$store.dispatch('add')"> actions+10 </button><br>
    <button @click="$store.dispatch('addAsync')"> actions 异步+10 </button>
</template>
<script>
export default {
    data () {
      return {
         uname : window.sessionStorage.getItem('user')
      }
    },
    computed: {
      count () {
        return this.$store.state.count
      }
    },
    methods: {
      myadd (){
        this.$store.commit('add', 10)//add 是 mutations 定义的方法
      },
      mysub (){
        this.$store.commit('sub', 5)
      }
    }
}
</script>
```

❼ 测试运行

在登录界面输入用户名 zhangsan，密码 123456。登录成功后打开主页面组件，如图 15.17 所示。

图 15.17　主页面

vuex-demo 目录是本节（15.5 节）的代码，读者可在目录下执行 npm install 命令自动安装所有的依赖，然后执行 npm run serve 命令启动服务。

15.6　本章小结

本章主要介绍了渲染函数、组合 API、webpack、Vue CLI、vue-router 插件以及 Vuex 插件。希望读者重点学习 Vue CLI、vue-router 插件以及 Vuex 插件的用法，为下一章综合项目实战夯实基础。

习 题 15

1．路由传参有几种方式？如何接收路由传递的参数？请举例说明。
2．什么是 h()函数？它有哪几个参数？
3．如何安装 Vue CLI？请使用 Vue CLI 的界面引导的方式创建项目。

视频讲解

第 16 章 人事管理系统的设计与实现（Spring Boot + Vue 3 + MyBatis）

学习目的与要求

本章以人事管理系统的设计与实现为综合案例，讲述如何使用 Spring Boot + Vue.js 3 + MyBatis 开发一个前后端分离的应用程序。通过本章的学习，掌握基于 Spring Boot + Vue.js 3 + MyBatis 的前后端分离的应用程序的开发流程、方法以及技术。

主要内容

- ❖ 系统设计
- ❖ 数据库设计
- ❖ 后台应用的实现
- ❖ 前端项目的实现

前后端分离的核心思想是前端页面通过 Ajax 调用后端的 RESTful API 进行数据交互。本章将使用 Spring Boot + MyBatis 实现后端系统的开发，使用 Vue.js 3 实现前端系统的开发，数据库采用的是 MySQL 5.x，后端集成开发环境为 IntelliJ IDEA，前端集成开发环境为 VSCode。

16.1 系统设计

系统总体目标是构建某单位的人力资源信息管理平台，不仅满足目前的业务需要，还要满足公司未来的发展，而且具备良好的可扩展性，形成公司未来人力资源管理信息化平台。

▶ 16.1.1 系统功能需求

人力资源部门的管理员成功登录系统后，具有如下功能。

（1）部门管理：主要用于描述组织的部门信息，以及部门的上下级关系，包括新建部门、修改部门、查询部门下的员工等功能。

（2）岗位管理：主要用于对组织内各岗位进行管理，包括增加、修改、删除岗位，以及查询岗位下的在职人员等功能。

（3）员工管理：主要用于员工基本信息录入与修改，包括员工部门、岗位、试用期及其他信息的录入。

（4）试用期管理：主要对试用期员工进行管理，包括试用期转正、试用期延期、试用期不通过、已转正员工信息查询等功能。

（5）岗位调动管理：主要对员工岗位调动进行管理，包括部门内岗位调动、部门间岗位调动、调动员工查询等功能。

（6）员工离职管理：主要对员工离职进行管理，包括确定离职员工、已离职员工信息查询等功能。离职的类型包括主动辞职、辞退、退休、开除、试用期未通过。

（7）报表管理：主要对给定时间段新聘员工报表、给定时间段离职员工报表、给定时间段岗位调动员工报表、人事月报等报表进行管理。

16.1.2　系统模块划分

系统包括部门管理、岗位管理、员工管理、试用期管理、岗位调动管理、员工离职管理、报表管理等功能模块。具体功能模块划分如图 16.1 所示。

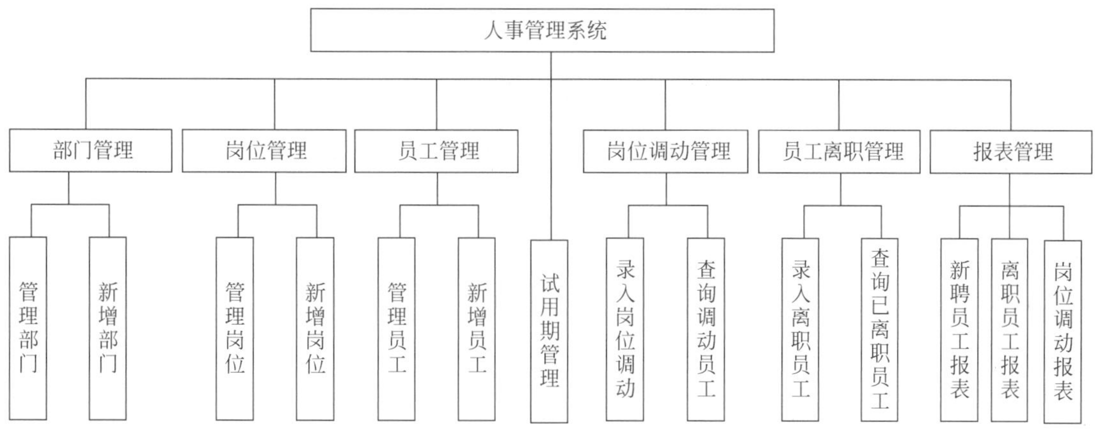

图 16.1　人事管理系统模块划分

16.2　数据库设计

在 MySQL 5.5 中创建数据库 personmis，并在 personmis 中创建 6 张与系统相关的数据表：ausertable、department、post、quit、staff 和 transfer。

16.2.1　数据库概念结构设计

根据系统设计与分析，可以设计出如下数据结构。

❶ 管理员

管理员包括 ID、用户名和密码。管理员的用户名和密码由数据库管理员预设，不需要注册。

❷ 部门

部门包括部门 ID、名称、类型、电话、传真、描述、上级部门以及成立日期。

❸ 岗位

岗位包括岗位 ID、岗位名称、岗位类型以及编制数。

❹ 员工

员工包括员工 ID（编号）、姓名、性别、出生日期、身份证号、所在部门、所在岗位、入职日期、参加工作日期、用工形式、人员来源、政治面貌、民族、籍贯、联系电话、电子邮件、身高、血型、婚姻状况、户口所在地、最高学历、最高学位、毕业院校、所学专业、毕业日期、试用期开始日期、试用期结束日期、状态等信息。其中，编号唯一，所在部门与“2. 部门 ID”关联；所在岗位与“3. 岗位 ID”关联。

❺ 离职记录

离职记录包括 ID、员工编号、员工名称、离职类型、离职日期、记录日期。其中，ID 唯一，员工编号与“4. 员工 ID”关联。

❻ 岗位调动记录

岗位调动记录包括 ID、员工编号、员工名称、调动前岗位、调动后岗位、调动类型、调动日期、记录日期。其中，ID 唯一，员工编号与“4. 员工 ID”关联；调动前岗位和调动后岗位与“3. 岗位 ID”关联。

▶ 16.2.2 数据库逻辑结构设计

将数据库概念结构设计转换为 MySQL 数据库所支持的实际数据模型，即数据库的逻辑结构。

管理员信息表（ausertable）的设计如表 16.1 所示。

表 16.1　管理员信息表

字段	含义	类型	长度	是否为空
id	管理员 ID（PK 自增）	int	11	no
aname	用户名	varchar	50	no
apwd	密码	varchar	50	no

部门信息表（department）的设计如表 16.2 所示。

表 16.2　部门信息表

字段	含义	类型	长度	是否为空
id	部门 ID（PK 自增）	int	11	no
dname	部门名称	varchar	50	no
dtype	部门类型	varchar	50	no
dtel	电话	varchar	50	
dfax	传真	varchar	50	
description	描述	varchar	500	
supdepartment	上级部门	int	11	
establishmentdate	创建日期	date		

岗位信息表（post）的设计如表 16.3 所示。

表 16.3　岗位信息表

字段	含义	类型	长度	是否为空
id	岗位 ID（PK 自增）	int	11	no
pname	岗位名称	varchar	50	no
ptype	部门类型	varchar	50	no
organization	岗位预设人数	int	11	

员工离职信息表（quit）的设计如表 16.4 所示。

表 16.4　员工离职信息表

字段	含义	类型	长度	是否为空
id	ID（PK 自增）	int	11	no
staff_id	员工 ID	varchar	50	no
sname	员工姓名	varchar	50	no
qtype	离职类型	varchar	50	no
qdate	离职日期	date	50	no
opdate	操作日期	date		no

员工信息表（staff）的设计如表 16.5 所示。

表 16.5 员工信息表

字段	含义	类型	长度	是否为空
id	员工 ID（PK 自增）	int	11	no
sname	姓名	varchar	50	no
sex	性别	varchar	10	no
birthday	生日	date		
sid	身份证号	varchar	50	
depart_id	部门	int	11	
post_id	岗位	int	11	
entrydate	入职日期	date		
joinworkdate	工作日期	date		
workform	用工形式	varchar	50	
staffsource	来源	varchar	50	
politicalstatus	政治面貌	varchar	50	
nation	国籍	varchar	50	
nativeplace	民族	varchar	50	
stel	电话	varchar	50	
semail	邮箱	varchar	100	
sheight	身高	decimal	12,2	
bloodtype	血型	varchar	50	
maritalstatus	婚姻状况	varchar	50	
registeredresidence	籍贯	varchar	50	
education	学历	varchar	50	
degree	学位	varchar	50	
university	毕业院校	varchar	50	
major	专业	varchar	50	
graduationdate	毕业时间	date		
startdate	试用期开始时间	date		
enddate	试用期结束时间	date		
status	状态	varchar	50	
peroidopdate	操作日期	date		

岗位调动信息表（transfer）的设计如表 16.6 所示。

表 16.6 岗位调动信息表

字段	含义	类型	长度	是否为空
id	ID（PK 自增）	int	11	no
staff_id	员工 ID	int	11	no
sname	员工姓名	varchar	50	no
beforepost_id	调动前部门	int	11	
afterpost_id	调动后部门	int	11	
ttype	调动类型	varchar	10	
tdate	调动日期	date		
opdate	操作日期	date		

16.2.3 创建数据表

根据 16.2.2 节的逻辑结构，创建数据表。由于篇幅受限，创建数据表代码请读者参考本书提供的源代码 personmis.sql。

16.3 后台应用的实现

16.3.1 使用 IntelliJ IDEA 构建后台应用

参考 5.2.4 节使用 IntelliJ IDEA 构建基于 Spring Web 和 MyBatis Framework 依赖的人事管理系统后台应用 personmis，如图 16.2 所示。

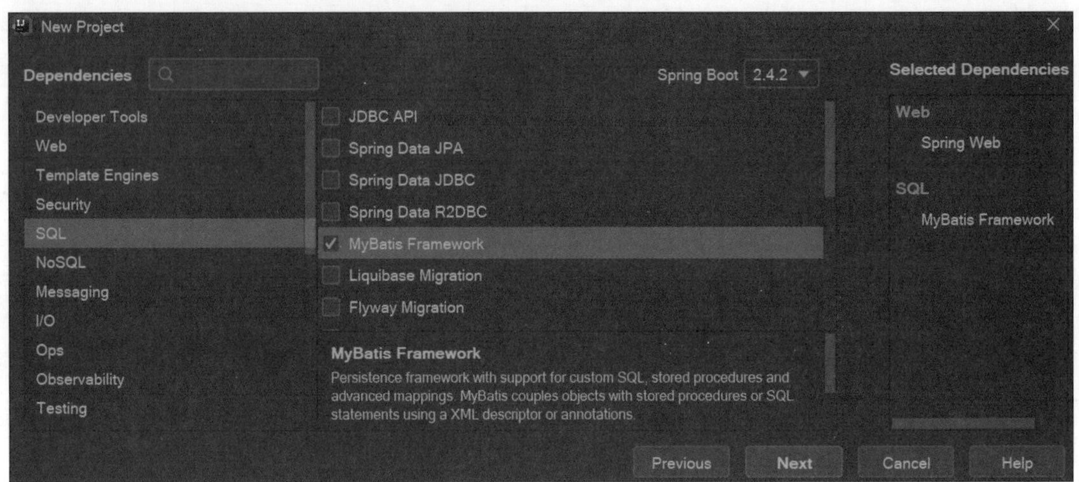

图 16.2 Spring Web 和 MyBatis Framework 依赖

16.3.2 修改 pom.xml

修改人事管理系统 personmis 的 pom.xml 文件，添加 MySQL 连接依赖。具体代码如下：

```
<dependency>
    <groupId>mysql</groupId>
    <artifactId>mysql-connector-java</artifactId>
    <version>5.1.45</version>
</dependency>
```

16.3.3 配置数据源等信息

在人事管理系统 personmis 的配置文件 application.properties 中，配置端口号、数据源等信息，具体内容如下：

```
server.port=8443
server.servlet.context-path=/personmis
spring.datasource.url=jdbc:mysql://localhost:3306/personmis?characterEncoding=utf8
#数据库用户名
```

```
spring.datasource.username=root
#数据库密码
spring.datasource.password=root
#数据库驱动
spring.datasource.driver-class-name=com.mysql.jdbc.Driver
#设置包别名（在 Mapper 映射文件中直接使用实体类名）
mybatis.type-aliases-package=com.ch.personmis.entity
#告诉系统到哪里去找 mapper.xml 文件（映射文件）
mybatis.mapperLocations=classpath:mappers/*.xml
#在控制台输出 SQL 语句日志
logging.level.com.ch.personmis.repository=debug
#让控制器输出的 JSON 字符串格式更美观
spring.jackson.serialization.indent-output=true
```

▶ 16.3.4　创建 CorsFilter 的 Bean 实例实现跨域访问

在 PersonmisApplication 主类中，创建 CorsFilter 的 Bean 实例实现跨域访问。PersonmisApplication 的具体代码如下：

```
package com.ch.personmis;
import org.mybatis.spring.annotation.MapperScan;
import org.springframework.boot.SpringApplication;
import org.springframework.boot.autoconfigure.SpringBootApplication;
import org.springframework.context.annotation.Bean;
import org.springframework.web.cors.CorsConfiguration;
import org.springframework.web.cors.UrlBasedCorsConfigurationSource;
import org.springframework.web.filter.CorsFilter;
@SpringBootApplication
@MapperScan(basePackages={"com.ch.personmis.repository"})
public class PersonmisApplication {
    public static void main(String[] args) {
        SpringApplication.run(PersonmisApplication.class, args);
    }
    //跨域设置
    private CorsConfiguration corsConfig() {
        CorsConfiguration corsConfiguration = new CorsConfiguration();
        //允许跨域请求的域名
        corsConfiguration.addAllowedOrigin("*");
        //允许发送的内容类型
        corsConfiguration.addAllowedHeader("*");
        //跨域请求允许的请求方式
        corsConfiguration.addAllowedMethod("*");
        corsConfiguration.setMaxAge(3600L);
        return corsConfiguration;
    }
    @Bean
    public CorsFilter corsFilter() {
        UrlBasedCorsConfigurationSource source = new UrlBasedCorsConfigurationSource();
        source.registerCorsConfiguration("/**", corsConfig());
        return new CorsFilter(source);
    }
}
```

▶ 16.3.5 管理员登录后台实现

每个功能模块的后台实现共有五部分内容：控制器层（com.ch.personmis.controller）、业务层（com.ch.personmis.service）、数据访问层（com.ch.personmis.repository）、实体层（com.ch.personmis.entity）以及 SQL 映射文件（resources/mappers）。由于篇幅受限，本书仅列出每个功能模块的核心实现，其他具体内容请参考本书提供的源代码。

❶ 控制器层

与管理员登录相关的控制器是 AdminController，具体代码如下：

```
package com.ch.personmis.controller;
import com.ch.personmis.entity.UserEntity;
import com.ch.personmis.service.AdminService;
import org.springframework.web.bind.annotation.PostMapping;
import org.springframework.web.bind.annotation.RequestBody;
import org.springframework.web.bind.annotation.RestController;
import javax.annotation.Resource;
import javax.servlet.http.HttpSession;
@RestController
public class AdminController {
    @Resource
    private AdminService adminService;
    @PostMapping(value = "/login")
    //前端提交的是 JSON 数据对象时，后端使用@RequestBody 解析
    public String login(@RequestBody UserEntity userEntity, HttpSession session) {
        return adminService.login(userEntity, session);
    }
}
```

❷ 业务层

与管理员登录相关的业务层包括 AdminService 接口和 AdminServiceImpl 实现类。AdminService 接口代码略，AdminServiceImpl 实现类的代码具体如下：

```
package com.ch.personmis.service;
import javax.annotation.Resource;
import javax.servlet.http.HttpSession;
import com.ch.personmis.entity.UserEntity;
import com.ch.personmis.repository.AdminRepository;
import org.springframework.stereotype.Service;
@Service
public class AdminServiceImpl implements AdminService{
    @Resource
    private  AdminRepository adminRepository;
    @Override
    public String login(UserEntity userEntity, HttpSession session) {
        UserEntity user = adminRepository.login(userEntity);
        if(user != null) {
            session.setAttribute("auser", user);
            return "ok";
        }else {
            return "no";
        }
    }
}
```

❸ 数据访问层

与管理员登录相关的数据访问层是 AdminRepository 接口，该接口代码略。

❹ SQL 映射文件

与管理员登录相关的 SQL 映射文件是 AdminMapper.xml，该文件的核心内容是：

```
<select id="login" resultType="UserEntity" parameterType="UserEntity">
    select * from ausertable where aname = #{uname} and apwd = #{upwd}
</select>
```

▶ 16.3.6 部门管理后台实现

部门管理功能模块包括增、删、改、查部门。

❶ 控制器层

与部门管理相关的控制器是 DepartController，核心代码如下：

```
@RestController
public class DepartController {
    @Resource
    private DepartService departService;
    @GetMapping("/getDepartment")
    public List<Depart> selectDepart() {
        return departService.selectDepart();
    }
    @GetMapping("/getDepartmentByPage")
    public Map<String, Object> selectDepartByPage(DepartByCon departByCon) {
        return departService.selectDepartByPage(departByCon);
    }
    @PostMapping("/addDepartment")
    public  String addDepartment(@RequestBody Depart depart){
        return  departService.addDepartment(depart);
    }
    @PostMapping("/selectDepartmentsByCon")
    public Map<String, Object> selectDepartmentsByCon(@RequestBody DepartByCon departByCon) {
        return departService.selectDepartmentsByCon(departByCon);
    }
    @GetMapping("/getDepartmentDetail")
    public Map<String, Object> getDepartmentDetail(int id){
        return departService.getDepartmentDetail(id);
    }
    @PostMapping("/updateDepartment")
    public  String updateDepartment(@RequestBody Depart depart){
        return  departService.updateDepartment(depart);
    }
    @PostMapping("/deleteDepartment")
    public  String deleteDepartment(int id){
        return departService.deleteDepartment(id);
    }
}
```

❷ 业务层

与部门管理相关的业务层包括 DepartService 接口和 DepartServiceImpl 实现类。DepartService 接口代码略，DepartServiceImpl 实现类的核心代码具体如下：

```
@Service
public class DepartServiceImpl implements DepartService {
    @Resource
    private DepartRepository departRepository;
    @Override
    public List<Depart> selectDepart() {
        return departRepository.selectDepart(null);
    }
    @Override
    public String addDepartment(Depart depart) {
        if(departRepository.addDepart(depart) > 0)
            return "ok";
        return "no";
    }
    @Override
    public Map<String, Object> selectDepartByPage(DepartByCon departByCon) {
        Map<String, Object> map = new HashMap<String, Object>();
        departByCon.setAct("byPage");
        List<Depart> departs = departRepository.selectDepart(departByCon);
        map.put("departs", departs);
        departByCon.setAct("byNoPage");
        map.put("total", departRepository.selectDepart(departByCon).size());
        return map;
    }
    @Override
    public Map<String, Object> selectDepartmentsByCon(DepartByCon departByCon) {
        Map<String, Object> map = new HashMap<String, Object>();
        departByCon.setAct("byPage");
        List<Depart> departs=departRepository.selectDepartmentsByCon(departByCon);
        map.put("departs", departs);
        departByCon.setAct("byNoPage");
        map.put("total", departRepository.selectDepartmentsByCon(departByCon).size());
        return map;
    }
    @Override
    public Map<String, Object> getDepartmentDetail(int id) {
        Map<String, Object> map = new HashMap<String, Object>();
        List<Depart> departs = departRepository.selectDepart(null);
        map.put("departs", departs);
        Depart aDepart = departRepository.getDepartmentDetail(id);
        map.put("aDepart", aDepart);
        return map;
    }
    @Override
    public String updateDepartment(Depart depart) {
        if(departRepository.updateDepartment(depart) > 0)
            return "ok";
        return "no";
    }
    @Override
    public String deleteDepartment(int id) {
        //先查询是否有关联数据
```

```
        List<Map<String, Object>> listMap =departRepository.selectAssociateDepart(id);
        if (listMap.size() <= 0) {
            if (departRepository.deleteDepart(id) > 0)
                return "ok";
        }
        return "no";
    }
}
```

❸ 数据访问层

与部门管理相关的数据访问层是 DepartRepository 接口，该接口代码略。

❹ 实体层

与部门管理相关的实体层有 Depart（部门实体）和 DepartByCon（条件查询实体），这两个实体的代码略。

❺ SQL 映射文件

与部门管理相关的 SQL 映射文件是 DepartMapper.xml，该文件的核心内容是：

```
    <!-- 查询所有部门 -->
    <select id="selectDepart"  resultType="Depart" parameterType="DepartByCon">
    select id,dname,dtype,dtel,dfax,description,supdepartment,date_format(establishmentdate,
'%Y-%m-%d') as establishmentdate1  from department
        <if test="act == 'byPage'" >
            limit #{startIndex}, #{pageSize}
        </if>
    </select>
    <!-- 条件查询部门 -->
    <select id="selectDepartmentsByCon"  resultType="Depart" parameterType="DepartByCon">
        select id,dname,dtype,dtel,dfax,description,supdepartment,date_format(establishmentdate,
'%Y-%m-%d') as establishmentdate1
        from department
        where 1=1
        <if test="dname !=null and dname!=''">
            and dname like concat('%',#{dname},'%')
        </if>
        <if test="dtype !=null and dtype!=''">
            and dtype = #{dtype}
        </if>
        <if test="act == 'byPage'" >
            limit #{startIndex}, #{pageSize}
        </if>
    </select>
    <!-- 添加部门 -->
    <insert id="addDepart"  parameterType="Depart">
       insert into department (id,dname,dtype,dtel,dfax,description,supdepartment,
establishmentdate)
        values  (null,  #{dname},  #{dtype},  #{dtel},  #{dfax},  #{description},
#{supdepartment}, now())
    </insert>
    <!-- 查询一个部门 -->
    <select id="getDepartmentDetail" resultType="Depart" parameterType="Integer">
        select d1.id,d1.dname,d1.dtype,d1.dtel,d1.dfax,d1.description,
```

```
        <if test="id != 1">
            d1.supdepartment, d2.dname as supdepartment1,
            date_format(d1.establishmentdate,'%Y-%m-%d') as establishmentdate1
            from department d1,department d2 where d1.id = #{id} and d1.supdepartment = d2.id
        </if>
        <if test="id == 1">
            d1.supdepartment, '顶级部门' as supdepartment1,
            date_format(d1.establishmentdate,'%Y-%m-%d') as establishmentdate1
            from department d1 where d1.id = #{id}
        </if>
    </select>
    <!-- 修改部门 -->
    <update id="updateDepartment" parameterType="Depart">
        update department
        <set>
            <if test="dname != null">
                dname = #{dname},
            </if>
            <if test="dtype != null">
                dtype = #{dtype},
            </if>
            <if test="dtel != null">
                dtel = #{dtel},
            </if>
            <if test="dfax != null">
                dfax = #{dfax},
            </if>
            <if test="description != null">
                description = #{description},
            </if>
            <if test="supdepartment != null">
                supdepartment = #{supdepartment}
            </if>
        </set>
        where id = #{id}
    </update>
    <!--查询关联部门-->
    <select id="selectAssociateDepart" resultType="map" parameterType="Integer">
        SELECT DISTINCT d.id, s.id
         FROM department d, staff s WHERE d.supdepartment = #{id} or s.depart_id  = #{id}
    </select>
    <!--删除部门-->
    <delete id="deleteDepart" parameterType="Integer">
        delete from department where id = #{id}
    </delete>
```

▶ 16.3.7 岗位管理后台实现

岗位管理功能模块包括增、删、改、查岗位。

❶ 控制器层

与岗位管理相关的控制器是 PostController，核心代码如下：

```
@RestController
public class PostController {
    @Resource
    private PostService postService;
    @GetMapping("/getPostByPage")
    public Map<String, Object> selectPostByPage(PostByCon postByCon) {
        return postService.selectPostByPage(postByCon);
    }
    @PostMapping("/addPost")
    public  String addPost(@RequestBody Post post){
        return  postService.addPost(post);
    }
    @PostMapping("/selectPostByCon")
    public Map<String, Object> selectPostByCon(@RequestBody PostByCon postByCon) {
        return postService.selectPostByCon(postByCon);
    }
    @GetMapping("/getPostDetail")
    public Post getPostDetail(int id) {
        return postService.getPostDetail(id);
    }
    @PostMapping("/updatePost")
    public  String updatePost(@RequestBody Post post){
        return  postService.updatePost(post);
    }
    @GetMapping("/getPost")
    public List<Post> getPost() {
        return postService.getPost();
    }
    @PostMapping("/deletePost")
    public  String deletePost(int id){
        return postService.deletePost(id);
    }
}
```

❷ 业务层

与岗位管理相关的业务层包括 PostService 接口和 PostServiceImpl 实现类。PostService 接口代码略，PostServiceImpl 实现类的核心代码具体如下：

```
@Service
public class PostServiceImpl implements  PostService{
    @Resource
    private PostRepository postRepository;
    @Override
    public String addPost(Post post) {
        if(postRepository.addPost(post) > 0)
            return "ok";
        return "no";
    }
    @Override
    public Map<String, Object> selectPostByPage(PostByCon postByCon) {
        Map<String, Object> map = new HashMap<String, Object>();
        postByCon.setAct("byPage");
```

```
        List<Post> posts = postRepository.selectPost(postByCon);
        map.put("posts", posts);
        postByCon.setAct("byNoPage");
        map.put("total", postRepository.selectPost(postByCon).size());
        return map;
    }
    @Override
    public Map<String, Object> selectPostByCon(PostByCon postByCon) {
        Map<String, Object> map = new HashMap<String, Object>();
        postByCon.setAct("byPage");
        List<Post> posts = postRepository.selectPostByCon(postByCon);
        map.put("posts", posts);
        postByCon.setAct("byNoPage");
        map.put("total", postRepository.selectPostByCon(postByCon).size());
        return map;
    }
    @Override
    public Post getPostDetail(int id) {
        return postRepository.selectAPost(id);
    }
    @Override
    public String updatePost(Post post) {
        if(postRepository.updatePost(post) > 0)
            return "ok";
        return "no";
    }
    @Override
    public List<Post> getPost() {
        return postRepository.selectPost(null);
    }
    @Override
    public String deletePost(int id) {
        //先查询是否有关联数据
        List<Map<String, Object>> listMap =postRepository.selectAssociatePost(id);
        if (listMap.size() <= 0) {
            if (postRepository.deletePost(id) > 0)
                return "ok";
        }
        return "no";
    }
}
```

❸ 数据访问层

与岗位管理相关的数据访问层是 PostRepository 接口，该接口代码略。

❹ 实体层

与岗位管理相关的实体层有 Post（岗位实体）和 PostByCon（条件查询实体），这两个实体的代码略。

❺ SQL 映射文件

与岗位管理相关的 SQL 映射文件是 PostMapper.xml，该文件的核心内容是：

```
<!-- 查询所有岗位 -->
<select id="selectPost"  resultType="Post" parameterType="PostByCon">
    select * from post
    <if test="act == 'byPage'" >
```

```
            limit #{startIndex}, #{pageSize}
        </if>
    </select>
    <!-- 添加岗位 -->
    <insert id="addPost"  parameterType="Post">
        insert into post (id,pname,ptype,organization) values (null, #{pname}, #{ptype},
#{organization})
    </insert>
    <!-- 条件查询 -->
    <select id="selectPostByCon" resultType="Post"  parameterType="PostByCon">
        select *
            from post where 1=1
            <if test="pname != null and pname != ''">
                and pname like concat('%',#{pname},'%')
            </if>
            <if test="ptype != null and ptype != ''">
                and ptype = #{ptype}
        </if>
        <if test="act == 'byPage'" >
            limit #{startIndex}, #{pageSize}
        </if>
    </select>
    <!-- 查询一个岗位 -->
    <select id="selectAPost" resultType="Post" parameterType="Integer">
        select * from  post where id = #{id}
    </select>
    <!-- 修改岗位 -->
    <update id="updatePost" parameterType="Post">
        update post
        <set>
            <if test="pname != null">
                pname = #{pname},
            </if>
            <if test="ptype != null">
                ptype = #{ptype},
            </if>
            <if test="organization != null">
                organization = #{organization}
            </if>
        </set>
        where id = #{id}
    </update>
    <!--查询关联岗位-->
    <select id="selectAssociatePost" resultType="map" parameterType="Integer">
        SELECT DISTINCT s.id, t.id  from staff s, transfer t WHERE
            s.post_id = #{id} or t.beforepost_id = #{id} or t.afterpost_id = #{id}
    </select>
    <!--删除职位-->
    <delete id="deletePost" parameterType="Integer">
        delete from post where id = #{id}
    </delete>
```

▶ 16.3.8 员工管理与试用期管理后台实现

员工管理功能模块包括增、删、改员工。试用期管理功能模块包括条件查询试用期员工、试用期转正、延期等。

❶ 控制器层

与员工管理和试用期管理相关的控制器是 StaffController，核心代码如下：

```
@RestController
public class StaffController {
    @Resource
    private StaffService staffService;
    @PostMapping("/addStaff")
    public  String addStaff(@RequestBody Staff staff ){
        return staffService.addStaff(staff);
    }
    @GetMapping("/getStaffByPage")
    public Map<String, Object> selectStaffByPage(StaffByCon staffByCon) {
        return staffService.selectStaffByPage(staffByCon);
    }
    @PostMapping("/selectStaffByCon")
    public Map<String, Object> selectStaffByCon(@RequestBody StaffByCon staffByCon){
        return staffService.selectStaffByCon(staffByCon);
    }
    @GetMapping("/getStaffDetail")
    public Staff getStaffDetail(int id){
        return staffService.getStaffDetail(id);
    }
    @PostMapping("/updateStaff")
    public  String updateStaff(@RequestBody Staff staff ){
        return staffService.updateStaff(staff);
    }
    //试用期管理
    @GetMapping("/getPeriods")
    public Map<String, Object> selectPeriodsByPage(PeriodByCon periodByCon) {
        return staffService.selectPeriodsByPage(periodByCon);
    }
    @PostMapping("/selectPeriodByCon")
    publicMap<String,Object>selectPeriodByCon(@RequestBody PeriodByConperiodByCon){
        return staffService.selectPeriodByCon(periodByCon);
    }
    @PostMapping("/periodOp")
    public String periodOp(int id, String status){
        return staffService.periodOp(id, status);
    }
    @PostMapping("/deleteStaff")
    public  String deleteStaff(int id){
        return staffService.deleteStaff(id);
    }
}
```

❷ 业务层

与员工管理和试用期管理相关的业务层包括 StaffService 接口和 StaffServiceImpl 实现类。

StaffService 接口代码略，StaffServiceImpl 实现类的核心代码具体如下：

```
@Service
public class StaffServiceImpl implements StaffService {
    @Resource
    private StaffRepository staffRepository;
    @Override
    public String addStaff(Staff staff) {
        if(staffRepository.addStaff(staff) > 0)
            return "ok";
        return "no";
    }
    @Override
    public Map<String, Object> selectStaffByPage(StaffByCon staffByCon) {
        Map<String, Object> map = new HashMap<String, Object>();
        staffByCon.setAct("byPage");
        List<Staff> staffs = staffRepository.selectStaff(staffByCon);
        map.put("staffs", staffs);
        staffByCon.setAct("byNoPage");
        map.put("total", staffRepository.selectStaff(staffByCon).size());
        return map;
    }
    @Override
    public Map<String, Object> selectStaffByCon(StaffByCon staffByCon) {
        Map<String, Object> map = new HashMap<String, Object>();
        staffByCon.setAct("byPage");
        List<Staff> staffs = staffRepository.selectStaffByCon(staffByCon);
        map.put("staffs", staffs);
        staffByCon.setAct("byNoPage");
        map.put("total", staffRepository.selectStaffByCon(staffByCon).size());
        return map;
    }
    @Override
    public Staff getStaffDetail(int id) {
        return staffRepository.selectAStaff(id);
    }
    @Override
    public String updateStaff(Staff staff) {
        if(staffRepository.updateStaff(staff) > 0)
            return "ok";
        return "no";
    }
    //试用期管理
    @Override
    public Map<String, Object> selectPeriodsByPage(PeriodByCon periodByCon) {
        Map<String, Object> map = new HashMap<String, Object>();
        periodByCon.setAct("byPage");
        List<Staff> periods = staffRepository.selectPeriodStaff(periodByCon);
        map.put("periods", periods);
        periodByCon.setAct("byNoPage");
        map.put("total", staffRepository.selectPeriodStaff(periodByCon).size());
        return map;
    }
```

```
    @Override
    public Map<String, Object> selectPeriodByCon(PeriodByCon periodByCon) {
        Map<String, Object> map = new HashMap<String, Object>();
        periodByCon.setAct("byPage");
        List<Staff>periods=staffRepository.selectPeriodStaffCon(periodByCon);
        map.put("periods", periods);
        periodByCon.setAct("byNoPage");
        map.put("total", staffRepository.selectPeriodStaffCon(periodByCon).size());
        return map;
    }
    @Override
    public String periodOp(int id, String status) {
        if (staffRepository.periodOp(id, status) > 0)
            return "ok";
        return "no";
    }
    @Override
    public String deleteStaff(int id) {
        //先查询是否有关联数据
        List<Map<String, Object>> listMap =staffRepository.selectAssociateStaff(id);
        if (listMap.size() <= 0) {
            if (staffRepository.deleteStaff(id) > 0)
                return "ok";
        }
        return "no";
    }
}
```

❸ 数据访问层

与员工管理和试用期管理相关的数据访问层是 StaffRepository 接口，该接口代码略。

❹ 实体层

与员工管理和试用期管理相关的实体层有 Staff（员工实体）、StaffByCon（条件查询员工实体）和 PeriodByCon（条件查询试用期实体），这三个实体的代码略。

❺ SQL 映射文件

与员工管理和试用期管理相关的 SQL 映射文件是 StaffMapper.xml，该文件的核心内容是：

```
<!-- 录入员工 -->
<insert id="addStaff"  parameterType="Staff">
    insert into staff values (null, #{sname}, #{sex}, #{birthday}, #{sid},
    #{depart_id}, #{post_id}, #{entrydate}, #{joinworkdate}, #{workform}, #{staffsource},
    #{politicalstatus}, #{nation}, #{nativeplace}, #{stel}, #{semail}, #{sheight},
    #{bloodtype}, #{maritalstatus}, #{registeredresidence}, #{education}, #{degree}, #{university},
    #{major}, #{graduationdate},#{startdate},#{enddate}
    <if test="startdate != null">
        ,'正常'
    </if>
    <if test="startdate == null">
        ,#{status}
    </if>
    ,#{peroidopdate}
    )
```

```
    </insert>
    <!-- 查询所有员工 -->
    <select id="selectStaff" resultType="Staff" parameterType="StaffByCon">
        select * from staff s, department d, post p where s.depart_id = d.id and s.post_id = p.id
        <if test="act == 'byPage'" >
            limit #{startIndex}, #{pageSize}
        </if>
    </select>
    <!-- 条件查询员工 -->
    <select id="selectStaffByCon" resultType="Staff" parameterType="StaffByCon">
        select * from staff s, department d, post p where s.depart_id = d.id and s.post_id = p.id
        <if test="sname != null and sname != ''">
            and s.sname like concat('%',#{sname},'%')
        </if>
        <if test="depart_id != 0">
            and s.depart_id = #{depart_id}
        </if>
        <if test="act == 'byPage'" >
            limit #{startIndex}, #{pageSize}
        </if>
    </select>
    <!-- 查询一个员工信息 -->
    <select id="selectAStaff" resultType="Staff" parameterType="Integer">
 select *,DATE_FORMAT(s.birthday,'%Y-%m-%d') as birthday1,DATE_FORMAT(s.entrydate,
'%Y-%m-%d') as entrydate1,DATE_FORMAT(s.joinworkdate,'%Y-%m-%d') as joinworkdate1,
DATE_FORMAT(s.graduationdate,'%Y-%m-%d') as graduationdate1
        ,DATE_FORMAT(s.startdate,'%Y-%m-%d') as startdate1 ,DATE_FORMAT(s.enddate,
'%Y-%m-%d') as enddate1  from staff s, department d, post p where s.id = #{id} and
s.depart_id = d.id and s.post_id = p.id;
    </select>
    <!-- 修改一个员工信息 -->
    <update id="updateStaff" parameterType="Staff">
        update staff
        <set>
            sname = #{sname},
            sex =  #{sex},
            birthday = #{birthday},
            sid = #{sid},
            depart_id =  #{depart_id},
            post_id = #{post_id},
            entrydate =  #{entrydate},
            joinworkdate = #{joinworkdate},
            workform =  #{workform},
            staffsource = #{staffsource},
            politicalstatus =  #{politicalstatus},
            nation = #{nation},
            nativeplace =  #{nativeplace},
            stel = #{stel},
            semail =  #{semail},
            sheight = #{sheight},
            bloodtype =  #{bloodtype},
            maritalstatus = #{maritalstatus},
```

```
            registeredresidence =  #{registeredresidence},
            education = #{education},
            degree =  #{degree},
            university = #{university},
            major =  #{major},
            graduationdate = #{graduationdate},
            startdate = #{startdate},
            enddate = #{enddate}
        </set>
        where id = #{id}
    </update>
    <!-- 试用期管理查询 -->
    <select id="selectPeriodStaff" resultType="Staff">
    select *,DATE_FORMAT(s.startdate,'%Y-%m-%d') as startdate1 ,DATE_FORMAT(s.enddate,
'%Y-%m-%d') as enddate1 from staff s, department d, post p where s.depart_id =
d.id and s.post_id = p.id and s.startdate IS NOT NULL  and s.enddate IS NOT NULL
        <if test="act == 'byPage'" >
            limit #{startIndex}, #{pageSize}
        </if>
    </select>
    <!-- 条件查询试用期管理 -->
    <select id="selectPeriodStaffCon" resultType="Staff">
     select *,DATE_FORMAT(s.startdate,'%Y-%m-%d') as startdate1 ,DATE_FORMAT(s.enddate,
'%Y-%m-%d') as enddate1  from staff s, department d, post p where s.depart_id =
d.id and s.post_id = p.id
        <if test="sname != null and sname != ''">
            and s.sname like concat('%',#{sname},'%')
        </if>
        <if test="depart_id != 0">
            and s.depart_id = #{depart_id}
        </if>
        <if test="post_id != 0">
            and s.post_id = #{post_id}
        </if>
        <if test="status != null and status != ''">
            and s.status = #{status}
        </if>
        <if test="startdate != null and enddate != null" >
            and date(s.startdate ) between #{startdate} and #{enddate}
            and date(s.enddate ) between #{startdate} and #{enddate}
        </if>
        <if test="act == 'byPage'" >
            limit #{startIndex}, #{pageSize}
        </if>
    </select>
    <!-- 试用期管理操作 -->
    <update id="periodOp">
        update staff
        <set>
            status = #{status},
            peroidopdate = now()
        </set>
```

```
    where id=#{id}
</update>
<!--查询关联员工-->
<select id="selectAssociateStaff" resultType="map" parameterType="Integer">
    SELECT DISTINCT q.id, t.id from quit q, transfer t where
        q.staff_id = #{id} or t.staff_id = #{id}
</select>
<!--删除员工-->
<delete id="deleteStaff" parameterType="Integer">
    delete from staff where id = #{id}
</delete>
```

▶ 16.3.9　岗位调动管理后台实现

岗位调动管理功能模块包括录入岗位调动信息、多条件查询岗位调动信息。

❶ 控制器层

与岗位调动管理相关的控制器是 TransferController，核心代码如下：

```
@RestController
public class TransferController {
    @Resource
    private TransferService transferService;
    @GetMapping("/getBeforePost")
    public Transfer getBeforePost(int id){
        return transferService.getBeforePost(id);
    }
    @PostMapping("/addTransfer")
    public  String addTransfer(@RequestBody Transfer transfer){
        return transferService.addTransfer(transfer);
    }
    @GetMapping("/getTransfer")
    public Map<String, Object> getTransfer(Transfer transfer) {
        return transferService.getTransfer(transfer);
    }
    @PostMapping("/selectTransfersByCon")
    public Map<String,Object>selectTransfersByCon(@RequestBody Transfer transfer){
        return transferService.selectTransfersByCon(transfer);
    }
}
```

❷ 业务层

与岗位调动管理相关的业务层包括 TransferService 接口和 TransferServiceImpl 实现类。TransferService 接口代码略，TransferServiceImpl 实现类的核心代码具体如下：

```
@Service
public class TransferServiceImpl implements  TransferService{
    @Resource
    private TransferRepository transferRepository;
    @Override
    public Transfer getBeforePost(int id) {
        return transferRepository.getBeforePost(id);
    }
```

```
    @Override
    public String addTransfer(Transfer transfer) {
        if (transferRepository.addTransferStaff(transfer) > 0
                &&transferRepository.updateStaff(transfer) > 0)
            return "ok";
        return "no";
    }
    @Override
    public Map<String, Object> getTransfer(Transfer transfer) {
        Map<String, Object> map = new HashMap<String, Object>();
        transfer.setAct("byPage");
        List<Transfer> transfers = transferRepository.selectTransfer(transfer);
        map.put("transfers", transfers);
        transfer.setAct("byNoPage");
        map.put("total", transferRepository.selectTransfer(transfer).size());
        return map;
    }
    @Override
    public Map<String, Object> selectTransfersByCon(Transfer transfer) {
        Map<String, Object> map = new HashMap<String, Object>();
        transfer.setAct("byPage");
        List<Transfer> transfers = transferRepository.selectTransfersByCon(transfer);
        map.put("transfers", transfers);
        transfer.setAct("byNoPage");
        map.put("total", transferRepository.selectTransfersByCon(transfer).size());
        return map;
    }
}
```

❸ 数据访问层

与岗位调动管理相关的数据访问层是 TransferRepository 接口，该接口代码略。

❹ 实体层

与岗位调动管理相关的实体层是 Transfer（封装岗位调动及条件查询信息），该实体的代码略。

❺ SQL 映射文件

与岗位调动管理相关的 SQL 映射文件是 TransferMapper.xml，该文件的核心内容是：

```
<!-- 查询调动一个员工 -->
<select id="getBeforePost" resultType="Transfer">
    select s.sname as sname, p.id as beforepost_id,
    p.pname as beforepost_name
     from staff s, post p where s.id = #{id} and s.post_id = p.id
</select>
<!-- 更新调动员工岗位 -->
<update id="updateStaff" parameterType="Transfer">
    update staff
    <set>
        post_id = #{afterpost_id} where id = #{staff_id}
    </set>
</update>
<!-- 记录调动员工 -->
```

```
    <insert id="addTransferStaff"  parameterType="Transfer">
insert into transfer values(null,#{staff_id},#{sname},#{beforepost_id},#{afterpost_id},
#{ttype},#{tdate},now())
    </insert>
    <!-- 查询调动员工 -->
    <select id="selectTransfer" resultType="Transfer" parameterType="Transfer">
        select t.staff_id, t.sname,t.ttype,
        ap.pname as afterpost_name,
        bp.pname as beforepost_name,
        DATE_FORMAT(t.tdate,'%Y-%m-%d') as tdate1,DATE_FORMAT(t.opdate,'%Y-%m-%d') as opdate1
         from transfer t, post bp, post ap
         where t.beforepost_id = bp.id and t.afterpost_id = ap.id
        <if test="act == 'byPage'" >
            limit #{startIndex}, #{pageSize}
        </if>
    </select>
    <!-- 条件查询调动员工 -->
    <select id="selectTransfersByCon" resultType="Transfer" parameterType="Transfer">
        select t.staff_id, t.sname,t.ttype,
        ap.pname as afterpost_name,
        bp.pname as beforepost_name,
        DATE_FORMAT(t.tdate,'%Y-%m-%d') as tdate1,DATE_FORMAT(t.opdate,'%Y-%m-%d') as opdate1
        from transfer t, post bp, post ap
        where t.beforepost_id = bp.id and t.afterpost_id = ap.id
        <if test="sname != null and sname != ''">
            and t.sname like concat('%',#{sname},'%')
        </if>
        <if test="staff_id != null and staff_id != 0">
            and t.staff_id = #{staff_id}
        </if>
        <if test="ttype != null and ttype != ''">
            and t.ttype = #{ttype}
        </if>
        <if test="startdate != null and enddate != null" >
            and date(t.tdate ) between #{startdate} and #{enddate}
        </if>
        <if test="act == 'byPage'" >
            limit #{startIndex}, #{pageSize}
        </if>
    </select>
```

▶ 16.3.10　员工离职管理后台实现

员工离职管理功能模块包括录入员工离职信息、查询及条件查询员工离职信息。

❶ 控制器层

与员工离职管理相关的控制器是 QuitController，核心代码如下：

```
@RestController
public class QuitController {
    @Resource
    private QuitService quitService;
```

```
    @PostMapping("/addQuit")
    public  String addQuit(@RequestBody Quit quit){
        return quitService.addQuit(quit);
    }
    @GetMapping("/getQuit")
    public Map<String, Object> getQuit(Quit quit){
        return quitService.getQuit(quit);
    }
    @PostMapping("/selectQuitsByCon")
    public Map<String, Object> selectQuitsByCon(@RequestBody Quit quit){
        return quitService.selectQuitsByCon(quit);
    }
}
```

❷ 业务层

与员工离职管理相关的业务层包括 QuitService 接口和 QuitServiceImpl 实现类。QuitService 接口代码略，QuitServiceImpl 实现类的核心代码具体如下：

```
@Service
public class QuitServiceImpl implements  QuitService {
    @Resource
    private QuitRepository quitRepository;
    @Override
    public String addQuit(Quit quit) {
        if(quitRepository.addQuit(quit) > 0)
            return "ok";
        return "no";
    }
    @Override
    public Map<String, Object> getQuit(Quit quit) {
        Map<String, Object> map = new HashMap<String, Object>();
        quit.setAct("byPage");
        List<Quit> quits = quitRepository.selectQuitStaff(quit);
        map.put("quits", quits);
        quit.setAct("byNoPage");
        map.put("total", quitRepository.selectQuitStaff(quit).size());
        return map;
    }
    @Override
    public Map<String, Object> selectQuitsByCon(Quit quit) {
        Map<String, Object> map = new HashMap<String, Object>();
        quit.setAct("byPage");
        List<Quit> quits = quitRepository.selectQuitsByCon(quit);
        map.put("quits", quits);
        quit.setAct("byNoPage");
        map.put("total", quitRepository.selectQuitsByCon(quit).size());
        return map;
    }
}
```

❸ 数据访问层

与员工离职管理相关的数据访问层是 QuitRepository 接口（该接口代码略）。

❹ 实体层

与员工离职管理相关的实体层是 Quit（封装离职和条件查询信息），该实体的代码略。

❺ SQL 映射文件

与员工离职管理相关的 SQL 映射文件是 QuitMapper.xml，该文件的核心内容是：

```
    <!-- 记录离职员工 -->
    <insert id="addQuit"  parameterType="Quit">
        insert into quit values(null,#{staff_id},#{sname},#{qtype},#{qdate},now())
     </insert>
    <!-- 查询离职员工 -->
    <select id="selectQuitStaff" resultType="Quit">
     select*,DATE_FORMAT(qdate,'%Y-%m-%d') as qdate1,DATE_FORMAT(opdate,'%Y-%m-%d')
as opdate1 from quit
        <if test="act == 'byPage'" >
           limit #{startIndex}, #{pageSize}
        </if>
     </select>
    <!-- 条件查询离职员工 -->
    <select id="selectQuitsByCon" resultType="Quit" parameterType="Quit">
        select*,DATE_FORMAT(qdate,'%Y-%m-%d') as qdate1,DATE_FORMAT(opdate,'%Y-%m-%d')
as opdate1 from quit
        where 1=1
        <if test="sname != null and sname != ''">
           and sname like concat('%',#{sname},'%')
        </if>
        <if test="staff_id != null and staff_id != 0">
           and staff_id = #{staff_id}
        </if>
        <if test="qtype != null and qtype != ''">
           and qtype = #{qtype}
        </if>
        <if test="startdate != null and enddate != null" >
           and date(qdate ) between #{startdate} and #{enddate}
        </if>
        <if test="act == 'byPage'" >
           limit #{startIndex}, #{pageSize}
        </if>
    </select>
```

▶ 16.3.11　报表管理后台实现

报表管理功能模块包括新聘员工报表、离职员工报表以及岗位调动报表的查询。

❶ 控制器层

与报表管理相关的控制器是 ReportController，核心代码如下：

```
    @RestController
    public class ReportController {
        @Resource
        private ReportService reportService;
        @GetMapping("/getNewStaffReport")
        public Map<String, Object> getNewStaffReport(Report report){
```

```
        return reportService.getNewStaffReport(report);
    }
    @PostMapping("/selectNewStaffReportByCon")
    public Map<String, Object> selectNewStaffReportByCon(@RequestBody Report report){
        return reportService.getNewStaffReport(report);
    }
    @GetMapping("/getQuitStaffReport")
    public Map<String, Object> getQuitStaffReport(Report report){
        return reportService.getQuitStaffReport(report);
    }
    @PostMapping("/selectQuitStaffReportByCon")
    public Map<String, Object> selectQuitStaffReportByCon(@RequestBody Report report){
        return reportService.getQuitStaffReport(report);
    }
    @GetMapping("/getTransferReport")
    public Map<String, Object> getTransferReport(Report report){
        return reportService.getTransferReport(report);
    }
    @PostMapping("/selectTransferStaffReportByCon")
    public Map<String, Object> selectTransferStaffReportByCon(@RequestBody Report report){
        return reportService.getTransferReport(report);
    }
}
```

❷ 业务层

与报表管理相关的业务层包括 ReportService 接口和 ReportServiceImpl 实现类。ReportService 接口代码略，ReportServiceImpl 实现类的核心代码具体如下：

```
@Service
public class ReportServiceImpl implements ReportService {
    @Resource
    private ReportRepository reportRepository;
    @Override
    public Map<String, Object> getNewStaffReport(Report report) {
        Map<String, Object> map = new HashMap<String, Object>();
        report.setAct("byPage");
        List<Map<String, Object>> newStaffReports = reportRepository.reportSelectNew(report);
        map.put("newStaffReports", newStaffReports);
        report.setAct("byNoPage");
        map.put("total", reportRepository.reportSelectNew(report).size());
        return map;
    }
    @Override
    public Map<String, Object> getQuitStaffReport(Report report) {
        Map<String, Object> map = new HashMap<String, Object>();
        report.setAct("byPage");
        List<Map<String, Object>> newQuitReports = reportRepository.reportSelectQuit(report);
        map.put("newQuitReports", newQuitReports);
        report.setAct("byNoPage");
        map.put("total", reportRepository.reportSelectQuit(report).size());
        return map;
    }
```

```
    @Override
    public Map<String, Object> getTransferReport(Report report) {
        Map<String, Object> map = new HashMap<String, Object>();
        report.setAct("byPage");
        List<Map<String, Object>> newTransferReports = reportRepository.reportSelectTransfer(report);
        map.put("newTransferReports", newTransferReports);
        report.setAct("byNoPage");
        map.put("total", reportRepository.reportSelectTransfer(report).size());
        return map;
    }
}
```

❸ 数据访问层

与报表管理相关的数据访问层是 ReportRepository 接口，该接口代码略。

❹ 实体层

与报表管理相关的实体层是 Report（封装条件查询信息），该实体的代码略。

❺ SQL 映射文件

与报表管理相关的 SQL 映射文件是 ReportMapper.xml，该文件的核心内容是：

```
<!-- 条件新员工 -->
<select id="reportSelectNew" resultType="map" parameterType="Report">
    select *,DATE_FORMAT(s.entrydate,'%Y-%m-%d') as entrydate
    from staff s, department d, post p
    where s.depart_id = d.id and s.post_id = p.id
    <if test="startdate != null and enddate != null" >
        and date(s.entrydate ) between #{startdate} and #{enddate}
    </if>
    order by entrydate desc
    <if test="act == 'byPage'" >
        limit #{startIndex}, #{pageSize}
    </if>
</select>
<!-- 条件离职员工 -->
<select id="reportSelectQuit" resultType="map" parameterType="Report">
    select *,DATE_FORMAT(q.qdate,'%Y-%m-%d') as qdate
    from staff s, department d, post p, quit q
    where s.depart_id = d.id and s.post_id = p.id and q.staff_id = s.id
    <if test="startdate != null and enddate != null" >
        and date(q.qdate ) between #{startdate} and #{enddate}
    </if>
    order by qdate desc
    <if test="act == 'byPage'" >
        limit #{startIndex}, #{pageSize}
    </if>
</select>
<!-- 条件调动员工 -->
<select id="reportSelectTransfer" resultType="map" parameterType="Report">
    select *,DATE_FORMAT(t.tdate,'%Y-%m-%d') as tdate, p1.pname as pname1, p2.pname as pname2
    from staff s, department d, post p1, post p2, transfer t
    where t.staff_id = s.id and s.depart_id = d.id and t.beforepost_id = p1.id and t.afterpost_id = p2.id
```

```
        <if test="startdate != null and enddate != null" >
            and date(t.tdate ) between #{startdate} and #{enddate}
        </if>
        order by tdate desc
        <if test="act == 'byPage'" >
            limit #{startIndex}, #{pageSize}
        </if>
    </select>
```

16.4 前端项目的实现

▶ 16.4.1 使用 Vue CLI 搭建前端项目

参考 15.4.2 节，使用 Vue CLI 搭建基于 Router 和 Vuex 功能的前端项目 personmis-vue。搭建成功后，使用 VSCode 打开 personmis-vue 目录即可进行前端项目的实现。

▶ 16.4.2 安装 axios

在前端界面组件中，通过 axios 模块向后端提交 Ajax 异步请求。所以需要打开 VSCode 的 Terminal 终端命令行窗口，执行 npm install --save axios 命令安装 axios 模块。

▶ 16.4.3 设置反向代理

成功安装 axios 模块后，首先在 personmis-vue 的入口文件 main.js 中，将 axios 挂载到 vue 实例上，示例代码如下：

```
//设置反向代理，前端请求默认发送到 http://localhost:8443/personmis
const axios = require('axios')//使用 axios 完成 ajax 请求
//全局注册，之后可在其他组件中通过 this.$axios 发送数据
axios.defaults.baseURL = 'http://localhost:8443/personmis'
//axios 挂载到 vue 实例
vapp.config.globalProperties.$axios = axios
```

然后，在 personmis-vue 根目录下，创建 Vue 的配置文件 vue.config.js，设置反向代理支持，示例代码如下：

```
module.exports = {
    //在本地会创建一个虚拟服务端，虚拟服务器访问后端的服务器不存在跨域
    devServer: {
        proxy: {
            '/personmis': {/*将所有以 /personmis 开头的请求自动代理到
http://localhost: 8443 后端的基准地址*/
                target: 'http://localhost:8443',
                //是否启用 websockets
                ws: true,
                /*开启代理：在本地会创建一个虚拟服务端，然后发送请求的数据，并同时接收请求的
数据，这样服务端和服务端进行数据的交互就不会有跨域问题*/
                changeOrigin: true,
                pathRewrite: {
```

```
                    '^/personmis': ''
                }
            }
        }
    }
}
```

▶ 16.4.4　配置页面路由

在 src/router/index.js 文件中配置页面路由。跳转登录界面的路由不需要登录权限验证，需要加上 meta:{auth:true}数据，以便在路由钩子函数中判断。路由配置内容具体如下：

```
import { createRouter, createWebHistory } from 'vue-router'
import Login from '../views/Login.vue'
import Department from '../views/Department.vue'
import AddDepartment from '../views/AddDepartment.vue'
import Post from '../views/Post.vue'
import AddPost from '../views/AddPost.vue'
import AddStaff from '../views/AddStaff.vue'
import Staff from '../views/Staff.vue'
import PeroidOp from '../views/PeroidOp.vue'
import AddTransferStaff from '../views/AddTransferStaff.vue'
import TransferStaff from '../views/TransferStaff.vue'
import AddQuit from '../views/AddQuit.vue'
import Quit from '../views/Quit.vue'
import NewStaffReport from '../views/NewStaffReport.vue'
import QuitStaffReport from '../views/QuitStaffReport.vue'
import TransferStaffReport from '../views/TransferStaffReport.vue'
const routes = [
    //打开程序直接跳转到登录页面
    {path: '/', redirect:'/login', meta:{auth:true}},//登录不需要验证权限
    {path: '/login', component: Login, meta:{auth:true}},
    //部门管理
    {path: '/department', component: Department},
    {path: '/adddepartment', component: AddDepartment},
    //岗位管理
    {path: '/post', component: Post},
    {path: '/addpost', component: AddPost},
    //员工管理
    {path: '/addStaff', component: AddStaff},
    {path: '/staff', component: Staff},
    //试用期管理
    {path: '/peroidOp', component: PeroidOp},
    //岗位调动管理
    {path: '/addTransferStaff', component: AddTransferStaff},
    {path: '/transferStaff', component: TransferStaff},
        //离职管理
    {path: '/addQuit', component: AddQuit},
    {path: '/quit', component: Quit},
    //报表管理
    {path: '/newStaffReport', component: NewStaffReport},
    {path: '/quitStaffReport', component: QuitStaffReport},
    {path: '/transferStaffReport', component: TransferStaffReport}
```

```
]
const router = createRouter({
    history: createWebHistory(process.env.BASE_URL),
    routes
})
export default router
```

▶ 16.4.5　安装 Element Plus

Element Plus 是一套为开发者、设计师和产品经理准备的基于 Vue 3.0 的桌面端组件库。我们使用 Element Plus 辅助开发人事管理系统的前端界面组件。所以，首先需要打开 VSCode 的 Terminal 终端命令行窗口，执行 npm install element-plus --save 命令安装 Element Plus 组件库。然后，在 src 目录下创建 plugins 目录，并在该目录下新建 Element Plus 的配置文件 element.js，配置内容如下：

```
import ElementPlus from 'element-plus'
import 'element-plus/lib/theme-chalk/index.css'
export default (app) => {
    app.use(ElementPlus)
}
```

最后，在 personmis-vue 的入口文件 main.js 中，将 Element Plus 组件库安装到 vue 实例上，示例代码如下：

```
import installElementPlus from './plugins/element'
installElementPlus(vapp)
```

上述操作流程如图 16.3 所示。

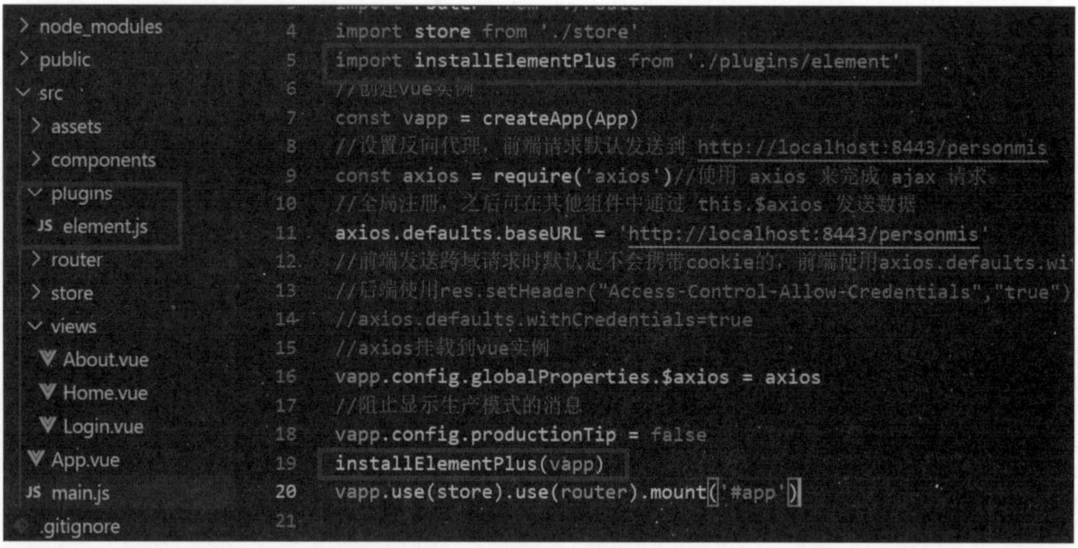

图 16.3　安装 Element Plus 组件库

▶ 16.4.6　管理员登录界面实现

前端项目首页路由默认跳转到登录界面。在 views 目录中，创建登录界面组件 Login.vue。

Login.vue 的运行效果如图 16.4 所示。

图 16.4　登录界面

界面组件 Login.vue 的代码如下：

```
    <template>
    <el-dialog title="管理员登录" v-model="dialogVisible"  width="30%">
      <div class="box">
      <el-form ref="loginForm" :model="loginForm" :rules="rules" style="width:100%;"
label-width="20%">
      <el-form-item label="用户名" prop="uname">
        <el-input v-model="loginForm.uname" placeholder="请输入用户名"></el-input>
      </el-form-item>
      <el-form-item label="密码" prop="upwd">
        <el-input v-model="loginForm.upwd" placeholder="请输入密码"></el-input>
      </el-form-item>
      <el-form-item>
        <el-button type="primary" @click="login(loginForm)" :loading="loadingbut">
{{loadingbuttext}}</el-button>
        <el-button type="danger" @click="cancel">重置</el-button>
      </el-form-item>
    </el-form>
      </div>
    </el-dialog>
    </template>
    <script>
      export default {
        name: 'Login',
        data () {
          return {
            loginForm: {},
            //验证规则
            rules: {
              uname: [{required: true, message: '请输入用户名', trigger: 'blur'}],
              upwd: [{required: true, message: '请输入密码', trigger: 'blur'}]
            },
          loadingbut: false,
```

```
      loadingbuttext: '登录',
      dialogVisible: true
    }
  },
  methods: {
    login (loginForm) {
      this.$refs['loginForm'].validate((valid) => {
        if (valid) {
          this.loadingbut = true;
          this.loadingbuttext = '登录中...';
          this.$axios
          .post('/login',{
              uname: loginForm.uname,
              upwd: loginForm.upwd
            })
            .then(successResponse => {
              if (successResponse.data === "ok") {
                //Message Box
                this.$alert('登录成功', {confirmButtonText: '确定' })
                this.$store.commit('changeLogin',this.loginForm.uname)
                let path = this.$route.query.redirect
                this.$router.replace({path:path==='/'||path===undefined?'/department':path})
              }else {
                this.$alert('用户名或密码错误！', {confirmButtonText: '确定' })
                this.loadingbut = false;
                this.loadingbuttext = '登录';
              }
            })
            .catch(failResponse => {
              this.$alert(failResponse.response.status,{confirmButtonText:'确定'})
            })
        }
        else {
            this.$alert('表单验证失败', {confirmButtonText: '确定' })
            return false;
        }
      })
    },
    cancel(){
        this.$refs['loginForm'].resetFields()
    }
  }
}
</script>
<style scoped>
.box{
    width: 100%;
    height: 180px;
  }
</style>
```

16.4.7 界面导航组件实现

登录成功后，进入部门管理界面。在所有管理界面中，都将引入界面导航组件，如图 16.5 所示。

人事管理系统

部门管理　岗位管理　员工管理　试用期管理　岗位调动管理　员工离职管理　报表管理

图 16.5　界面导航组件

所以，首先在 components 目录中，创建界面导航组件 NavMain.vue。然后，在各个管理界面组件中，使用如下代码引入界面导航组件：

```
import NavMain from '@/components/NavMain.vue'
…
    components:{
        NavMain
    }
```

界面导航组件 NavMain.vue 的代码如下：

```
<template>
    <h2>人事管理系统</h2>
<el-menu
    :default-active="activeIndex"
    class="el-menu-demo"
    mode="horizontal"
    background-color="#545c64"
    text-color="#fff"
    active-text-color="#ffd04b">
    <el-submenu index="1">
        <template #title>部门管理</template>
        <el-menu-item index="1-1"><router-link to="/department" class="a">管
理部门</router-link></el-menu-item>
        <el-menu-item index="1-2"><router-link to="/adddepartment" class="a">
新增部门</router-link></el-menu-item>
    </el-submenu>
    <el-submenu index="2">
        <template #title>岗位管理</template>
        <el-menu-item index="2-1"><router-link to="/post" class="a">管理岗位
</router- link></el-menu-item>
        <el-menu-item index="2-2"><router-link to="/addpost" class="a">新增岗
位</router-link></el-menu-item>
    </el-submenu>
    <el-submenu index="3">
        <template #title>员工管理</template>
        <el-menu-item index="3-1"><router-link to="/staff" class="a">管理员工
</router- link></el-menu-item>
        <el-menu-item index="3-2"><router-link to="/addStaff" class="a">新增
员工</router-link></el-menu-item>
    </el-submenu>
    <el-menu-item index="4"><router-link to="/peroidOp" class="a">试用期管理
</router-link></el-menu-item>
    <el-submenu index="5">
        <template #title>岗位调动管理</template>
        <el-menu-item index="5-1"><router-link to="/addTransferStaff" class="a">
录入岗位调动</router-link></el-menu-item>
        <el-menu-item index="5-2"><router-link to="/transferStaff" class="a">
查询调动员工</router-link></el-menu-item>
    </el-submenu>
```

```
        <el-submenu index="6">
            <template #title>员工离职管理</template>
            <el-menu-item index="6-1"><router-link to="/addQuit" class="a">录入离职员工</router-link></el-menu-item>
            <el-menu-item index="6-2"><router-link to="/quit" class="a">查询已离职员工</router-link></el-menu-item>
        </el-submenu>
        <el-submenu index="7">
            <template #title>报表管理</template>
            <el-menu-item index="7-1"><router-link to="/newStaffReport" class="a">新聘员工报表</router-link></el-menu-item>
            <el-menu-item index="7-2"><router-link to="/quitStaffReport" class="a">离职员工报表</router-link></el-menu-item>
            <el-menu-item index="7-3"><router-link to="/transferStaffReport" class="a">岗位调动报表</router-link></el-menu-item>
        </el-submenu>
    </el-menu>
    </template>
    <script>
        export default {
            name: 'NavMain'
        }
    </script>
    <style>
        .a{
            text-decoration: none;
            color: aliceblue;
        }
    </style>
```

▶ 16.4.8　部门管理界面实现

部门管理界面有两个界面组件，一个是新增部门界面组件，一个是部门管理界面组件。在 views 目录中，创建新增部门界面组件 AddDepartment.vue，运行效果如图 16.6 所示；创建

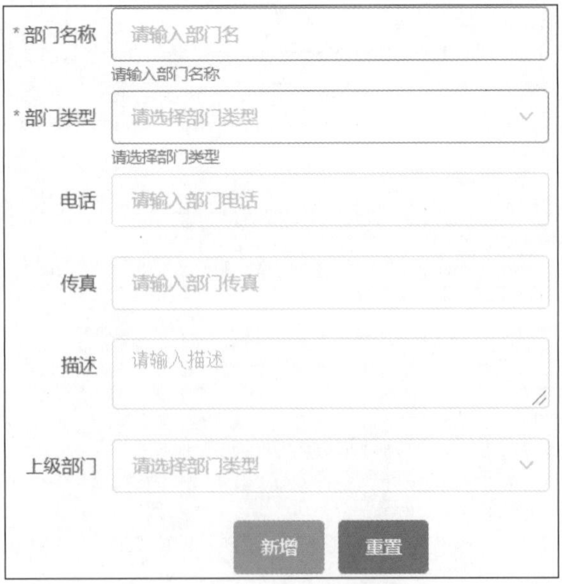

图 16.6　新增部门界面

部门管理界面组件 Department.vue，运行效果如图 16.7 所示。

图 16.7　部门管理界面

在图 16.6 所示的新增部门界面中，“部门名称”和“部门类型”是必需输入项，“上级部门”是从数据库查询出来，供管理员选择。

在图 16.7 所示的部门管理界面中，可以根据部门名称和部门类型进行查询，并可对部门进行编辑、详情以及删除（不能删除有数据关联的部门）操作。

新增部门界面组件 AddDepartment.vue 的具体代码如下：

```
<template>
  <NavMain></NavMain>
  <div class="box">
    <br>
   <el-form ref="addForm" :model="addForm" :rules="rules" style="width:50%;"
   label-width="50%" >
  <el-form-item label="部门名称" prop="dname">
    <el-input v-model="addForm.dname" placeholder="请输入部门名"></el-input>
  </el-form-item>
 <el-form-item label="部门类型" prop="dtype">
    <el-select v-model="addForm.dtype" placeholder="请选择部门类型">
      <el-option v-for="(item,index) in dtypes" :key="index" :label="item" :
      value="item"></el-option>
    </el-select>
  </el-form-item>
  <el-form-item label="电话" prop="dtel">
    <el-input v-model="addForm.dtel" placeholder="请输入部门电话"></el-input>
  </el-form-item>
  <el-form-item label="传真" prop="dfax">
    <el-input v-model="addForm.dfax" placeholder="请输入部门传真"></el-input>
  </el-form-item>
  <el-form-item label="描述" prop="description">
    <el-input v-model="addForm.description" type="textarea" placeholder="
    请输入描述"></el-input>
  </el-form-item>
  <el-form-item label="上级部门" prop="supdepartment">
    <el-select v-model="addForm.supdepartment" placeholder="请选择部门类型">
      <el-option v-for="(item,index) in supdepartments" :key="index" :label=
      "item.dname" :value="item.id"></el-option>
```

```
      </el-select>
    </el-form-item>
    <el-form-item>
      <el-button type="primary" @click="add(addForm)" :loading="loadingbut">
      {{loadingbuttext}}</el-button>
      <el-button type="danger" @click="cancel">重置</el-button>
    </el-form-item>
  </el-form>
  </div>
</template>
<script>
import NavMain from '@/components/NavMain.vue'
export default {
  components:{
    NavMain
  },
  data() {
    return {
      dtypes:[ '公司', '部门', '车间', '生产线', '班组'] ,
      supdepartments:[ {id: '', dname: ''}] ,//定义空数组接收后台数据
      addForm: {},
      //验证规则
     rules: {
        dname: [{required: true, message: '请输入部门', trigger: 'blur'}],
        dtype: [{required: true, message: '请选择部门类型', trigger: 'change'}]
     },
     loadingbut: false,
     loadingbuttext: '新增'
    }
  },
  created: function() {
     this.loadSupdepartment()
  },
  methods: {
    add(){
      this.$refs['addForm'].validate((valid) => {
        if (valid) {
          this.loadingbut = true;
          this.loadingbuttext = '添加中...';
          this.$axios
            .post('/addDepartment', this.addForm)//直接提交表单
            .then(successResponse => {
             if (successResponse.data === "ok") {
               this.$alert('添加成功', {confirmButtonText: '确定' })
               this.$router.replace({path: '/department'})
             }else {
               this.$alert('添加失败', {confirmButtonText: '确定' })
               this.loadingbut = false;
               this.loadingbuttext = '新增';
             }
            })
            .catch(failResponse => {
```

```
                this.$alert(failResponse.response.status,{confirmButtonText:'确定'})
              })
          }
          else {
                this.$alert('表单验证失败', {confirmButtonText: '确定' })
                return false;
          }
        })
      },
      loadSupdepartment(){
          this.$axios
            .get('/getDepartment')
            .then(successResponse => {
              this.supdepartments = successResponse.data
            })
            .catch(failResponse => {
              this.$alert(failResponse.response.status)
            })
      },
      cancel(){
            this.$refs['addForm'].resetFields()
        }
    }
  }
</script>
<style scoped>
.box{
     width: 80%;
     height: 200px;
     margin-left: 20%;
   }
.el-select{
   width:100%
}
</style>
```

部门管理界面组件 Department.vue 的具体代码如下：

```
<template>
   <NavMain></NavMain>
<div class="box1">
<br>
<el-form ref="selectForm" :model="selectForm" style="width:50%;" label-width="50%" >
  <el-form-item label="部门名称"  prop="dname">
    <el-input v-model="selectForm.dname"  placeholder="请输入部门名称"></el-input>
  </el-form-item>
 <el-form-item label="部门类型" prop="dtype">
    <el-select v-model="selectForm.dtype" placeholder="请选择部门类型">
      <el-option v-for="(item,index) in dtypes" :key="index" :label="item" :
      value="item"></el-option>
    </el-select>
  </el-form-item>
  <el-form-item>
```

```
      <el-button type="primary" @click="selectDepartmentsByCon">查询</el-button>
    </el-form-item>
  </el-form>
  </div>
  <div class="box2">
    <el-table
      :data="tableData"  :header-cell-style="headClass"
       :cell-style="{ textAlign: 'center' }"
      :default-sort = "{prop: 'id', order: 'descending'}">
      <el-table-column prop="id" label="ID" sortable></el-table-column>
      <el-table-column prop="dname" label="名称"></el-table-column>
      <el-table-column prop="dtype"  label="类型"></el-table-column>
      <el-table-column prop="establishmentdate1" label="成立日期" sortable></el-
      table-column>
      <el-table-column label="操作">
        <template #default="scope">
          <el-button size="mini"  type="success"
            @click="handleEdit(scope.$index, scope.row, 'update')">编辑</el-button>
         <el-button size="mini" type="primary"
            @click="handleEdit(scope.$index, scope.row, 'detail')">详情</el-button>
          <el-button size="mini" type="danger"
            @click="handleDelete(scope.$index, scope.row)">删除</el-button>
        </template>
      </el-table-column>
    </el-table>
    </div>
   <div class="block">
      <el-pagination
        @current-change="handleCurrentChange"
        v-model:currentPage="currentPage"
        :page-size="pageSize"
        layout="total, prev, pager, next"
        :total="total">
      </el-pagination>
    </div>
  <!--部门编辑和详情与新增界面基本一样，不再赘述-->
  </template>
  <script>
  import NavMain from '@/components/NavMain.vue'
  export default {
    components:{
      NavMain
    },
    created: function () {
      this.loadDepartments()
    },
    data() {
       return {
         dtypes:[ '', '公司', '部门', '车间', '生产线', '班组'] ,
         selectForm: {
           currentPage: 1,
           pageSize: 1,
```

```
          act: ''
        },
        currentPage: 1,
        tableData: [{}],//定义空数组接收数据
        pageSize: 1,
        total: 0,
        dialogVisible: false,//详情对话框是否显示
        dialogVisibleDetail: false,
        detailData: {},
        supdepartments:[ {id: '', dname: ''}] ,//定义空数组接收后台数据
      }
    },
    methods: {
      loadDepartments(){
        this.$axios
            .get('/getDepartmentByPage?currentPage=' + this.currentPage +
            '&&pageSize=' + this.pageSize)
            .then(successResponse => {
                this.tableData = successResponse.data.departs
                this.total = successResponse.data.total
            })
            .catch(failResponse => {
              this.$alert(failResponse.response.status)
            })
      },
      //条件查询
      selectDepartmentsByCon(){
          this.selectForm.act = "byCon"
          this.$axios
            .post('/selectDepartmentsByCon', this.selectForm)//直接提交表单
            .then(successResponse => {
              this.tableData = successResponse.data.departs
              this.total = successResponse.data.total
            })
            .catch(failResponse => {
              this.$alert(failResponse.response.status, {confirmButtonText: '确定' })
            })
      },
       // 表头样式设置
      headClass() {
          return 'text-align: center;background:rgb(242,242,242);color:rgb(140,138,140)'
      },
      //页码变换
      handleCurrentChange(val) {
        this.currentPage = val
        if(this.selectForm.act === 'byCon'){
          this.selectForm.currentPage = this.currentPage
          this.selectForm.pageSize = this.pageSize
          this.selectDepartmentsByCon()
        }else{
           this.loadDepartments()
        }
```

```
    },
    //编辑与详情
    handleEdit(index, row, act) {
      console.log(index, row);
      this.$axios
          .get('/getDepartmentDetail?id=' + row.id )
          .then(successResponse => {
             this.detailData = successResponse.data.aDepart
             this.supdepartments = successResponse.data.departs;
          })
          .catch(failResponse => {
            this.$alert(failResponse.response.status)
          })
      if(act === 'update')
        this.dialogVisibleDetail = true
      else
        this.dialogVisible = true
    },
    handleDelete(index, row) {
      console.log(index, row);
      this.$confirm('删除部门, 是否继续?', '提示', {
          confirmButtonText: '确定',
          cancelButtonText: '取消',
          type: 'warning'
        }).then(() => {
          this.$axios
          .post('/deleteDepartment?id=' + row.id)
          .then(successResponse => {
            if (successResponse.data === "ok") {
              this.$message({
                type: 'success',
                message: '删除成功!'
              });
              //删除成功后重新加载
              this.loadDepartments()
            }else {
              this.$alert('不能删除有关联数据! ', {confirmButtonText: '确定' })
            }
          })
          .catch(failResponse => {
            this.$alert(failResponse.response.status, {confirmButtonText: '确定' })
          })
        }).catch(() => {
          this.$message({
            type: 'info',
            message: '已取消删除'
          });
        });
    },
    cancel(){
         this.$refs['detailData'].resetFields()
    },
```

```
    update(){
        this.$axios
          .post('/updateDepartment', this.detailData)//直接提交表单
          .then(successResponse => {
            if (successResponse.data === "ok") {
              this.$alert('修改成功', {confirmButtonText: '确定' })
              this.dialogVisibleDetail = false
              //修改成功后重新加载
              this.loadDepartments()
            }else {
              this.$alert('修改失败', {confirmButtonText: '确定' })
              this.dialogVisibleDetail = false
            }
          })
          .catch(failResponse => {
            this.$alert(failResponse.response.status, {confirmButtonText: '确定' })
          })
      }
    }
  }
</script>
<style scoped>
  .box1{
      width: 80%;
      height: 200px;
      margin-left: 20%;
    }
  .box2{
    margin-left: 5px;
    margin-right: 5px;
  }
 .el-select{
    width:100%
  }
</style>
```

▶ 16.4.9　岗位管理界面实现

岗位管理界面有两个界面组件，一个是新增岗位界面组件，一个是管理岗位界面组件。在 views 目录中，创建新增岗位界面组件 AddPost.vue，运行效果如图 16.8 所示；创建管理岗

* 岗位名称	请输入部门名称
* 岗位类型	请选择岗位类型
编制人数	请输入编制人数
	新增　重置

图 16.8　新增岗位界面

位界面组件 Post.vue，运行效果如图 16.9 所示。

图 16.9　管理岗位界面

在图 16.8 所示的新增岗位界面中，岗位名称和岗位类型是必需输入项。

在图 16.9 所示的管理岗位界面中，可以根据岗位名称和岗位类型进行查询，并可对岗位进行编辑、详情以及删除（不能删除有数据关联的岗位）操作。

新增岗位界面组件 AddPost.vue 和管理岗位界面组件 Post.vue 的实现与部门管理类似，因篇幅受限，具体代码不再赘述，请参考本书提供的源代码。

16.4.10　员工管理界面实现

员工管理界面有两个界面组件，一个是新增员工界面组件，一个是管理员工界面组件。在 views 目录中，创建新增员工界面组件 AddStaff.vue，运行效果如图 16.10 所示；创建管理员工界面组件 Staff.vue，运行效果如图 16.11 所示。

图 16.10　新增员工界面

在图 16.11 所示的管理员工界面中，可以根据员工名称和部门名称进行查询，并可对员工进行编辑、详情以及删除（不能删除有数据关联的员工）操作。

新增员工界面组件 AddStaff.vue 和管理员工界面组件 Staff.vue 的实现与部门管理类似，因篇幅受限，具体代码不再赘述，请参考本书提供的源代码。

图 16.11　管理员工界面

16.4.11　试用期管理界面实现

在 views 目录中，创建试用期管理界面组件 PeroidOp.vue，运行效果如图 16.12 所示。在试用期管理界面中，可以对正常状态的试用期人员进行转正、延期、不录用等操作。同时，也可以根据多个条件查询处于试用期的员工。

图 16.12　试用期管理界面

试用期管理界面组件 PeroidOp.vue 的实现与部门管理类似，因篇幅受限，具体代码不再赘述，请参考本书提供的源代码。

16.4.12　岗位调动管理界面实现

岗位调动管理界面有两个界面组件，一个是录入岗位调动界面组件，一个是查询调动员工界面组件。在 views 目录中，创建录入岗位调动界面组件 AddTransferStaff.vue，运行效果如图 16.13 所示；创建查询调动员工界面组件 TransferStaff.vue，运行效果如图 16.14 所示。

在图 16.13 所示的录入岗位调动界面中，可以根据员工编号自动带入员工姓名和之前岗位。

录入岗位调动界面组件 AddTransferStaff.vue 和查询调动员工界面组件 TransferStaff.vue 的实现与部门管理类似，因篇幅受限，具体代码不再赘述，请参考本书提供的源代码。

16.4.13　员工离职管理界面实现

员工离职管理界面有两个界面组件，一个是录入离职员工界面组件，一个是查询已离职员工界面组件。在 views 目录中，创建录入离职员工界面组件 AddQuit.vue，运行效果如图 16.15 所示；创建查询已离职员工界面组件 Quit.vue，运行效果如图 16.16 所示。

* 员工编号 请输入员工编号
员工姓名
之前岗位
* 调动后岗位 请选择岗位类型
* 调动类型 请选择岗位类型
* 调动日期 选择调动日期
新增 重置

图 16.13　录入岗位调动界面

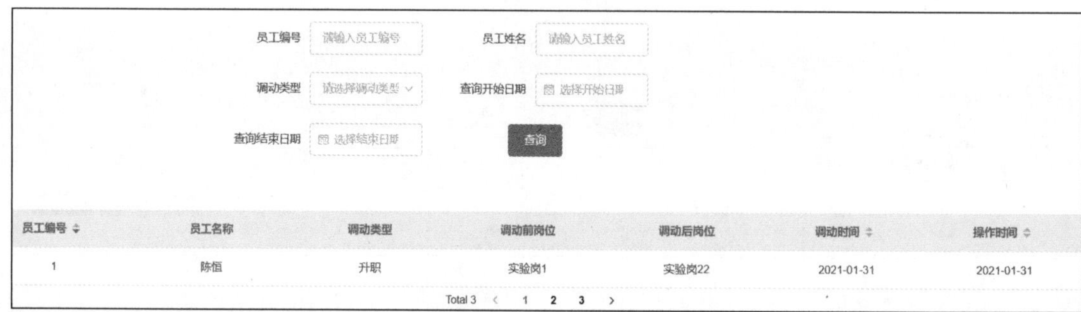

图 16.14　查询调动员工界面

* 员工编号 请输入员工编号
* 员工姓名
* 离职类型 请选择离职类型
* 离职日期 选择离职日期
新增 重置

图 16.15　录入离职员工界面

图 16.16 查询已离职员工界面

在图 16.15 所示的录入离职员工界面中，可以根据员工编号自动带入员工姓名。

录入离职员工界面组件 AddQuit.vue 和查询已离职员工界面组件 Quit.vue 的实现与部门管理类似，因篇幅受限，具体代码不再赘述，请参考本书提供的源代码。

▶ 16.4.14 报表管理界面实现

报表管理界面有三个界面组件，一个是新聘员工报表界面组件，一个是离职员工报表界面组件，一个是岗位调动报表界面组件。在 views 目录中，创建新聘员工报表界面组件 NewStaffReport.vue，运行效果如图 16.17 所示；创建离职员工报表界面组件 QuitStaffReport.vue，运行效果如图 16.18 所示；创建岗位调动报表界面组件 TransferStaffReport.vue，运行效果如图 16.19 所示。

查询开始日期 选择开始日期 查询结束日期 选择结束日期

查询

ID	员工名称	性别	所在部门	岗位名称	入职日期
4	王五	男	采购部333	实验岗22	2021-02-11

Total 3 1 2 3

图 16.17 新聘员工报表界面

查询开始日期 选择开始日期 查询结束日期 选择结束日期

查询

ID	员工名称	性别	所在部门	岗位名称	离职日期	离职类型
2	李四	女	采购部333	实验岗1	2021-01-12	辞退

Total 2 1 2

图 16.18 离职员工报表界面

新聘员工报表界面组件 NewStaffReport.vue、离职员工报表界面组件 QuitStaffReport.vue 和岗位调动报表界面组件 TransferStaffReport.vue 与查询已离职员工界面组件 Quit.vue 的实现类似，因篇幅受限，具体代码不再赘述，请参考本书提供的源代码。

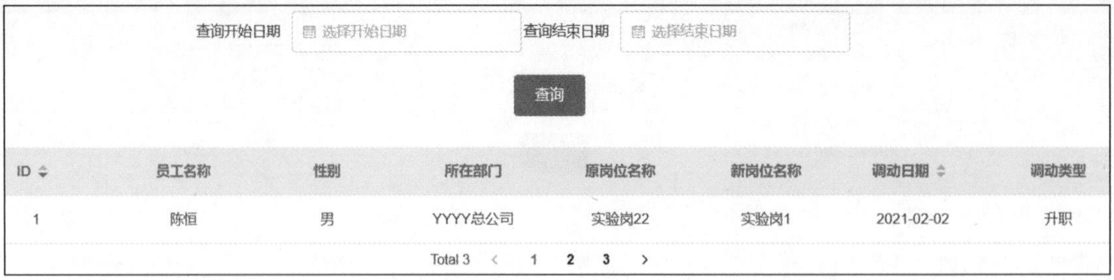

查询开始日期 选择开始日期 查询结束日期 选择结束日期

查询

ID	员工名称	性别	所在部门	原岗位名称	新岗位名称	调动日期	调动类型
1	陈恒	男	YYYY总公司	实验岗22	实验岗1	2021-02-02	升职

Total 3 ‹ 1 2 3 ›

图 16.19　岗位调动报表界面

16.4.15　使用钩子函数实现登录权限认证

前端登录权限认证的具体实现步骤如下。

❶ 定义状态变量

在 vuex 的配置文件 src/store/index.js 中，新建状态变量 isLogin，判断用户是否已登录，并在 mutations 中定义改变状态变量的方法 changeLogin。具体配置如下：

```
import { createStore } from 'vuex'
export default createStore({
  state: {
    //初始时给一个 isLogin='0' 表示用户未登录
    isLogin:window.sessionStorage.getItem('user')==null?'0':window.sessionStorage.
    getItem('user')
  },
  mutations: {
    changeLogin(state, data) {
      state.isLogin = data;
      window.sessionStorage.setItem('user', data)
    }
  }
})
```

❷ 修改状态变量

当管理员登录成功后，调用 this.$store.commit('changeLogin',this.loginForm.uname)修改状态变量 isLogin，标记为已登录状态。代码见 16.4.6 节管理员登录界面实现。

❸ 使用前置路由钩子函数 beforeEach 判断是否登录

在 personmis-vue 的入口文件 main.js 中，使用前置路由钩子函数 beforeEach，进入每个路由导航之前（登录路由除外），判断是否已登录。完整的 main.js 代码如下：

```
import { createApp } from 'vue'
import App from './App.vue'
import router from './router'
import store from './store'
import installElementPlus from './plugins/element'
//创建 vue 实例
const vapp = createApp(App)
//设置反向代理，前端请求默认发送到 http://localhost:8443/personmis
const axios = require('axios')//使用 axios 完成 ajax 请求
//全局注册，之后可在其他组件中通过 this.$axios 发送数据
```

```
axios.defaults.baseURL = 'http://localhost:8443/personmis'
//axios 挂载到 vue 实例
vapp.config.globalProperties.$axios = axios
//阻止显示生产模式的消息
vapp.config.productionTip = false
installElementPlus(vapp)
vapp.use(store).use(router).mount('#app')
//验证是否登录
router.beforeEach((to,from,next)=>{
    //如果路由器需要验证
    if(!to.matched.some(m=>m.meta.auth)){
      //对路由进行验证
      if (store.state.isLogin == '0') {
        alert("您没有登录，无权访问！", {confirmButtonText: '确定' })
        //未登录则跳转到登录界面，query:{ redirect: to.fullPath}表示把当前路由信息
        //传递过去方便登录后跳转回来
        next({
          path: 'login',
          query: {redirect: to.fullPath}
        })
      } else { // 已经登录
        next() //正常跳转到设置好的页面
      }
    }else{
      next()
    }
  }
)
```

16.5　测试运行

首先，运行后台应用 personmis 的主类 PersonmisApplication，启动 personmis，如图 16.20 所示。

```
Run:  PersonmisApplication
No active profile set, falling back to default profiles: default
Devtools property defaults active! Set 'spring.devtools.add-properties' to 'false' to disable
For additional web related logging consider setting the 'logging.level.web' property to 'DEBUG'
Tomcat initialized with port(s): 8443 (http)
Starting service [Tomcat]
Starting Servlet engine: [Apache Tomcat/9.0.41]
Initializing Spring embedded WebApplicationContext
Root WebApplicationContext: initialization completed in 2273 ms
Initializing ExecutorService 'applicationTaskExecutor'
LiveReload server is running on port 35729
Tomcat started on port(s): 8443 (http) with context path '/personmis'
Started PersonmisApplication in 3.947 seconds (JVM running for 4.753)
```

图 16.20　启动 personmis

然后，在打开 personmis-vue 的 VSCode 的 Terminal 终端命令行窗口中，执行 npm run serve 命令启动前端项目 personmis-vue。

最后，可通过 http://localhost:8080/测试运行。

personmis-vue 目录是本章的代码，读者可在目录下执行 npm install 命令自动安装所有的依赖，然后执行 npm run serve 命令启动服务。

16.6 本章小结

本章主要介绍了基于 Spring Boot + Vue.js 3 + MyBatis 的人事管理系统的设计与开发。通过本章的学习，读者应该重点掌握前后端分离项目的跨域访问。

习 题 16

1. 在基于 Spring Boot + Vue.js 3 的前后端分离项目中，如何实现跨域访问？
2. 在 Vue.js 3 的前端项目中，如何实现登录权限验证？

图书资源支持

感谢您一直以来对清华版图书的支持和爱护。为了配合本书的使用，本书提供配套的资源，有需求的读者请扫描下方的"书圈"微信公众号二维码，在图书专区下载，也可以拨打电话或发送电子邮件咨询。

如果您在使用本书的过程中遇到了什么问题，或者有相关图书出版计划，也请您发邮件告诉我们，以便我们更好地为您服务。

我们的联系方式：

地　　址：北京市海淀区双清路学研大厦 A 座 714

邮　　编：100084

电　　话：010-83470236　010-83470237

客服邮箱：2301891038@qq.com

QQ：2301891038（请写明您的单位和姓名）

资源下载：关注公众号"书圈"下载配套资源。

资源下载、样书申请

书圈

获取最新书目

观看课程直播